Earth and Environmental Sciences Library

Earth and Environmental Sciences Library (EESL) is a multidisciplinary book series focusing on innovative approaches and solid reviews to strengthen the role of the Earth and Environmental Sciences communities, while also providing sound guidance for stakeholders, decision-makers, policymakers, international organizations, and NGOs.

Topics of interest include oceanography, the marine environment, atmospheric sciences, hydrology and soil sciences, geophysics and geology, agriculture, environmental pollution, remote sensing, climate change, water resources, and natural resources management. In pursuit of these topics, the Earth Sciences and Environmental Sciences communities are invited to share their knowledge and expertise in the form of edited books, monographs, and conference proceedings.

Makarand M. Ghangrekar · Shalini Yadav ·
Ram Narayan Yadava
Editors

Biological and Hybrid Wastewater Treatment Technology

Recent Developments in India

Springer

Editors
Makarand M. Ghangrekar
Department of Civil Engineering
Indian Institute of Technology Kharagpur
Kharagpur, West Bengal, India

Shalini Yadav
Department of Civil Engineering
Rabindranath Tagore University
Raisen, Madhya Pradesh, India

Ram Narayan Yadava
Research and International Affairs Division
Mansarovar Global University
Bhopal, Madhya Pradesh, India

ISSN 2730-6674 ISSN 2730-6682 (electronic)
Earth and Environmental Sciences Library
ISBN 978-3-031-63045-3 ISBN 978-3-031-63046-0 (eBook)
https://doi.org/10.1007/978-3-031-63046-0

This Springer imprint is published by the registered company Springer Nature Switzerland AG
The registered company address is: Gewerbestrasse 11, 6330 Cham, Switzerland

Contents

Integrated Hybrid Biological Systems in Wastewater Treatment and Recycling

Arun Kumar Thalla and R. S. Devika

Abstract Population growth and climate change are identified as the primary cause of water scarcity at global as well as regional levels, and the final outcome of this scarcity or crisis is that a large portion of the world's population is under severe water stress and does not have adequate freshwater resources for satisfying present and future demands. Adding to this, the current water supply practices wherein freshwater is being supplied even for secondary usages like flushing toilets, gardening, street cleaning, etc., aggravate the water stress. It has become imperative to judiciously use our water resources to have sustainable progress. In this context, the world across, policymakers, researchers, and industries started thinking about treating the wastewater and effectively recycling the same to meet the secondary requirement, which accounts for about 50% of the total water requirement. The present chapter discusses the role of integrated hybrid biological systems in addressing wastewater treatment to recyclable levels, specifically in handling volumetric and organic shock loadings.

Keywords AS-Biofilm · Shock loading · Hybrid systems · Multistage reactors · Nutrient removal

1 Introduction

The history of wastewater management dates to 4000–3000 BC, when Mesopotamians and the people of Indus valley civilisation first designed and implemented drainage to carry wastewater away from residential regions and urban centres. On later periods, evidences have been found pointing towards the hygiene practices adopted by the Roman empire to manage storm water. However, the emphasis that

A. K. Thalla (✉) · R. S. Devika
Department of Civil Engineering, NITK Surathkal, Mangalore, Karnataka 575025, India
e-mail: arunkumar@nitk.edu.in

R. S. Devika
e-mail: devikars.222en009@nitk.edu.in

M. M. Ghangrekar et al. (eds.), *Biological and Hybrid Wastewater Treatment Technology*, Earth and Environmental Sciences Library,
https://doi.org/10.1007/978-3-031-63046-0_1

was given to wastewater management went on diminishing as human population progressed over the centuries. The consequences of this ignorance towards public hygiene and waste management were several contagious diseases including Cholera which killed over 2.5 lakhs people in Europe in the nineteenth century.

With the advent of industrial revolution, huge quantities of water were consumed for industrial applications. The advancement in chemical industries propelled the use of chemicals manifold. But no significant improvement was made in the field of treatment of wastewater. Only after the microbiology branch's development, awareness was made on the pathogenic microorganisms present in untreated water (https://organicabiotech.com/a-peek-into-the-history-of-wastewater-treatment/, accessed on 23 January 2024). In the early 1900s, treatment of wastewater started to be explored in terms of treatment units like filtration, settling and septic tanks. Disinfection methods by the use of chlorine were also prevalent. In 1916, Activated sludge process (ASP) was first implemented in the USA, marking the onset of secondary treatment research [1]. This also necessitated the need for further studies in sludge treatment and the importance of dissolved oxygen in wastewater treatment. Later, parameters like Chemical Oxygen Demand (COD), Biochemical Oxygen Demand (BOD), pH etc. and modifications to ASP like Sequence Batch Reactor (SBR) were introduced.

The 1950s witnessed vast changes globally, with and accelerated industrial development. This marked the progress of advanced research in processes and devices for wastewater treatment. This was the period when membrane technology was developed. Regulatory authorities framed and imposed various stricter regulations and standards due to growing concerns about the effects of wastewater treatment plant effluents on the environment [1]. Additionally, existing plants frequently need to be upgraded. Post 1970, computerisation fuelled automation in sample collection and analysis. Standardised methods were formulated for monitoring individual parameters.

The focus of research has been more inclined towards energy efficient and practical treatment methods, in the past few decades. As per UN Habitat, in 2015, only 32% (data from 42 countries, representing 18% of global population) of the domestic wastewater generated received some forms of treatment. It was 30% (data from 14 countries, corresponding to 4% of world population) for industrial wastewater. In 2020, with a wider range of information available from 128 countries (80% of global population), 56% of household wastewater was treated to safety standards (UN Habitat and WHO, 2021). Globally, more research articles on the area of wastewater treatment have been from countries like China (19.66%), USA (9.78%), and India (7.98%). Based on the bibliographic study conducted by Mao et al. (2020), biological treatment processes like plain ASP might reach a stage of saturation in the near future, however the modifications or upgradations of ASP to hybrid systems is being looked at a viable alternative to upgrade the existing systems to enhance the efficacy of the same. A significant amount of studies that have been conducted on biofilm and anaerobic processes have contributed to the rapid maturity of biological treatment technology.

More than just finding ways to reduce the environmental pollution caused by wastewater, we live in an era where the perspective has been changed, considering wastewater as a resource. Due to the rising concern of water scarcity, wastewater reuse after appropriate treatment has become necessary. It is predicted that by 2050, 40% of the world population will be facing water scarcity, given the tremendous increase in water consumption [2]. The concept of circular economy, wherein bioresource are recovered from the process of wastewater treatment in the form of biogas etc. has also gained much interest over the last few years [3]. This chapter aims to consolidate the opportunities that integrated biological systems offer in wastewater treatment.

2 Aerobic Systems for Wastewater Treatment

Aerobic systems treat and stabilize wastewater by using microorganisms to break down pollutants like organic matter and nutrients like nitrogen and phosphorous in the presence of oxygen. The use of mechanical aerators or mixers ensures the sufficient supply of oxygen in these systems. Aerobic systems are most suitable for treating diluted effluents. The end product of aerobic digestion is mostly carbon dioxide, nitrates etc. Aerobic systems can be of suspended or attached growth. In suspended growth, microorganisms will remain floating freely in the media. While in attached growth processes, the microorganisms grow on an inert media, forming a biofilm.

2.1 Aerobic Suspended Growth Process

The activated sludge process maintains an active sludge containing juvenile living biomass, bacteria, and other microorganisms, along with organic and inorganic particles, suspended in wastewater through aeration or stirring. This biomass utilizes organic material as an energy source, effectively removing organic substances and generating more biomass, leading to faster rates of organism stabilization. Control of suspended material is typically achieved by incorporating a sedimentation tank at the process's end, allowing gravity settling to transport biomass to the tank bottom. The purified wastewater is withdrawn from the top of the tank for further treatment or direct release into a water body, while excess sludge is recirculated or removed.

2.1.1 Activated Sludge Process (ASP)

The aerobic biological wastewater treatment method that is most frequently used and well studied is the activated sludge process. It is essentially a biological process where suspended aerobic microorganisms in the reactor degrades the organic matter in the wastewater in the presence of oxygen. The system basically consists of an aeration tank, where organic matter is oxidized, and a secondary settling tank where the sludge formed in the aeration tank is allowed to settle. The sludge thus settled is rich in microbial population, hence recycled back to the aeration tank.

ASPs provide excellent versatility for treating industrial and municipal wastewater. Despite decades of extensive research, the activated sludge method still has many practical drawbacks, including low resistance, low stability, low nitrification rate, and high rate of extra sludge. Either raising the system's total biomass content or promoting the growth of specialized organisms within the system will improve the process's efficiency. However, as elevated MLSS concentrations are typically accompanied by shortcomings in the viscosity, oxygen transfer, and overall activity of the biomass, the total biomass concentration in the system cannot be increased deliberately. Usually, the speciality organisms are grown on stationary surfaces inside the reactor. Therefore, to meet both the present and future treatment needs, a strategy for improving the current activated sludge process must address the structure of the microbial activated sludge aggregates themselves and combine the benefits of such a biofilm system with a suspended growth system.

2.2 *Attached Growth Grocess*

A more thorough description would be a layer of prokaryotic and eukaryotic organisms embedded in an organic matrix with biological origins that is anchored to a substratum surface. The bacteria become more resilient to changes in their environment by adhering to surfaces, forming little ecosystems, and adapting to withstand shock loads. In the field of environmental biotechnology, especially in the treatment of water and wastewater, biofilms are very critical.

Because of rapid urbanization, expanding treatment units in highly populated areas can be difficult from an economic and physical standpoint. Therefore, engineers and researchers must prioritize high-rate systems that can remove carbon and nitrogen while having small environmental footprints and offering the most economical solutions. Nitrifying biofilm reactors have been discovered to be an effective alternative for upgrading when compared to the activated sludge process. Biofilm units frequently require less area than typical basins with suspended bacteria because they can operate the reactors at high biomass concentrations without requiring the biomass and treated effluents to be separated. The process's controllability and restricted flexibility, however, are its clear drawbacks. Trickling Filters and rotating biological contractors are the most popular configurations among the attached growth systems.

(a) Trickling Filter (TF)

Trickling filters are aerobic attached growth systems, where the microorganism grow on packed beds (diameter upto 200 ft) made of inert material, while the wastewater flows over it in a downward direction. By attaching themselves to the medium as a biological film or slime layer (about 0.1 to 0.2 mm thick), a population of microorganisms (aerobic, anaerobic, and facultative bacteria; fungi; algae; and protozoa) can assimilate organic material from the wastewater. Microorganisms in the wastewater gradually attach to the media and form a biofilm as the wastewater flows over the medium. The aerobic bacteria in the slime layer's outermost layer then break down the organic material. Anaerobic microorganisms also proliferate because oxygen cannot pass through the medium face as the the layer become thicker due to microbial growth. The bacteria close to the surface of the biological film become unable to stick to it as it grows, and some of the slime layer slides off the filter. This process is termed as sloughing. The sloughed solids are removed from the wastewater by the underdrain system, which then carries them to a clarifier. The conventional materials used as the media in shallow filters (for diluted effluents) for are broken rocks, aggregates, coke, PVC, polyurethane foam, geotextiles, etc. Deep filters use corrugated PVC plastic sheets stacked up to allow the influent to flow in a criss-cross manner. A void content of 95% characterises this type of media and this together with the elongated pathway provides the maximum retention time for the wastewater in the system, leading to better treatment efficiency [4], simple process, autocleaning system, and effective ammonia removal. When the BOD load of the influent is so high, TFs can be provided in two stages for better removal rates of organic matter and nitrogen. Large sludge output and a high clogging rate are major drawbacks.

(b) Rotating Biological Contactor (RBC)

These are attached growth systems where large closely spaced circular discs are mounted over a horizontal shaft in one or more basins. These slowly rotating discs are partially submerged in wastewater, with biofilm formed on them. The biofilm, when it comes in contact with air takes up oxygen required for aerobic digestion. Apart from the bacteria in the biofilm, the system also has a consortium of microorganisms in suspension, by the sloughing of biofilm from the disc due to rotating motion. In general, partially submerged RBCs are used for carbonaceous BOD removal, combined carbon oxidation, and nitrification. For de-nitrification, completely submerged RBCs can be used. This treatment option is widely recognized for its simple design, low energy requirement, high reliability and efficiency in handling fluctuating loads. They can be used for various applications ranging from small scale decentralised systems to large scale municipal wastewater treatment plants [5].

RBCs are found to be effective in removing organic content as well as total nitrogen from the effluent under lower HRTs. Ashfaq [6] conducted studies showing that after a period of 1.5 h, there was no further improvement in effluent quality while treating grey water for organic content removal. Also, these systems are very effective in removing NH_4^+-N and total nitrogen from wastewater even at lower C/N ratios and lower temperatures [7].

2.3 Aerobic Hybrid Systems/Processes

(a) Activated Sludge-Biofilm Process

The AS-Biofilm Process combines attached and suspended growing biomass in a single unit by integrating biofilm carriers with traditional activated sludge units. The reactor can be categorized as either a fixed media or movable media type depending on how the biofilm media are firmly placed or permitted to float inside. In order to help maintain a higher concentration of biomass inside the activated sludge unit, biomass carriers offer a sizable surface area for the development of biofilm. Inert support medium is used by attached growth processes to promote biomass growth. There are dozens of media formats available. A vital and crucial part of an aerobic hybrid process is bio-media. The following qualities should be present in a quality media: High void fraction (ii) and high specific surface area (iii) cheap cost per unit surface area (iv) inert construction material (v) resistance to blockage or clogging (vi) robust mechanical strength (vii) flexible overall shape (viii) low maintenance (ix) low energy consumption (x) light attenuation (since nitrifying bacteria are light-sensitive, the packing should be opaque and the shape should prevent light from penetrating into the interior of the packing), and (xi) wettability (bacteria must attach to and grow on the media's surface in order for it to be hydrophilic.

The advantages of movable media over fixed media hybrid processes include high specific biofilm surface area, no need for backwashing, no filter bed channelling; low head loss, excellent mixing of reactor contents leading to an efficient mass transfer and resistance to plugging and clogging. However, the moving bed reactor has certain drawbacks when it comes to biological nutrient removals. They require high DO in the system to produce greater nitrification rates. Other drawbacks are high capital costs for the carrier material, and separation and recycling of media. Buhari et al. (2023) studied the application of co-biofilm, where biofilm develops on the root surface of water hyacinth plant for the removal of ammonia and Bisphenol A from wastewater. Bisphenol A (BPA) is an endocrine disrupter, considered as an emerging contaminant and ammonia–nitrogen (NH_3-N) is one of the most common pollutants in wastewater that uses dissolved oxygen in water and promotes eutrophication in surface water bodies. Co-biofilm treatment is a hybrid wastewater treatment method that combines root uptake by water hyacinth plant and moving bed biofilm reactors (MBBR) and was found effective for removing of BPA and ammonia.

(b) Membrane Bioreactor (MBR)

Membrane Bioreactors can be defined as combinations of membrane separation processes like ultrafiltration, microfiltration etc. and activated sludge process, which is essentially a biological process. The system basically consists of an aeration tank, which is the bioreactor, with a membrane (mostly immersed) placed either along with the bioreactor or in a separate compartment. The membrane acts as an inert barrier, which separates the sludge and gives a highly clarified effluent. Hence, the need for a secondary settling tank is eliminated. Compared to conventional biological treatments, the MBR has a number of advantages, such as stable effluent, less footprint, less sludge production, and flexibility for future development. High nitrification rates characterise them along with high removal rates of organic materials compared to a conventional ASP. However, the rate of denitrification is usually less, owing to the lack of anoxic zones in the system [8] the retention of microorganisms in the bioreactor, leading to better oxidation and sludge retention. This eventually allows for more control over the biological process. Based on the configuration MBRs can be of three types: (i) Submerged/immersed membrane bioreactor (iMBR) as shown in Fig. 1, where membrane is immersed directly into the bioreactor. A suction pump is used in conjunction with it to draw the effluent. Advantages are lesser energy consumption and low fouling potential. (ii) Cross flow membrane bioreactor (cMBR) as shown in Fig. 2, in which the membrane is incorporated into the aeration tank as a single unit. There is no requirement for a settling tank and the solids retention is higher. (iii) Hybrid membrane bioreactor as shown in Fig. 3, which is identical to an iMBR, but contains some carriers in the reactor which stabilises the treatment process and reduces the fouling of membrane [9].

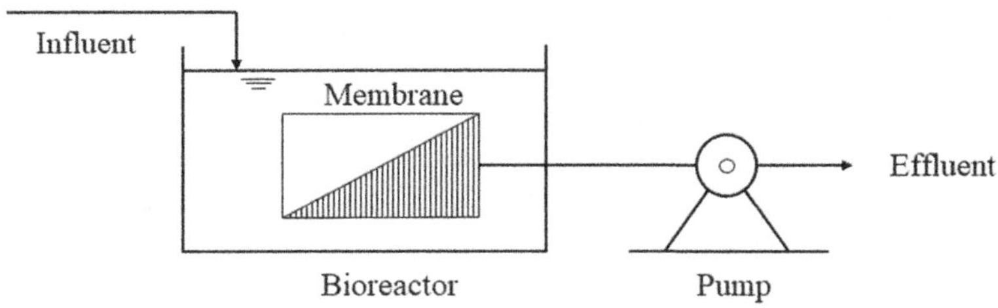

Fig. 1 Schematic diagram of submerged/ immersed membrane bioreactor. *Source* Wen et al. [9]

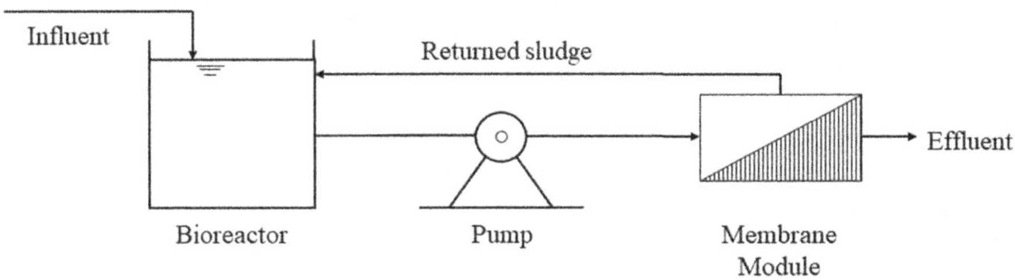

Fig. 2 Schematic diagram of cross flow membrane bioreactor. *Source* Wen et al. [9]

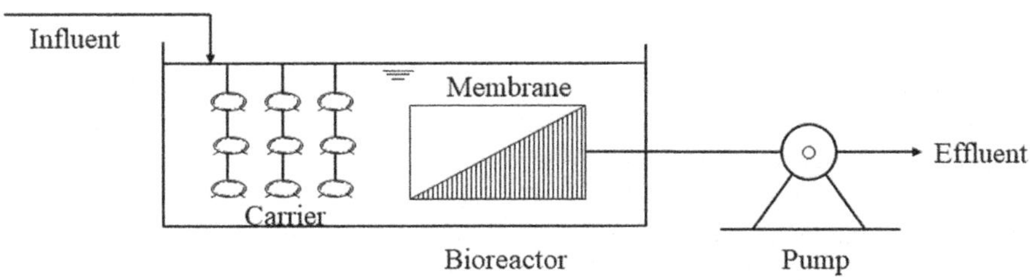

Fig. 3 Schematic diagram of hybrid membrane bioreactor. *Source* Wen et al. [9]

The membranes which separates the solids and liquid effluents in an MBR, can be of three geometries: (i) hollow fibre (ii) flat sheet or (iii) tubular (or multitubular). Hollow fibres are the most commonly ones used in water and wastewater treatment in the form of microfilters. The pores in the membrane are in the shape of long hollow tubes. While, flat sheet membranes are mostly rectangular, mounted on a frame, with a planar geometry. Tubular membranes have a porous tube-like structure, which are suitable for the separation of effluents containing high amounts of solids. This kind of membrane has a very high filtration surface, making it less prone to clogging. Among the three configurations, hollow fibre and flat sheet membranes are the most commonly used ones for wastewater treatment processes, with polyvinylidene difluoride (PVDF) being the most commonly used material.

3 Anaerobic Systems for Wastewater Treatment

Anaerobic systems are those which digests the organic fraction of wastewater by four basic processes, hydrolysis, acidogenesis, acetogenesis and methanogenesis. All of these phases take place under the absence of oxygen, unlike in aerobic processes. In the hydrolysis phase which is the rate limiting step, larger molecules (carbohydrates, proteins, lipids) are broken down into soluble simpler molecules (sugars, amino acids, fatty acids) in the presence of water. *Streptococcus* and *Enterobacterium* are some of the hydrolysis microorganisms. The acidogenesis phase involves the conversion of these hydrolysed organic molecules into short chain volatile fatty acids and alcohols by acid phase microorganisms like *Clostridium, Peptococcus Anaerobus, Lactobacillus*, and *Actinomyces*. Due to the production of organic acids in this phase pH in the system drops, which is not favourable for the further steps. The optimum pH for acidogenesis is 5.5–6.5. In the third step, volatile fatty acids are being converted to substrates for methanogenesis like acetates, hydrogen and carbon dioxide by acetogenic bacteria like *Syntrophomonas, Syntrophobacter* and *Smithella*. The final step is the conversion of acetates and hydrogen generated in the acetogenesis phase into end products which are gases like methane, carbon dioxide etc. by the aid of methanogens like *Crenarchaeota*, and *Euryarchaeota* belonging to *archaebacteria* genera. This step is characterised by a rise in the pH of the system to 6.8–8. A pH less than 6.5 is often not favourable for the process of methanogenesis to take place [10].

Apart from pH, the other major factors that influence the process of anaerobic digestion are organic loading rate (OLR), hydraulic retention time (HRT), and temperature. Based on the temperature at which digestion takes place, anaerobic digestion can be psychrophilic (10–30 °C), mesophilic or thermophilic. Mesophilic process takes place in a temperature range of 30–40 °C, ideally 37 °C. Thermophilic range is generally from 55 to 60 °C. Thermophilic digestion is usually recommended for wastewater with high organic contents, even though they require additional energy for maintaining the temperature. The major advantage of thermophilic digestion is more biogas production, under a shorter start-up time [11]. The treated effluents from anaerobic systems usually need to be polished in aerobic systems to meet the standards for disposal or reuse. Popular configurations among anaerobic systems are Up flow Anaerobic Sludge Blanket reactors and anaerobic baffled reactors,

(a) UASB Reactors

Upflow Anaerobic Sludge Blanket (UASB) Reactors are anaerobic systems for treating effluents that contain even complex compounds as well. They are high rate reactors characterised by high organic loading rates and a shorter retention time. They were developed in 1980 by Lettinga et al. [11] and have proven efficient in treating industrial and domestic effluents. Most advanced anaerobic systems rely on biofilms formed when microbial consortia get attached to an inert surface. While, UASB reactors primarily make use of granules of biomass that have good settling characteristics for the treatment of wastewater [11]. The granules along with the suspended matter forms the dense sludge bed under appropriate retention time, which is the most essential component of the system. Being dense, the biomass concentration in an UASB reactor has very minor chances of getting washed out, making it an efficient treatment option. The major advantages of UASB reactors are as follows:

- Higher removal of COD can be obtained with effluents of greater organic loads. Can handle shock loads as well in mesophilic range.
- Generation of biogas in the reactor by the anaerobic digestion of organic matter, that could be utilized for energy production.
- Lesser susceptibility of microbial consortia (both suspended and accumulated) to get washed out with effluent, reduces the volume requirements and retention time.
- Does not require primary sedimentation tanks, sludge digestors and secondary settling tanks.
- Like in other anaerobic systems, microbial (methanogens) growth is slower, since only a small fraction of organic materials is being utilized for biomass generation. Hence, the generated sludge is less in quantity and more stabilized in nature than aerobic systems.
- Ideal for treating effluents with high organic matter like industrial effluents, aquaculture wastewater, slaughterhouse wastewater etc.

(b) Anaerobic Baffled Reactors

Anaerobic Baffled Reactors were developed by Bachmann et al. [12] in the 1980s in the Stanford University. The distinct feature of these reactors is the longitudinal separation between the phases of anaerobic digestion provided by a number of vertical baffles. Theoretically, the hydrolytic and acidogenic bacteria are separated from the methanogenic bacteria which require a comparatively higher pH, in different compartments. This ensures appropriate growth conditions for each species of bacteria. Wastewater is forced to flow in upward and downward directions through the longitudinally placed baffles, ensuring high retention time. They are multi-stage system which require no mechanical mixing. The system is found to be efficient in treating high strength wastewater like that from food processing units and industrial sources, since the chances of biomass getting washed out are even lesser compared to UASB reactors, which are essentially single chambered. Among the various compartments, the first compartment is observed to have higher potential in removing organic matter and nutrients. Pre-treatment for removing suspended solids can be provided for better performance of the reactor. For the treated effluent to attain disposal standards, polishing must be provided througherobic systems [12].

4 Integrated Anaerobic-Aerobic Systems for Wastewater Treatment

Wastewater treatment plants consists of a number of unit operations, each one targeted to remove a particular type/ group of pollutants. The preliminary and primary treatment units remove most of the inorganic and larger particles, by means of physical processes. The secondary treatment is biological process aimed at removing the organic matter and nutrients present in wastewater. The average concentration of COD municipal wastewater falls approximately in the range of 250–1000 mg/L and that of BOD does in the range of 110–400 mg/L. However, the concentration of BOD and COD in industrial effluents is in the range of several hundreds or thousands, depending on the nature of the industry [13]. In addition to the higher organic loads, effluents from various sources contain compounds that have harmful effects on environment, unless treated properly. There has been an emerging concern over the Pharmaceutical and Personal Care Products (PPCPs), that have been detected in wastewater in recent years. They, along with several other chemical compounds that are present in effluents are basically organic compounds, but of unnatural origin. The degradation of such compounds sometimes becomes a complex process, as these compounds can be toxic to the microorganisms that are present in secondary treatment units. In such cases were biological treatment units need to be chosen but must deal with these compounds or very high organic loads, integrated systems are used for treatment. Integrated systems consist of two or more unit processes (aerobic or anaerobic) in a certain configuration that enables the whole system to manage the contaminants in a more efficient manner.

Anaerobic systems are usually preferred for the digestion of highly concentrated effluents, as they produce sludge with better settleability that often require no further treatment, unlike aerobic systems. In addition to that, the sludge generated by dead biomass in anaerobic systems are much lesser compared to aerobic systems. In aerobic systems the half of the organic matter consumed will be converted to energy, while the other half is converted into biomass (cell growth and multiplication). At the same time in anaerobic systems, only a small part of the organic matter is utilized for cellular growth, while the major part is converted into energy. Furthermore 60–70% of the biogas generated by anaerobic digestion will be methane, which can be utilized as fuel for producing heat energy [14]. Hence, anaerobic systems can be viewed as sustainable options that reduce the treatment process's overall carbon footprint. Although, anaerobic systems are effective in bringing down the organic content of high COD effluents, it becomes necessary to provide aerobic digestion systems followed by anaerobic reactors to meet the standards.

Various configurations of integrated treatment systems have been designed according to the required quality of treated effluent and the degree of pollution that is to be addressed. Satyendra and Vijay [15] studied a system consisting of anaerobic baffled reactor followed by an up-flow constructed wetland for treating municipal wastewater. The study reports the efficiency of an anaerobic baffle reactor (ABR) for treating organic matter and settling down inorganic materials. The treated water was found fit for activities like flushing, cleaning, gardening, and floriculture. In another study, anaerobic baffled reactors and aerobic trickling filters were combined to create a combination system for the treatment of lipid-rich wastewater. The ABR converted 57.8% of the COD in the inflow to methane, allowing for bioresource recovery as well [16]. A summary of other reactor configurations reviewed for the municipal wastewater are given in Table 1 and the integrated systems for treating industrial effluents are given in Table 2.

(a) Pharmaceutical Contaminant Removal

Pharmaceutical contaminates have been detected in surface water bodies all over the globe in recent times in alarming concentrations and harmful effects on

Table 1 Integrated biological systems for the treatment of domestic wastewater and their removal efficiencies

System	Type of effluent	Target pollutant	Removal rate (%)	Remarks	References
Integrated anaerobic/anoxic/oxic activated sludge biofilm (A^2O-AS-biofilm) reactor	Domestic wastewater	COD	95.5	C/N ratio = 4 Suspended biomass = 3 to 3.5 g/L HRT = 10 h DO = 1.5 to 2.5 mg/L	[25]
		Total Phosphorous	93.1		
		Total Ammonia Nitrogen	98		
		Total Nitrogen	80		
Aerobic Baffled Filter followed by Up-flow CW	Municipal wastewater	TSS	96		[15]
		COD	84		
		TN	68		
		TP	65.5		
Hybrid anaerobic/anoxic digester—constructed wetland	Municipal wastewater	TSS	98	Recirculation of nitrified effluent from wetland to digestor for denitrification	[26]
		COD	94		
		BOD	97		
		NH_4^+-N	86		

Table 2 Integrated systems for the treatment of different industrial effluents

System	Type of effluent	Target pollutant	Removal rates (%)	Remarks	References
Anaerobic digestor (AD) + UASB + MFC	Swine wastewater	COD	80.5% in AD, 61.1% in MFC	MFC enhanced TAN removal from the effluent from AF, which improved the COD/TAN ratio, leading to better COD removal in MFC	[22]
		Total Ammonia Nitrogen (TAN)	5.8% in AD, 77.5% in MFC		
Integrated anaerobic–aerobic sequence batch reactor	Slaughterhouse wastewater	COD	97 ± 2%	Tolerance to shock loads, Sludge with more settleability	[27]
		NH3-N	95 ± 3%		
		TSS	96 ± 3%		
Packed bed UASB-biological aerated filter-inclined plate settler	Synthetic wastewater	Enteric viruses (rotavirus, norovirus, Hepatitis E virus, and adenovirus)	95% TSS, 90% COD, 90% BOD, 66% TKN, 71% TP	Flow rate = 7 m^3/d, OLR = 2.64 kg BOD_5 / m^3/d	[28]
AS-Biofilm reactor followed by photocatalysis by TiO_2	Synthetic wastewater	COD	72%	Recovery of photocatalyst	[29]
		MPN	95%		
		Phosphorous	52%		
Hybrid anaerobic–aerobic SBR	Textile wastewater	COD	99.5%	Biogas generation from Anaerobic SBR, Optimum HRT: 54 h	[30]
		TKN	99.3%		
		Dyes	78.4%		
UASB followed by aerobic reactor	Industrial wastewater	2-Naphthol Red dye	98.4%	HRT = 6 h, OLR = 12.97 g COD L^{-1} d^{-1}, Maximum methane yield = 13.3 mmol CH_4 g COD^{-1} d^{-1}	[31]
		COD	85.6% in UASB, 92.4% in aerobic reactor		

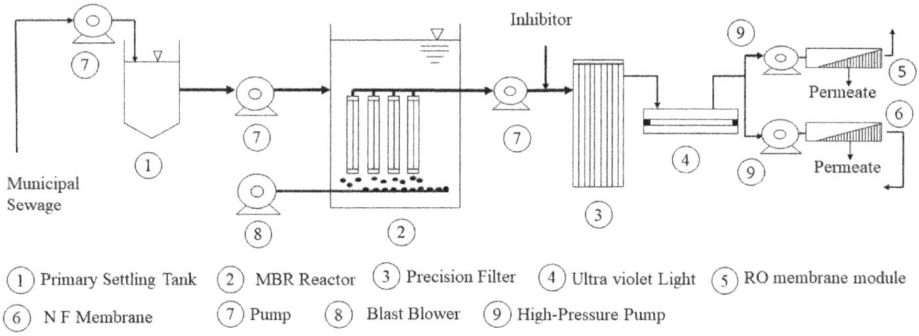

Fig. 4 MBR-RO/NF pilot plant. *Source* Gulamhussein et al. [17]

ecosystem. Gulamhussein et al. [17] investigated the possibility of using membrane bioreactors for the removal of pharmaceutical contaminants. Conventional methods of treating contaminants include activated carbon and chlorination. The MBR can be fitted with extra components, such as an external loop airlift membrane, which has been shown to eliminate all acetaminophen completely in less than 24 h. Additionally, it can be coupled with a reverse osmosis or nano-filtration membrane, as shown in Fig. 4 which has been shown to remove about 20 and 13 contaminants, with removal efficiencies > 90%. Increasing a few other parameters, including the sludge retention period, has reported to have notable removal efficiency while improving the bioreactor's performance.

The removal and fate of pharmaceuticals during wastewater treatment was investigated by Iliopoulou et al. [18] using an integrated lab-scale wastewater treatment system that consists of an aerobic Membrane Bioreactor (AeMBR) and an anaerobic Moving Bed Biofilm Reactor (AnMBBR) in series. Batch studies were carried out to calculate the sorption and biotransformation rates, while continuous-flow tests were carried out by applying different temperatures to the AnMBBR (35 and 20 °C). The AnMBBR's temperature did not impact the degree to which the principal contaminants and target medicines were removed. Ammonium nitrogen (NH_4–N), BOD, and dissolved COD were removed on average 86%, 91%, and 96% at 35 °C and 91%, 96%, and 96% at 20 °C, respectively in the system. Compounds like metronidazole, trimethoprim, sulfamethoxazole, and valsartan showed removal efficiency ranging from 65 to 100%, but carbamazepine and diclofenac showed little (< 30%) or no removal at all. The biodegradation of trimethoprim, sulfamethoxazole, and valsartan depended on the AeMBR, but the AnMBBR significantly boosted the removal of metronidazole. Over time. Bacteroidetes was the predominant phylum identified in both bioreactors. Bacteroides and Cloacibacterium were observed more in the biocarriers of the AnMBBR than in the suspended biomass.

(b) Nutrient Removal

Nutrient removal in wastewater treatment has evolved into a crucial process in preventing the eutrophication of surface water bodies. The ecological effects of excess quantities of nitrogen and phosphorous reaching the water bodies are much drastic. Unlike carbonaceous organic matter which require either aerobic or anaerobic systems for degradation, nitrogen and phosphorous in wastewater require combinations of various processes for their proper removal. In case of nitrogen, the first product formed by the decomposition of organic matter is ammonia nitrogen, which needs to undergo nitrification to form the intermediate compound nitrite that is extremely harmful and then the final product, nitrates. These nitrification processes take place in aerobic conditions. However, the non-polluting end products are not nitrates, but nitrogen gas. The process of conversion of nitrates to nitrogen gas is termed as denitrification, which takes place ideally under anoxic conditions. Similarly, for phosphorous removal also anaerobic digestion is to be provided at first, followed by aerobic. In most cases phosphorous removal in wastewater treatment plants is done by physicochemical methods like coagulation with ferric chloride. Sreekumar et al. [19] used adsorption by activated carbon made from activated sludge and thermally activated laterite soil for the removal of phosphorous.

The potential of integrated anoxic filter- aerobic rotary biological contactor for the treatment of rural domestic sewage was investigated by Cheng et al. [20]. The system featured great shock resistance, low sludge formation, and good efficiency have one pump-driven retention anoxic filter (RAF) and a hydraulic rotary biological contactor (HRBC). The variables that led to enhanced denitrification (85.9% nitrate removal) and deodorization (89.3% sulphide removal) included a reflux ratio of 175%, 5 h HRT in RAF and 2,5 h HRT in HRBC. The process of sulphide removal and denitrification was accelerated in the RAF by the selective enrichment of autotrophic sulphide-oxidating nitrate-reducing bacteria and heterotrophic sulphide-oxidating nitrate-reducing bacteria. The first stage RBC in the two-stage HRBC system primarily consisted of up of organic degraders, and the second stage RBC was mostly made up of ammonium oxidizers.

Wastewater with a high organic and nutritional content can be treated on-site using the sequential batch movable media biofilm reactor (SB-MBBR), as per Tsitouras [21]. In order to degrade organic matter as well as ammonia in the system, consortia of both autotrophic and heterotrophic bacteria need to be maintained. The competition among the two bacteria population will be higher if they are present in a single reactor treating high C/N effluents. This is avoided by a configuration with two SB-MBBRs connected in series to divide the two different bacteria. The first reactor (AN/AE) cycled between anaerobic and aerobic stages while breaking down carbon and phosphorus for the selection for organisms that accumulated polyphosphate. To encourage nitrification, the second reactor was run in an aerobic environment. The performance of the system was measured in terms of morphology and thickness of the biofilm, growth kinetics, dry weight and density. The efficiencies of removal that the AN/AE reactor attained were as follows: COD—$97.5 \pm 1.7\%$, total nitrogen—$58.8 \pm 5\%$, and total phosphorous—$49 \pm 1\%$. A high rate of $98.7 \pm 2.4\%$ total ammonia nitrogen was removed by the second reactor, suggesting this technology can be integrated into an anammox system.

Hybrid systems with microbial fuel cells (MFCs) were investigated by [22] for their potential use in the removal of total ammonia nitrogen (TAN) and residual COD in the effluent that was previously treated in an anaerobic digester (AD) designed to treat swine wastewater in batch process. The AD had a cumulative COD removal of 80.5%, but only 5.8% of TAN was removed, which resulted in a drop in the COD/TAN ratio from 14.5 to 3.0. In the followed MFC system, the COD/TAN ratio increased from 4.6 to 8.1 after 36 days, with 77.5% of the TAN eliminated. Consequently, the COD in the anode decreased even more, removing 61.1%. Based on these findings, removal of TAN in MFC was efficient in improving the COD/TAN ratio, which further enhanced the COD removal in MFC. Additionally, suppressing the activity of methanogens in the MFC can significantly prevent substrate loss and enhance the power output. Pre-treatment of anodic inoculum with of marine algae, Chaetoceros was found to enhance the Coulombic efficiency by 45% [23]. Another method of supressing the methanogenic bacteria is by heating the sludge inoculum at a temperature of 100 °C for 15 min [24].

(c) Shock Loads

Shock loads are created when high volumes or high strength wastewater flows into the treatment plant, causing disruptions to the functioning. They are usually short-term changes but can be twice as high as the average values. Systems with fixed film can perform well under shock load compared to suspended growth processes [32]. Not a lot of studies have been conducted on the effects of shock loads. Some of the works dealing with shock loads in biological systems are added here.

In a two stage Activated Sludge- biofilm reactor when shock loads were applied for an HRT of 6 h, with influent COD raised 2–4 times the average value, it was observed that the COD removal rate was proportional to the effective OLR. The recovery time was also proportional to the shock load applied [32].

Senturk et al. [33] studied how organic shock loads affected the performance of a anaerobic contact reactor operating in thermophilic conditions for treating potato processing wastewater. To the reactor operating at a constant loading rate of 5.5 kg COD/m^3.d, an acetate shock load of 1 g/L was applied. It was found out that the shock load had an adverse effect on the performance of the system. However, by a period of 24 h the functioning of the system reached back to a steady state, indicating the resistance of thermophilic systems in handling shock loads.

Saharia and Chakraborty [34] compared the effects of thiocyanate shock load on two continuous and batch fed moving bed bioreactors having anaerobic, anoxic and aerobic reactors in series. The thiocyanate load was increased from 600 to 1000 mg/L and then to 1200 mg/L in an interval of 3 days each. In anaerobic continuous reactor, increase of effluent COD due to shock load was found to be only 2%, while in fed batch reactor it was 14%. The damage and the reduction in ammonia removal efficiency observed in aerobic reactors were significant and irreversible, compared to anaerobic and anoxic reactors. In case of anoxic reactors, partial recovery was possible for fed batch systems, whereas complete recovery was possible for continuous systems. Even though fed batch system showed higher efficiency in substrate removal, it is more susceptible to the effects of organic shock loads.

Saharia et al. [35] analysed the effect of phenol shock loads on two sequential moving bed bioreactor systems operating in anaerobic–anoxic–aerobic conditions-continuous moving bed bioreactor and semi-continuous batch fed moving bed bioreactor respectively. Both the systems were applied a constant phenol loading of 2500 mg/L and it was increased to 3000 and 3500 mg/L on day 3 of the total 6 days HRT. The overall performance of the continuous system was found to be superior to that of the semi-continuous fed system. The load of 3000 mg/L had not much effect on the performance of the system, but the load of 3500 mg/L resulted in significant reduction in COD removal rates in the anaerobic reactors.

Zhang et al. [36] studied the effect of shock loads while treating leachate on UASBRs, which usually have poor resistance to organic shock loads. In order to improve the stability of anaerobic performance, pumice was added along with the leachate. When the influent COD load was increased to approximately 6 times, irreversible acidification and a reduction in methane content were observed in the absence of pumice. When pumice filler was added, the COD removal rate reached up to 91.8% and a 42.3% rise in methane output were also obtained. It was concluded that the pumice filler had a significant stabilising effect on the methanogens and thus on improving the resistance of the system to organic shock loads.

5 Natural Purification Syatems as a Sustainable Option: Constructed Wetlands (CWs)

Constructed wetland is a natural purification process wherein mechanisms such as anaerobic and/or aerobic digestion, and phytoremediation together play a significant role in the pollutant remediation. Wetland systems are these days widely used as it is a sustainable technology that meets SDG goal (SDG-6) and in addition it gives a reliable treatment quality and these CWs are proved to be capable of reducing the bacterial loads, which otherwise in any conventional biological treatment has to be met by additional disinfection systems. CWs has additional advantages viz. lesser energy consumption, aesthetic appeal, low operation and maintenance costs. Constructed wetlands are suitable especially for tropical climate, as there is more availability of sunlight. Lower temperatures and lesser sunlight can as adversely affect the performance. Apart from all the advantages, there are some drawbacks for constructed wetlands also, like large area requirement, emission of greenhouse gases, and potential for clogging and constant management.

The right plant species and sensible planting techniques are required to improve the effectiveness of phytoremediation in CWs. Numerous techniques [37] have been developed to monitor and assess CW clogging, including solid sampling, hydraulic conductivity testing, tracer tests, bio-electrochemical sensors, and geophysical methods. Pretreating wastewater, replacing the matrix, adding chemical oxidants, biosurfactants, and enzymes, adding earthworms or degrading bacteria, and other techniques are some of the primary anticlogging treatments. During the wastewater treatment process, CWs frequently release greenhouse gasses (such as CH_4 and N_2O), which can be managed using both conventional and bio-electrochemical techniques.

Conventional techniques include flow patterns, vegetation kinds, electron receptor supplementation, and the use of specific fillers. When it comes to shock loads in wetlands, it can come into picture in various events of changes in weather as well as seasons. The type of units and certain fluctuating parameters (such as inflow concentrations, seasonal variations in the vegetation, and water levels) have an impact on the overall system performance in such multistage CWs. In contrast to single CW, a multi-stage CW system has the potential to recover faster with respect to DO and ammonia nitrogen. However, the recovery is slower on BOD and total phosphorous. It was found that for handling the shock loads from BOD and total phosphorous, a pre-processing and buffering section were required. Compared to single-stage CWs, it is more difficult for multi-stage CWs to return to their initial water levels, although both recover swiftly [38]. To achieve the desired results CW [39].

The studies conducted by Thalla et al. [40] at the sewage treatment plant of the National Institute of Technology Karnataka (NITK), INDIA revealed that Vertical Flow Constructed Wetland (VFCW) showed improved mass removal efficacy with respect to Horizontal Sub Surface Flow Constructed Wetland (HSSFCW).

Büngener [41] studied the potential retention of microplastic, which have been a pollutant of emerging concern in recent times. Microplastics in the size range of 25–1000 μm were examined in inflow and outflow water as well as atmospheric deposition in a full-scale reed-based surface flow wetland that serves as a polishing stage for municipal wastewater treatment. It was observed that the passage of secondary treated effluent through the wetland in rainy season increased the microplastic concentration by 92%, while in the dry season it was 43%. The rise in the concentration during rainy season was attributed to atmospheric deposition (590 MPs m^{-2} $week^{-1}$).

Sánchez et al. [26] incorporated a hybrid anaerobic/ anoxic digestor (HD) as a pre-treatment for a vertical subsurface built wetland (VF) for the first time. To encourage the simultaneous removal of nitrogen along with organic matter, a portion of the already nitrified VF effluent was recycled back to the HD, accelerating the process of denitrification in the HD and the nitrification process in the VF. The findings show that HD primarily eliminated organic materials and suspended particles from the VF up to amounts that were consistently below the discharge limits. The elimination of nitrate increased from 64 to 94%, and the removal of ammonium from 64 to 89%. The combined HD-VF system reached an average of 98% for TSS, 94% for COD, 97% for BOD_5, and 86% for $NH_4{}^+$-N, indicating a good overall removal efficiency. The amount of methane released was less than what was documented in the literature for CWs. The VF didn't appear to be clogged.

The design approaches for CW are either area-based, which evaluates pollution reduction using the total wetland areas, or volume-based, which takes into consideration the hydraulic retention time to guarantee optimal pollutant removal. According to the study of Vymaza [42] that analysed various configurations of CW, the design to be used and applied is highly reliant on a number of criteria, including accessible assets, size, cost, environmental conditions, wastewater quality, treatment degree, and purpose of treatment. A significant barrier to the widespread use of CW is its enormous land area requirement, particularly in hilly and heavily inhabited places. In these areas, where land area is limited, and cost of the land is high, the demand for a compact footprint is greatest. The cost and availability of land may restrict the system's applicability because CW treatment might only be a financially viable choice in areas with inexpensive land. In order to get around this restriction, some academics have proposed deepening the system without expanding its footprint and stacking stages in a way that minimises footprint. Furthermore, a great deal of study has shown that when there is enough oxygen in the system, microorganisms may finish biodegradation, which increases the nitrogen and organic matter removal efficiency of the system and reduces system footprint. According to Zhang et al. [43] the removal of BOD was > 75 and 90% in Horizontal Flow type Wetland (HFCW) and Vertical Flow type Wetland (VFCW), the removal of COD was > 65%, and that of total suspended solids (TSSs) was < 80 and > 85% in VFCW and HFCW each. However, it was shown that HFCWs were more successful in removing TSS (> 90%), BOD (> 85%), and COD (> 85%) respectively.

According to studies, CWs might need a big footprint in order to guarantee high-quality effluent. However, each country has a different CW footprint. Reducing the footprint to 1 m^2 per Population Equivalent (PE) for tertiary treatment and 0.5 m^2 PE^{-1} for storm water treatment is advised by the Environmental Protection Agency (EPA). The variation with nations and climate zones is also there in the case of footprint. Recent findings reveal that an area of 4 ft^2 per person should be the adequate land requirement for constructed wetlands in wastewater treatment.

5.1 Recent Advancements in CW

5.1.1 Hybrid Constructed Wetland

According to the findings of Sakurai et al. [44] in order to improve treatment efficiency, vertical and horizontal flow wetland systems are joined to build hybrid wetlands. Hybrid CWs are preferred when removal of nitrogen (simultaneous nitrification and denitrification) is desired and being implemented across the world. Hybrid CW compensates for the shortcomings of each CW; for example, horizontal flow CW performs poorly in nitrification since there is less oxygen available. On the other hand, nitrification by vertical flow can convert ammonia to nitrate. A few of the numerous possible combinations are Horizontal flow followed by vertical flow and vice versa, and other stages, including the flow pattern alterations i.e. staged flows.

Leachate from landfills is a serious problem everywhere. Pesticides and a variety of personal care items are present in landfill leachate. Leachate pesticides such as dichlorodiphenyltrichloroethane (DDT) and hexachlorocyclohexane (HCH) are prohibited.

In the experiments of Liu et al. aand Zhang et al. [39, 43] to treat the HCH and DDT from landfill leachate, a recirculating subsurface three stage vertical flow CW (connected in series) was used. It is reported that the removal of HCH and DDT organics were 38.6% and 82.7% respectively. The various mechanisms involved in the removal of HCH and DDT in the CW's were said to be Phyto uptake (6.87–21.56%), bacterial remediation (20.38–27.89%) and medium adsorption (50.55–72.74%) respectively.

5.1.2 Baffled Subsurface Flow Constructed Wetland

To enhance the removal of contaminants, a novel design for the wetland built by HSF has been developed, which successively integrates turbulent flow [45]. In a built wetland, the design allows for the sequential treatment of pollutants in a variety of aerobic, anoxic, and anaerobic conditions. By installing vertical baffles throughout the wetland's width, the wastewater flows in vertical channels rather than horizontal from inlet to outlet. By allowing for up-and down-flow movement, baffles lengthen the track and maintain the wastewater's interaction with the root zone and surrounding microenvironment. Highway runoff is a significant non-point source that contributes heavy metals, polyaromatic hydrocarbons, nitrogen, and phosphorus, among other contaminants. The ability of baffled flow-constructed wetlands to create successively anaerobic, anoxic, and aerobic environments has made them popular recently.

5.1.3 Tidal Flow Constructed Wetlands

The most significant developments in CWs during the past 10 years have been tidal flow CWs (TFCWs), which surpass the insufficient O_2 source of traditional CWs to greatly improve BOD and ammonia nitrogen treatment capability. The filling and draining of wetland units with wastewater is the fundamental idea behind TFCWs. Saeed et al. [46] found that when TFCW is unsaturated, its pollutant removal rate is higher. When building waste, such as gravel and concrete blocks, is employed as media or substrate in TFCW, the removal of phosphate is maximised. The abundance of Ca, Al, and Fe aids in the phosphate adsorption process. Utilising organic waste as a substrate improves the CW's denitrification process.

5.1.4 CW-MFC

There are two zones in the CW: anaerobic and aerobic. It has mostly been determined that the upper layer of the CW is aerobic while the lower layer is anaerobic. The constructed wetland-microbial fuel cell (CW-MFC) is an integrated treatment system that was created taking into account the potential of the CW to create a redox environment. The CW-MFC is a bio-system consisting of bioprocess and biocontrol. The plant, which uses photosynthetic processes to produce voltage, is a component of the biocontrol framework. The the bioprocess structure implies bacterial population that consumes the oozes produced by the root to release power outputs through microbial metabolism. As per Ebrahami et al. [47] scalability is the primary flaw of CW-MFC's. The costs associated with design, operation, and maintenance increase when it is intended for field application, while the power output decreases as a result of an increase in internal resistance.

6 Summary/Conclusions

Based on the different systems discussed in the above sections, it is evident that though there are technologies that can treat wastewater to the desired recyclable quality standards for various purposes, the real challenge is to stitch together the appropriate technology that satisfies the needs of a particular environment and purpose with reliable performance and reasonable cost. In the present scenario where demand for clean water is rising rapidly and water is continuously getting polluted due to industrial and anthropogenic activities, it was proved that hybrid treatment units are more efficient in remediating wastewater than conventional treatment systems. It is also worthy to note that constructed wetlands are showcased as a natural and efficient method of treatment. It has also been observed that the cost of treatment of sewage using constructed wetlands have lower carbon footprint and accounts to about 52% lower expenditure per square meters compared to systems such as waste stabilization ponds, integrated anaerobic aerobic treatment plant and activated sludge plant.

References

1. Ranade VV, Bhandari VM (2014) Industrial wastewater treatment, recycling, and reuse-past, present and future. In: Industrial wastewater treatment, recycling and reuse. Elsevier Inc., pp 521–535. https://doi.org/10.1016/B978-0-08-099968-5.00014-3
2. Kumar V, Bilal M, Ferreira LFR (2022) Editorial: recent trends in integrated wastewater treatment for sustainable development. Front Microbiol 13. https://doi.org/10.3389/fmicb.2022.846503
3. Villarín MC, Merel S (2020) Paradigm shifts and current challenges in wastewater management. J Hazard Mater 390. https://doi.org/10.1016/j.jhazmat.2020.122139
4. Wastewater Technology Fact Sheet Trickling Filters 2000
5. Waqas S, Harun NY, Sambudi NS et al (2023) A review of rotating biological contactors for wastewater treatment. Water (Switzerland). 15(10). https://doi.org/10.3390/w15101913
6. Ashfaq A (2011) Preliminary study of greywater treatment through rotating biological contactor. https://www.researchgate.net/publication/264849826
7. Chen X, Zhang Q, Zhu Y, Zhao T (2022) Response of rotating biological contactor started up by heterotrophic nitrification-aerobic denitrification bacteria to various C/N ratios. Chemosphere 291(133048):0045–8574
8. Moktadir MdA, Maliha M, Tujjohra F et al (2024) Treatment of tannery wastewater by different membrane bioreactors: a critical review. Environ Adv 15:100478. https://doi.org/10.1016/j.envadv.2023.100478
9. Wen G, Ma J, Zhang L, Yu G (2010) 4.07 Membrane bioreactor in water treatment the design of bioreactor 198 4.07.2.1.2 membrane module selection and design 198 4.07.2.1.3 aeration system design 199 4.07.2.2 operation of membrane bioreactor

10. Zamri MFMA, Hasmady S, Akhiar A et al (2021) A comprehensive review on anaerobic digestion of organic fraction of municipal solid waste. Renew Sustain Energy Rev 137. https://doi.org/10.1016/j.rser.2020.110637
11. Latif MA, Ghufran R, Wahid ZA, Ahmad A (2011) Integrated application of upflow anaerobic sludge blanket reactor for the treatment of wastewaters. Water Res 45(16):4683–4699. https://doi.org/10.1016/j.watres.2011.05.049
12. Lau PL, Trzcinski AP (2022) A review of modified and hybrid anaerobic baffled reactors for industrial wastewater treatment. Water Sci Eng. 15(3):247–256. https://doi.org/10.1016/j.wse.2022.06.004
13. Bielefeldt AR (2009) Water treatment, industrial. http://www.epa.gov/
14. Guo H, Wei J, Ma Y et al (2022) Facilitated transport membranes for CO_2/CH_4 separation—state of the art. Adv Membr. 2. https://doi.org/10.1016/j.advmem.2022.100040
15. Satyendra VR (2024) Anaerobic baffled reactor followed by up-flow constructed wetland for treatment of municipal wastewater. J Water Process Eng. 58. https://doi.org/10.1016/j.jwpe.2024.104810
16. Tanikawa D, Yokote N, Nakahara K, Yamaguchi T (2016) Evaluation of process performance for lipid-rich wastewater treatment using a combination system of an anaerobic baffled reactor and an aerobic trickling filter. J Water Environ Technol 14(2):90–95. https://doi.org/10.2965/jwet.15-064
17. Gulamhussein MA, Saini B, Dey A (2023) Removal of pharmaceutical contaminants through membrane bioreactor. Mater Today Proc 77:260–268. https://doi.org/10.1016/j.matpr.2022.11.299
18. Iliopoulou A, Arvaniti OS, Deligiannis M et al (2023) Combined use of strictly anaerobic MBBR and aerobic MBR for municipal wastewater treatment and removal of pharmaceuticals. J Environ Manage 343(118211), 0301-4797
19. Sreekumar U, Thalla AK, Nair VV (2019) Comparative evaluation of phosphate abatement using laterite soil and activated carbon. Int J Environ Sci Technol 16(8):4493–4502. https://doi.org/10.1007/s13762-018-1926-1
20. Cheng H, Lee W, Wen C, Dai H, Cheng F, Lu X (2023) A sustainable integrated anoxic/aerobic bio-contactor process for simultaneously in-situ deodorization and pollutants removal from decentralized domestic sewage. Heliyon 9(11). https://doi.org/10.1016/j.heliyon.2023.e22339
21. Tsitouras A (2021) Sequencing batch moving bed biofilm reactors for treatment of cheese production wastewater
22. Kim T, An J, Jang JK, Chang IS (2015) Coupling of anaerobic digester and microbial fuel cell for COD removal and ammonia recovery. Bioresour Technol 195:217–222. https://doi.org/10.1016/j.biortech.2015.06.009
23. Rajesh PP, Jadhav DA, Ghangrekar MM (2015) Improving performance of microbial fuel cell while controlling methanogenesis by Chaetoceros pretreatment of anodic inoculum. Bioresour Technol 180:66–71. https://doi.org/10.1016/j.biortech.2014.12.095
24. Jadhav DA, Jain SC, Ghangrekar MM (2017) Simultaneous wastewater treatment, algal biomass production and electricity generation in clayware microbial carbon capture cells. Appl Biochem Biotechnol 183(3):1076–1092. https://doi.org/10.1007/s12010-017-2485-5
25. Manu DS, Thalla AK (2018) The combined effects of carbon/nitrogen ratio, suspended biomass, hydraulic retention time and dissolved oxygen on nutrient removal in a laboratory-scale anaerobic–anoxic–oxic activated sludge biofilm reactor. Water Sci Technol 77(1):248–259. https://doi.org/10.2166/wst.2017.537
26. Sánchez M, Ruiz I, Soto M (2023) Sustainable wastewater treatment using a new combined hybrid digester—constructed wetland system. J Environ Chem Eng 11(5). https://doi.org/10.1016/j.jece.2023.110861
27. Rajab AR, Salim MR, Sohaili J, Anuar AN, Salmiati, Lakkaboyana SK (2017) Performance of integrated anaerobic/aerobic sequencing batch reactor treating poultry slaughterhouse wastewater. Chem Eng J. 313:967–974. https://doi.org/10.1016/j.cej.2016.10.144
28. El-Senousy WM, Abou-Elela SI (2017) Assessment and evaluation of an integrated hybrid anaerobic-aerobic sewage treatment system for the removal of enteric viruses. Food Environ Virol. 9(3):287–303. https://doi.org/10.1007/s12560-017-9286-4

29. Manu DS, Thalla AK (2019) Influence of various operating conditions on wastewater treatment in an AS-biofilm reactor and post-treatment using TiO_2-based solar/UV photocatalysis. Environ Technol (UK) 40(10):1271–1288. https://doi.org/10.1080/09593330.2017.1420697
30. Shoukat R, Khan SJ, Jamal Y (2019) Hybrid anaerobic-aerobic biological treatment for real textile wastewater. J Water Process Eng. 29. https://doi.org/10.1016/j.jwpe.2019.100804
31. Gadow SI, Li YY (2020) Development of an integrated anaerobic/aerobic bioreactor for biodegradation of recalcitrant azo dye and bioenergy recovery: HRT effects and functional resilience. Bioresour Technol Rep 2020;9. https://doi.org/10.1016/j.biteb.2020.100388
32. Seetha N, Bhargava R, Kumar P (2010) Effect of organic shock loads on a two-stage activated sludge-biofilm reactor. Bioresour Technol 101(9):3060–3066. https://doi.org/10.1016/j.biortech.2009.12.055
33. Senturk E, Ince M, Engin GO (2014) The effect of shock loading on the performance of a thermophilic anaerobic contact reactor at constant organic loading rate. J Environ Health Sci Eng 12(1). https://doi.org/10.1186/2052-336X-12-84
34. Sahariah BP, Chakraborty S (2015) Comparative study on response of thiocyanate shock load on continuous and fed batch anaerobic-anoxic-aerobic sequential moving bed reactors. Environ Eng Res 20(1):65–72. https://doi.org/10.4491/eer.2014.049
35. Sahariah BP, Anandkumar J, Chakraborty S (2018) Stability of continuous and fed batch sequential anaerobic–anoxic–aerobic moving bed bioreactor systems at phenol shock load application. Environ Technol (UK) 39(15):1898–1907. https://doi.org/10.1080/09593330.2017.1343388
36. Zhang Y, Ji F, Hu Q et al (2023) Effect of organic shock loading on anaerobic performance of pumice-reinforced up-flow anaerobic sludge bed for incineration leachate treatment. Braz J Chem Eng 40(2):417–426. https://doi.org/10.1007/s43153-022-00264-0
37. Licciardello F, Sacco A, Barbagallo S, Ventura D, Cirelli GL (2020) Evaluation of different methods to assess the hydraulic behavior in horizontal treatment wetlands. Water (Switzerland). 12(8). https://doi.org/10.3390/w12082286
38. Jiang L, Chui TFM (2023) Performance and sustainability of a multi-stage free water surface constructed wetland under seasonal change and shock loading. Ecol Eng 186(106834):0925–8574
39. Liu S, Zhang Y, Feng X, Pyo SH (2024) Current problems and countermeasures of constructed wetland for wastewater treatment: a review. J Water Process Eng 57. https://doi.org/10.1016/j.jwpe.2023.104569
40. Thalla AK, Devatha CP, Anagh K, Sony E (2019) Performance evaluation of horizontal and vertical flow constructed wetlands as tertiary treatment option for secondary effluents. Appl Water Sci 9(6). https://doi.org/10.1007/s13201-019-1014-9
41. Büngener L, Postila H, Löder MGJ, Laforsch C, Ronkanen AK, Heiderscheidt E (2023) The fate of microplastics from municipal wastewater in a surface flow treatment wetland. Sci Total Environ 903. https://doi.org/10.1016/j.scitotenv.2023.166334
42. Vymaza J (2022) The historical development of constructed wetlands for wastewater treatment. Land (Basel). 11(2). https://doi.org/10.3390/land11020174
43. Zhang DQ, Jinadasa KBSN, Gersberg RM, Liu Y, Ng WJ, Tan SK (2014) Application of constructed wetlands for wastewater treatment in developing countries—a review of recent developments (2000–2013). J Environ Manage 141:116–131. https://doi.org/10.1016/j.jenvman.2014.03.015
44. Sakurai KSI, Pompei CME, Tomita IN, Santos-Neto ÁJ, Silva GHR (2021) Hybrid constructed wetlands as post-treatment of blackwater: An assessment of the removal of antibiotics. J Environ Manage 278. https://doi.org/10.1016/j.jenvman.2020.111552
45. Muduli M, Choudharya M, Ray S (2023) A review on constructed wetlands for environmental and emerging contaminants removal from wastewater: traditional and recent developments. Environ Dev Sustain. Published online 2023. https://doi.org/10.1007/s10668-023-04190-0
46. Saeed T, Miah MJ, Khan T, Ove A (2020) Pollutant removal employing tidal flow constructed wetlands: media and feeding strategies. Chem Eng J 382. https://doi.org/10.1016/j.cej.2019.122874

47. Ebrahimi A, Sivakumar M, McLauchlan C (2021) A taxonomy of design factors in constructed wetland-microbial fuel cell performance: a review. J Environ Manage 291. https://doi.org/10.1016/j.jenvman.2021.112723

An Overview on Recent Developments in Biological and Hybrid Wastewater Treatment Technology in India

Sunil Jayant Kulkarni

Abstract Conventional biological wastewater treatments are broadly classified as suspended growth and attached growth treatments. The biomass is suspended in wastewater in suspended growth processes. It is attached as a slime layer to packings or rotating discs in attached growth processes. In attached growth processes microorganisms have large surface area for growth. Attached growth processes are costly and have high oxygen requirements. Activated sludge processes are suspended growth processes for biological treatment of water. Microorganisms consume the organic matter to purify the wastewater. The biological methods can also be aerobic and anaerobic. Anaerobic water treatment is utilized for very high organic loading and produces methane. Due to odour problems, this method is underutilized. Advancement in technology to intensify the processes is a key research area in wastewater treatment technology. Reductions in cost, area and time have been the driving forces for innovations in biological treatment technologies. Combining the advantages of one biological treatment process with other technologies such as membrane separation, advanced oxidation with UV light, ozone treatment, and biosorption for wastewater treatment are widely investigated research topics in wastewater engineering. Investigations of biological treatment in recent past are focused on optimization of process parameters and combining the biological treatment processes to obtain high percent removal of the pollutants along with modelling of the removal mechanism. This book chapter sheds light on the recent investigations on biological treatments along with an overview of treatment methodologies with a focus on research, investigations and initiatives in India.

Keywords Biomaterials · Membrane · Oxidation · Microorganisms · Contaminants

---

S. J. Kulkarni (✉)
Chemical Engineering Department, Gharda Institute of Technology, Lavel, Maharashtra, India
e-mail: suniljayantkulkarni@gmail.com

M. M. Ghangrekar et al. (eds.), *Biological and Hybrid Wastewater Treatment Technology*, Earth and Environmental Sciences Library,
https://doi.org/10.1007/978-3-031-63046-0_2

1 Introduction

Treatment of sewage wastewater is a major environmental concern across the world along with industrial effluent. Only 28% of urban wastewater is properly treated [1]. Unaware of its consequences, the large population of India is still being exposed to unsafe water resources. The rivers like Ganga and Yamuna are used to dump the effluent from industries. The industries such as dairy, dye, agrochemicals and textile are becoming cause of concern due to environmental foot prints through water contamination. By 2050, 90 and 81 km^3 of water will be required for urban population and Industrial use respectively which is 30% more than current requirement according to Central pollution control board [2]. According to Frost and Sullivan report, the water and wastewater treatment in industry is expected to cross 2 billion dollars which is 1.3 billion dollars today [3]. In India, approximately 6.2 billion litres of industrial effluent is generated every day [4]. Reuse of gray water has the potential to reduce water scarcity by 50% [5]. About 80% of water used in domestic activities becomes wastewater of which 62% is greywater [5].

Treatment of domestic and industrial wastewater focuses on removal of pollutants, not only below the standards of regulatory bodies but also for reuse and recycle. Increased water scarcity issues have made the government, housing societies and industry sector to facilitate reuse of water. Three Rs[reuse, recycle and reduce] are becoming important aspects of sustainable development. Reducing the water usage by avoiding unnecessary wastage, recycling or reusing of the water after treatment; are common practices that should be followed considering water scarcity and pollution problems. Figure 1 depicts the technologies used for sewage treatment in India [6]. The data is obtained from NITI Ayog report, 'Reuse of Treated Wastewater in Urban/Peri-Urban Agriculture in India'; published in June 2023 [6].

Wastewater treatment technologies are classified as physical, biological and chemical treatments. Most of the treatment plants are treating the effluent or sewage in three stages named as primary, secondary and tertiary treatments. Biological treatments are considered as secondary treatments. Conventional water treatment plants use activated sludge process with or without modification. The biological wastewater treatment can be suspended growth or attached growth. The biological sludge is suspended in wastewater in the former and attached to support in the later. Activated sludge process (ASP) is suspended growth process whereas biotowers and trickling filters are widely used attached growth processes. Moving bed biofilm reactors (MBBR) exploits advantages of attached and suspended growth processes. The wastewater treatment can be based on aerobic or anaerobic decomposition. Highly thick sludge is normally subjected to aerobic-anaerobic or anaerobic treatment. Anaerobic treatment provides value added products like methane and biogas. Odour is a major problem with anaerobic treatments. Aerobic treatment is facilitated by bubbling air or oxygen in the reactor to oxidize the organic matter and produce settleable sludge. Bionanotechnology is an emerging area with huge potential in wide range of applications and sectors. Application of bionanotechnology has a potential

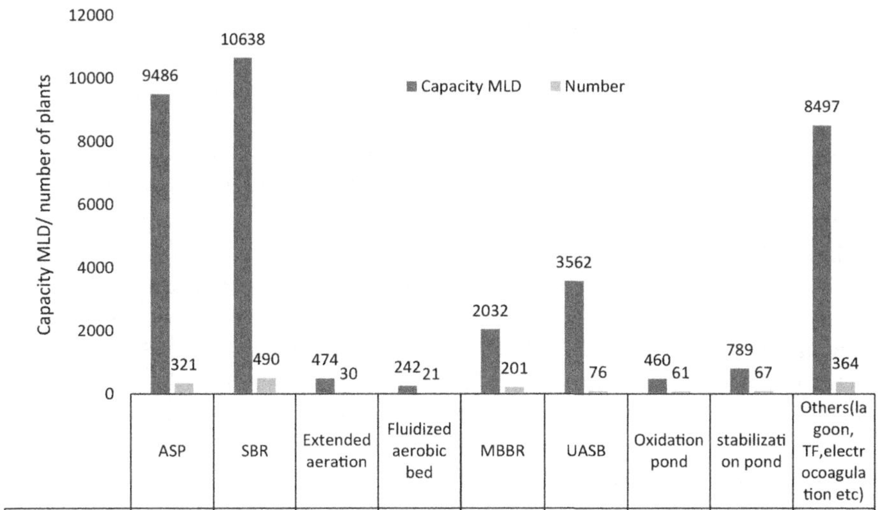

	ASP	SBR	Extended aeration	Fluidized aerobic bed	MBBR	UASB	Oxidation pond	stabilization pond	Others(lagoon, TF,electrocoagulation etc)
Capacity MLD	9486	10638	474	242	2032	3562	460	789	8497
Number	321	490	30	21	201	76	61	67	364

Fig. 1 Technologies used for sewage treatment in India [6]

to provide solution to many wastewater treatment issues. Also, conventional wastewater treatment technologies are inadequate to deal with 'Emerging Contaminants'. Bionanotechnological approaches can be explored to eliminate these new pollutants from waster. This chapter aims to sheds light on the recent developments in biological wastewater treatments and their combinations with other treatment technologies to provide advanced and effective treatment methods for wastewater.

1.1 Objective

Investigations on biological treatments are focused on studies of parameters affecting the treatment, removal of specific pollutant, treatment of emerging pollutants, possibility of combining the aerobic and anaerobic mechanisms, exploiting the advantages of membrane separation, filtration, cavitation, advanced oxidation to enhance performance of the biological treatment. Recent investigations are focused on advanced treatment methods and modelling of the kinetics and mechanism of the treatment processes. This chapter contains insight into the studies and investigations on biological treatments in India. The book chapter explains basic principles of the wastewater treatment method, thrust areas of investigations for the technology followed by India centric insight into research and studies on the particular treatment method. The book chapter aims to shed light on recent investigations while explaining basic principles for better understanding of research and investigations on biological treatments in India.

1.2 Outline of the Chapter

In the book chapter, Sect. 2 is divided into 10 subsections based on the treatment methods. Each subsection contains two parts. The first part contains the working principle, important aspects of the treatment process with citation of some significant research, advantages and disadvantages. In the second subsection, some significant investigations and studies on the particular treatment technology, mainly from the last 3–4 years in India are discussed. The treatment technologies considered in the chapter are selected based on their use in the treatment plants in India. These treatment methods are, activated sludge process (ASP), Trickling beds (biotowers), Rotating biological contactors (RBC), Biosorption, Advanced oxidation process (AOP), Extended aeration, Membrane bioreactors (MBRs), Moving bed biofilm reactors (MBBR), Sequential batch reactor (SBR) and Upflow anaerobic sludge blanket (UASB) reactor. The last section of the chapter is devoted to hybrid wastewater treatments followed by the conclusion.

2 Insight into Developments on Wastewater Treatment Methods with Focus on Research, Studies and Initiatives in India

2.1 Activated Sludge Process

Reducing the aeration cost is one of the important topics for investigators. Aeration efficiency can be increased by using microbubble and nanobubble technologies. Use of bacterial microalgae can also help in increasing treatment efficiency. Increasing nutrient removal is also very important aspect of modern-day investigations. Advanced activated sludge processes are being explored keeping these two aspects as main focus along with recovery of valuable materials [7]. Materials like cellulose, bioplastics, biomass, clean water and other valuable materials present in effluent or sewage water can be recovered. Activated sludge process, thus can be incorporated into the circular economy of waste material. More and more detailing of the wastewater composition and the mechanism can help in more effective isolation of useful materials. The conventional BOD/COD ratio for selection and design of ASP can be inaccurate as it depends on oxygen consumption. The oxygen independent organic matter and inorganic matter is not considered in this ratio. Aspects like this needs to be relooked into while designing activated sludge processes. Figure 2 depicts advantages and disadvantages of activated sludge process.

Aziz et al. [8] used conventional design method for stage wise activated sludge treatment. Reusing the wastewater was one of the objectives of their work. They used 23 parameters including chemical and biological oxygen demand as quality parameters. Their studies indicated that the COD to BOD ratio of 0.5 was good

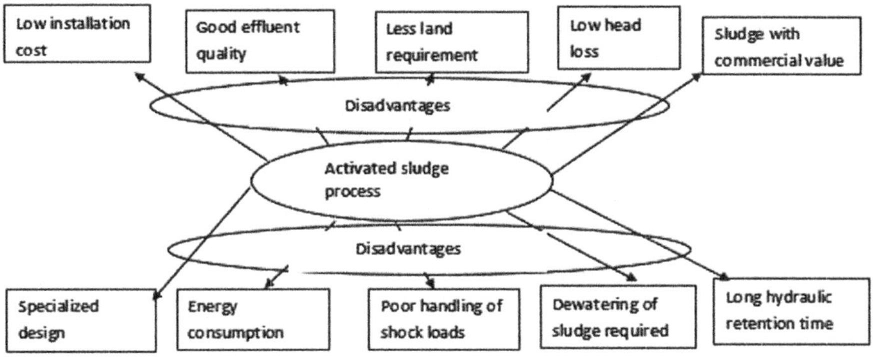

Fig. 2 Advantages and disadvantages of activated sludge process

enough to treat the wastewater by biological treatment. They used screening, grit chamber followed by flow equalization. Flow equalization avoids operational difficulties like flow fluctuations and helps improve downstream processing performance. The equalization basins can be in line or offline.

Gernaey et al. [9] discussed various aspects related to modelling of activated sludge process. Initial steps in the stepwise model development include definition, model and parameter identification. Characterization of wastewater and data reconciliation are next steps. Calibration of parameters and model falsification is done after data reconciliation. Accuracy of the model is determined in model unfalsification step. The last step includes scenario evaluations. For design of activated sludge process, International water association proposed models named activated sludge model ASM1, ASM2, ASM2D and ASM3 [10, 11]. While modelling, Environmental effects and biodegradation kinetics were important aspects in ASM1. The models evolved at later stages considered these parameters for further developments. Temperature dependence can be studied by using temperature correction for kinetics. Some investigators have reported Arrhenius like correlation for temperature effects (ASM2). ASM2 temperature coefficients are valid for 10 to 25 °C. Cell growth limitations were included in ASM2 model.

According to Demoulin et al. [12], cyclic ASP technology and sequencing batch reactors belong to the group of activated sludge process. Deepak and Rustum [13] reviewed nature inspired algorithm applications in activated sludge systems. Particle swarm optimizations and genetic algorithms were employed by them. There are three main types of modelling approaches namely Deterministic, stochastic and hybrid. According to them, ASM1 can be used for simulation of carbon removal and denitrification. ASM2 can be used for phosphorus removal, decarbonation and denitrification. According to Lowe et al. [14], limited academic transparency is a road block in further development of models for ASP. Small size of data used in academic research is also a limitation [15].

Conventional activated sludge processes are not energy efficient. To overcome this issue, investigators have suggested high-rate activated sludge treatment. Most of the sludge in these systems with low sludge retention time, is retained and can be used for biogas synthesis [16]. Calibrated model was used by Latif [17] for eliminating the effect of primary settling from activated sludge process. Chemical and biological oxygen demand (COD and BOD) was not affected by elimination of primary settling. After removal of primary settling tank, total suspended solid removal decreased and NH4-N removal increased.

2.1.1 India Centric Insight into Research on ASP

Narayan C. and Narayan V. [18] have reviewed biological wastewater treatment related literature [18]. According to them, it is necessary to optimize the amount of activated sludge that is recycled. Bardenpho process employs four stirred tanks in series. These serve the purpose of pre-denitrification, aeration, denitrification and sedimentation. Activated sludge plants from northern India were studied for removal of Coliforms by Kazmi et al. [19]. They proposed that high mixed liquor suspended solid mechanism with high sludge retention time was capable of removing the residual fecal and total coliforms. Tyagi et al. [20] have investigated the effect of ciliated protozoan population density on the effluent quality in two modes of activated sludge processes in India. They identified ciliated protozoa in effluent at sewage treatment plant at Haridwar. Few protozoan communities were identified from a plant in Delhi. Their studies indicated that *Aspidisca*, *Euplotes* and *Chilodonella* species were more frequent in the sludge samples. These species were indicators of effluent quality as these have strong association with low effluent concentration of BOD. Sarkar et al. [21] carried out dynamic simulation of the wastewater from Titagadh wastewater treatment plant. They proposed modified processes for treatment of the sludge by using the software 'STOAT'. By using modified process, they were able to reduce the COD as low as 38.8 mg/l. Pillai and Subramanyan [22] explained the role of protozoa in activated sludge process. Six or more species of Epistylis have been found active in the activated sludge processes across India. Normally these species flourish in an aeration tanks and its walls. In India, activated sludge process is also used effectively to treat textile and dye industry wastewater [23, 24].

2.2 Trickling Beds [Biotowers]

Trickling filters [TF] are attached growth biological processes. The trickling filters consists of a bed containing packing materials, on which biological slim layers are allowed to grow. Filter media supports can be rocks, gravels, plastic media or foam of polyurethane. The wastewater is sprinkled on the trickle bed. The biological slime grows on the media support. The microbes on the packings absorb dissolved organic compounds from the effluent. The trickling filters are easy to operate, high filter loading capacity, and efficient in ammonia removal. Tricking filters are capable of removing 85% of organic matter from wastewater [25]. TFs are appropriate for small

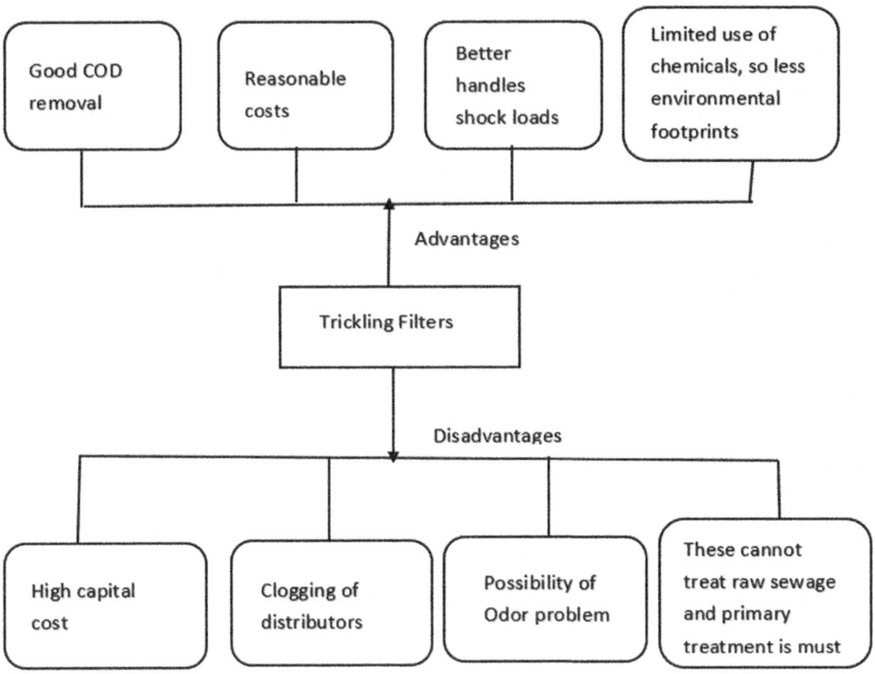

Fig. 3 Advantages and disadvantages of trickling filters

and medium scale enterprises [26]. Greater void space and porous nature of media allows more hydraulic loading [27]. Figure 3 depicts advantages and disadvantages of trickling filters. Bio tower is a generic term used for the processes where biofilm is attached to the media that is used as filter material for wastewater.

2.2.1 India-Centric Insight into Research on Trickling Filters and Biotowers

Aslam et al. [28] used cotton sticks as a filter media in trickling media. By using this trickling filter, BOD removal of 75% on an average was obtained. COD removal was 65 to 80%. Total suspended solid removal was 20 to 36%. Effluent samples from Barapullah drain, New Delhi were treated by trickling filters for emerging pollutants by Shukla and Ahammad [29]. They modified the trickling filter to contain sequential aerobic and anaerobic zones. Polyeurethin foam was used as filling material in the trickling filter. The aerobic zone of Modified trickling filter was more effective for removal of emerging pollutants. Bhat et al. [30] used three bed trickling filter for the secondary treatment of dairy effluent. Up to 4 trials, a steady performance was obtained in terms of COD, BOD and nitrogen removal. Investigators used various agricultural materials as a media support in tricking beds by investigators [31–35]. Investigations on kinetics of the pollutant removal from wastewater for biofilms are also reported in the literature [36, 37]. Optimization of trickling bed processes is focused on the parameters namely filter media, filter media thickness and flowrates [38]. Optimum parameters for the trickling filter system depend on the water's strength and composition [39–41]. Recirculation of the secondary clarified effluent was studied by Hanumanulu [39] for deep tricking filter (5.5 m deep). It was observed that the recirculation of the secondary effluent enables seeding throughout the depth of the filter and dampen the variations in the loading. Performance of trickling filters can be improved by using various packing materials namely stone, rubber, polystyrene and plastic [40]. By using these filter media material, 80 to 90% COD removal was obtained [40]. Kanwar et al. [41] reported an investigation on use of locally available agricultural materials namely maize cob and date palm fibre as packing material in trickling filter for treating municipal sewage wastewater [41]. Pathogen [total count, faecal coliforms and *Escherichia coli]* removal with 61 to 76.3% efficiency was obtained for these two trickling filters. For all pollution indicators, maize cob trickling filter showed 8 to 15% higher efficiency than the date palm fiber trickling filter.

Trickling filters and biotowers are used for the treatment of domestic as well as industrial effluent. The research on tricking filters and biotowers is mainly focused on the use of different media or packing materials that are a waste of agricultural activities. Also, investigations of plastic and polyethylene support in biotowers are reported with encouraging results. Investigations on the kinetics of pollutant uptake have been carried out by many investigators [26, 34, 36].

2.3 Rotating Biological Contactors

Rotating biological contactors (RBC) are attached growth methodology for treating effluent with high organic loading. High contact time enables RBCs for 8 to 10 times higher treatment than the trickling filters. Rotation of RBC and its exposure to the atmosphere enable the uptake of the food, oxygen and mixing needed for attached growth organisms to multiply and thrive. This method is limited in terms of use, by factors like the need for continuous electricity, high initial and operating costs, need for skilled manpower, need to protect from sunlight and possible odour problems. Low-stage production, high stability, low space requirement, short contact period and noiseless treatments are attractive attributes of the process.

2.3.1 India Centric Insight into Research on RBC

Radha [42] explored the application of rotating biological contactors for the treatment of vegetable oil wastewater. She achieved the maximum BOD removal above 95% and COD removal near to 90 by using RBC. RPM of 8, 30% submergence of the

8 discs were the optimum used conditions for the treatment. The modified Stover–Kincannon model fitted the kinetic data with almost same experimental values. Powar et al. [43] obtained nearly 89–90% removal of COD and BOD in RBC. They treated the dairy industry effluent for its reuse in agriculture. An investigation carried out by Subhadarsini and Dash [44] indicated that, with polypropylene foam as biofilm medium, it is possible to removal the BOD in the range of 85 to 90%. Many investigations are reported on effluent treatment by RBCs for dairy and dye industry wastewater [45, 46]. Rotating biological contactors are also useful for removal specific pollutants like phenol from wastewater [47]. Anaerobic RBCs can also be used for heavy metal removal under sulphate reducing conditions [48, 49]. Many times, lack of know-how and technical knowledge become hindrance in the use of RBCs for wastewater treatment [50].

2.4 Biosorption

Microorganisms like algae, yeast and bacteria have attracted the attention of experts due to their ability to biosorb heavy metals [51]. Biosorption is energy independent process. It requires strongly bonded solid phase for removal of pollutant from liquid phase [52]. Biosorption is a quick process and hence can treat large volume of water. It is a complex process that involves different mechanisms for removing pollutants from wastewater [52]. Not only microorganisms like bacteria, fungi, algae but also waste of dead animals are capable of adsorbing pollutants. The mechanism to remove pollutants can be metabolically mediated or physiobiological. According to some literature, Biosorption is metabolically inactive process as against bioaccumulation in which the biosorbent is metabolically active. Barange et al. [53] investigated use of microalgae for removal of heavy metal from wastewater. According to them, removal of pollutants by biosorption depends on pretreatment, pH and temperature. They obtained lead removal more than 400 mg per gram of biosorbent for microalgae. Figure 4 depicts mechanisms involved in biosorption.

Hussein et al. [54] reported use of walnut shell for biosorption of lead, chromium, mercury and sodium ions. Rao and Bhargavi [55] used alkali-treated biomass for the removal of heavy metals from wastewater. Their study indicated that pretreated biomass of Aspergillus Niger was effective in removal of lead and nickel from wastewater. Maximum removal of 97% was observed at optimum pH value of 6. Moringa oleífera was used for biosorption of pollutants from pharmaceutical waste by De Olivera [56]. They were able to remove 88% Diclofenac (DCF) and 50% Oxytetracycline [OTC] from wastewater at pH 2 and 10 respectively.

2.4.1 India Centric Insight into Research on Biosorption

Oves et al. [57] carried out an investigation to estimate metal biosorption by bacterial strain OSM29 recovered from rhizosphere of cauliflower grown in soil irrigated

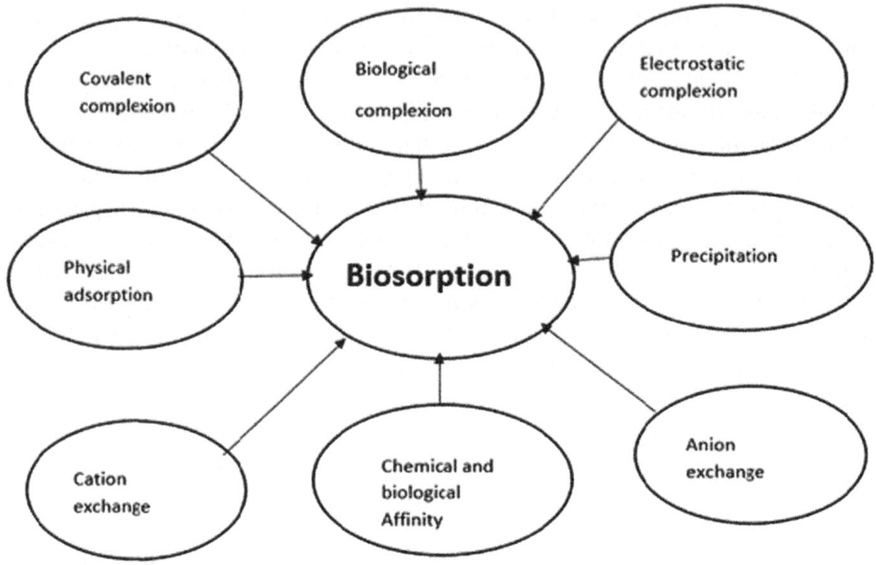

Fig. 4 Biosorption mechanisms

by industrial effluent. They identified the bacteria as Bacillus thuringiensis. They conducted experiments for heavy metal removal and observed that optimum pH for nickel and chromium biosorption was 7. For cadmium, copper and lead, optimum pH was 6. Biosorption capacity of the biomass was highest for nickel (94%). Tyagi et al. [58] isolated parapedobacter sp. ISTM3 strain from the Mawamai caves in Meghalaya. They used the Extracellular polymeric substances [EPS] obtained from these strains for removal of chromium from wastewater. They carried out experimentation with 20 mg/l concentrations of heavy metal mixture containing Zn^{2+}, Cu^{2+}, Pb^{2+}, Cr^{6+}, Fe^{2+}, and Cd^{2+}. The highest removal was obtained for chromium-VI. The adsorption capacity of 19 mg/g and 95.1% removal of Cr^6 was obtained by them. Langmuir isotherm was a better fit for metal uptake indicating adsorption capacity of 33.7 mg/g. Green marine microalgae from the gulf of Kutch (Okhla port) was investigated for biosorption by Kumar et al. [59]. They carried out experimentation for removal of cadmium, mercury and lead from wastewater. Mercury biosorption was maximum at initial concentration of 20 mg/l, 40 mg/l and 60 mg/l depending on the microalgae used. For different microalgae different parameter values exhibited maximum removal. Lead and Cadmium removal was maximum at 20 mg/l for almost all algae. Two hours contact time was optimum in all the cases.

Tanvi et al. [60] investigated indigenous microbiota for arsenic removal from the industrial sewage from Davangere district in Karnataka. *Aspergillus* spp APR-1 and APR-2 indicated arsenic removal above 50%. Patel et al. [61] used marine seaweeds from north west coast of India for biosorption of copper. At pH 5 and adsorption dosage of 10 g/l, they obtained maximum adsorption 60.97 mg per gram of adsorbent. Three hours contact time was sufficient for biosorption of 100 ml of sample. Sajenbam et al. [62] isolated aqueous Streptomyces VITSVK9 sp. from marine sediments from the bay of Bengal, India. They used these for biosorption of lead and mercury from wastewater. Their investigation indicated that the isolates had maximum tolerance concentration of 1800 mg/l for lead while for mercury it was 200 mg/l. Biosorption of phenolic compounds on various biological substances is also reported [63–66]. Rao and Viraraghwan [66] were able to remove 66% phenol from effluent at pH 5.1 and initial concentration of 1000 μg per litre by Aspergillus niger biomass. Studies carried out by Aravindan et al. [64] indicated that green microalgae has the capacity of 20 mg/g phenol loading. Adsorption and biotreatment combination can be used for cyanide removal along with phenol [65].

2.5 Advanced Oxidation Process

Strong oxidising agents are used in advanced oxidation processes (AOPs). Generation of transitory species can be facilitated by using nanomaterials as catalysts [67]. Large scale design can reduce the cost of AOP. The AOP processes can be classified into various categories based on method used for advanced oxidation. Various types of AOPs are depicted in Fig. 5. Various nanoparticles can be used as catalyst in catalytic oxidation to target specific pollutant.

In advanced oxidation process some pollutants that cannot be oxidized by using normal oxidising agent, are oxidized by using non-selective chemical oxidant, hydroxyl radical [68]. Homogeneous AOPs with Ultraviolet light are used for materials that absorb UV light. In heterogeneous AOPs, catalysts are used for degrading the pollutants. Combination of different AOPs can complement each other in terms of removing each other's drawbacks. Amount of OH radicals produced is important parameters for evaluating efficacy of the processes [69]. By-product formation

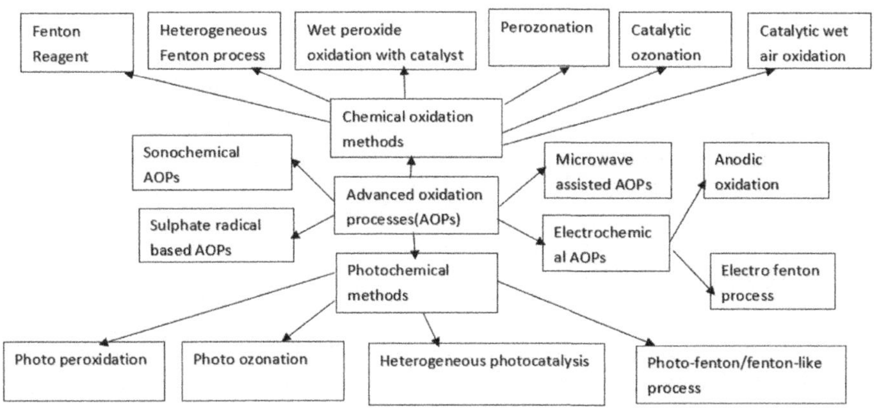

Fig. 5 Various types of advanced oxidation processes

is drawback of chemical methods. UV/hydrogen peroxide processes are ineffective for high concentrations and also are costly. In sonolysis, uncontrolled products and parameter versatility can pose problems [70, 71].

2.5.1 India Centric Insight into Research on AOPs

The contaminants with emerging concerns are causing novel difficulties in their removal. Trihalomethane like contaminants need to be investigated for removal from wastewater [71]. These are major disinfection by-products of chlorine treatment. Tak and Vellanki [72] carried out investigation on Trihalomethane in the treatment plant from Mathura and a moving bed biofilm based non-conventional water treatment plant at Agra. They observed that the treatment plants were unable to remove Trihalomethane from wastewater. They found that UV-L/H_2O_2 treatment was most efficient for removal of Trihalomethanes. Kalra et al. [73] treated textile wastewater by using different advanced oxidation processes. According to them, the efficacy of the treatment various according to the effluent characteristics. AOPs were able to degrade the dye by more than 90%. Mandal et al. [74] investigated Tambla Tributary [Durgapur, India] industrial wastewater for removal of pollutants by Fenton oxidation method. They also optimised parameters namely pH, temperature, H_2O_2 and $FeSO_4$. Their investigation indicated that pH and temperature plays very important role in the treatment. With optimum $FeSO_4$ quantity, 60% COD removal was possible. 95% COD reduction was obtained with optimum amount of hydrogen peroxide. By using microbial treatment following $FeSO_4$ treatment, the COD removal of 97% was obtained by them. Fenton treatment and biological treatment have synergetic effect resulting in enhanced treatment. Advanced oxidation processes are very useful for removal toxicity from wastewater. AOPs are capable of treating arsenic toxicity by converting arsenite to arsenate [75]. Sivagami et al. [76] reported advanced oxidation treatment for tannery wastewater. Their research indicated that ozone treatment preceded by coagulation and aeration was able to remove 90% COD. Sathya et al. [77] have explored the use of AOP in combination with membrane bioreactor. They treated textile wastewater from Tiruppur. Submerged membrane bioreactor was used with combination of ozonation and photocatalysis. They used Tungsten oxide in the photoreactor. Ozone dosage also increased biodegradability of the wastewater. 93% COD reduction was obtained in the integrated treatment method.

2.6 Extended Aeration

Extended aeration process is dealt here separately though it is modification of activated sludge process. In the extended aeration the aeration time is extended to 15 to 24 h. These systems are better equipped to handle higher organic loading and also fluctuations in flow. Advantages and disadvantages of extended aeration are depicted in Fig. 6.

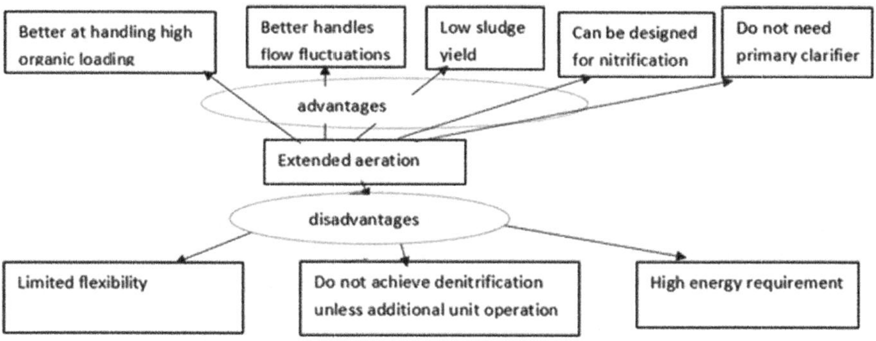

Fig. 6 Advantages and disadvantages of extended aeration system

Extended aeration can be used an intermittent cycle (ICEAS) or continuous flow systems (SBR) [78]. ICEAS systems show high removal efficiencies for organic matter and nitrogen. Investigations on treatment of leachate, and hospital wastewater indicate the potential of extended aeration systems in eliminating of pathogens and emerging pollutants from wastewater [79–81]. Extended aeration has been explored by the investigators for removal of phenol like compounds from wastewater [82, 83]. The literature outlines wide variety of pollutants that can be treated using extended aeration systems.

2.6.1 India Centric Insight into Research on Extended Aeration

Mutiyar and Mittal [84] have carried out investigation on the occurrence and removal of an antibiotic, Amoxicillin, at a sewage treatment plant (STP) located in Delhi. They carried out sampling exercise for almost 6 months to get proper representative samples. Average removal efficiency of amoxicillin was observed to be 49.7%. Singh et al. [85] carried out studies on performance evaluation of extended aeration in sewage treatment plant in Delhi. The samples were collected from plants in Vasant Kunj(02), Meharauli(01) and Najfgarh(01). COD removal obtained was from 78% at Najfgargh to 86% at Vasantkunj. TSS removal in the range of 87% to 91% was observed.

Gupta et al. [86] carried out an investigation on physicochemical and toxicity profiling of sludge and wastewater samples from extended aeration facilities of municipal wastewater treatment plants. The COD reduction in the treatment was above 90%. Cytotoxicity and genotoxicity were present in the samples; hence further treatment was necessary. Few other investigations confirm the effectiveness of extended aeration in wastewater treatment [87, 88]. Extended aeration can be made further effective with the help of artificial neural networks to predict faecal coliform and total coliform removal for an intermittent cycle extended aeration-sequential batch reactor ([ICEAS-SBR) [89].

2.7 *Membrane Bioreactors*

Membrane bioreactors (MBRs) explore the advantages of membrane processes like microfiltration and ultrafiltration with suspended growth bioreactors for efficient treatment of wastewater. There are many advantages of membrane bioreactors. Possibility of retaining almost all the suspended matter in the bioreactor and bacteria, viruses results in sterile and pure water [90]. Membrane also acts to retain the extracellular enzymes, suspended matter and biological species. Problems associated with settling are eliminated in MBR. The MBR systems are more compact than the conventional treatment plants. The main disadvantages of MBRs are related to cost and the energy. Cost of high-quality membranes affects the economy and there is need for high pressure drop because of membranes increases energy cost. MBRs can be integrated or submerged and external (Fig. 7). Different types of membrane modules are used for MBR namely hollow fiber, spiral wound and tubular [91].

Membrane fouling is one of the major problems in MBRs. Fouling results in increase in hydraulic resistance. This results in reducing permeate flux at constant pressure. About 95% removal of COD, BOD and TSS is reported in MBR [91]. These MBRs can be based on aerobic or anaerobic principles for biological digestion.

2.7.1 India Centric Insight into Research on MBRs

Garg and Chaudhari [92] treated food industry wastewater by using membrane bioreactor. They were able to remove about 95% suspended solids from the effluent. Initial COD (1650 mg/l) was reduced by 90% after the treatment. They were able to reduce the pollution parameters well below the Indian norms. COD removal efficiency was marginally more than BOD removal efficiency. In the recent past, the micropollutant removal from wastewater is emerged as a challenging task for researchers. Membrane bioreactor find widespread application in removal of micropollutants.

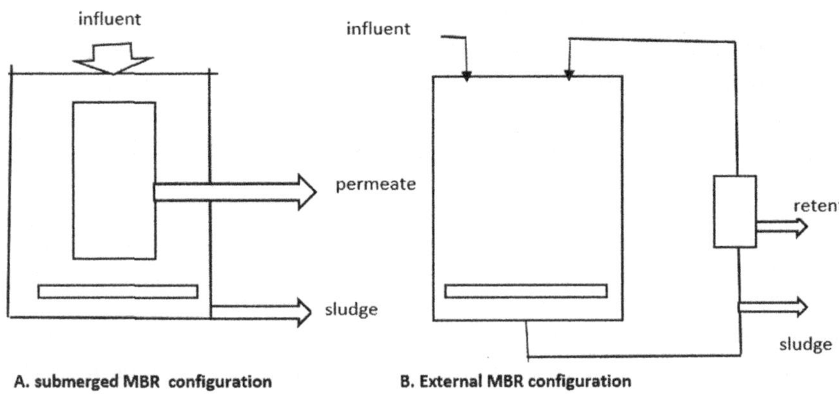

Fig. 7 Types of MBR

European Union has listed more than 100,000 chemicals as micropollutants [93]. Around 50% of these chemicals are of daily use. According to Dohare and Trivedi [94], compound annual growth of MBR technology was more than 14%. This is more than any other wastewater treatment technology. Gupta et al. [95] provided overview on use of submerged MBRs for municipal wastewater treatment. According to their studies ammonia removal more than 90% is possible in municipal treatment plants in India. MBR can achieve 30% more removal of organic matter and other pollutants than activated sludge process [95]. Normally washing is done when permeability reduces to 10% of initial permeability. About 82% of nitrogen was denitrified in MBR. Mukherjee et al. [96] developed ceramic membrane for application in membrane bioreactor. This membrane was tested for the wastewater containing toxic pollutant like atrazine. The ultrafiltration membrane was developed by them using iron oxide nanoparticles synthesized by green route from *Aloe vera* leaf extract, on the microporous support tubes with chitosan as matrix and glutaraldehyde as cross-linker. The atrazine removal in the clay based UF membrane was 15% more than lay-alumina based ceramic microfiltration membranes (MF-MBR). Chandrashekhar et al. [97] designed integrated membrane bioreactor for treatment of complex coffee industry effluent. Treatment of food industry wastewater poses a serious challenge due to high organic loading rate, strong odour and colour. Initial turbidity of 1100 formazin attenuation unit (FAU) was reduced to 0 after treatment, whereas initial COD of 20,000 reduced to 40 mg/l after the treatment. The scale up ratio of 10 (from 2 to 20 L) was found to comply the treatment effectiveness. Keerthi et al. [98] developed membrane bioreactor for treatment of tannery wastewater. They treated the wastewater from the common effluent treatment plant (CETP) located at Pallavaram, Chennai. They coupled membrane bioreactor with electrocoagulation. The hybrid membrane bioreactor showed 11% less fouling than MBR. For treating high strength wastewater, anaerobic MBRs are used because of low energy requirement and biogas as value added material. Aerobic MBR is normally used for water reclamation only rather than recovery of material and energy [99]. Various investigations on dairy wastewater, domestic wastewater treatment by aerobic and anaerobic membrane bioreactors have been reported in the recent past [100–102]. The investigations on membrane modules are also reported [103, 104]. Different types of coatings are explored for membrane to reduce the fouling. Investigations are reported on use of silicon carbide coatings on membranes. The coating were found to be useful for reducing fouling without compromising on performance. Recent investigations are also focused on energy intensification of MBRs. Factors such as membrane characteristics, operating conditions and feed and biomass characteristics affect membrane fouling [100–104]. Measures such as membrane scouring, permeate extraction and sludge transfer can considerably reduce energy costs.

2.8 *Moving Bed Biofilm Reactors*

Moving bed biofilm reactors (MBBR) are used for treatment of dairy, beverages, automobile pulp and paper industry waster. In moving bed reactors plastic media is used for increasing surface area. The biofilm is attached to the plastic media that is suspended in water. Increased process stability, mass transfer and settling characteristics help in reducing footprints of the treatment process. MBBR process is completely mix, continuous flow wastewater treatment operation. In MBBR, the principles of activated sludge process and biofilm are combined. It is combination of attached growth and suspended growth process. Modelling of MBBR is more complicated than the activated sludge processes. There are no models existing that can predict the substrate flux, biofilm thickness and most importantly the effective bio-film thickness [105]. MBBR processes can be used for BOD/COD-removal, nitrification and denitrification in municipal and industrial wastewater treatment [106]. MBBR can be used as effective secondary treatment with hydraulic retention time less than one hour. Johnson et al. [107] reported application of MBBR for treatment of refinery and slaughter house treatment effluent. Tak et al. [108] carried out research on identification of emerging contaminants and their transformational products in drinking water plants near Yamuna river. According to their investigation, pharmaceutical industries are major sources of emerging contaminants, followed by plasticizers and insecticides. Their study indicated presence of some toxic pollutants even towards the end of the treatment processes [108].

2.8.1 India Centric Insight into Research on MBBRs

Singh et al. [109] carried out lifecycle assessment of MBBR plant in India. Their study showed that the main bottleneck in the used of MBBR was energy consumption. Efficient collection of wastewater and optimization of energy requirements and use can increase the acceptability of MBBRs across the country. In their work on wastewater treatment, Thakur and Khedikar [110] carried out evaluation of Moving bed bio-film reactor [MBBR] for treatment of domestic wastewater. They carried out experimentation for hydraulic retention time of 24 to 48 h. The treatment indicated the organic matter removal in the range of 90 to 95%. Solid removal was 60 to 65% in the treatment. Osmani et al. [111] reported work on upgradation of conventional MBBR for nitrogen removal. Changed norms of regulatory bodies and the need for non-potable utilization of wastewater to decrease the environmental footprints were major motivations for upgrading. They were able to optimize and upgrade the plant to achieve 50 to 90% nitrification and 40% denitrification. They converted the conventional MBBR into PVA gel based MBBR. Simultaneous sludge minimization, pollutant and nitrogen removal using integrated MBBR configuration for tannery wastewater treatment was reported by Sodhi et al. [112]. They used anoxic-aerobic moving bed biofilm reactor configuration. Using this configuration, they obtained 94% COD removal and 88% nitrogen removal. Barwal and Chaudhari

[113] carried out an investigation to the performance of biocarriers in moving bed biofilm reactor (MBBR). Also, they studied the kinetics of biofilm for retrofitting the existing aerobic treatment systems. According to them, performance of MBBR depends on the percent of media provided in the reactor, surface area of the biocarrier, dissolved oxygen and the organic loading. MBBRs are also useful in removing phenol from the effluent [114]. Phenol removal of 99% is reported in aerobic MBBR [115].

2.9 *Sequential Batch Reactor [SBR]*

Sequential batch reactors can be used for aerobic and anaerobic wastewater treatments. Generally, the SBR application can be explained in five steps [116]. First step is filling of the reactor. In second step the reaction takes place followed by third step in which the sludge is allowed to settle. Then in next step the supernant is decanted. Final step is named as idle just before starting next cycle with filling step. Advantage of SBR lies in choice of the steps and flexibility in step operation. The cycle time, hydraulic retention time can be changed as per requirement and effluent quality. The SBRs can be modified for anaerobic treatments also. Studies reveal that decrease in operating temperature decreases the nitrification capacity of the plant [117]. Decrease in the cycle time causes increase in effluent concentration. Oxygen concentration in the SBR should be close to 5 mg/l so that thinner aerobic layer and thicker anaerobic layer is formed facilitating elimination of higher NO_3-N loading. The major factors affecting the operation of SBR are organic loading, hydraulic retention time, sludge retention time, dissolved oxygen and influent characteristics like COD, BOD and C/N ratio [118]. SBR can be designed to perform specific functions such as nitrification, denitrification and phosphorus removal. Figure 8 depicts advantages and disadvantages of SBR.

2.9.1 India Centric Insight into Research on SBR

Ruhela et al. [119] studied efficacy of sequential batch reactor for treating effluent. The effluent discharge after treatment and its effect on the reservoir was also studied. Their studies indicated that the maximum BOD removal was 79–80%. SBR was found effective for treating the petroleum industry wastewater [120]. COD and TOC removal efficiencies of 80 and 84% were observed during the treatment. Sharma et al. investigated effect of various parameters on removal of resorcinol from wastewater in SBR treatment [121]. Mixed liquor suspended solid, initial concentration of pollutant and ratio of aeration to fill phase were identified as crucial parameters by them. Optimum time for react and settle phases were 1.75 and 0.75 h respectively for 5 h cycle time. For initial concentration of 50 mg/l, resorcinol removal of 81% was obtained. An investigation by Shaukat and Najar [122] indicated that SBR treatment was able to remove objectionable contaminants from wastewater. However

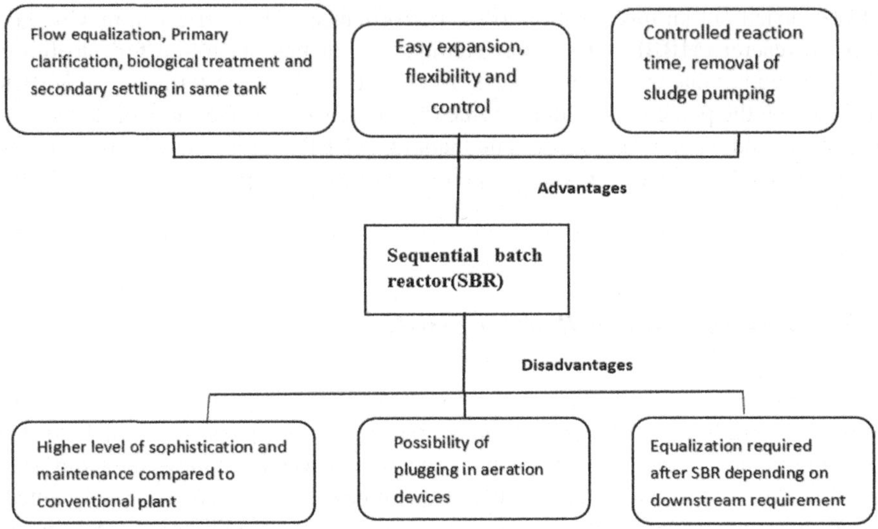

Fig. 8 Advantages and disadvantages of SBR

low reduction in ortho phosphates, suspended solid and dissolved solid was still a problem. An investigation on treatment of slaughter house effluent with sequential batch reactor was carried out by Kundu et al. [123]. They prepared SBR from perpex sheets. They treated 20 L of effluent in a reactor to which, 2.5 L of pre-acclimatized mixed seed with carbonaceous bacteria was added. They performed the experiment in three combinations of aeration and anoxic react periods, 4 + 4, 5 + 3 and 3 + 5 h. The removal efficiency of chemical oxygen demand depends on COD: TKN ratio, 10–12 being optimum ratio. In this investigation it was 11.24 [123]. The total react period of 8 h with 4 h aerobic and 4 h of anoxic process yielded maximum removal of carbonaceous matter, nitrification and denitrification. Many investigations on treatment of wastewater by aerobic and anoxic pathways for treatment of wastewater are reported [124–126]. Domestic wastewater, tannery and dairy industry wastewater treatment experiments yielded promising results [126–129]. Most of these investigations were focused on the parameters like MLSS concentrations, optimum cycle time of aerobic and anoxic phases, sludge settling ability and removal efficiency for carbonaceous and nitrogen based materials.

2.10 *Upflow Anaerobic Sludge Blanket Reactor*

UASB is a reactor to treat the wastewater by anaerobic mechanism. In this process, methane is produced. The effluent flows upward through the dense sludge containing microorganisms while undergoing anaerobic microbial activity. The UASB reactor handles high organic loading with good COD removal efficiency. Also, it has short hydraulic retention time and low energy consumption. Disadvantages being, sensitivity to shock loads and difficulty in granulation process.

2.10.1 India Centric Insight into Research on UASB

UASB reactor was used for treating dairy wastewater by Shirule et al. [130]. They found that the gas generated was having 80% methane. Also, the COD removed had straight proportionality with methane generated. Kasaudhan et al. [131] studied operation of UASB reactor at Bharwara, Lucknow. They suggested use of algicide for algal cells [131]. In important schemes like the Yamuna action plan, 16 out of 28 plants had UASB technology [132]. 83% of the total treated effluent was treated in UASB plants. Recent developments to reduce cost, increase durability and simple operation and maintenance are salient features of the modifications in the process. Yadav et al. [133] analysed flow and quality parameters for the UASB treatment plants in Uttar Pradesh (Bharwara STP). By using machine learning algorithms, they proposed a model to predict influent and effluent parameters. Hasan et al. [134], carried out investigation on performance of seven sewage treatment plants in different cities from North India for two years. Their studies indicated that the treatment plants based on SBR, MBBR and combined UASB with downflow hanging sponge treatment were able to reduce the quality parameters as per regulatory norms. Bandhavya and Prashanth [135] carried out investigation for treatment of low strength wastewater by using UASB. They investigated effect of hydraulic loading and sludge occupancy on the performance. Till sludge loading rate alteration up to 0.41, there was no change in the COD removal which was 84 to 85%. At the sludge loading rate of 0.51, the COD reduction dropped to 81%. Studies carried out for the effluent treatment across middle stretches of Ganga river basin indicated that UASB based treatment plants were able to remove 87% of the microplastics [136]. Treatment of synthetic slaughterhouse effluent by using UASB was experimented by Loganath and Mazumder [137]. They used hybrid UASB with enhanced clarifier having 13.5 L volume. They were able to remove 87% COD with biogas generation of 41.5 L/d. Optimum COD loading rate was 8.00 kg COD/m^3 /day. Investigations on the treatment of effluent various industries namely dairy wastewater, slaughterhouse wastewater were carried out to explore the UASB treatment for the effluent with reasonable success [138–141]. Most of these investigations reported the COD removal above 80%. Biogas generation is value added side application of UASB. UASB technology is playing important role in energy intensification by virtue of biogas production [142–144].

3 Hybrid Wastewater Treatment Technologies

The sewage treatment plants can be classified as natural treatment systems and mechanized treatment systems. In India only 37% of the effluent gets the treatment [145]. In a city like Mumbai, 1600 MLD/day of effluent gets only primary treatment. Investment in Municipal treatment plants can enhance the tourism, avoid health costs and increase fish catches [146]. Study indicated that these can be beneficial in long term [146]. Industrial wastewater is another concern which is the responsibility of generator. Mechanized treatment plants contain ASP, SBR, MBBR, UASB kind of technologies for treatment of effluent. Natural treatment systems like stabilization pond are limited by their land requirements. In India, the treatment plants needs to be upgraded considering their efficacies and need to reuse the water for non-potable use. Emergence of new contaminates have posed a new challenge before the environmentalists and experts. New norms for COD, BOD, TSS have necessitated the urgent overhaul of treatment plants [147]. Considering these aspects, there is need to set up new treatment plants or upgrade the existing ones. These challenges are not limited to India but are at global level. There is a need for more efficient and sustainable infrastructure for treatment of wastewater to meet the limitations of land requirement and need for reuse of wastewater along with less energy consumption and environmental foot prints. In hybrid wastewater treatment (HTS, hybrid treatment systems); Natural Treatment systems (NTS) and Mechanized Treatment Systems (MTS) are combined [145]. Hybrid wastewater treatment plants have potential to solve the pretreatment of wastewater from industries and domestic sources by overcoming drawbacks of conventional treatment methodologies [148, 149].

3.1 India Centric Insight into Research on Hybrid Wastewater Treatments

Most of the hybrid plants use biological treatments followed by chemical or other advanced treatments. Removal of pharmaceuticals can be done with hybrid systems. The hybrid systems combining activated sludge process with ultrafiltration, reverse osmosis can remove trace organic contaminants [149]. Soni et al. [150] proposed a solar powered water purification system for the households [150]. Wind and solar energy is utilized for applying vacuum inside the vessel and solar energy is used to run the plant. According to the studies carried out by Kalbar [145], hybrid treatment systems can overcome the limitations of natural and mechanized treatment systems. His studies revealed that the hybrid systems are more economical than mechanized treatments for urban, peri-urban and rural wastewater treatment. He expressed need to adopt the hybrid technologies across the country [145, 151]. Mechanized treatments are very high on operation and maintenance costs but low in land requirement as against the Natural treatment plants. Neither Natural treatment systems nor Mechanised treatment systems are individually capable to remove the pollutants effectively from wastewater [145, 151]. Hybrid treatment systems are combination of natural and mechanized approaches for treatment of wastewater. Land requirement for hybrid systems are greater than mechanized treatments but since mechanised systems can hardly be used without any of the natural treatment systems, the Hybrid treatment has advantage. Also, lower operational and maintenance cost is advantage of Hybrid treatment system. In an attempt to improve the performance of activated sludge biofilm reactor system treatment, Manu treated the wastewater from this treatment by titanium dioxide based photocatalysis with UV radiations [152]. They were able to remove 72% COD and 52% phosphorus from the treated effluent by UV and 71% and 50% by solar radiations. Gajbhiye and Yadav [153] treated the industrial wastewater by using a combined system based on ultrasound, ozonation and catalyst loading. Sonicator and ozonation as isolated treatment yielded inferior results with 26% and 46% COD removal respectively. These three processes when combined together, yielded 75% COD reduction.

4 Future Scope and Recommendations

Considering the water scarcity in different regions and cities in India, the wastewater treatment for meeting only the disposal standards will be history in near future. The reuse and recycle of wastewater has become essential component of the sustainability. Modification of the existing facilities is required to upgrade them to meet the requirements for agricultural, domestic and potable applications. Three Rs namely reuse, recycle and reduce are important buzzwords for water treatment. With increase in population, sewage generation has increased. The sewage treatment plants need to be upgraded in terms of technology and capacity. The data from the literature indicates that only 50% of the urban sewage is treated through centralized facilities [154]. Increasing number of non-operational facilities and low compliance rate of many facilities calls for urgent upgrade in view of zero discharge policy. Artificial intelligence (AI) models can be used to optimize the treatment facilities by reducing the amounts of chemicals and adsorbents used for the treatment [155]. AI long with Internet of Things (IoT) can be used for better design and operation of the plants by acquiring and processing the data for more economical and effective wastewater treatment.

5 Conclusion

Wastewater treatment technologies in India need to be upgraded to meet the requirements of regulatory norms as per the Green Tribunal. The limits for COD, BOD and other parameters are becoming more stringent. The increase in the population, results in the increase in the demand for potable and nonpotable water. The treatment plants needs to cater the non-potable water requirement by making the water recyclable and reusable. Three types of treatments, physical, biological and chemical are used in primary, secondary and tertiary treatment plants. Advanced technologies like ozonation, ultrasound-assisted treatments and solar radiations etc. have potential to reduce the contaminants from the wastewater. Natural treatment systems like stabilization ponds have limitations of land requirements and mechanized treatment systems are limited by high operation and maintenance costs. Neither of these two systems can be used in isolation. Hybrid systems combining these two have the potential to solve the problem by reducing these two costs considerably. Investigations in India, on biological treatment processes are focused on activated sludge processes and trickling filters along with their modifications. Processes using combined principles of suspended and attached growth processes have added advantages but have limitations in terms of expertise and technical know-how. Hybrid wastewater systems combining natural treatments, mechanized systems, and advanced treatments can provide a solution to the upgradation of wastewater treatment plants. Investigations on modelling and simulation of wastewater treatment facilities indicate that there are ample opportunities for exploring computational facilities and automation of the wastewater treatment plants.

6 Availability of Data and Materials

Author declare that, no additional data and materials are associated with this article.

Acknowledgements The author acknowledges the affiliating institute, Gharda Institute and the Gharda Foundation for their support and encouragement. Also, author is grateful to his colleagues for cooperation.

Author's Contributions This book chapter is conceptualized, prepared and compiled by Dr. Sunil Jayant Kulkarni.

Funding Not applicable.
Not applicable.

Declarations Ethics Approval and Consent to Participate

Consent for Publication Not applicable.

Competing Interests The author declares no competing interests.

References

1. Times of India (2023) Waste is winning: Cities' effluent management is way behind target and a health hazard. Eco-friendly, inexpensive solutions exist. https://timesofindia.indiatimes.com/blogs/toi-editorials/waste-is-winning-cities-effluent-management-is-way-behind-target-and-a-health-hazard-eco-friendly-inexpensive-solutions-exist/

2. Minhas PS, Saha JK, Dotaniya ML, Sarkar A, Saha M (2022) Wastewater irrigation in India: current status, impacts and response options. Sci Total Environ 808:152001. https://doi.org/10.1016/j.scitotenv.2021.152001
3. The International Trade Administration (2023) India water and wastewater treatment industry. https://www.trade.gov/market-intelligence/india-water-and-wastewater-treatment-industry
4. India Infrastructure Magazine (2021) Working with waste-key trends and developments in India Effluent Management. https://indianinfrastructure.com/2021/12/31/73992/#:~:text=Approximately%2013%2C500%20mld%20of%20industrial,generated%20every%20day%20across%20India
5. Anees U, Kumar P, Prasad J (2023) World Water Week 2023: reuse of greywater can be an answer to India's shortage problem. https://www.downtoearth.org.in/news/water/world-water-week-2023-reuse-of-greywater-can-be-an-answer-to-india-s- shortage-problem-91223
6. Mishra A, Arunlala K, Anand S, Goel S (2023) Reuse of treated wastewater in urban/peri-urban agriculture in India. National institution for transforming India (NITI) Aayog, New Delhi. Water & Land Resources vertical, NITI Aayog, Govt. of India. https://www.niti.gov.in/sites/default/files/2023-08/Revised_Strategy-Paper-on-Reuse-of-Treated-wastewater-in-peri-urban-agriculture-in-India.pdf
7. Lee DJ, Taherzadeh MJ, Tyagi RD, Chen C (2022) Advanced activated sludge processes toward circular bioeconomy. Biores Technol 368:128325. https://doi.org/10.1016/j.biortech.2022.128325
8. Aziz SQ, Omar IA, Bashir MJ, Mojiri A (2020) Stage by stage design for primary, conventional activated sludge, SBR and MBBR units for residential wastewater treatment and reusing. Adv Environ Res 9(4):233–249. https://doi.org/10.12989/aer.2020.9.4.233
9. Gernaey KV, Van Loosdrecht MC, Henze M, Lind M, Jorgensen SB (2004) Activated sludge wastewater treatment plant modelling and simulation: state of the art. Environ Model Softw 19(9):763–783. https://doi.org/10.1016/j.envsoft.2003.03.005
10. Henze M, Gujer W, Mino T, van Loosdrecht M (2000) Activated sludge models ASM1, ASM2, ASM2D, ASM3. 5. https://doi.org/10.2166/9781780402369
11. Hassanpour M (2014) Activated sludge model 1 (ASM 1). Int J Eng Sci Res Technol 3(1):587–597
12. Demoulin G, Rüdiger A, Goronszy MC (2001) Cyclic activated sludge technology-recent operating experience with a 90,000 pe plant in Germany. Water Sci Technol 43(3):331–337. https://doi.org/10.2166/wst.2001.0154
13. Deepak M, Rustum R (2022) Review of latest advances in nature-inspired algorithms for optimization of activated sludge processes. Processes 11(1):77. https://doi.org/10.3390/pr11010077
14. Lowe M, Qin R, Mao X (2022) A review on machine learning, artificial intelligence, and smart technology in water treatment and monitoring. Water 14(9):1384. https://doi.org/10.3390/w14091384
15. Zhao L, Dai T, Qiao Z, Sun P, Hao J, Yang Y (2020) Application of artificial intelligence to wastewater treatment: a bibliometric analysis and systematic review of technology, economy, management, and wastewater reuse. Process Saf Environ Prot 133:169–182. https://doi.org/10.1016/j.psep.2019.11.014
16. Van Winckel T (2014) Development of high-rate activated sludge processes for energy-efficient wastewater treatment (Doctoral dissertation, MSc. thesis, Gent University, Belgium). https://libstore.ugent.be/fulltxt/RUG01/002/166/682/RUG01-002166682_2014_0001_AC.pdf
17. Latif EF (2022) Applying novel methods in conventional activated sludge plants to treat low-strength wastewater. Environ Monit Assess 194(5):323. https://doi.org/10.1007/s10661-022-09968-9
18. Narayanan CM, Narayan V (2019) Biological wastewater treatment and bioreactor design: a review. Sustain Environ Res 29:33. https://doi.org/10.1186/s42834-019-0036-1
19. Kazmi AA, Tyagi VK, Trivedi RC, Kumar A (2008) Coliforms removal in full-scale activated sludge plants in India. J Environ Manage 87(3):415–419. https://doi.org/10.1016/j.jenvman.2007.01.017

20. Tyagi VK, Subramaniyan S, Kazmi AA, Chopra AK (2008) Microbial community in conventional and extended aeration activated sludge plants in India. Ecol Ind 8(5):550–554
21. Sarkar U, Dasgupta D, Bhattacharya T, Pal S, Chakroborty T (2010) Dynamic simulation of activated sludge based wastewater treatment processes: case studies with Titagarh sewage treatment plant, India. Desalination 252(1–3):120–126. https://doi.org/10.1016/j.desal.2009.10.014
22. Pillai SC, Subrahmanyan V (1942) Role of protozoa in the activated sludge process. Nature 150(3809):525–525. http://eprints.iisc.ac.in/id/eprint/16334
23. Kumar K, Singh GK, Dastidar MG, Sreekrishnan TR (2014) Effect of mixed liquor volatile suspended solids (MLVSS) and hydraulic retention time (HRT) on the performance of activated sludge process during the biotreatment of real textile wastewater. Water Resour Ind 5:1–8. https://doi.org/10.1016/j.wri.2014.01.001
24. Kanekar P, Sarnaik S (1991) An activated sludge process to reduce the pollution load of a dye-industry waste. Environ Pollut 70(1):27–33. https://doi.org/10.1016/0269-7491(91)90129-k
25. Rezai B, Allahkarami E (2021) Wastewater treatment processes—techniques, technologies, challenges faced, and alternative solutions. In: Soft computing techniques in solid waste and wastewater management. Elsevier, pp 35–53. https://doi.org/10.1016/B978-0-12-824463-0.00004-5
26. Dubey SK, Kashyap AK (2022) Biofilm: a doable microbial continuum for the treatment of wastewater. In: Microbial consortium and biotransformation for pollution decontamination. Elsevier, pp 321–345. https://doi.org/10.1016/B978-0-323-91893-0.00013-4
27. Buchanan JR (2014) Decentralized wastewater treatment. In: Ahuja S, Larsen MC, Eimers JL, Patterson CL, Sengupta S, Schnoor JL (eds) Comprehensive water quality and purification, vol 1. Elsevier, Amsterdam, pp 44–45. https://doi.org/10.1016/B978-0-12-382182-9.00050-5
28. Aslam MMA, Khan ZM, Sultan M, Niaz Y, Mahmood MH, Shoaib M, Shakoor A, Ahmad, M (2017) Performance evaluation of trickling filter-based wastewater treatment system utilizing cotton sticks as filter media. Polish J Environ Stud 26(5). https://doi.org/10.15244/pjoes/69443
29. Shukla R, Ahammad SZ (2023) Performance assessment of a modified trickling filter and conventional activated sludge process along with tertiary treatment in removing emerging pollutants from urban sewage. Sci Total Environ 858:159833. https://doi.org/10.1016/j.scitotenv.2022.159833
30. Ramananda Bhat M, Nayak S, Pariti A, Dhawan S (2017) Secondary treatment of dairy effluents with trickle bed. In: Materials, energy and environment engineering: select proceedings of ICACE 2015. Springer Singapore, pp 309–316. https://doi.org/10.1007/978-981-10-2675-1_36
31. Zylka R, Dąbrowski W, Gogina E, Yancen O (2018) Trickling filter for high efficiency treatment of dairy sewage. J Ecol Eng 19(4):269–275. https://doi.org/10.12911/22998993/89657
32. Arsalan M, Khan ZM, Sultan M, Ali I, Shakoor A, Mahmood MH, Ahmad M, Shashiri R Imran M, Khalid MU (2021) Experimental investigation of a wastewater treatment system utilizing maize cob as trickling filter media. Fresenius Environ Bull 30:148–157
33. Rathore P, Parde D, Killedar DJ (2022) A study of LCCA-based wastewater treatment technologies in the context of an Indian scenario. https://doi.org/10.21203/rs.3.rs-2098736/v1
34. Dhokpande SR, Kulkarni SJ, Kaware JP (2014) A review on research on application of trickling filters in removal of various pollutants from effluent. Int J Eng Sci Res Technol 3(7):359–365
35. Naz I, Saroj DP, Mumtaz S, Ali N, Ahmed S (2015) Assessment of biological trickling filter systems with various packing materials for improved wastewater treatment. Environ Technol 36(4):424–434
36. Kulkarni S, Dhokpande SR, Kaware DJ (2015) Modeling for biological wastewater treatment facilities-a review. Int J Sci Res Sci, Eng Technol 1(2):104–106

37. Vijayagopal V, Sabarathinam PL (2008) Kinetics of biological treatment of phenolic wastewater in a three phase draft tube fluidized bed bioreactor containing biofilm. Afr J Biotechnol 7(6):834–837. https://www.ajol.info/index.php/ajb/article/view/58555
38. Shrivastava V, Verma AK, Chandra P, Roy SD (2015) Optimization of trickling bio filter at different water flow rate and filter media thickness. Poll Res 34(3):135–143. http://www.envirobiotechjournals.com/article_abstract.php?aid=6135&iid=194&jid=4
39. Hanumanulu V (1969) Effect of recirculation on deep trickling filter performance. J (Water Pollution Control Federation):1803–1806. https://www.jstor.org/stable/25039137
40. Naz I, Saroj DP, Mumtaz S, Ali N, Ahmed S (2015) Assessment of biological trickling filter systems with various packing materials for improved wastewater treatment. Environ Technol 36(4):424–434. https://doi.org/10.1080/09593330.2014.951400
41. Kanwar RMA, Khan ZM, Farid HU (2021) Investigation of municipal wastewater treatment by agricultural waste materials in locally designed trickling filter for peri-urban agriculture. Water Supply 21(5):2298–2312. https://doi.org/10.2166/ws.2021.075
42. Radha KV (2023) Performance and modeling studies of rotating biological contactor for treatment of vegetable oil wastewater: (Studies on modeling of fabricated rotating biological contactor). Environ Qual Manage 32(3):301–312. https://doi.org/10.1002/tqem.21909
43. Powar P, Nejkar V, Patil P, Naik V (2023) Dairy wastewater treatment with rotating biological contactor as secondary unit and use of treated effluent for agriculture. In: E3S web of conferences 405: 01006. EDP Sciences. https://doi.org/10.1051/e3sconf/202340501006
44. Subhadarsini L, Dash RR (2020) Treatment of industrial waste water using single-stage rotating biological contactor. IOP Conf Ser: Mater Sci Eng 970(1):012025. https://doi.org/10.1088/1757-899X/970/1/012025/meta
45. Palakshappa K, Sneha MK, Chiranjeevi MR (2019) Feasibility studies on the treatment of dairy wastewater using rotating biological contactor (RBC) under variable experimental conditions. In: Proceedings of the second international conference on emerging trends in science & technologies for engineering systems (ICETSE-2019). https://doi.org/10.2139/ssrn.3511456
46. Vairavel P, Murty VR (2020) Decolorization of Congo red dye in a continuously operated rotating biological contactor reactor. Desalin Water Treat 196:299–314. https://doi.org/10.5004/dwt.2020.25931
47. Kulkarni SJ, Kaware JP (2013) Review on research for removal of phenol from wastewater. Int J Sci Res Publ 3(4):1–5. https://www.ijsrp.org/research-paper-0413.php?rp=P161089
48. Kiran MG, Pakshirajan K, Das G (2017) A new application of anaerobic rotating biological contactor reactor for heavy metal removal under sulfate reducing condition. Chem Eng J 321:67–75. https://doi.org/10.1016/j.cej.2017.03.080
49. Gopi Kiran M, Das R, Behera SK, Pakshirajan K, Das G (2021) Modelling a rotating biological contactor treating heavy metal contaminated wastewater using artificial neural network. Water Supply 21(5):1895–1912. https://doi.org/10.2166/ws.2020.304
50. Murugesan S (2017) Performance study on rotating biological contactor. https://doi.org/10.13140/RG.2.2.25602.17607
51. Srivastava I, Pal A, Sahu S, Singh KR (2023) A Study of the suitability of microbial cells for the biosorption and bioaccumulation of heavy metal removal. Int J Curr Sci Rev 6(5):2914–2923. https://ijcsrr.org/wp-content/uploads/2023/05/29-20-2023.pdf
52. Gouda SA, Taha A (2023) Biosorption of heavy metals as a new alternative method for wastewater treatment: a review. Egypt J Aquat Biol Fish 27(2):135–153. https://doi.org/10.21608/ejabf.2023.291671
53. Barange M, Srivastava A, Srivastava JK, Palsania J (2014) Biosorption of heavy metals from wastewater by using microalgae. Int J Chem Phys Sci 3(6):67–81
54. Hussain SA, Hasan NK, Al-Abodi EE (2021) Biosorption to removing heavy metals from wastewater. In: J Phys: Conf Ser 1853(1):012012. https://doi.org/10.1088/1742-6596/1853/1/012012/pdf
55. Rao PR, Bhargavi C (2013) Studies on biosorption of heavy metals using pre-treated biomass of fungal species. Int J Chem Chem Eng 3(3):171–180

56. De Olivera AR (2018) Biosorption of pharmaceuticals from wastewater using Moringa oleífera as biosorbent (Doctoral dissertation, Instituto Politecnico de Braganca (Portugal)). https://www.proquest.com/openview/ef90a0808952be7529897f6a01d18f69/1?pq-origsite=gscholar&cbl=2026366&diss=y
57. Oves M, Khan MS, Zaidi A (2013) Biosorption of heavy metals by Bacillus thuringiensis strain OSM29 originating from industrial effluent contaminated north Indian soil. Saudi J Biol Sci 20(2):121–129. https://doi.org/10.1016/j.sjbs.2012.11.006
58. Tyagi B, Gupta B, Thakur IS (2020) Biosorption of Cr (VI) from aqueous solution by extracellular polymeric substances (EPS) produced by Parapedobacter sp. ISTM3 strain isolated from Mawsmai cave, Meghalaya, India. Environ Res 191:110064. https://doi.org/10.1016/j.envres.2020.110064
59. Kumar JN, Oommen C, Kumar RN (2009) Biosorption of heavy metals from aqueous solution by green marine macroalgae from Okha Port, Gulf of Kutch, India. Am Eurasian J Agric Environ Sci 6(3):317–323. https://citeseerx.ist.psu.edu/document?repid=rep1&type=pdf&doi=bcc714fb36a7ce91f4b0e53c98588aea5a4b07a6
60. Tanvi DA, Pratam KM, Lohit RT, Vijayalakshmi BK, Devaraja TN, Vasudha M, Ramesh A, Chakra PS, Gayathri D (2020) Biosorption of heavy metal arsenic from Industrial Sewage of Davangere District, Karnataka, India, using indigenous fungal isolates. SN Appl Sci 2:1–7. https://doi.org/10.1007/s42452-020-03622-0
61. Patel GG, Doshi HV, Thakur MC (2016) Biosorption and equilibrium study of copper by marine seaweeds from north west coast of India. IOSR J Environ Sci, Toxicol Food Technol 10(7):54–65. https://www.iosrjournals.org/iosr-jestft/papers/vol10-issue7/Version-2/H1007025464.pdf
62. Sanjenbam P, Saurav K, Kannabiran K (2012) Biosorption of mercury and lead by aqueous Streptomyces VITSVK9 sp. isolated from marine sediments from the Bay of Bengal, India. Front Chem Sci Eng 6:198–202. https://doi.org/10.1007/s11705-012-1285-2
63. Kulkarni SJ (2016) Biosorption for phenol removal from wastewater: an insight into studies and investigations. Int J Sci Healthc Res 1(4):15–18. https://ijshr.com/IJSHR_Vol.1_Issue.4_Oct2016/IJSHR0003.pdf
64. Aravindhan R, Rao JR, Nair BU (2009) Application of a chemically modified green macro alga as a biosorbent for phenol removal. J Environ Manage 90(5):1877–1883. https://doi.org/10.1016/j.jenvman.2008.12.005
65. Singh N, Agarwal B, Balomajumder C (2016) Simultaneous treatment of phenol and cyanide containing aqueous solution by adsorption, biotreatment and simultaneous adsorption and biotreatment (SAB) process. J Environ Chem Eng 4(1):564–575. https://doi.org/10.1016/j.jece.2015.11.041
66. Rao JR, Viraraghavan T (2002) Biosorption of phenol from an aqueous solution by Aspergillus niger biomass. Bioresour Technol 85(2):165–171. https://doi.org/10.1016/S0960-8524(02)00079-2
67. Kurian M (2021) Advanced oxidation processes and nanomaterials-a review. Clean Eng Technol 2:100090. https://doi.org/10.1016/j.clet.2021.100090
68. Al Mayyahi A, Al-Asadi HAA (2018) Advanced oxidation processes (AOPs) for wastewater treatment and reuse: a brief review. Asian J Appl Sci Technol 2:18–30. https://ajast.net/data/uploads/6002.pdf
69. Garrido-Cardenas JA, Esteban-García B, Agüera A, Sánchez-Pérez JA, Manzano-Agugliaro F (2020) Wastewater treatment by advanced oxidation process and their worldwide research trends. Int J Environ Res Public Health 17(1):170. https://doi.org/10.3390/ijerph17010170
70. Wang JL, Xu LJ (2012) Advanced oxidation processes for wastewater treatment: formation of hydroxyl radical and application. Crit Rev Environ Sci Technol 42(3):251–325. https://doi.org/10.1080/10643389.2010.507698
71. Pandis PK, Kalogirou C, Kanellou E, Vaitsis C, Savvidou MG, Sourkouni G, Zorpas AA, Argirusis C (2022) Key points of advanced oxidation processes (AOPs) for wastewater, organic pollutants and pharmaceutical waste treatment: a mini review. ChemEngineering 6(1):8. https://doi.org/10.3390/chemengineering6010008

72. Tak S, Vellanki BP (2019) Applicability of advanced oxidation processes in removing anthropogenically influenced chlorination disinfection byproduct precursors in a developing country. Ecotoxicol Environ Saf 186:109768. https://doi.org/10.1016/j.ecoenv.2019.109768
73. Kalra S S, Mohan S, Sinha A, Singh G (2011) Advanced oxidation processes for treatment of textile and dye wastewater: a review. In: 2nd International conference on environmental science and development, vol 4. IACSIT Press Singapore, pp 271–275
74. Mandal T, Maity S, Dasgupta D, Datta S (2010) Advanced oxidation process and biotreatment: their roles in combined industrial wastewater treatment. Desalination 250(1):87–94. https://doi.org/10.1016/j.desal.2009.04.012
75. Babu DS, Srivastava V, Nidheesh PV, Kumar MS (2019) Detoxification of water and wastewater by advanced oxidation processes. Sci Total Environ 696:133961. https://doi.org/10.1016/j.scitotenv.2019.133961
76. Sivagami K, Sakthivel KP, Nambi IM (2018) Advanced oxidation processes for the treatment of tannery wastewater. J Environ Chem Eng 6(3):3656–3663. https://doi.org/10.1016/j.jece.2017.06.004
77. Sathya U, Nithya M, Balasubramanian N (2019) Evaluation of advanced oxidation processes (AOPs) integrated membrane bioreactor (MBR) for the real textile wastewater treatment. J Environ Manage 246:768–775. https://doi.org/10.1016/j.jenvman.2019.06.039
78. Dan NH, Phe TTM, Thanh BX, Hoinkis J, Le Luu T (2021) The application of intermittent cycle extended aeration systems (ICEAS) in wastewater treatment. J Water Process Eng 40:101909. https://doi.org/10.1016/j.jwpe.2020.101909
79. Takdastan A, Kordestani B, Nisi A, Jalilzadeh Yangjeh R (2016) Study of operational and maintenance problems and parameters of extended aeration activated sludge process in Golestan hospital wastewater treatment plant, Ahvaz, and their solutions. J Environ Health Eng 3(4):270–279. https://doi.org/10.18869/acadpub.jehe.3.4.270
80. Hashemzadeh B, Geravandi S, Mohammadi MJ, Yari AR, Charkhloo E, Omidi Khaniabadi, Y, Takdastan A, Niri MV, Fazizadehdavild M, Khoshgoftar M (2017) Performance of extended aeration biological system in removal of organic matter from Razi hospital wastewater during 2015, Iran. Arch Hyg Sci 6(3):244–249. https://doi.org/10.29252/ArchHygSci.6.3.244
81. Nikmanesh M, Eslami H, Momtaz SM Biabani R, Mohammadi A, Shiravand B, Zarei Mahmoudabadi T (2018) Performance evaluation of the extended aeration activated sludge system in the removal of physicochemical and microbial parameters of municipal wastewater: a case study of Nowshahr wastewater treatment plant. J Environ Health Sustain Dev 3(2):509–517. http://dorl.net/dor/20.1001.1.24766267.2018.3.2.5.8
82. Mareai BM, Fayed M, Aly SA, Elbarki WI (2020) Performance comparison of phenol removal in pharmaceutical wastewater by activated sludge and extended aeration augmented with activated carbon. Alex Eng J 59(6):5187–5196. https://doi.org/10.1016/j.aej.2020.09.048
83. Ramírez-Morales D, Masís-Mora M, Montiel-Mora JR, Cambronero-Heinrichs JC, Briceño-Guevara S, Rojas-Sánchez CE, Mendez-Rivera M, Arias-Mora V, Tormo-Budowski R, Brenes-Alfaro L, Rodríguez-Rodríguez CE (2020) Occurrence of pharmaceuticals, hazard assessment and ecotoxicological evaluation of wastewater treatment plants in Costa Rica. Sci Total Environ 746:141200. https://doi.org/10.1016/j.scitotenv.2020.141200
84. Mutiyar PK, Mittal AK (2013) Occurrences and fate of an antibiotic amoxicillin in extended aeration-based sewage treatment plant in Delhi, India: a case study of emerging pollutant. Desalin Water Treat 51(31–33):6158–6164. https://doi.org/10.1080/19443994.2013.770199
85. Singh SK, Kapoor V, Siriya K, Vashisth M (2021) Performance evaluation of extended aeration-based sewage treatment plants at NCT of Delhi, India. Shanghai Ligong Daxue Xuebao. J Univ Shanghai Sci Technol 23(5):515–525. https://doi.org/10.51201/JUSST/21/05126
86. Gupta A, Kumar M, Ghosh P, Thakur IS (2022) Risk assessment of a municipal extended aeration activated sludge treatment plant using physico-chemical and in vitro bioassay analyses. Environ Technol Innov 26:102254. https://doi.org/10.1016/j.eti.2021.102254

87. Sharma P, Sood S, Mishra SK (2020) Development of multiple linear regression model for biochemical oxygen demand (BOD) removal efficiency of different sewage treatment technologies in Delhi, India. Sustain Water Resour Manage 6:1–13. https://doi.org/10.1007/s40899-020-00377-9
88. Suneethi S, Keerthiga G, Soundhar R, Kanmani M, Boobalan T, Krithika D, Philip L (2015) Qualitative evaluation of small scale municipal wastewater treatment plants (WWTPs) in South India. Water Pract Technol 10(4):711–719. https://doi.org/10.2166/wpt.2015.087
89. Khatri N, Khatri KK, Sharma A (2020) Artificial neural network modelling of faecal coliform removal in an intermittent cycle extended aeration system-sequential batch reactor based wastewater treatment plant. J Water Process Eng 37:101477. https://doi.org/10.1016/j.jwpe.2020.101477
90. Cicek N (2003) A review of membrane bioreactors and their potential application in the treatment of agricultural wastewater. Can Biosyst Eng 45:6–37. https://library.csbe-scgab.ca/docs/journal/45/c0219.pdf
91. Kootenaei FG, Aminirad H (2014) Membrane biological reactors (MBR) and their applications for water reuse. Int J Adv Biol Biomed Res 2(7):2208–2216. https://citeseerx.ist.psu.edu/document?repid=rep1&type=pdf&doi=76f6efabe09ef249b3c91eb067d2acdcb8d14730
92. Garg S, Chaudhry S (2017) Treatment of wastewater of food industry by membrane bioreactor. Int Adv Res J Sci, Eng Technol 4(6):153–157
93. Goswami L, Kumar RV, Borah SN, Manikandan NA, Pakshirajan K, Pugazhenthi G (2018) Membrane bioreactor and integrated membrane bioreactor systems for micropollutant removal from wastewater: a review. J Water Process Eng 26:314–328. https://doi.org/10.1016/j.jwpe.2018.10.024
94. Dohare D, Trivedi R (2014) A review on membrane bioreactors: an emerging technology for industrial wastewater treatment. Int J Emerg Technol Adv Eng 4(12):226–236
95. Gupta N, Jana N, Majumder CB (2008) Submerged membrane bioreactor system for municipal wastewater treatment process: an overview. Indian J Chem Technol 15:604–612. https://nopr.niscpr.res.in/handle/123456789/2871
96. Mukherjee D, Bhattacharya P, Jana A, Bhattacharya S, Sarkar S, Ghosh S, Majumdar S, Swarnakar S (2018) Synthesis of ceramic ultrafiltration membrane and application in membrane bioreactor process for pesticide remediation from wastewater. Process Saf Environ Prot 116:22–33. https://doi.org/10.1016/j.psep.2018.01.010
97. Chandrasekhar SS, Vaishnavi D, Sahu N, Sridhar S (2020) Design of an integrated membrane bioreactor process for effective and environmentally safe treatment of highly complex coffee industrial effluent. J Water Process Eng 37:101436. https://doi.org/10.1016/j.jwpe.2020.101436
98. Keerthi Suganthi V, Mahalakshmi M, Balasubramanian B (2013) Development of hybrid membrane bioreactor for tannery effluent treatment. Desalination 309:231–236. https://doi.org/10.1016/j.desal.2012.10.014
99. Liu W, Song X, Huda N, Xie M, Li G, Luo W (2020) Comparison between aerobic and anaerobic membrane bioreactors for trace organic contaminant removal in wastewater treatment. Environ Technol Innov 17:100564. https://doi.org/10.1016/j.eti.2019.100564
100. Sushmitha MB, Chanakya HN, Khuntia HK (2019) Efficient grey water treatment and reuse options for India—a review. In: Waste water recycling and management: 7th IconSWM–ISWMAW 2017, vol 3, pp 143–149. https://doi.org/10.1007/978-981-13-2619-6_12
101. Praneeth K, Moulik S, Vadthya P, Bhargava SK, Tardio J, Sridhar S (2014) Performance assessment and hydrodynamic analysis of a submerged membrane bioreactor for treating dairy industrial effluent. J Hazard Mater 274:300–313. https://doi.org/10.1016/j.jhazmat.2014.04.030
102. Yogalakshmi KN, Joseph K, Visvanathan C (2007) Response of membrane bioreactor to feed starvation shock load. Asian J Water Environ Pollut 4(2):81–86
103. Jadhao RK, Dawande SD (2012) Feasibility study of hollow fiber submerged membrane bioreactor. Int J Biotechnol, Chem Environ Eng 1(1):12–21

104. Surappanahalli Rajanna S, Madhu GM, Madhusoodana CD, Govindarajan A (2019) Silicon carbide-coated ceramic membrane bioreactor for sustainable water purification. Membranes 9(4):47. https://doi.org/10.3390/membranes9040047
105. Goswami S, Mazumder D (2019) Modelling and process design of moving bed bioreactor (MBBR) for wastewater treatment—a review. J Indian Chem Soc 96:215–229
106. Odegaard H (2006) Innovations in wastewater treatment:–the moving bed biofilm process. Water Sci Technol 53(9):17–33. https://doi.org/10.2166/wst.2006.284
107. Johnson CH, Page MW, Blaha L (2000) Full scale moving bed biofilm reactor results from refinery and slaughter house treatment facilities. Water Sci Technol 41(4–5):401–407. https://doi.org/10.2166/wst.2000.0472
108. Tak S, Tiwari A, Vellanki BP (2020) Identification of emerging contaminants and their transformation products in a moving bed biofilm reactor (MBBR)–based drinking water treatment plant around River Yamuna in India. Environ Monit Assess 192:1–23. https://doi.org/10.1007/s10661-020-08303-4
109. Singh A, Kamble SJ, Sawant M, Chakravarthy Y, Kazmi A, Aymerich E, Starkl M, Ghangrekar M, Philip L (2018) Technical, hygiene, economic, and life cycle assessment of full-scale moving bed biofilm reactors for wastewater treatment in India. Environ Sci Pollut Res 25:2552–2569. https://doi.org/10.1007/s11356-017-0605-y
110. Thakur SA, Khedikar IP (2015) Performance evaluation of moving bed bio-film reactor (MBBR) for treatment of domestic wastewater. Int J Sci Res 6(4):973–976. https://doi.org/10.13140/RG.2.2.14199.04008
111. Osmani SA, Rajpal A, Kazmi AA (2021) Upgradation of conventional MBBR into aerobic/anoxic/aerobic configuration: a case study of carbon and nitrogen removal based sewage treatment plant. J Water Process Eng 40:101921. https://doi.org/10.1016/j.jwpe.2021.101921
112. Sodhi V, Singh C, Cheema PPS, Sharma R, Bansal A, Jha MK (2021) Simultaneous sludge minimization, pollutant and nitrogen removal using integrated MBBR configuration for tannery wastewater treatment. Bioresour Technol 341:125748. https://doi.org/10.1016/j.biortech.2021.125748
113. Barwal A, Chaudhary R (2014) To study the performance of biocarriers in moving bed biofilm reactor (MBBR) technology and kinetics of biofilm for retrofitting the existing aerobic treatment systems: a review. Rev Environ Sci Bio/Technol 13:285–299. https://doi.org/10.1007/s11157-014-9333-7
114. Kulkarni SJ (2014) An insight into research and studies on modeling and simulation-biological treatment methods for wastewater. Int J Eng Res Gen Sci 2(5):35–38. http://ijergs.org.managewebsiteportal.com/files/documents/REMOVAL-5.pdf
115. Nakhli SAA, Ahmadizadeh K, Fereshtehnejad M, Rostami MH, Safari M, Borghei SM (2014) Biological removal of phenol from saline wastewater using a moving bed biofilm reactor containing acclimated mixed consortia. Springerplus 3:1–10. https://doi.org/10.1186/2193-1801-3-112
116. Mane SS, Munavalli GR (2012) Sequential batch reactor-application to wastewater–a review. In: Proceeding of international conference SWRDM, pp 121–128
117. Wichern M, Lübken M, Horn H (2008) Optimizing sequencing batch reactor (SBR) reactor operation for treatment of dairy wastewater with aerobic granular sludge. Water Sci Technol 58(6):1199–1206. https://doi.org/10.2166/wst.2008.486
118. Mahvi AH (2008) Sequencing batch reactor: a promising technology in wastewater treatment. Iran J Env. Hdealth Sci. Engg. 5(2):79–90
119. Ruhela M, Wani AA, Ahamad F (2020) Efficiency of sequential batch reactor (SBR) based sewage treatment plant and its discharge impact on Dal Lake, Jammu & Kashmir, India. Arch Agric Environ Sci 5(4):517–524. https://doi.org/10.26832/24566632.2020.0504013
120. Thakur C, Srivastava VC, Mall ID (2014) Aerobic degradation of petroleum refinery wastewater in sequential batch reactor. J Environ Sci Health Part A49(12):1436–1444. https://doi.org/10.1080/10934529.2014.928557
121. Sharma V, Srivastava VC, Kushwaha JP, Mall ID (2010) Studies on biodegradation of resorcinol in sequential batch reactor. Int Biodeterior Biodegradation 64(8):764–768. https://doi.org/10.1016/j.ibiod.2010.08.009

122. Showkat U, Najar IA (2019) Study on the efficiency of sequential batch reactor (SBR)-based sewage treatment plant. Appl Water Sci 9(1):2. https://doi.org/10.1007/s13201-018-0882-8
123. Kundu P, Debsarkar A, Mukherjee S (2013) Treatment of slaughter house wastewater in a sequencing batch reactor: performance evaluation and biodegradation kinetics. BioMed Res Int. https://doi.org/10.1155/2013/134872
124. Dutta A, Sarkar S (2015) Sequencing batch reactor for wastewater treatment: recent advances. Curr Pollut Rep 1:177–190. https://doi.org/10.1007/s40726-015-0016-y
125. Suresh S, Tripathi RK, Gernal M (2011) Review on treatment of industrial wastewater using sequential batch reactor. Int J Sci Technol Manage 2(1):64–84
126. Suresh S (2011) Biodegradation of hydroquinone using sequential batch reactor: a preliminary study of industrial effluent. Res J Chem Environ 15(2):48–56
127. Durai G, Rajasimman M, Rajamohan N (2011) Kinetic studies on biodegradation of tannery wastewater in a sequential batch bioreactor. J Biotech Res 3:19–26
128. Durai G, Rajasimman M, Rajamohan N (2011) Aerobic digestion of tannery wastewater in a sequential batch reactor by salt-tolerant bacterial strains. Appl Water Sci 1:35–40. https://doi.org/10.1007/s13201-011-0006-1
129. Acharya N, Kumar V, Gupta V, Thakur C, Chaudhari PK (2021) Aerobic sequential batch reactor for domestic sewage treatment: parametric optimization and kinetics studies. Int J Chem Reactor Eng 20(6):609–617. https://doi.org/10.1515/ijcre-2021-0094
130. Shirule PA, Mulik MNM, Sangore VP (2013) Treatment of dairy waste water using UASB reactor and generation of energy. Pratibha: Int J Sci Spirit Bus Technol (IJSSBT) 2(1), 2277-7261
131. Kasaudhan GK, Mumtaz N, Raj V (2013) Optimization of municipal wastewater treatment by UASB reactor and polising pond. Biolife 1(4):200–207
132. Khalil N, Sinha R, Raghav AK, Mittal AK (2008) UASB technology for sewage treatment in India: experience, economic evaluation and its potential in other developing countries. In: Twelfth international water technology conference, vol 12, pp 1411–1427
133. Yadav P, Chaudhary A, Keshari A, Chaudhary NK, Sharma P, Kumar S, Yadav BS (2022) Data visualization of influent and effluent parameters of UASB-based wastewater treatment plant in Uttar Pradesh. Int J Adv Comput Sci Appl 13(2). https://www.proquest.com/openview/f3b59a22308f6d1a06d90d3e8385e9d4/1?pq-origsite=gscholar&cbl=5444811
134. Hasan MN, Khan AA, Ahmad S, Lew B (2019) Anaerobic and aerobic sewage treatment plants in northern India: two years intensive evaluation and perspectives. Environ Technol Innov 15:100396. https://doi.org/10.1016/j.eti.2019.100396
135. Bandhavya GB, Prashanth S (2020) Effect of hydraulic loading rate and sludge occupancy on the performance of UASB reactor treating low strength wastewater. Appl J Environ Eng Sci 6(4):356–362. https://revues.imist.ma/index.php/AJEES/article/view/23175
136. Parashar N, Hait S (2023) Abundance, characterization, and removal of microplastics in different technology-based sewage treatment plants discharging into the middle stretch of the Ganga River, India. Sci Total Environ 905:167099. https://doi.org/10.1016/j.scitotenv.2023.167099
137. Loganath R, Mazumder D (2020) Treatability study of synthetic slaughterhouse wastewater using enhanced clarifier hybrid UASB reactor. J Indian Chem Soc 97(9A):1361–1369. http://indianchemicalsociety.com/portal/uploads/journal/September%207.pdf
138. Bhaduri M (2021) Available techniques of wastewater management practices in Indian tanneries: a literature review. J Environ Treat Tech 9(2):491–498. https://www.academia.edu/download/83960472/85.pdf
139. Shroff KC, Shah NG (2023) The performance evaluation and process optimization of anaerobic co-digestion of MSW with bio-flocculated sludge from secondary settling tank: a key to integrated solid–liquid waste management. Waste Biomass Valorization 15:405–415. https://doi.org/10.1007/s12649-023-02176-7
140. Arunadevi PS, Saravanaraja M (2020) Two phase upflow anaerobic sludge blanket (UASB) reactor on the reduction of chemical oxygen demand in dairy effluent. Asian J Adv Res 3(1):86–94

141. Mittal S, Ahlgren EO, Shukla PR (2019) Future biogas resource potential in India: a bottom-up analysis. Renewable Energy 141:379–389. https://doi.org/10.1016/j.renene.2019.03.133
142. Vaishnavi M, Gopinath KP, Ghodke PK (2023) Membrane bioreactor (MBR) technologies for treatment of tannery waste water and biogas production. In: Biorefinery for water and wastewater treatment. Springer International Publishing, Cham, pp 217–247. https://doi.org/10.1007/978-3-031-20822-5_11
143. Wagh M, Nemade P, Biradar A (2021) Potential of recalcitrant effluent to generate biogas by up-flow anaerobic sludge blanket and mechanical vapour recompression (MVR)-a case study. https://doi.org/10.21203/rs.3.rs-394660/v1
144. Sundararajan M M, Thirunavukkarasan S (2023) Enhanced copper removal and biogas production through intermittent operation of coupled modified upflow with static granular bed reactor treating copper-containing wastewater. Environ Eng Manage J 22(7). https://doi.org/10.1016/j.renene.2019.03.133
145. Kalbar PP (2021) Hybrid treatment systems: a paradigm shift to achieve sustainable wastewater treatment and recycling in India. Clean Technol Environ Policy 23:1365–1373. https://doi.org/10.1007/s10098-021-02034-x
146. Gupta I, Vachasiddha L, Kumar R (2017) Evaluation of the costs and benefits of Mumbai sewage disposal project, India. Indian J Geo-Marine Sci. 46:1539–1545. https://nopr.niscpr.res.in/handle/123456789/42530
147. NGT (2019) National Green Tribunal, Principal Bench, New Delhi, India Order dated 30 April 2019, Original Application No. 1069/2018. http://www.indiaenvironmentportal.org.in/files/file/revised-standards-STPsNGT-Order.pdf. Accessed 02 Feb 2020
148. Hai FI, Yamamoto K, Fukushi K (2007) Hybrid treatment systems for dye wastewater. Crit Rev Environ Sci Technol 37:315–377. https://doi.org/10.1080/10643380601174723
149. Dhangar K, Kumar M (2020) Tricks and tracks in removal of emerging contaminants from the wastewater through hybrid treatment systems: a review. Sci Total Environ 738:140320. https://doi.org/10.1016/j.scitotenv.2020.140320
150. Soni A, Stagner JA, Ting DSK (2017) Adaptable wind/solar powered hybrid system for household wastewater treatment. Sustain Energy Technol Assess 24:8–18. https://doi.org/10.1016/j.seta.2017.02.015
151. Kalbar PP (2021) Need for adopting hybrid treatment systems to reduce O&M costs of sewage treatment and recycling in India. J Indian Water Works Assoc 53(4):247–252
152. Manu DS (2018) Hybrid biological systems for wastewater treatment (Doctoral dissertation, National Institute of Technology Karnataka, Surathkal). https://idr.nitk.ac.in/jspui/handle/123456789/14088
153. Gajbhiye P, Yadav AM (2020) Industrial wastewater treatment using hybrid techniques and data modeling using RSM-ANOVA approach. J Indian Chem Soc 97(11 B):2495–2500
154. Vaidya R, Verma K, Kumar M, Hoysall, C, Rao L (2023) Assessing wastewater management challenges in developing countries: a case study of India, current status and future scope. Environ Dev Sustain, pp 1–28. https://doi.org/10.1007/s10668-023-03540-2
155. Uwamungu JY, Kumar P, Alkhayyat A, Younas T, Capangpangan RY, Alguno AC, Ofori I (2022) Future of water/wastewater treatment and management by industry 4.0 integrated nanocomposite manufacturing. J Nanomaterials, pp 1–11. 2022. https://doi.org/10.1155/2022/5316228

Biosorption for Wastewater Treatment and Post-sorption Utilization of Treated Wastewater and Spent Biosorbent

K. Man Mohan and S. Gajalakshmi

Abstract Biosorption is one of the potent strategies for pollutant removal from aqueous solutions employing living or non-living biological materials. The mass transfer results in accumulating and concentrating the adsorbate in/on the biomass adsorbent (biosorbent), mostly through complexation and chelation. Generally, biosorbents such as bacteria, fungi, algae, chitin, and chitosan, plant biomass from agricultural and horticultural activities have been widely explored for the pollutant removal efficacy and for their comparative performance with activated carbon, zeolite, silicates and ion-exchange materials. Principally, biosorption is solely a mass transfer method, in which degradation or destruction of the pollutant does not happen. Only interphase transfer of mass (pollutant) and leaching of chemicals from biosorbent take place during the process which ultimately provides treated wastewater and pollutant loaded post-sorbent material. Management of post-sorption products is relatively less explored domain and a few options i.e., recovery of adsorbates by desorption/elution, land fill, storage, incineration, animal feed additive, crop irrigation (treated water) etc., have been reported. With this backdrop, the present chapter aims to provide a comprehensive account on the following aspects related to biosorption process:

1. Mechanism of biosorption
2. Factors influencing the biosorption process
3. Techniques used in biosorption research
4. Post-sorption exploration such as regeneration of spent adsorbent and recovery of adsorbates
5. Key issues that restrict the field application of biosorption process
6. SWOT analysis of biosorption process
7. Conclusion
8. Recommendations

K. Man Mohan · S. Gajalakshmi (✉)
Bioremediation Lab, Centre for Pollution Control and Environmental Engineering, Pondicherry University, Kalapet, Puducherry 605014, India
e-mail: gajalakshmis@pondiuni.ac.in

K. Man Mohan
e-mail: mannmohan3009@gmail.com

M. M. Ghangrekar et al. (eds.), *Biological and Hybrid Wastewater Treatment Technology*, Earth and Environmental Sciences Library,
https://doi.org/10.1007/978-3-031-63046-0_3

Keywords Biosorption · Post-sorption · Wastewater · Treatment · Sustainability · Regeneration · Recovery

1 Introduction

Biological methods of wastewater treatment have been undeniably outperforming from the point of view of economic feasibility and environmental impacts. In line with other treatment methods, biosorption is one of the novel and effective approach for the pollutant elimination from aqueous solutions which is defined as the removal or decrease in the solute/adsorbate concentration in the aqueous solution applying biological materials (living/non-living). The mass transfer results in accumulating and concentrating the adsorbate in/on the biomass (adsorbent) mostly through the complexation and chelation mechanisms. There are two subcategories of biosorption process based on the *vital* status of biomass i.e., biosorption of contaminants using either alive or non-living biomass. Non-living biomass-based adsorption method have been extensively explored for the decontamination of wastewater due to its inherent virtues such as inexpensiveness, eco-friendliness, comparative pollutant uptake capacity, abundant availability, recyclability and low sludge formation. Better adsorbate selectivity of biosorbent is an added benefit compared to activated charcoal and ion-exchange materials [1]. Generally, biosorbents such as bacteria, fungi, algae, chitin, chitosan, plant biomass from agricultural and horticultural activities have been widely explored for the pollutant removal research works for their comparative performance with activated carbon, zeolite, silicates and ion exchange materials. Majority of the research work deals with the plant biomass based biosorbents acquired from the agriculture, weeds and horticulture waste biomasses. Therefore, further sections have been oriented towards biosorption of pollutant such as dyes using plant biomass.

2 Mechanism of Biosorption

The efficiency of the plant biomass based biosorbents for wastewater treatment entirely depend on the physico-chemical properties, interactions between the biomass surface and the pollutant molecules (dye, metals, organics etc.) on given experimental conditions. Plant biomass typically contains macromolecules such as cellulose, hemicellulose, lignin, xylan, tannin etc., in unique composition which is specific for a plant species and its body parts (roots, leaf, stem etc.). Cellulose and hemicellulose are the main structural components of plant cell wall made of linear chain and branched chain polysaccharides, polymers respectively. Hydrogen bonding and dispersion forces in the β-(1,4) linked D-Glucose chains have been referred for the crystallinity, insolubility and resistance to hydrolysis of the cellulose structure [2]. On the other hand, hemicellulose, a heteropolymer compound made of β-(1,4) and α-(1,2) linked xylose to 4-O-methyl ether of glucuronic acid, result in branched matrix that gives amorphous, soluble and easy to hydrolysable nature to it [3, 4]. In softwood α-(1,3)-glycosidic bond prevails between xylan backbone and arabinose, furanose branched units [5]. Lignin, major constituent and known as cementing agent of plant cell wall, is a complex aromatic polymer which contains aryl alcohol (Phenylpropane) as main component unlike polysaccharides in cellulose and hemicellulose [6]. It attaches via covalent bonding to hemicellulose and cellulosic matrices forming a protective cover around the cell wall against structural damage, hydrolysis and microbial degradation [7]. Another complex compound, tannin is formed as a part of defence mechanism in the plants which irreversibly complexes the proteins of the attacking organisms i.e., bacteria, fungi, insects etc. It is made of polyphenolic compounds and it has multiple industrial applications such as leather tanning, pharmaceuticals, adhesive, liquor additive and fire-insulation [8]. The richness and diversity of aforementioned plant biopolymers suggests abundant availability of multiple functional groups in the lignocellulosic matrix. For instance, the complex matrix of lignin contains high number of –OH, CH_3–O–, –COOH and –C=O that gives polarity to the matrix [5, 9]. Tannin have polyhydroxy polyphenol groups as an active component. These functional moieties play vital role in the adsorbate-adsorbent interactions and can be exploited for biosorption of pollutant species from wastewater. The cellulose, hemicellulose, polyphenolics from the lignin and tannin can be employed as the binding agent for the wastewater dyes [1].

The complexity of biomaterials can give rise to diverse adsorption mechanisms that often create difficulty in the biosorption process evaluation and characterization at particular set of experimental conditions. Biosorption can be classified on the basis of the dependency of biosorption process on the cellular metabolism of the biosorbent under investigation. There are two types of biosorption process and described as (a) metabolism dependent and (b) non-metabolism dependent. Figure 1 provides an overview of the mechanisms involved in biosorption process. The metabolism dependent biosorption chiefly involves the uptake/accumulation of pollutant molecules into the living microbial cells (e.g., bacteria) as part of their immune response mechanism [10]. On the other hand, the non-metabolism dependent biosorption process occurs mostly due to the physico-chemical interactions between the pollutant molecules and the surface of dead biomass. In both types of biosorption, physico-chemical interactions such as ion-exchange, adsorption, complexation and precipitation etc., may take place at varied rates and intensities which solely depend on the nature of biosorbent and pollutant (adsorbate) as well as the experimental conditions. However, the metabolism dependent biosorption involves transmembrane and intracellular transport of pollutants in living microbial cells. Independency from metabolism gives rapid kinetics and reversibility in non-metabolism dependent biosorption processes [11].

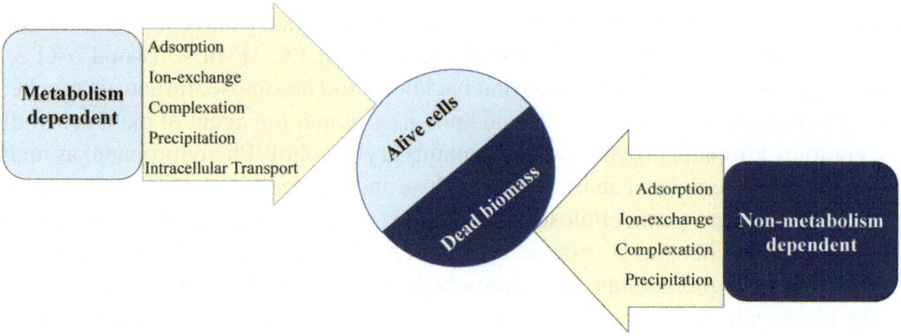

Fig. 1 Mechanisms involved in two types of biosorption processes

2.1 *Ion Exchange*

Ion-exchange is a complex process which involves adhesion of adsorbate ionic species by replacing the existing ionic moieties on the adsorbent surface while maintaing the electro-neutrality of the sorbate-sorbent system environment [12]. It is a reversible and stoichiometrically defined process which can be influenced from the existence of competitive ionic species, pH of the solution and intraparticle diffusion etc. [13].

2.2 *Precipitation*

Precipitation is the transition of chemicals from dissolved to undissolved compounds on spiking up the ionic concentration of precipitating agents in saturated solution phase. It is highly influenced by the solute concentration (adsorbate) and the pH of the solution [14, 15]. The biological surface functional moieties such as hydroxyl, carboxyl, sulfahydryl, and carbonyl facilitate cationic precipitation at specific pH [16]. In the case of metal biosorption, the surface bound metal ions can serve as precipitating loci and result in high degree of adsorption capacity. However it can advesrly affect the desorption process at the time of recovery [17].

2.3 *Complexation/Coordination*

Complexation is the formation of complex macromolecules through sharing of electrons from ligands (atoms/molecules/ions) to a central metal atom via Lewis acid–base reaction. Unlike double salts, coordinate complexes do not readily dissociate in aqueous solutions. Biosorption of pollutants may involve significant proportion of complexation due to the availability of diverse functional groups on the surface of biosorbents and charged adsorbent molecules along with the surface adsorption, precipitation and ion exchange. Complexation occurs within a particular pH range where the emergence of specific charge on the biosorbent surface as well as adsorbate molecules take place according to the dissociation constants (pKa) values of their functional moieties that enhance the electrostatic interactions which ultimately result in high biosorption efficiency [18]. Increase in metal binding to surface biomolecule moieties in metabolism dependent biosorption process is possibly due to the complexation reactions in presence of specific surface functional groups, e.g. –SH, $-NH_2$ etc. [17]. Presence of many ligands on the biomass surfaces facilitate chelation of metallic pollutants [19].

2.4 *Surface Adsorption*

Adsorption occurs on the biosorbent surface's active sites when adsorbate molecules arrive by crossing the boundary layer resistance of biosorbent particles. It mainly involves the Van der Waal's forces, dipole–dipole and electrostatic interactions. These interactions result from the high functional group diversity on the biosorbent surface and have been cosidered the major contributor for the wide variety of pollutant removal by biosorbents [20]. The interactions between adsorbent and adsorbate are heavily dependent on the solution pH [21]. At lower pH values, the suface active sites becomes highly protonated that produce positively charged biosorbent surface which facilitate the adsorption of anionic molecules and retards the cationic molecues approching the surface via electrostatic repulsive forces. Apart from pH of the solution, multiple factors impact the overall performance of the biosorption process, which are briefly disscussed in the following sections.

3 Factors Influencing the Biosorption Process

The rate and the mechanism of biosorption process are often governed by the nature of adsorbate and adsorbent materials as well as the environmental factors of sorbate-sorbent system. In general, most affecting factors are the pH of the solution, temperature of the system, size of biosorbent particles, biosorbent dosage, composition and initial concentration of adsorbate, and contact time of biosorption process.

3.1 *pH of the Solution*

The pH of the adsorption system most precisely governs the biosorption process by modulating the chemical properties of the biosorbent surface as well as the adsorbate molecules. Changes in solution pH can alter the surface charge of the particles, the activity of functional groups, the chemical nature of adsorbate and the competition with coexisting ionic moieties in the solution [22]. Solution pH defines the overall charge on the surface of the biosorbent that ensures accessibility of adsorbate molecules towards specific functional groups on biosorbent surface [23]. The pH directly impacts on the dissociation of dye molecules affecting its solubility in aqueous phase. It can also cause structural changes in dye molecules. Lower pH values often decrease the uptake of cationic pollutants (e.g., metals, dyes etc.) due to the competition with protons to occupy the active sites on biosorbent surface [17]. However, anionic dyes favourably biosorbed at the lower pH of solution, due to the protonation of surface functional groups which enhances the electrostatic attractions among the dye molecules and adsorbent surface moieties. On the other hand, increased pH values enhance the surface's deprotonation, resulting in the increased negative charge which poses electrostatic repulsion towards the anionic dyes but attraction for the cationic dyes [22].

3.2 *Temperature*

The temperature of the adsorption system is intimately related to the energetics and thermodynamics of the adsorption process. Changes in the system temperature can transform the reaction from exothermic to endothermic or vice-versa by altering the kinetic energy, mobility of the adsorbate molecules and adsorptive power between the adsorbate molecules and adsorbent surface. Adsorption capacity of a biosorbent is a function of system temperature. In general, the increase in temperature causes increase in kinetic energy of adsorbate molecules and decreases the viscosity of the aqueous phase simultaneously which lowers the boundary layer resistance for the enhanced mass transfer in biosorbent surface. There is optimum temperature range for particular sorbate-sorbent system where it functions at its maximum and the rate of the adsorption decreases at both extremes of temperature range (lower and higher) [24]. Higher temperature can cause structural damage to adsorbent particles, leading to fragmentation and dissolution. In case of chemical adsorption, the increase in system temperature enhances the adsorption capacity by increasing the rate of reaction. However, the increase in the kinetic energy at increased temperature reduces the probability of electrostatic attraction among the adsorbent-adsorbate molecules and increases the desorption rate in a physical adsorption process [25]. On the basis of energy exchange, there are two types of adsorption process: endothermic and exothermic. Increase in temperature causes decline in the adsorption efficiency in the exothermic adsorption process due to the weakening of the adsorptive power among the adsorbate and active sites of adsorbent particle. On the other hand, endothermic adsorption process efficiency improves on increase in temperature due to the elevated mobility of heavy dye molecules and thermal activation mediated increase in number of active sites on adsorbent surface. It can be simplified as, better uptake at higher temperature indicates endothermic adsorption while better uptake at lower temperature indicates exothermic nature of biosorption process. Major thermodynamic parameters such as enthalpy, entropy and Gibbs free energy related information can be obtained by conducting the batch biosorption experiments varying the system temperature and keeping all other process parameter at constant level. The enthalpy and entropy of the sorbate-sorbent system might be elevated or declined during the adsorption process [22].

3.3 Initial Concentration of the Adsorbate

The initial concentration of pollutant (adsorbate) in the aqueous system has significant impact on the adsorption process as it acts as driving force against the mass transfer resistance from aqueous phase to solid phase as well as it increases the collision frequency (probability) of adsorbate molecules with adsorbent particles during the competition for occupying the active sites [22]. Initial dye concentration can have negative, positive, and neutral impacts on the biosorption efficiency. In most of the cases, adsorbate removal percentage decreases on the increased initial concentration which is inferred due to the saturation of active sites of biosorbent. But increased removal at higher concentrations indicates the increased driving force of adsorbate molecules towards the biosorbent surface [23]. At lower concentrations the ratio of adsorbate molecules to active surface area is lower so the process is independent from the effect of concentration; but at higher concentrations the active sites become limited which poses the competition among the adsorbate molecules thus the process becomes concentration dependent [26].

3.4 Quantity of Biosorbent

The amount of biosorbent poses significant impact on the adsorption efficiency (removal percentage) and uptake capacity which is directly linked to economic feasibility of a biosorbent. Optimization experiments are often carried out for achieving maximum uptake capacity at minimum biosorbent wastage [22]. Most of the previous studies reported, the dosage of biosorbent material (at constant initial concentration, temperature, pH and agitation speed etc.) enhances the adsorption efficiency but declines the uptake capacity of the material [27]. Two main reasons that have been suggested for the decline in capacity are: (i) prevalence of unoccupied active sites at equilibrium and (ii) reduced surface area due to clumping/agglomeration of biosorbent material [28].

3.5 Contact Time

The duration of interaction between adsorbent and adsorbate containing liquid phase is a key parameter to evaluate the rate of the adsorption and the point of time where the system attains equilibrium. It is crucial in designing the batch adsorption reactor system for optimum wastewater treatment. In most of the research work, favourable adsorption process shows high rate of pollutant removal at early stages of operation which declines and subsequently attain the equilibrium with increase in contact time that indicate the saturation of active sites on adsorbent surface [29, 30]. Increase in contact time indirectly increases and maximises the probability of contact between adsorbate and adsorbent material. This probability can also be enhanced by increasing the agitation or mixing speed of the sorbate-sorbent system, which facilitates the adsorbate to overcome the boundary layer resistance and to diffuse into the adsorbent surface. However, it can create structural damage to the adsorbent particles due to increase in sheering forces of liquid phase [22].

3.6 Particle Size

The effect of the size of biosorbent particles has been less studied but considered to be an important parameter for the efficient removal of the pollutant from the aqueous phase. Size reduction of biosorbent provides high surface area that improves the pollutant removal efficiency by increasing the rate of adsorption process. But smaller particle size may have negative impact on adsorption efficiency due to the insufficient contact between adsorbate and adsorbent caused by the floating of biosorbent on the surface of solution [22]. From the various inferences, particle size of biosorbent has been found to be inversely proportional to the efficiency of biosorption process. Large sized particles facilitate attachment of a greater number of adsorbate molecule to the active sites by proper mixing in solution [23]. However, for the column biosorption studies, the particle size should be maintained smaller to maximize the mass transfer of dye molecules by minimizing the resistance. The small sized biosorbents suffer from the durability and economic feasibility related drawback [31]. To rectify the ambiguities and denote the particle size range (smaller-larger), optimization experiments should be carried out for the biosorbent and adsorbate under investigation.

4 Techniques Used in Biosorption Research

Over the long period of research in the biosorption domain, various physical and chemical analysis techniques have been evolved and employed by the researchers to understand the underlying mechanisms and effects of the adsorbent material modifications on the efficacy of biosorption process. Major characterization techniques include microscopy, spectroscopy, titrimetric and bulk properties analysis and so on.

The surface of the biosorbent is the vital space where mass transfer takes place. Its structural and chemical nature governs entire process of pollutant uptake. The study of surface morphology by electron microscopy (Scanning Electron Microscopy) provides information on the topography of the surface, structure of pores, distribution of pores, presence of specific microstructures and surface elemental composition of biosorbent [32]. It helps to elucidate the impact of biosorption process and concentrated adsorbate molecules on the surface. However, elemental analysis is qualitative approach which is performed using SEM coupled Energy dispersive X-ray analysis. On the other hand, surface area and porosity of the biosorbent material are known to be the key players in the biosorption process by providing the active sites for the attachment. The surface area of the biosorbents can be determined by methylene blue number estimation and Branueur-Emmet-Teller (BET) analysis etc. The BET technique is widely employed and considered to be more reliable for the specific surface area and pore dimension characterisation [32]. It is performed using nitrogen adsorption–desorption on properly degassed biosorbent material [33]. However, BET surface area values should not be taken alone for the purpose of biosorbent uptake capacity comparison. It should be considered as supporting information as there are multiple factors affecting the complex biosorption process i.e., pH, temperature, initial concentration, and surface chemistry of the biosorbent etc. It is well known that the biosorbent possess lower surface area than conventional adsorbents such as activated carbon, biochar, silica etc., but their comparable adsorption capacity is the resultant of the abundance and diversity of available functional groups on the surface. The chemistry of biosorbent surface can be examined by point of zero charge analysis, Boehm or potentiometric titration and Fourier transform infrared spectroscopy (FT-IR) analysis. Point of zero charge study demonstrate the overall electrostatic charge on the surface of biosorbent at particular pH of the solution that can be useful for the suitability of adsorbate uptake at desired pH. Titrimetric analysis provides quantitative information of the surface functional groups such as phenolics, acidic and sulfonic which aid to determine the ion-exchange capacity of the biosorbents. The FT-IR spectroscopy demonstrates the involvement of surface functional moieties and bond formations in the biosorption process. The shift in the band position and change in the frequency patterns can be compared taking FT-IR spectra of pre- and post-sorption materials to identify the involved surface functional groups. To gain deeper insight into the chemical composition, state of oxidation of surface elements and biosorption mechanism, X-ray photoelectron spectroscopy (XPS) can be used. It provides electronic configuration, quantity and ratio of elements of compounds on the biosorbent surface that is crucial to understand the biosorption mechanism on the surface [16, 34].

5 Key Issues that Restrict the Field Application of Biosorption Process

Although biosorption has been considered as low-cost and eco-friendly approach in most of the previous laboratory studies, yet the pilot and field scale applicability have not gained much attention; most probably due to its stability and feasibility related challenges. However, few studies on the pilot scale and semi-industrial scale applications of biosorbent for pollutant removal (mostly heavy metal removal) have been reported in the past [16, 35], but there is severe dearth of report on pilot or industrial scale biosorption process [36]. Key issues prompting to restrict the field application of biosorption have been listed below.

- Lack of instantly usable biosorbent materials [37].
- Biosorbent material collection, storage and logistics may impart overall high cost, which may lead to non-feasibility of process implementation [38].
- Most of the biosorbents are physically weak materials and unable to withstand high pressure (in column process), agitation speed (in batch process), longer contact time and higher temperature.
- Many biosorbents have affinity towards a particular adsorbate (mostly metal) that decrease their applicability for multiple pollutant containing complex real-time effluents.
- Usually, biosorbents show lower sorption capacity in their natural forms. Due to this reason, many biosorbents require pre-treatment to enhance their efficiency which escalates overall cost of the treatment process [39, 40].
- Chemical and physical pre-treated biosorbents have excellent removal efficiency, but they lose the virtues of cost-effectivity and eco-friendliness by releasing hazardous chemical into the environment after pre-treatments and high energy consumption during the physical pre-treatments [41].
- Leaching of chemicals and incorporation of color during the biosorption process often lead to decline in aesthetics as well as quality parameters of the treated water [38, 42].
- The persisting problems related to the management and safe disposal of adsorbate loaded hazardous spent biosorbents are challenging [17, 43].

The emphasis on the enhancement of sorption capacity of biosorbents by pre-treatment are being explored for more than two decades. However, many research works have proved significant improvement in the performance on pre-treatment of biosorbents but with increased overall cost and environmental unfavorability.

6 Post-sorption Exploration

Principally, biosorption is solely a mass transfer method, in which degradation or destruction of the pollutant does not happen. Only interphase transfer of mass (pollutants) and leaching of chemicals from biosorbent take place during the process which ultimately provides treated wastewater as well as pollutant loaded post-sorbent material. By definition, post-sorbent material is just adsorbate loaded biomass while spent or exhausted biosorbent is the residue which remains after recovery of adsorbate (e.g., metal, dyes, other organic compounds etc.) from post-sorbent material. Management of post-sorption products is relatively less explored domain and a few options i.e., recovery of adsorbates by desorption/elution, land fill, storage, incineration, animal feed additive, crop irrigation (treated water) etc., are discussed in previous reports. A brief overview of commonly employed and proposed methods for post-sorbent management is given in subsequent sections.

6.1 Regeneration of Spent Adsorbent and Recovery of Adsorbates

In the wake of sustainable development goals, majority of the industrial as well as domestic activities are being leaned towards the "3Rs—Reduce, Recycle and Reuse" approach [44]. Although, wastewater treatment employing adsorption process is feasible, adaptable, and efficient, yet the adsorbate loaded toxic solid waste management issues create substantial repression at large scale. Therefore, many research laboratories have oriented their studies towards the recycling and reuse of spent adsorbents by various methods [45]. Regeneration of spent adsorbent can be carried out by various physico-chemical and biological processes. Figure 2 illustrates commonly employed regeneration methods at lab scale as well as industrial level. Each of the regeneration process has its own advantages and limitations. Though, having high efficiency, the regeneration process lacks attention on ground due to higher cost of operation, secondary pollution and significant material loss on regeneration [46].

6.2 Desorption of Adsorbates

Thermal regeneration is mostly carried out on exhausted activated carbon materials either onsite or offsite. It relies on high energy input for three stage heating of exhausted adsorbent i.e., drying (150 °C), desorption/decomposition (500 °C – 800 °C) and gasification (800 °C + CO_2 gas) of charred pollutant compound to re-exfoliate the active sites on the exhausted carbon columns. It is envisaged that the higher temperature can pose significant risk of destruction of adsorbent column

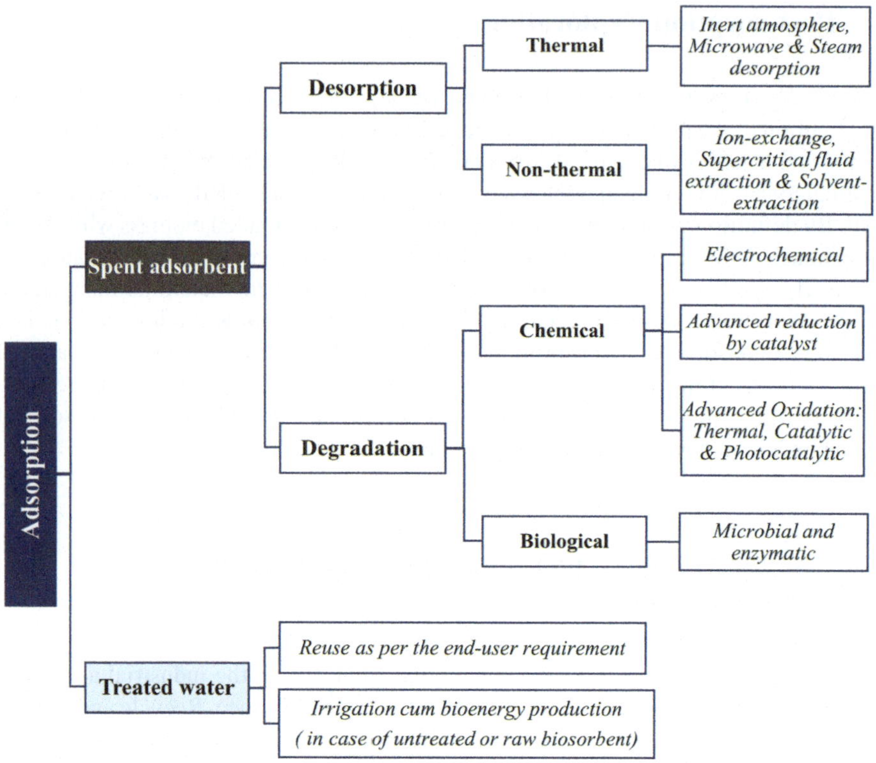

Fig. 2 Common adsorbent regeneration strategies (adopted from Omorogie [46])

material as well as toxic flue gas release in atmosphere. Therefore, thermal regeneration must be carried out after understanding the thermal behaviour (physical stability, combustion and pyrolysis) of adsorbate—adsorbent complex. Microwave mediated desorption also gained interest due to distinctive mode of uniform and fast heat generation at molecular level which enables selective desorption of a particular adsorbate from multiple adsorbates loaded matrix of exhausted adsorbent material [47]. Polar adsorbate molecules and carbonaceous adsorbent materials have been found to be apt for the microwave mediated desorption method due to higher tendency to absorb microwave radiation. It should be noted that the fine-tuned microwave mediated heating at molecular scale helps to increase the specific surface area and improve the pore structures with minimal destructive loss of adsorbent material [48]. Although, microwave desorption method offers low time, energy and space consumption merits for the adsorbent regeneration, but evolution of noxious by-products i.e., hydrochloric acid, carbon dioxide, nitrogen etc., during the process hinders its wide application on field [47]. On the other hand, chemical mediated regeneration method is independent of energy input. It exploits the nature of adsorbate—solvent interactions to desorb or elute the exhausted adsorbent material. It mostly depends on the solubility of the adsorbate in selected solvent system (extractant). Principally, there are three ways to attempt chemical regeneration i.e., alteration of acidity and alkalinity, complexation with suitable organics, ligands or metal ions, and oxidative degradation of adsorbed pollutants. Acids and alkali mediated desorption have been successfully demonstrated in many research works by using diverse solvent strengths. In this method, the desorption of adsorbate is accomplished due to the ionic exchange which is resulted from the concentration gradient across the solvent and spent adsorbent surface [46]. Reports on desorption of organic pollutants such as dyes, aryl alcohols and pesticides from the live or dead microbial biomass based biosorbents (bacterial, fungal) reveal that the wide range of eluents i.e., deionised distilled water, alkaline solutions, $CaCl_2$ solution, surfactants, and organic solvents etc., can be applied effectively for the regeneration of biosorbents and recovery of adsorbate consequently. The desorption process is highly influenced by the pH of the eluent [49]. Alteration in adsorption capacity (increase or decrease) may occur on the regeneration of adsorbents, possibly due to the alteration in hydrophilic / hydrophobic interactions. However, this desorption process often impacts positively on the uptake capacity of biosorbent after regeneration, but the possibility of structural disintegration of biosorbent materials must not be ruled out [17]. Recently, Himanshu Patel [45] reviewed a few comprehensive studies on the elution of azo dyes and heavy metals loaded biosorbents i.e., neem and flamboyant leaf powders using various types and strength of acidic, alkaline and organic solvent systems in batch as well as column desorption experiments for seven consecutive sorption–desorption cycles. The study revealed that the desorption efficiency of the solvent system mostly depended on the pH of solvent system. The attractive interactions among the oppositely charged adsorbate and eluent molecules result in high desorption efficiency at particular pH of eluent system [45]. These type of studies are desirable in the selection of appropriate eluent and pH for optimum regeneration efficiency. Furthermore, it was concluded that the column mode of elution-based regeneration process provides more feasible operation than the batch mode in terms of quantity of required eluents and operation time.

The supercritical fluid extraction method has been widely employed to desorb attached adsorbates and regenerate adsorbent materials for more than three decades. It takes the advantage of tuneable dissolution capacity (enhanced solubilisation of adsorbed molecules by varying temperature and pressure) of a supercritical fluid. Basically, any matter attains a unique state of indistinguishable phase (neither liquid nor gas), if pushed above the critical point of temperature and pressure and known as supercritical fluid. The supercritical fluid presents lower viscosity than liquid phase and higher density than gaseous phase which gives it special abilities of rapid, high and desirable solute dissolution of extractable components from many matrices [50]. Supercritical carbon dioxide has been profoundly employed to desorb organic compounds for its superior mass-transfer kinetics, flexible dissolution capacity, economic feasibility and non-hazardous nature [46]. However, the supercritical fluid regeneration method imposes lower operational cost compared to conventional thermal desorption methods, but higher initial capital investments and requirement of highly skilled human resources leads to restricted large scale application [51].

The sonication mediated desorption is an inexpensive method of regeneration, which exploits the effect of ultrasound frequency on liquid such as nucleation mediated micro-turbulence, micro-streaming, shock wave in solvent system which enables diffusion-convection of adsorbate from the adsorbent to the medium. The operational frequency (below 40,000 Hz) of sonication mediated desorption process should not cross the threshold of adsorbent stability and structural integrity to avoid the destructive loss of expensive adsorbent materials. Desorption of pollutant using this method chiefly depends on the nature of adsorbate, adsorbent and the interactions among them. Therefore, the adsorption isotherm and kinetics patterns could be employed to modulate and achieve maximum desorption capacity in case of favourable adsorption process [46].

6.3 Degradation of Adsorbates

Regeneration of spent adsorbent material through adsorbate degradation should be carried out when the adsorbate possesses degradability (chemical, biological, physical) and its recovery has no substantial importance over the regeneration of spent adsorbent. In this context, advanced oxidation processes such as Fenton, modified Fenton, photo-Fenton, permanganate oxidation etc., have been successfully employed for the regeneration of organic pollutant loaded spent adsorbents in previous studies. The Fenton method exploits the extraordinary oxidising ability of highly reactive HO• free radicals generated from Hydrogen peroxide in presence of iron as catalyst in acidic environment. These free radicals can degrade organic matter either by snatching hydrogen or inserting hydroxyl groups to the adsorbed organic compounds in violent attack. Despite the fact that the Fenton process provides safe and low-cost solution for regeneration of loaded adsorbents, but self-hunting cum neutralisation of free radicals by Hydrogen peroxide as a scavenger and formation of iron containing hazardous sludge poses performance as well as environmental challenges in large scale applications [46, 52]. However, elevated levels of Hydrogen peroxide and input of additional chemicals significantly improves the quantity and diversity of the free radicals in the reaction mixture which results in enhanced oxidation of diverse organic compounds. Though, higher strength of Hydrogen peroxide may lead to increased self-hunting activity towards HO• free radicals, yet this modified Fenton process shows better capacity of oxidation, fast reaction kinetics and degradation of highly recalcitrant adsorbates [52, 53]. It must be noted that the progression of the Fenton reactions leads to the consumption of ferrous ions and simultaneous accumulation of ferric ions in the reaction mixture that poses inhibitory effect and ultimately ceasing of overall process. To address this issue, radiation mediated ferric to ferrous ion conversion mechanism have been proposed and successfully applied to Fenton process for last few decades in high strength wastewater treatment [54]. Oxidative degradation efficiency of Fenton process increases tremendously, when the Fenton process is combined with the ultraviolet or visible radiation mediated ferrous to ferric conversion mechanism [55]. Efficiency of photo-Fenton process increases with the increase in the acidity of solution due to the formation of H_2CO_3 which significantly decreases the interference of CO_3^- and HCO_3^- ions towards activity of HO• free radicals. In addition to photo-Fenton, many other Fenton associated processes i.e., sono-Fenton, electro-Fenton etc., have been developed and explored for the highly recalcitrant organic pollutant degradation from various polluted matrices as well as to achieve high economic and environmental feasibility [56]. Permanganate mediated degradation of organic compounds have been widely employed in many pollutants loaded matrices such as soil and water. It exploits higher oxidative nature of $KMnO_4$ for the degradation of organic materials to CO_2 along with MnO_2 production. Although, the process is highly effective towards many organic pollutants such as petroleum hydrocarbons, pesticides, organic acids etc., but its reactivity towards CN^- and S^- groups can cause significant alteration in the chemical nature of the regenerated adsorbent material [46].

Ozone mediated degradation of adsorbed organic compounds such as nitrobenzene, aryl alcohols etc., has been explored by the researchers in previous studies. Organic compounds react with O_3 either directly or HO• free radical (decomposition product of O_3 in alkaline conditions) which result in breakdown of larger molecules and mineralization. The molecular O_3 interacts specifically to particular organic molecule, therefore it has been employed for the selective degradation processes. Although, ozonation provides benefit of pollutant selectivity and high degradation efficiency, yet pre-treatment step such as calcination involved in the process demands high energy input and alteration of adsorbent surface chemistry after regeneration are considered to be unfavourable for large-scale application [46, 57].

Regeneration of spent adsorbent employing electrolysis phenomenon has virtue of achieving desorption as well as degradation of adsorbate materials. It is carried out in electrolytic cells where spent adsorbent material is subjected to electrical current flow. The current induced change in pH and conductivity of the microenvironment of spent adsorbent surface causes dissociation of adsorbates from saturated active sites (cathodic accumulation of inorganic pollutants) or degradation of adsorbate compounds in smaller intermediate compounds (oxidative destruction of organic pollutants at anodic side). The efficiency of the process is mainly governed by the type of electrode material, current flow, electrolyte concentration, reaction time, pH of the solution and bed height of the spent adsorbent material [58]. However, having apparent efficiency of regeneration via desorption as well as degradation, electrolytic regeneration suffers from the inhibitory impacts of salt formation as by-products, longer regeneration time due to mass transfer resistance in electrodes and higher energy requirements [59].

Biological agent mediated regeneration of spent adsorbents provides a plausible solution by degrading recalcitrant pollutants into harmless smaller compounds. Principally, the process employs microbes such as bacteria and fungi to eliminate the degradable adsorbates which are attached to the surface of spent adsorbent. Microbe mediated regeneration is accomplished through simultaneous desorption and degradation of adsorbate molecules in synthetic media, that has been termed as off-line mode of microbe mediated regeneration. The desorption of adsorbate molecules occurs due to the existence of concentration gradient of adsorbate on the solid and liquid phase boundary in the microenvironment as well as the activity of exoenzymes from microbial secretions. The concentration gradient exists due to the continuous depletion of adsorbate molecules in the liquid media by bacterial degradation activity which provides driving force for the desorption process [60]. It should be noted that the Gibb's free energy differences among the adsorbed pollutant molecules (on the surface of adsorbent) and the free molecules (desorbed molecules) have direct impact on the desorption process during the microbial regeneration process [61]. On the other hand, exoenzymatic secretions enable desorption cum degradation of deeply attached adsorbate molecules in the porous matrix of adsorbent which are inaccessible by the bacterial cells. Various physico-chemical, biological and environmental factors have direct impact on the efficiency of the microbe mediated regeneration processes. Few governing factors of microbe mediated regeneration process are listed as follows:

i. Quantity, surface morphology and physico-chemical property of spent adsorbent.
ii. Nature and structure of adsorbate molecules.
iii. Concentration of adsorbates on the adsorbent surface.
iv. Type of interactions among the adsorbate—adsorbent.
v. Selected microbial species.
vi. Chemical composition of nutrient media.
vii. Operating conditions i.e., pH, temperature, and availability of oxygen to the microbes etc.

Chemical structure of adsorbate has significant influence on the spent adsorbent regeneration efficiency. For instance, microbial regeneration is found to be better in phenolic group deficient compounds (e.g., 3-chlorbenzoic acid) and lower in phenolic group containing compounds (e.g., 2-methyl phenol) adsorbed on the spent adsorbent surface [62]. Additionally, the type of substitution in adsorbate molecules can alter the degradation efficiency, such as p-Cresol, 4-Ethyl phenol, 4-Isopropyl phenol etc., containing spent adsorbent show lower regeneration efficiency than the phenol loaded adsorbent material. Saturated functional groups ($-CH_3$, $-C_2H_5$ etc.) containing phenols show higher microbial degradation over $-COOH$, $-NO_2$ functional group containing phenols. It may be attributed to the hindrance in unsaturated functional group containing phenols that suppresses bacterial metabolism mediated oxidation [62, 63]. Furthermore, the adsorbates which normally resist to undergo degradation even provided with prolonged contact time on adsorbent surface have also reported with increased degradability on microbial degradation process implementation [46]. According to previous reports, in terms of biodegradability, fungi outperform bacteria in microbe mediated regeneration of active carbons, mostly due to its tolerance towards the high strength of adsorbates in solution as well as ease of separation from the solution [64]. It is well evident that the mechanism of the microbial degradation mediated regeneration profoundly depends on the nature of adsorbate, adsorbent and their interactions. Ng et al. (2009–2010) [65, 66] group made the initial attempt of understanding and modelling of degradation mediated regeneration kinetics. They proposed first order kinetic model which is implied on both subsequent processes i.e., desorption of adsorbate from the spent adsorbents and microbial degradation of desorbed adsorbents in the solution. Biological regeneration process has been considered to be highly effective but suffering from the slow operation speed, alteration of efficiency on regeneration, material loss and selective and limited applicability (only on biodegradable pollutants).

6.4 Application of Spent Adsorbent as Animal Feed Additive, Soil Amendment in Agriculture, Nanoparticle Preparation and Gasification

Biosorption results in the adsorbate loaded biomass which can be further utilized as inexpensive alternative nutrient sources, especially in case of micronutrient (e.g., Cu, Zn, Mn, Cr etc.,) loaded biosorbents. The process of biosorption can be optimized according to the required amount of targeted adsorbate in the post-sorbent biomass. In this regard, the process would be called 'bio-removal', if the targeted equilibrium uptake capacity of adsorbate is minimized and 'bio-binding' if the equilibrium uptake capacity is maximized (Fig. 3) [67].

Numerous studies have been conducted to establish post-biosorbents as alternative sources of micronutrients in animal feedstocks [69]. In an early investigation by Chojnacka [67], qualitative and quantitative analysis of *Lemna minor* biosorbent after biosorption (bio-removal as well as bio-binding) of Cr, Mn, Zn, and Cu was carried

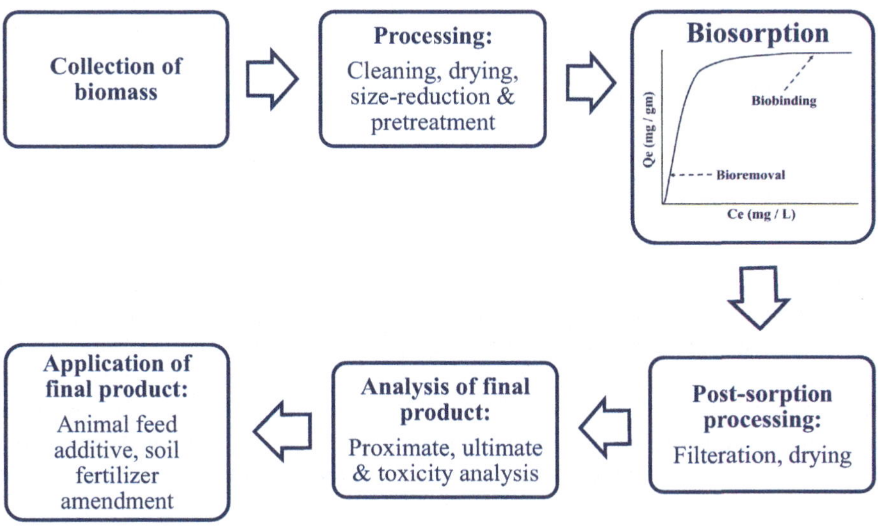

Fig. 3 Schematics of post-sorbent applications (adopted from Michalak et al. [68] and Chojnacka [67])

out for feedstock supplement applicability. In another study by Saeid et al. [70], post-sorption applicability of copper loaded *Spirulina* biosorbent on pigs showed increase in absorption and decrease in excretion of copper by the animals which suggested the post-biosorption formulation as a promising animal feed supplement. A comprehensive study demonstrated the positive impacts of post-sorption utilization of microelements (i.e., Cu, Cr, Zn, Mn) loaded soya meal feed for poultry hen's productivity and eggs quality. The post-sorption-based soya meal fed animals did not present much deviation or decline in the quality as well as organoleptic parameters [71, 72]. Another study from the same laboratory reported enhanced micronutrient uptake i.e., Fe, Zn, Cu and fortification of eggs with elevated levels of Fe, Zn and Cu in comparison to inorganic salt fortification feedstock [73]. In a swine feeding experiments, the impact of the micronutrient loaded *Enteromorpha* sp. biomass revealed better bioavailability of micronutrients than the commercial inorganic salt-based feed supplements [74].

Another sustainable approach to manage the post-sorption solid waste material is to utilize it as soil amendment as source of micronutrients and alternative of fertilizers. Sorption based soil amendments offer benefits of biocompatibility, low soil health risk, optimum and sustained release of nutrient to the plants. In this context, spent biochar have been explored as soil fertilizer for few decades. Biochar from various sources such as plant biomass, sludge and other organic waste have significant application for the biosorption of pollutants such as heavy metals, organic compounds etc. [75]. Studies on the utilization of spent biochar as soil amendment indicate lack of consistent performance due to numerous factors such as diverse surface morphology (porosity), chemical composition of source of origin (carbon and other mineral concentrations), parameters of biochar preparation (temperature, atmosphere and duration etc.) and the nature of the amended soil (pH, mineral composition, ion-exchange capacity etc.) [76]. The fidelity of post-sorbent biochar as soil amendment can be enhanced and optimized by physico-chemical modifications for the particular set of conditions. For instance, applicability of biochar as post-sorption amendment agent for PO_4^{3-} and NO_3^- delivery in soil is restricted due to the similar electrostatic nature of majority of biochar surface (mostly anionic). However, the problem of electrostatic incompatibility can be overcome by the formation of biochar-cationic (Mg, Al, Fe etc.) composites and magnetic composites [77, 78]. Other than biochar, few post-sorbents such as feather waste, plant biomass, mushroom waste and algae have been exploited for the biosorption of minerals, followed by the soil amendment applications [79–81]. Biosorptive removal of zinc using biomass of marine algae and subsequent utilization of spent biosorbent as fertilizer showed biofortification and increased dry weight of plants under study. Furthermore, the application of mushroom waste as biosorbent for copper, zinc and manganese followed by spent utilization experiment for the soil amendment in *Zea mays* plants revealed the positive impacts on the plant's productivity (grain yield), micronutrient and macronutrient composition.

Another fascinating application of spent adsorbent is to utilize it as shielding agent for stable metal nanoparticle synthesis it is also known as carbon-encapsulated nanoparticles. Ideally, naive metal nanoparticles are highly reactive as well as labile in nature due to the elevated surface area to volume ratio. They are prone to oxidative decomposition thus losing magnetic as well as other features easily and it has been considered as one of the main factors which restrains their large-scale application. However, there are methods to avoid nanoparticle degradation by developing protective shields on their exterior surface using various organic, inorganic and non-reactive materials [82]. Application of carbon for metal nanoparticle encapsulation has been found to be more attractive as well as feasible option in terms of electric, magnetic and chemical reactivity as well as stability of the final product [76]. It provides ease of operations as well as eco-friendly and economic desirability with compared to the conventional carbon–metal nano-composites manufacturing process. Additionally, the carbon nano-composites prepared employing spent adsorbent (metal loaded adsorbent/biosorbents) have better metal nanoparticle holding strength than the conventional carbon nano-composites which cease the leaching of metal nanoparticles in the aqueous phase during the wastewater treatment process. It also results in high surface area, reduced agglomerating and oxidizing tendency bearing nanocomposites [83].

Biomass gasification is one of the promising strategies to manage post-sorption solid waste. Generally, biomass gasification is carried out at high temperature and in presence or absence of air (combustion, reduction and pyrolysis) to obtain value added products such as syngas, hydrogen fuel, charcoal, minerals and other organic chemicals (e.g., unsaturated cyclic hydrocarbons, tar etc.). Though, gasification virtually provides low-cost and energy efficient alternative for post-sorbent management, yet the formation of excessive tar, emission of harmful gases into the atmosphere poses the challenge of sustainability. However, the issue regarding to tar formation can be rectified using catalytic cracking but it may incur the extra cost of the entire process. Additionally, the metal loaded spent biosorbents provide superior yield on gasification compared to another biomass [84]. Recent review, by Rial [85], proposes the possible applicability of dye loaded post-adsorbent material for the subsequent removal of cationic pollutants (Ca^{2+}, Mg^{2+} and metal ions) from wastewater. Furthermore, the metal loaded adsorbent could be employed as doping agent in the ceramics and construction materials [85]. It has been suggested that the exploitation of appropriate biosorbent with the optimization of multistage biosorption cum post-sorption strategies can help to eliminate the barrier between laboratory scale to industrial applicability of biosorption process in textile and other industries.

7 SWOT Analysis of Biosorption Process

Biosorption of hazardous waste surrounds with the complex intricacies of technical, economic and environmental impact. Majority of the reviewed literature suggest that nullifying the negative shortcomings of the biosorption process with the elevated affirmative outcomes can provide justifiable space in the upcoming commercial market. SWOT (Strength, Weakness, Opportunities and Threat) analysis of any process provides overall outlook and futuristic directions to the researchers. Table 1 provides key SWOT points suggested in recently published review articles.

8 Conclusion

The present chapter walks through the basic roadmap of biosorption mechanism and impact of various parameters on its performance followed by the realm of post-sorption strategies to attain circular economy relatable solution. Despite obtaining humongous lab-scale evidences, biosorption technique has not found space in the real-world commercial market. The major lacunae to overcome are research leading to pilot scale and commercial popularization of well performing biosorbents among the stakeholders i.e., farmers, wastewater professionals, environmentalist etc. There is no exaggeration to say that biosorption can overpower many existing high carbon footprint technologies in future if the major shortcomings are dealt collaboratively and in standardized manner.

9 Recommendations

Biosorption is a well-studied phenomena throughout the globe to achieve various goals such as pollutant removal from wastewater, biomass fortification for cattle feed, and soil conditioning on biochar application etc. Majority of the biosorption related work have been carried out at lab-scale for academic purposes which is poorly carried forward to the next levels such as pilot and large-scale explorations. One of the main obstacles in moving forward is unavailability of a globally standardized methodology of testing, reporting of results and repository of biosorbent related database. A common platform consisting essential details such as chemical and physical characteristics of biosorbent, structural properties, pollutant removal efficiencies, adsorption kinetics and probable mechanism with a model adsorbate molecule, etc., can be designed and maintained for the stakeholders i.e., researchers, industrialist, farmers and policy makers etc., which can accelerate the process of designing and selecting suitable biosorbent for targeted application. It will provide ease of comparing and deciding modification strategies to develop most effective biosorbent as well as optimum conditions for the process. On the other hand, use of effective biosorbents should be incentivised for field application to promote the technology. More focus should be given to develop and promote locally available biosorbent materials for financial and environmental sustainability. Extensive research must be carried out to develop strategies to deal with post-biosorption by-products in cleaner manner. Circular bioeconomy approach should be adopted via integrating biosorption process with existing infrastructures to reduce carbon footprint on capital expenditure (CAPEX) and operating expenditure (OPEX).

Table 1 SWOT analysis of biosorption process

Review article	Strengths	Weaknesses	Opportunities	Threats
Elgarahy et al. [86]	• Biosorption minimizes the concentration of water pollutants to the acceptable limits recommended by different federal regulations • Coincides with the green chemistry principles • Low operating and manufacturing cost with high efficiency • Agriculture biomass waste, biopolymers and biochar: Possession of diverse functional groups make them potential agents for the removal of wide range of pollutants	• Instability of biosorbents in terms of durability and performance on varying pH of the solution	• Improvement in textural properties can enhance the characteristics of biosorbents • Exploration of synergistic effects of pollutant interaction in aqueous solution on the biosorption efficiency can be useful in capacity enhancement • Sustainability evaluation of pollutant removal by biosorption in terms of environment and human health risk must be accomplished to ensure the future path	• Ion exchange resins are highly competitive in terms of physical strength, durability and porosity
Skrzypczak et al. [69]	• Post-extraction biomass (waste biomass residue after extraction of valued bioactive compounds) carries high diversity of functional groups on its surface, which can act as chelating agent for heavy metals from wastewater stream • Biosorbent based chelating feed additives can be cheaper than the synthetic feed additives • Biomass based fertilizers provide comparable bioavailability sustained release of micronutrients to the plants which minimizes the nutrient loss due to leaching in mineral based fertilizers	• Insufficient attempts on pilot and field level optimization • No popularity of post-sorption biomass valorization as feed additives and fertilizers among the industrial producers and consumers	• Application of biosorption for the production of micronutrient (Ni, Cu, Mn, Mo, Cl, B etc.) loaded animal feeds using plant biomass as carrier • Production of highly bioavailable micronutrient enriched biofertilizers • Pilot to large scale optimization and implementation • Economics analysis for real time application • Awareness and popularization of these product among the farmers must be carried out	• Synthetic fertilizer granules with soluble coating which provides slow and controlled nutrient release are strong competitors

(continued)

Table 1 (continued)

Review article	Strengths	Weaknesses	Opportunities	Threats
Younas et al. [87]	• Forestry and agriculture waste biomass i.e., pine needles, bark, saw dust, coffee waste, teawaste etc., are abundant in polysaccharides and polyphenols which have carboxyl and hydroxyl passing ions to carryout chelation and ion exchange process • Cost of carbonaceous biosorbent ($\leq$ $ 0.1) is much less than the commercial activated carbons (~$3.0) for industrial application	• Absence of pilot-scale experimental studies on biosorption • Lack of systematic studies on the regeneration and recycling of biosorbents for industry level application	• Regeneration and recycling of agriculture/forestry based biosorbents can significantly bring down the economic burden of wastewater treatment systems in ecofriendly way • Localities with high biomass waste generation can venture in the pilot and industrial scale application	• Substantial loss of biomass integrity and adsorption capacity on regeneration/ recycling compared to ion-exchangers and activated carbons

(continued)

Table 1 (continued)

Review article	Strengths	Weaknesses	Opportunities	Threats
Carreira et al. [88]	• Energy independent way of metal removal from aqueous solutions • Biosorption widens the metal removal mechanism from just sorption to internalization and bioaccumulation of metals • Non-living biomass are not affected by toxic metal pollutants and interacts only through surface • This feature enables desorption of metals and recovery and regeneration of biosorbents for multiple cycle of use • Non-living algal biosorbents are easy to handle, biologically compatible, enhanced kinetics and easy availability	• Living biosorbents are sensitive to the concentration related toxicity of pollutants • Manual cultivation of algal biosorbent is time consuming and laborious • Bio-photoreactors are energy consuming and costly • Cultivation of algae is highly water demanding • High algae to metal ratio suppress the performance of the biosorbent • More focus is given to wastewater decontamination instead of metal recovery and recyclability of algal biomass	• Algal biosorbent scale up and algae harvesting • Understanding the ion-metal competition for the adsorption sites • Experimental data deficit for the real time wastewater • Hyper proliferators such as algal blooms can turn into useful biosorbents for metal removal • Optimization of algae-to-metal ratio for best performance • Valorization of metal loaded algal biomass such as biochar, super capacitor and catalyst • Multi-elemental studies will provide crucial information of large-scale application challenges	• Largescale cultivation of algal species must be thoroughly studied to avoid cross contamination and ecological disasters • Metal loaded algal biomass is a potential ecological hazard if not managed sustainably • Cost performance studies suggest that algae based biosorption strategy is not competitive with other waste-based metal removal strategies

(continued)

Table 1 (continued)

Review article	Strengths	Weaknesses	Opportunities	Threats
Sah et al. [89]	• Significant cost reduction on the use of locally available biomass resources • Agriculture waste and nanoparticles produced from waste biomass are cheap and effective biosorbents	• Alkali and acid pretreatment are required to improve porous structure • Unreliability of performance in natural conditions • pH adjustment for better outcome • Sustainable management of spent material	• Need to extrapolate lab scale results in field scale by conducting pilot scale studies in adequate number • Think globally, act locally approach can help researcher to resolve local issues using locally available waste material	• Production of secondary hazards from biosorbent activation and regeneration processes
Rai [90]	• Biosorption coupled phytoremediation can significantly improve the wastewater treatment process • Micro-remediation by extracellular polymeric substance (EPS) from microbes improves overall metal removal efficiency • Cost benefit study suggest the biosorption process is economically viable in comparison to other commercial adsorbents • Opportunity to generate biofertilizer, biochar and biofuels as byproducts	• Biosorption is relatively slower than chemical adsorption process	• Plant leaves waste should be explored for economic feasibility to incorporate in large scale application • Constructed wetland based biosorption of waste can increase the removal efficiency which ultimately help to achieve SDGs • Expansion of biosorption process is sought in analytical, molecular and sustainability analysis • Work on the selectivity should be explored via polymer science integration	• Preference to synthetic chemical adsorbents for heavily contaminated sites

(continued)

Table 1 (continued)

Review article	Strengths	Weaknesses	Opportunities	Threats
Kumar et al. [91]	• Batch biosorption of metal pollutant shows higher removal efficiency • Low investment, ease of maintenance, simple and reliable outcomes • Improvement in the efficiency can be achieved after modelling and understanding the biosorption process	• Extremely pH dependent process • Lack of real time industrial adoption of well performing biosorbents • Consumption of chemicals during regeneration leads to secondary pollution • Less studies on organics and other pollutants compared to metal biosorption	• Development of hybrid biosorbent materials for higher endurance and removal efficiency • A worldwide cooperation required for comprehensive optimization studies and database development for standardized process • Total life cycle assessment and evaluation of environmental fate of hybrid biosorbents	• Hazards from spent hybrid materials

(continued)

Table 1 (continued)

Review article	Strengths	Weaknesses	Opportunities	Threats
Nathan et al. [92]	• Flexibility of process design and operation • Metal ions recovery through desorption • Theoretically well described process via modelling and lab scale parameter optimization	• Very high diversity and elevated concentration of heavy *metals in real time* effluents poses great optimization challenges in multi-ion biosorption • Costly analytical techniques i.e., ICP-MS, TEM,NMR etc. limits the exploratory research in laboratories • Extreme emphasis on the theoretical explanation rather than addressing the practical translatability issues • Modified isotherm and kinetic models do not work well with strong interaction between multiple ions in the wastewater	• Food and vegetable waste have huge potential for large scale *application in biosorption* process • Multi-ion biosorption studies need to be conducted in sufficient number for real world application • Effect of wastewater volume should be tested for large volume treatment in real time	• Structural weakness compared to other *commercial* adsorbents

(continued)

Table 1 (continued)

Review article	Strengths	Weaknesses	Opportunities	Threats
Pham et al. [93]	• Economical and ergonomic way of bacterial biomass implementation for wastewater remediation • Cocktail of heavy metals can be processed in a single shot • No extra nutritional supply • Large quantities of wastewater can be treated • Extended range of operation • Regeneration possible	• Low performance at real world scenarios • Active site blockage due to metal binding ligands • High energy demand • Bacterial culture identification, operation and maintenance is tedious and high skill demanding process	• Immobilization of microbes on carriers for enhanced removal efficiency can be explored in more detail • Extremophiles can be assessed in depth for various pollutant removal • Development of multifunctional bacterial system for the removal of multiple pollutants simultaneously	• Inadequate understanding of microbial process led to negative outcomes • Low bacterial productivity under stress conditions
Hashmi et al. [94]	• Biosorbents have tendency to replace traditional adsorbents • Abundant biomass for utilization • Surface modification enhances the adsorption capacity	• Presence of multiple pollutants in a mixed system • Unreliability of regeneration cost possess challenge of recycling of used material • Most of the studies done on batch mode at lab scale	• Higher emphasis on lab to field translatability studies required • Increase commercial popularity using various media • Low-cost surface modification methods development	• Incineration of adsorbed hazardous pollutant cause ecological damage • Land filling spent biosorbent can lead to severe odour and environmental issues • Production cost, maintenance, and commercial market competition with existing solutions

(continued)

Table 1 (continued)

Review article	Strengths	Weaknesses	Opportunities	Threats
Harshala et al. [95]	• Use of agriculture waste as biosorbents supports recovering the environmental damage caused by other activities • Appropriate surface modification strategies can significantly improve the metallic pollutant removal efficiency of biosorbents	• Pollutant specificity of biosorbents limits the wide range application in real time scenarios • Effective regeneration of used biosorbents hinders due to very high eluent specific desorption	• Composite biosorbents can be designed and employed for multiple types of pollutant removal • Development of universal eluent system for regeneration of biosorbents as well as recovery of heavy metals	• Less competitive than other physicochemical processes in terms of temperature and pH sensitivity
Krishnani et al. [96]	• Environmentally viable and energy efficient approach for contaminants of environmental and emerging concern (CEECs) removal from waste streams • Agriculture, plantation crops, cash crops, medicinal-aromatic plants, weeds, ethanol fermentation waste, algae, animal waste residues perform better for the heavy metal (single or multiple) removal on surface modification • Cation exchange capacity of the biosorbents can be improved by surface modifications • Microbial biosorption such as mycoremediation poses comparable efficacy due to bioaccumulation as well as exopolysaccharides formation	• Many biomasses in nature have very low pollutant binding efficiency • Surface modification techniques lead to increase in overall cost as well as non-ecofriendly production of unused chemicals • Identification of effective biosorbent for a targeted pollutant is a significant challenge due to high diversity of functional groups on their surface	• Nanomodification of biochar has potential to elevate the pollutant removal capacity with very high economic feasibility • Effective remediation of tannery-effluent (Cr) contaminated soil using biochar coupled with sulphur opens up new dimension in phytoremediation coupled biosorption • Micronutrient biosorption on vegetable dietary fibres has potential to manage nutrient deficiency in human • Combined bioremediation techniques • Holistic technoeconomic evaluation of pilot scale operations for industrial applicability	• Chitosan, zeolites and silica based advanced adsorption processes are strong competitors • Co-contamination of nontargeted pollutants hinder the largescale applicability of biosorption

(continued)

Table 1 (continued)

Review article	Strengths	Weaknesses	Opportunities	Threats
Tofan et al. [97]	• Green and ecofriendly option for wastewater treatment • Broad range of biosorbents with distinct properties • Abundance of lignocellulosic waste biomass in nature • Chitosan possesses biodegradability, low density, excellent reactivity, biocompatibility, and ease of use • Hydrogel biosorbents are biocompatible and having reversibility with sterilization • Versatility and adequacy of method • Column continuous mode of operation is simple to handle, holds great concentration gradient to derive the driving force for biosorption, high operational yield, easy scaling up, suitable for automatization	• Majority of the work done on synthetic wastewaters in batch mode • Lack of continuous mode biosorption process information • Chitosan has low endurance, high solubility in organic solvents, and high expansion on absorption • Continuous flow modelling results in empirical or semiempirical formulation which provide insufficient information for process design • Column adsorption–desorption studies for reuse and recovery are rarely conducted	• More emphasis is needed on continuous mode operation data collection and optimization • Exploration of biopolymers by cross-linking or chemical reactions for advanced biosorbent development • Computational simulation studies on different biosorption processes • Extensive research on recyclability of biosorbents in continuous flow column biosorption systems • More studies on real time effluents • Sustainability assessment of continuous flow column systems	• Less popularity and repulsion from largescale application studies of biosorption than other competitive processes
Bibbins-Martínez et al. [98]	• Fungi are capable to metabolize and mineralize pollutants after biosorption • Biodegradation of pollutants by fungi is very effective in wastewater treatment	• Quantitative analysis is extremely tedious exercise • Very few studies on industrial application of fungal biosorbent	• Exploration of process mechanism of fungal biosorbent for lab and pilot scale application • Incorporation of nanotechnological and other material science ideas in fungal biosorption process for performance enhancement	• Technical hurdles in the development and implementation of fugal bioremediation process

References

1. Crini G (2006) Non-conventional low-cost adsorbents for dye removal: a review. Bioresour Technol 97:1061–1085. https://doi.org/10.1016/j.biortech.2005.05.001
2. Nishiyama Y (2018) Molecular interactions in nanocellulose assembly. Philos Trans R Soc London 376. https://doi.org/10.1098/rsta.2017.0047
3. Saha BC (2003) Hemicellulose bioconversion. J Ind Microbiol Biotechnol 30:279–291. https://doi.org/10.1007/s10295-003-0049-x
4. Rajesh Banu J, Preethi KS et al (2021) Lignocellulosic biomass based biorefinery: a successful platform towards circular bioeconomy. Fuel 302:121086. https://doi.org/10.1016/j.fuel.2021.121086table
5. Rangabhashiyam S, Anu N, Nandagopal MSG, Selvaraju N (2014) Relevance of isotherm models in biosorption of pollutants by agricultural byproducts. J Environ Chem Eng 2:398–414. https://doi.org/10.1016/j.jece.2014.01.014
6. Xu F (2010) Structure, ultrastructure, and chemical composition, 1st edn. Elsevier
7. Nair A, Ithnin NB, Sim HL, Appleton DR (2016) Energy crops, 2nd edn. Elsevier
8. Pizzi A (2019) Tannins: prospectives and actual industrial applications. Biomolecules 9. https://doi.org/10.3390/biom9080344
9. El Mansouri N-E, Salvado J (2007) Analytical methods for determining functional groups in various technical lignins. Ind Crops Prod 26:116–124. https://doi.org/10.1016/j.indcrop.2007.02.006
10. Ahalya, N., T.V. Ramachandra RDK (2003) Biosorption of heavy metals. Res J Chem Environ 7:71–79. https://doi.org/10.1002/0470050594.ch11
11. Kuyucak N, Volesky B (1988) Biosorbents for recovery of metals from industrial solutions. Biotechnol Lett 10:137–142
12. Ray SS, Gusain R, Kumar N (2020) Adsorption in the context of water purification
13. Schiewer S, Volesky B (1995) Modeling of the proton-metal ion exchange in biosorption. Environ Sci Technol 29:3049–3058. https://doi.org/10.1021/es00012a024
14. Rodriguez-Freire L, Gonzalez-Estrella J, Li G (2020) Technologies for fractionation of wastewater and resource recovery. Elsevier Inc.
15. Halling-Sorensen B, Jorgensen SEBT-S in ES (1993) Precipitation. In: Studies in environmental science. Elsevier, pp 355–392
16. Michalak I, Chojnacka K, Witek-Krowiak A (2013) State of the art for the biosorption process—a review. Appl Biochem Biotechnol 170:1389–1416. https://doi.org/10.1007/s12010-013-0269-0
17. Gadd GM (2009) Biosorption: Critical review of scientific rationale, environmental importance and significance for pollution treatment. J Chem Technol Biotechnol 84:13–28. https://doi.org/10.1002/jctb.1999
18. Hisada M, Tomizawa Y, Kawase Y (2019) Removal kinetics of cationic azo-dye from aqueous solution by poly-γ-glutamic acid biosorbent: contributions of adsorption and complexation/precipitation to Basic Orange 2 removal. J Environ Chem Eng 7:103157. https://doi.org/10.1016/j.jece.2019.103157
19. Ramachandra TV, Ahalya N KR (2006) Technical Report—110: biosorption: teachniques & mechanisms. Cent Ecol Sci (CES), Idian Inst Sci Bangalore 1–91
20. Elgarahy AM, Elwakeel KZ, Mohammad SH, Elshoubaky GA (2021) A critical review of biosorption of dyes, heavy metals and metalloids from wastewater as an efficient and green process. Clean Eng Technol 4:100209. https://doi.org/10.1016/j.clet.2021.100209
21. Won SW, Han MH, Yun YS (2008) Different binding mechanisms in biosorption of reactive dyes according to their reactivity. Water Res 42:4847–4855. https://doi.org/10.1016/j.watres.2008.09.003
22. Rápó E, Tonk S (2021) Factors affecting synthetic dye adsorption; desorption studies: a review of results from the last five years (2017–2021). Molecules 26. https://doi.org/10.3390/molecules26175419

23. Saravanan P, Josephraj J, Pushpa Thillainayagam B (2021) A comprehensive analysis of biosorptive removal of basic dyes by different biosorbents. Environ Nanotechnol, Monit Manag 16:100560. https://doi.org/10.1016/j.enmm.2021.100560
24. Yeow PK, Wong SW, Hadibarata T (2021) Removal of azo and anthraquinone dye by plant biomass as adsorbent—a review. Biointerface Res Appl Chem 11:8218–8232. https://doi.org/10.33263/BRIAC111.82188232
25. Adegoke KA, Bello OS (2015) Dye sequestration using agricultural wastes as adsorbents. Water Resour Ind 12:8–24. https://doi.org/10.1016/j.wri.2015.09.002
26. Hema M, Arivoli S (2008) Adsorption thermodynamics of malachite green dye onto acid activated low cost carbon. J Appl Sci Environ Manag 12:43–51
27. Gokulan R, Prabhu GG, Jegan J (2019) Remediation of complex remazol effluent using biochar derived from green seaweed biomass. Int J Phytoremediation 21:1179–1189. https://doi.org/10.1080/15226514.2019.1612845
28. Popa S, Radulescu-Grad ME, Perdivara A, Mosoarca G (2021) Aspects regarding colour fastness and adsorption studies of a new azo-stilbene dye for acrylic resins. Sci Rep 11:1–9. https://doi.org/10.1038/s41598-021-85452-7
29. Low KS, Lee CK, Tan KK (1995) Biosorption of basic dyes by water hyacinth roots. Bioresour Technol 52:79–83. https://doi.org/10.1016/0960-8524(95)00007-2
30. Ravikumar M, King P (2020) Application of response surface optimization on biosorption of congo red dye onto spathodea campanulata leaves. Desalin Water Treat 182:342–350. https://doi.org/10.5004/dwt.2020.25140
31. Volesky B (2001) Detoxification of metal-bearing effluents: biosorption for the next century. Hydrometallurgy 59:203–216. https://doi.org/10.1016/S0304-386X(00)00160-2
32. Ighalo JO, Adeniyi AG (2020) A mini-review of the morphological properties of biosorbents derived from plant leaves. SN Appl Sci 2:1–16. https://doi.org/10.1007/s42452-020-2335-x
33. Alencar WS, Acayanka E, Lima EC et al (2012) Application of Mangifera indica (mango) seeds as a biosorbent for removal of Victazol Orange 3R dye from aqueous solution and study of the biosorption mechanism. Chem Eng J 209:577–588. https://doi.org/10.1016/j.cej.2012.08.053
34. Pethkar AV, Kulkarni SK, Paknikar KM (2001) Comparative studies on metal biosorption by two strains of Cladosporium cladosporioides. Bioresour Technol 80:211–215. https://doi.org/10.1016/S0960-8524(01)00080-3
35. Calderón OAR, Abdeldayem OM, Pugazhendhi A, Rene ER (2020) Current updates and perspectives of biosorption technology: an alternative for the removal of heavy metals from wastewater. Curr Pollut Reports Biol Pollut 6:8–27. https://doi.org/10.1007/s40726-020-00135-7
36. Legorreta-Castañeda AJ, Lucho-Constantino CA, Beltrán-Hernández RI et al (2020) Biosorption of water pollutants by fungal pellets. Water (Switzerland) 2. https://doi.org/10.3390/W12041155
37. Lim AP, Aris AZ (2014) A review on economically adsorbents on heavy metals removal in water and wastewater. Rev Environ Sci Biotechnol 13:163–181. https://doi.org/10.1007/s11157-013-9330-2
38. Dhir B (2014) Potential of biological materials for removing heavy metals from wastewater. Environ Sci Pollut Res 21:1614–1627. https://doi.org/10.1007/s11356-013-2230-8
39. Ahmad T, Danish M, Rafatullah M et al (2012) The use of date palm as a potential adsorbent for wastewater treatment: a review. Environ Sci Pollut Res 19:1464–1484. https://doi.org/10.1007/s11356-011-0709-8
40. Man mohan K, Rajeev A, Gajalakshmi S (2022) Azo Dye Removal from Aqueous Phase by Surface Modified Agroindustrial Waste: Equilibrium and Kinetics studies
41. Fomina M, Gadd GM (2014) Biosorption: current perspectives on concept, definition and application. Bioresour Technol 160:3–14. https://doi.org/10.1016/j.biortech.2013.12.102
42. Man mohan K, Kiran Kumar V, Sreevidya VH, Gajalakshmi S (2021) Biosorptionally treated dye wastewater employed in "Olla-pot coupled microbial fuel cell" for tomato (*Solanum lycopersicum L.*) irrigation and bio-electricity production. Bioresour Technol Rep 13:100626. https://doi.org/10.1016/j.biteb.2021.100626

43. Man mohan K, Rajeev A, Gajalakshmi S (2023) Biosorption of AB-193 dye using rice straw biomass and post-biosorption application of treated wastewater in olla-pot coupled microbial fuel cells (OPMFCs). Waste and Biomass Valorization 14(5):1539-1549. https://doi.org/10.1007/s12649-022-01993-6
44. Mensah J (2019) Sustainable development: meaning, history, principles, pillar, and implications for human action: literature review. Cogent Soc Sci 5:1–21. https://doi.org/10.1080/23311886.2019.1653531
45. Patel H (2021) Review on solvent desorption study from exhausted adsorbent. J Saudi Chem Soc 25:101302. https://doi.org/10.1016/j.jscs.2021.101302
46. Omorogie MO, Babalola JO, Unuabonah EI (2016) Regeneration strategies for spent solid matrices used in adsorption of organic pollutants from surface water: a critical review. Desalin Water Treat 57:518–544. https://doi.org/10.1080/19443994.2014.967726
47. Yuen FK, Hameed BH (2009) Recent developments in the preparation and regeneration of activated carbons by microwaves. Adv Colloid Interface Sci 149:19–27. https://doi.org/10.1016/j.cis.2008.12.005
48. Srivastava A, Gupta B, Majumder A et al (2021) A comprehensive review on the synthesis, performance, modifications, and regeneration of activated carbon for the adsorptive removal of various water pollutants. J Environ Chem Eng 9:106177. https://doi.org/10.1016/j.jece.2021.106177
49. Aksu Z (2005) Application of biosorption for the removal of organic pollutants: a review. Process Biochem 40:997–1026. https://doi.org/10.1016/j.procbio.2004.04.008
50. Yousefi M, Rahimi-Nasrabadi M, Mirsadeghi S, Pourmortazavi SM (2021) Supercritical fluid extraction of pesticides and insecticides from food samples and plant materials. Crit Rev Anal Chem 51:482–501. https://doi.org/10.1080/10408347.2020.1743965
51. Sunarso J, Ismadji S (2009) Decontamination of hazardous substances from solid matrices and liquids using supercritical fluids extraction: a review. J Hazard Mater 161:1–20. https://doi.org/10.1016/j.jhazmat.2008.03.069
52. Verma M, Kumar V, Ekka A, Ghosh P (2020) Application of biological and advanced oxidation processes (AOPs) for the remediation of wastewater laden with toxic pollutants. Elsevier B.V.
53. Kakarla PK, Andrews T, Greenberg RS, Zervas DS (2002) Mdified Fenton's processes for effective In-sit chemical oxidation—Laboratory and Field evaluation. Remediation 23. https://doi.org/10.1002/rem.10043
54. Bauer R, Fallmann H (1997) The Photo-Fenton oxidation—a cheap and efficient wastewater treatment method. Res Chem Intermed 23:341–354. https://doi.org/10.1163/156856797X00565
55. Kim SM, Vogelpohl A (1998) Degradation of organic pollutants by the photo-Fenton-process. Chem Eng Technol 21:187–191
56. Ameta R, Chohadia AK, Jain A, Punjabi PB (2018) Fenton and Photo-Fenton Processes
57. Álvarez PM, Beltrán FJ, Gómez-Serrano V et al (2004) Comparison between thermal and ozone regenerations of spent activated carbon exhausted with phenol. Water Res 38:2155–2165. https://doi.org/10.1016/j.watres.2004.01.030
58. Simon RG, Stöckl M, Becker D et al (2018) Current to clean water—electrochemical solutions for groundwater, water, and wastewater treatment. Chem-Ing-Tech 90:1832–1854. https://doi.org/10.1002/cite.201800081
59. Narbaitz RM, Cen J (1994) Electrochemical regeneration of granular activated carbon. Water Res 28:1771–1778. https://doi.org/10.1016/0043-1354(94)90250-X
60. Silva M, Fernandes A, Mendes A et al (2004) Preliminary feasibility study for the use of an adsorption/bio-regeneration system for molinate removal from effluents. Water Res 38:2677–2684. https://doi.org/10.1016/j.watres.2004.03.016
61. Gu X, Zhou J, Zhang A et al (2008) Feasibility study of the treatment of aniline hypersaline wastewater with a combined adsorption/bio-regeneration system. Desalination 227:139–149. https://doi.org/10.1016/j.desal.2007.06.021
62. de Jonge RJ, Breure AM, van Andel JG (1996) Biodegradation of pawdered activated carbon (PAC) loaded with aromatic compounds. Pergamon 30:875–882

63. Ha SR, Vinitnantharat S (2000) Competitive removal of phenol and 2,4-dichlorophenol in biological activated carbon system. Environ Technol (UK) 21:387–396. https://doi.org/10.1080/09593332108618107
64. Yang L, Zhou Z, Xiao L, Wang X (2003) Chemical and Biological regeneration of HDTMA-modified Montmorillonite after sorption with phenol. Environ Sci Technol 37:5057–5061. https://doi.org/10.1021/es0342493
65. Ng SL, Seng CE, Lim PE (2009) Quantification of bioregeneration of activated carbon and activated rice husk loaded with phenolic compounds. Chemosphere 75:1392–1400. https://doi.org/10.1016/j.chemosphere.2009.02.049
66. Ng SL, Seng CE, Lim PE (2010) Bioregeneration of activated carbon and activated rice husk loaded with phenolic compounds: kinetic modeling. Chemosphere 78:510–516. https://doi.org/10.1016/j.chemosphere.2009.11.041
67. Chojnacka K (2006) The application of multielemental analysis in the elaboration of technology of mineral feed additives based on Lemna minor biomass. Talanta 70:966–972. https://doi.org/10.1016/j.talanta.2006.05.063
68. Michalak I, Witek-Krowiak A, Chojnacka K, Bhatnagar A (2015) Advances in biosorption of microelements—the starting point for the production of new agrochemicals. Rev Inorg Chem 35:115–133. https://doi.org/10.1515/revic-2015-0003
69. Skrzypczak D, Ligas B, Mikula K et al (2021) Valorization of post-extraction biomass residues as carriers of bioavailable micronutrients for plants and livestock. Biomass Convers Biorefinery 11:3037–3052. https://doi.org/10.1007/s13399-019-00586-z
70. Saeid A, Chojnacka K, Korczyński M et al (2013) Biomass of Spirulina maxima enriched by biosorption process as a new feed supplement for swine. J Appl Phycol 25:667–675. https://doi.org/10.1007/s10811-012-9901-6
71. Witkowska Z, Chojnacka K, Korczyński M et al (2014) Soybean meal enriched with microelements by biosorption—a new biological feed supplement for laying hens. Part I. Performance and egg traits. Food Chem 151:86–92. https://doi.org/10.1016/j.foodchem.2013.11.023
72. Janeczek M, Korczynski M, Basinska K et al (2015) Yumurtacı tavuklarda (lohman brown) in vivo körbarsak fermentasyonunda sindirilebilir çinko (II) şelat kompleksi ve çinko (II) ile zenginleştirilmiş soya fasulyesi küspesi ilavesinin seçilen parametrelere etkisi. Kafkas Univ Vet Fak Derg 21:49–54. https://doi.org/10.9775/kvfd.2014.11679
73. Witkowska Z, Swiniarska M, Korczyński M et al (2018) Biofortification of hens' eggs with microelements by innovative bio-based dietry supplement. Anim Physiol Anim Nutr 103:485–492. https://doi.org/10.1111/jpn.13027
74. Michalak I, Chojnacka K, Korniewicz D (2015) New feed supplement from macroalgae as the dietary source of microelements for pigs. Open Chem 13:1341–1352. https://doi.org/10.1515/chem-2015-0149
75. Seow YX, Tan YH, Mubarak NM, et al (2022) A review on biochar production from different biomass wastes by recent carbonization technologies and its sustainable applications. J Environ Chem Eng 10:107017. https://doi.org/10.1016/j.jece.2021.107017
76. Harikishore Kumar Reddy D, Vijayaraghavan K, Kim JA, Yun YS (2017) Valorisation of post-sorption materials: opportunities, strategies, and challenges. Adv Colloid Interface Sci 242:35–58. https://doi.org/10.1016/j.cis.2016.12.002
77. Yao Y, Gao B, Zhang M et al (2012) Effect of biochar amendment on sorption and leaching of nitrate, ammonium, and phosphate in a sandy soil. Chemosphere 89:1467–1471. https://doi.org/10.1016/j.chemosphere.2012.06.002
78. Jung KW, Jeong TU, Hwang MJ et al (2015) Phosphate adsorption ability of biochar/Mg-Al assembled nanocomposites prepared by aluminum-electrode based electro-assisted modification method with $MgCl_2$ as electrolyte. Bioresour Technol 198:603–610. https://doi.org/10.1016/j.biortech.2015.09.068
79. Tuhy Ł, Samoraj M, Michalak I, Chojnacka K (2014) The application of biosorption for production of micronutrient fertilizers based on waste biomass. Appl Biochem Biotechnol 174:1376–1392. https://doi.org/10.1007/s12010-014-1074-0

80. Tuhy Ł, Samoraj M, Witkowska Z et al (2015) Using spent mushroom substrate as the base for organic-mineral micronutrient fertilizer-dield test on maize. BioResources 10:5709–5719
81. Tuhy Ł, Samoraj M, Basladynska S, Chojnacka K (2015) New micronutrient fertilizer biocomponents based on seaweed biomass. Pol J Environ Stud 24:2213–2221. https://doi.org/10.15244/pjoes/39552
82. Shen Y (2015) Carbothermal synthesis of metal-functionalized nanostructures for energy and environmental applications. J Mater Chem A 3:13114–13188. https://doi.org/10.1039/c5ta01228g
83. Zhang S, Zeng M, Li J et al (2014) 2014.pdf. J Mater Chem A 2:4391–4397. https://doi.org/10.1039/c3ta14604a
84. Richardson Y, Drobek M, Julbe A et al (2015) Biomass gasification to produce syngas
85. Rial JB, Ferreira ML (2022) Potential applications of spent adsorbents and catalysts: re-valorization of waste. Sci Total Environ 823:153370. https://doi.org/10.1016/j.scitotenv.2022.153370
86. Elgarahy AM, Elwakeel KZ, Mohammad SH, Elshoubaky GA (2021) A critical review of biosorption of dyes, heavy metals and metalloids from wastewater as an efficient and green process 4
87. Younas F, Mustafa A, Farooqi ZUR et al (2021) Current and emerging adsorbent technologies for wastewater treatment: trends, limitations, and environmental implications. Water (Switzerland) 13:1–25. https://doi.org/10.3390/w13020215
88. Carreira ARF, Passos H, Coutinho JAP (2023) Metal biosorption onto non-living algae: a critical review on metal recovery from wastewater. Green Chem 25:5775–5788. https://doi.org/10.1039/d3gc01993d
89. Sah MK, Edbey K, EL-Hashani A et al (2022) Exploring the biosorption of methylene blue dye onto agricultural products: a critical review. Separations 9:1–26. https://doi.org/10.3390/separations9090256
90. Rai PK (2022) Novel adsorbents in remediation of hazardous environmental pollutants: progress, selectivity, and sustainability prospects. Clean Mater 3:100054. https://doi.org/10.1016/j.clema.2022.100054
91. Kumar P, Prajapati KB, Mahajan AK et al (2022) Evaluating feasibility of biosorption technique for heavy metals removal: limitations and future perspective. Int J Environ Anal Chem 00:1–25. https://doi.org/10.1080/03067319.2022.2145196
92. Nathan RJ, Jain AK, Rosengren RJ (2022) Biosorption of heavy metals from water: mechanism, critical evaluation and translatability of methodology. Environ Technol Rev 11:91–117. https://doi.org/10.1080/21622515.2022.2078232
93. Pham VHT, Kim J, Chang S, Chung W (2022) Bacterial biosorbents, an efficient heavy metals green clean-up strategy: prospects, challenges, and opportunities. Microorganisms 10:1–16. https://doi.org/10.3390/microorganisms10030610
94. Hashmi Z, Jatoi AS, Nadeem S et al (2022) Comparative analysis of conventional to biomass-derived adsorbent for wastewater treatment: a review. Springer, Berlin Heidelberg
95. Harshala K, Wagh ND (2022) Use of agricultural waste-based biosorbents for the removal of heavy metals from aqueous solution: a review. Nat Environ Pollut Technol 21:1003–1014. https://doi.org/10.46488/NEPT.2022.v21i03.007
96. Krishnani KK, Boddu VM, Singh RD et al (2023) Plants, animals, and fisheries waste-mediated bioremediation of contaminants of environmental and emerging concern (CEECs)—a circular bioresource utilization approach. Springer, Berlin Heidelberg
97. Tofan L, Suteu D (2023) Renewable resource biosorbents for pollutant removal from aqueous effluents in column mode. Separations 10. https://doi.org/10.3390/separations10020143
98. Bibbins-Martínez M, Juárez-Hernández J, López-Domínguez JY, et al (2023) Potential application of fungal biosorption and/or bioaccumulation for the bioremediation of wastewater contamination: a review. J Environ Biol 44:135–145. https://doi.org/10.22438/JEB/44/2/MRN-5093

Membrane Bioreactors and Other Emerging Membrane Technologies

Geophry Wasonga Anyango, Gourav Dhar Bhowmick, and Veeramani Rajasekar

Abstract As fascinating as it sounds, membrane technologies have become very important for various applications in the present world. In the realm of wastewater treatment and resource recovery, the relentless pursuit of sustainable and efficient solutions has led to remarkable advancements in membrane technologies. Among these transformative innovations, membrane bioreactors (MBRs) stand out as a pioneering approach that marries the principles of biotechnology and filtration. Harnessing the power of semi-permeable membranes, MBRs, which were introduced in the 1960s, have revolutionized the field, offering unparalleled treatment efficiency, reduced footprint, and enhanced water quality. This process combines the ultrafiltration or microfiltration technique of membrane with biological treatments to separate sludge generated from the initial biological activities, thus eliminating the conventional gravity-enhanced clarifier/secondary sedimentation process in purifying wastewater. However, the journey of membrane technologies continues after MBRs. Exciting emerging membrane technologies are pushing the boundaries even further, promising even more remarkable achievements in water treatment, desalinization, and beyond. In exploring MBRs and emerging membrane technologies, we embark on a captivating journey through the intricacies of these cutting-edge solutions, revealing the immense possibilities they possess in shaping a more environmentally friendly tomorrow.

Keywords Ultrafiltration · Immersed membrane bioreactor · Side-stream membrane bioreactor · Membrane fouling · Membrane filtration

Abbreviation

MBR	Membrane bioreactor
UV	Ultraviolet

G. W. Anyango · G. D. Bhowmick (✉) · V. Rajasekar
Agricultural and Food Engineering Department, IIT Kharagpur, Kharagpur, India
e-mail: gourav@agfe.iitkgp.ac.in

M. M. Ghangrekar et al. (eds.), *Biological and Hybrid Wastewater Treatment Technology*, Earth and Environmental Sciences Library,
https://doi.org/10.1007/978-3-031-63046-0_4

TMP	Transmembrane pressure
BOD	Biochemical Oxygen Demand
COD	Chemical Oxygen Demand
HRT	Hydraulic Retention Time
MD	Membrane Distillation
SRT	Solid Retention Time
NTU	Nephelometric Turbidity Unit
OLR	Organic Loading Rate
MLSS	Mixed Liquor Suspended Solids
PVDF	Polyvinylidene fluoride
PES	Polyethersulfone
PP	Polypropylene
HF-MBR	Hollow Fibre Membrane bioreactor
FS-MBR	Flat Sheet Membrane bioreactor
FO	Forward Osmosis
T-MBR	Tubular Membrane Bioreactor
AI	Artificial Intelligence
ANN	Artificial Neural Networks
CST	Capillary Suction Time
PCA	Principal Component Analysis
TSS/VSS	Total Suspended and Volatile Solids
ATP	Adenosine-50-triphosphate
PLS	Projection to latent structures
MSPC	Multivariate Statistical Process Control
SDWC	Sequential days without cleaning
RO	Reverse Osmosis
MOF	Metal-organic frameworks

1 Introduction

Membranes have become a captivating contrivance of the present world. In the wastewater treatment and resource recovery domain, the relentless pursuit of sustainable and efficient solutions has led to remarkable advancements in membrane technologies [1]. Among these transformative innovations, membrane bioreactors (MBRs) stand out as a pioneering approach that marries the principles of biotechnology and filtration [2]. Harnessing the power of semi-permeable membranes, MBRs, which were introduced in the 1960s [1], have revolutionized the field, offering unparalleled treatment efficiency, reduced footprint, as well as enhanced water condition.

In the novel practice, MBR process combines membrane technology with aerobic or anaerobic treatment (biological treatments) in purifying wastewater [1]. The membranes utilize ultrafiltration or microfiltration techniques to separate sludge generated from the initial biological activities, thus eliminating the conventional gravity-enhanced clarifier/secondary sedimentation process [3]. This way, the overall treatment process consumes a smaller space, generates a better permeate quality, and enhances resource recovery [1, 4]. Therefore, the MBR is credited as a more sustainable and efficient wastewater treatment approach, making it a better alternative to conventional techniques.

However, the journey of membrane technologies continues after MBRs. Exciting emerging technologies are pushing the boundaries further, promising even more remarkable achievements in managing emerging wastewater contaminants, desalinization, and beyond [5]. In exploring MBRs and emerging membrane technologies, we embark on a captivating journey through the intricacies of these cutting-edge solutions, uncovering their vast potential for shaping a sustainable future.

1.1 The Relevance of Membrane Bioreactor in Wastewater Treatment

MBRs have become a highly significant mechanism in sewage treatment plant, primarily owing to their multitude of advantages and their remarkable capability to tackle the critical challenges encountered in conventional treatment methodologies. In the previous decade, MBR has attracted much attention in municipal wastewater treatment, including the widespread 175 full-scale leachate treatments across China, indicating a rapid growth of MBR in this field [6]. By combining biological treatment with membrane filtration [7], MBRs offer superior treatment efficiency, resulting in enhanced removal of contaminants such as suspended solids, pathogens, and nutrients, leading to higher effluent quality [8, 9]. Additionally, MBRs exhibit a smaller footprint than traditional activated sludge processes [10, 11], making them particularly suitable for space-limited areas or decentralized wastewater treatment applications.

Implementing MBRs allows process intensification by achieving high sludge concentrations, enhancing treatment efficiency, and optimizing resource utilization [12]. Furthermore, if integrated with the biofilm [13], this process can effectively retain biomass, fostering the development of a stable and highly active microbial population, thereby enabling the effective treatment of complex or high-strength wastewaters. Their adaptability to fluctuating conditions makes MBRs well-suited for handling variable or challenging wastewater streams [14].

Notably, MBRs produce high-quality effluent that meets the requirements for water reuse applications while also facilitating the recovery of valuable resources from wastewater [15, 16]. With advanced monitoring and control capabilities, MBRs ensure stable operation, minimize downtime, and optimize performance. Importantly, MBRs enable compliance with stringent effluent discharge regulations, which is crucial in reducing the environmental impact of discharged effluent in wastewater treatment processes [8].

Altogether, the relevance of membrane bioreactors in wastewater treatment lies in their ability to provide high-quality treatment, process efficiency, flexibility, and opportunities for resource recovery and water reuse. With ongoing advancements in membrane technology and research, MBRs continue to evolve as promising solutions for addressing the challenges of modern wastewater treatment.

1.2 Membrane Bioreactors: Design, Process Description, and Applications

For a cost-effective process, the recommended mixed liquor suspended solids (MLSS) range in designing an MBR is 8000–10,000 mg/L, solid retention time (SRT) ranging between 5–20 days and a 4–6 h hydraulic retention time (HRT) specific to sewage treatment. The permeate flux rate is manufacture-specified but often ranges between 600–1200 L/m^2. While organic loading rate (OLR) for industrial wastewater treatment is laboratory experimental-based, a daily 1–3 kg biochemical oxygen demand (BOD) unit per m^3 is appropriate for a sewage treatment design [17].

MBRs combine the principles of biological wastewater treatment and membrane filtration to achieve efficient and advanced water purification [18]. The system has two main components: a membrane filtration system and biological treatment process. In this case, biological treatment process is like traditional activated sludge processes where water is introduced into a bioreactor containing microorganisms like bacteria and protozoa, which breaks down organic matter and removes nutrients like nitrogen and phosphorous, thus lowering the organic and nutrient component of the wastewater.

The key feature of membrane filtration sets MBRs apart from conventional wastewater treatment systems (Fig. 1). Instead of using clarifiers or settling tanks after the biological process in separating the cleaned water from the activated sludge, the technique employ membrane filters as physical obstacle to bacteria, suspended solids and other contaminants. This process brings several advantages: enhanced treatment efficiency, integration of biotechnology and filtration, membrane configurations, reduced footprints, improved solids retention, membrane fouling and maintenance, and resource recovery [19].

The assembly of MBR itself consists of membranes, inlet and outlet tubes, and support structures all in one casing to form the main bioreactor. The permeates outside the bioreactor is sucked from the mixed liquor by approximately 50 kPa vacuum while retaining the biosolids. Due to the biofouling effect, compressed air is jetted onto the membrane surface to scrub off the solids attached to the membrane cartridge, along with periodic external cleaning with water or chemical membrane cleaner products [17]. Acid and alkaline chemical products such as citric acid, sodium hypochlorite, sodium hydroxide, hydrogen peroxide, or hydrochloric acid have been applied in Seelscheid, Nordkanal, and Woffelsbach wastewater treatment operations in Germany for cleaning membrane surfaces [17, 20].

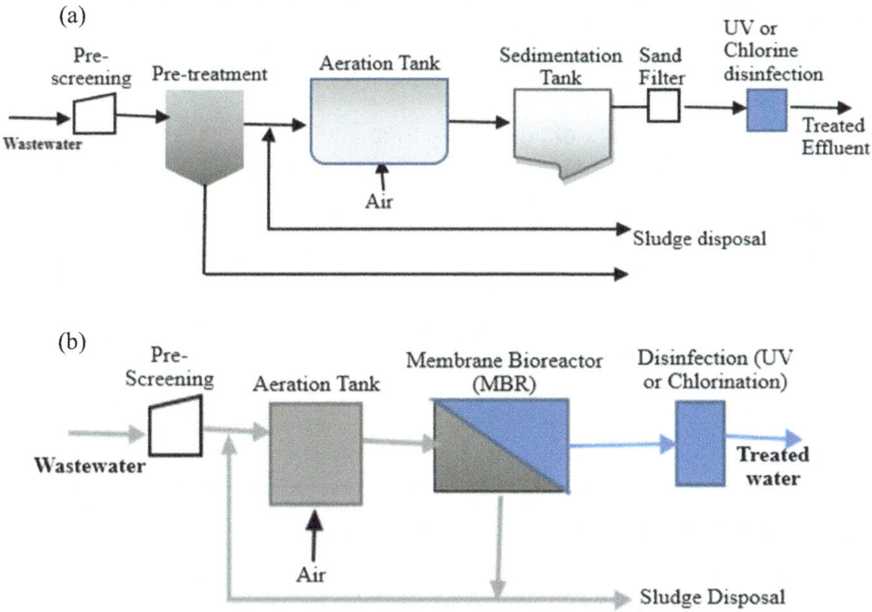

Fig. 1 Wastewater treatment by **a** Traditional system and **b** MBR

In the integration process, the MBRs integrate the power of microorganisms and semi-permeable membranes through a biological process involving microorganisms, like bacteria and fungi, to break down organic matter and eliminate pollutants from wastewater [21]. The filtration process, Fig. 2, utilizes membranes with tiny pores to physically separate solids, microorganisms, and dissolved substances from the water [22]. Simultaneous nitrification and denitrification are achievable using the MBR in performance situations with the presence of lower dissolved oxygen [17].

Permeability of the membrane is considered a controlling principle of membrane operation in an MBR. It offers an inseparable membrane property applicable in establishing membrane performance. It relies on both transmembrane pressure (TMP) and generated permeate volume per unit membrane area in a given time, also referred to as permeate flux rate (J). The TMP, in this case, is attributed to membrane's total driving pressure required to force water across the membrane material and often determined by rectifying pressure losses in the piping alongside reading of the filtrate pressure to gauge height in the membrane tank (Eq. 1). The permeability is therefore considered as the ratio of the flux rate and TMP (Eq. 2) [23].

$$\text{TMP} = \text{P}_{\text{feed}} - \text{P}_{\text{filtrate}} \quad (1)$$

$$\text{Permeability} = \frac{\text{J}}{\text{TMP}} \quad (2)$$

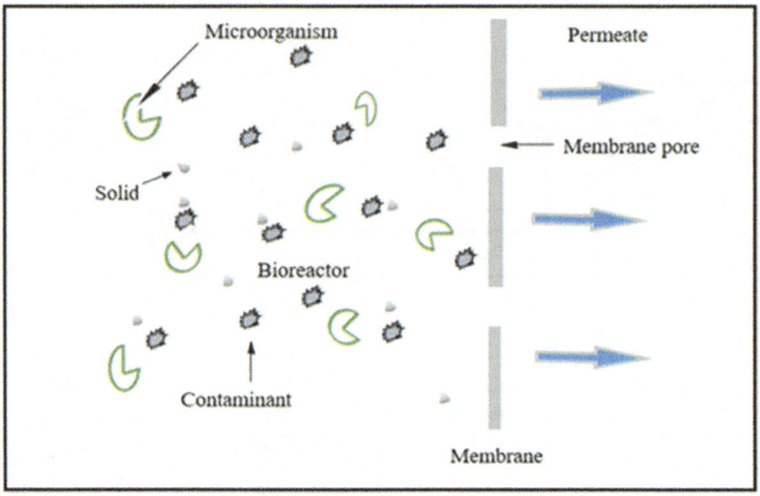

Fig. 2 Integrated microorganisms and semi-permeable membranes

where TMP is Transmembrane pressure in Pa, P_{feed} is static water head above the membrane, $P_{filtrate}$ is corrected filtrate gauge to membrane top, and J is the flux rate in $m^3/m^2.s$ to generate permeability values in m/Pa.s.

Compared to conventional wastewater treatment methods, MBRs offer superior treatment efficiency as they are physical obstacle to bacteria, suspended solids and other contaminants, resulting in high-quality effluent with low Turbidity and reduced levels of contaminants [24]. MBR configurations can be performed differently, including immersed and external systems. In immersed systems, membranes are submerged directly into the mixture, while in external systems, membranes are placed outside the bioreactor and separated by a membrane tank. Both configurations have advantages and can be tailored to specific treatment requirements [19].

A major constraint in MBR operation is membrane fouling. Due to this obstacle, inorganic and organic matter pile up on the membrane surface, lowering its permeability. Frequent membrane polishing and maintenance are critical to avoid fouling and ensure optimal performance. Techniques like chemical cleaning, backwashing and air scouring mitigate fouling effects. The concentrated biomass retained by the membranes can be further processed to recover energy as biogas or produce valuable byproducts like fertilizers to enhance resource recovery. Additionally, MBRs can be integrated with advanced treatment technologies like membrane distillation for water reuse or desalination.

1.3 *Performance Evaluation and Comparison of MBRs with Conventional Biological Treatment*

Generally, the levels of effluent measured regarding biochemical oxygen demand (BOD), suspended solids, and turbidity in MBR are much better than those measured with the conventional technique, justifying its relevance in wastewater treatment. Unlike the conventional sludge treatment, the MBR technique employs microfiltration in sludge separation, whereas the conventional method utilizes clarifiers/settling tanks. Therefore, the layout offers a compact and modular design, making it ideal for municipalities with limited space.

Treatment efficiency is further affected by the processing time required for the solid fraction of the wastewater, which is termed hydraulic retention time (HRT) and solid retention time (SRT). In conventional method, the SRT is lower than MBR. In comparison, the HRT is higher in the conventional method than in the MBR process [1]. Determining SRT is by establishing quantity of the matter maintained in the reactor and the one coming out daily (Eq. 3).

$$\mathrm{SRT} = \frac{\mathrm{V} \times \mathrm{C_s}}{\mathrm{Q_{out}} \times \mathrm{C_{out}}} \tag{3}$$

where V is the digester volume (L), C_s is the solids concentration (mg/L), Q_{out} is the volume wasted each day (L/day), C_{out} is the effluent's solid concentration in the treatment unit (mg/L) and SRT is the solid retention time (in days).

1.4 *Applications of MBRs in Different Wastewater Treatment Scenarios*

Membrane bioreactors have applications in various wastewater treatment scenarios because of their high-quality effluent and compact design. Some treatment scenarios in which its applications are performed in municipal, industrial, and decentralized wastewater are discussed below.

1.4.1 Performance of MBR in Treating Municipal Wastewater

Investigative operation on domestic wastewater treatment using MBRs revealed a satisfactory effluent quality of chemical oxygen demand (COD) removal efficiencies > 96%, Turbidity > 99.3%, TSS consistently below 1.1 mg/L with Turbidity below 0.3 NTU [25]. An MBR system of integrated post denitrification and membrane filtration processes for concurrent nitrogen and phosphorous elimination reported removal efficiencies of 85% (effluent concentrations less than 6.0 mg/L) and 87% (concentrations less than 1.0 mg/L) of total nitrogen and phosphorous respectively in less than 10 h [26]. This is crucial for protecting water bodies from eutrophication and ensuring compliance with stringent effluent discharge regulations.

1.4.2 Performance of MBR in Treating Industrial Wastewater

Industrial wastewater often contains complex and high-strength contaminants that require specialized treatment. MBRs can effectively handle various industrial wastewaters with their commercial applications in pulp, paper, food processing, textile and tannery, pharmaceutical, and petrochemical facilities. Toxic pollutants and high organic strength of extreme physicochemical nature characterize industrial wastewater. Consequently, the conventional biological treatment method encounters difficulties; MBR technology is resorted to as a prospective alternative [27].

A review on investigation of the industrial application of MBR in a fish canning factory reported a five-day 92% efficiency of COD removal from an initial 18,000 mg/L concentration [17]. They also outlined an MBR use on piggery wastewater treatment that facilitated 97.5%, 97.2%, and 98.9% removal efficiencies of COD, BOD, and NH_3-N, respectively, from their respective influent of 2,050 mg/L, 1,198 mg/L, and 248 mg/L. Furthermore, their review indicated above 99% decolorization of azo dye from wastewater under steady 2.5 days of hydraulic retention time (HRT) and between 57.5 to 63.1% COD removal efficiency from pharmaceutical wastewater.

1.4.3 Decentralized and Cluster Application of MBR

Reported studies in remote locations of Northern Canada demonstrated significant total nutrient removal by both conventional MBR and electronically enhanced MBR, which indicated > 99% and > 95% removal of ammonium-N and dissolved COD, respectively [28]. MBR is preferable in remote locations of cold climatic conditions, mining camps, military bases, and offshore platforms. These areas often need access to municipal wastewater treatment systems, and MBRs can provide a compact and reliable solution for treating and managing wastewater on-site.

2 Types of Membrane Bioreactors and Membrane Materials

Immersed membrane bioreactor (iMBR) and external or side-stream membrane bioreactor (eMBR) are the examples of different types of MBRs [1]. Both iMBRs and eMBRs shown in Fig. 3 have their respective advantages and disadvantages. The choice between the two types depends on factors such as the scale of the treatment system, desired treatment objectives, space availability, and operational requirements. In iMBRs, the membranes are submerged directly into the mixture,

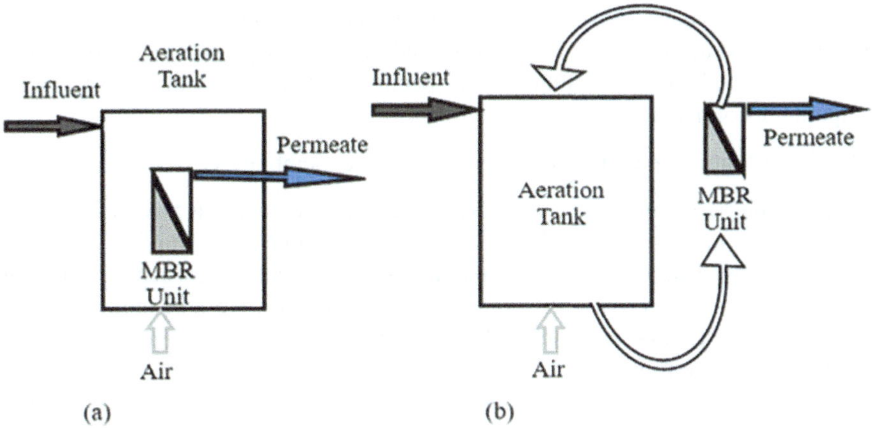

Fig. 3 **a** Immersed MBR, and **b** external MBR

forming a filtration barrier within the bioreactor. The membranes can be arranged in a flat-sheet or hollow fiber configuration. iMBRs are known for their simplicity and compact design, making them suitable for various applications, including small-scale systems.

On the other hand, the eMBRs are positioned outside the bioreactor and separated by a membrane tank. Pumps circulate the mixture across the bioreactor and its membrane vessel. The eMBRs provide easier access for membrane inspection, cleaning, and replacement and are commonly used in larger-scale applications.

2.1 Overview of Membrane Materials Used in MBR Systems

Selecting a suitable membrane material is based on various factors such as the type of wastewater, desired treatment objectives, budget, and operational conditions. Each membrane material has unique characteristics, and selecting the appropriate membrane is crucial for optimizing MBR performance and ensuring long-term sustainability [29].

i. Polymeric Membranes (Organic membranes)

Polymeric membranes are frequently used membranes in an MBRs. They are typically made from materials like polyvinylidene fluoride (PVDF), polyethersulfone (PES), and polypropylene (PP). Polymeric membranes are flexible, cost-effective, and have good chemical and mechanical stability. They can be manufactured as flat-sheet or hollow fiber membranes, offering a range of options for MBR applications.

ii. Ceramic Membranes (inorganic membranes)

Ceramic membranes have patronizing chemical and thermal resistance. They are made from inorganic alumina, zirconia, or titania. Ceramic membranes have smaller pore sizes, allowing for excellent separation efficiency and resistance to fouling. However, ceramic membranes are costly and less flexible compared to polymeric membranes, making them suitable for specific applications with high operating temperatures or harsh chemical environments.

iii. Mixed Matrix Membranes (inorganic–organic membranes)

Mixed matrix membranes are a relatively new development in MBR technology. They are a combination of polymers and inorganic materials, aiming to harness the advantages of both types. These membranes offer fouling resistance, improved permeability and selectivity. Assorted matrix membranes are still under research and development. However, they hold a promising capability for enhancing MBR performance in the future.

Membrane technologies are effective in diverse applications across different industrial sectors. Specifically, these sectors include water reclamation and recycling, landfill leachate treatment, water treatment, desalination pre-treatment, greywater treatment, and industrial process water treatment [19].

2.2 Membrane Configuration Options

MBRs offer different arrangements and setups of the membranes within the bioreactor based on the type of membranes used. These configurations determine how the membranes are positioned, how the inflowing wastewater interacts with the membranes, and how the permeate (filtered or separated component) is collected. The selection of a specific membrane configuration depends on factors such as space availability, desired treatment capacity, membrane performance requirements, operational considerations, and cost-effectiveness. The choice should also be based on the application's specific needs since each configuration has advantages and limitations.

2.2.1 Hollow Fiber MBR

Hollow fibre MBR (HF-MBR) utilizes bundles of hollow fiber membranes that were formerly handmade and progressively prototyped for today's fully automated mass production. Reinforced hollow fiber technology is preferable for medium-sized to substantial wastewater treatment facilities [30]. The membranes are hollow fibers bundled together, and the wastewater is circulated outside the fibers. At the same time, the treated water is collected inside the fibers and withdrawn under pressure. HF-MBRs have a high packing density, effective solids separation, and backwashing capacity but they have the drawback of binding together filaments [31]. However, an additional design by rotating the hollow fiber module can eliminate the fouling cake from the membranes [32].

2.2.2 Flat Sheet MBR

Flat Sheet MBR (FS-MBR) consists of flat sheet membranes stacked together to form modules. Wastewater is passed over the surface of the membrane, and the treated water extracted on the opposite end. As compared to hollow fiber configurations, FS-MBRs provide good solids removal efficiency and are easier to clean due to low sludge sedimentation [31], and have been observed to operate at 0.2 bar transmembrane pressure (TMP) value, which is below the suggested extent of 2 to 3 times the chemical cleaning [33]. This arrangement is mentioned as favourable in treatment of high-strength dyeing wastewater to about 62% COD removal rate from a 7800 mg/L influent [34].

2.2.3 Tubular

A tubular membrane bioreactor (T-MBR) is designed incorporate membrane benefits in filtration with the biological breakdown of organic pollutants. While the configuration nearly resembles a hollow membrane, the two should be distinct. They exhibit a distinctive flow direction for wastewater and treated water collection besides the configuration. Wastewater flows through the inner diameter of the tubes as treated water collects from the tubes' center in T-BMR. This configuration helps minimize fouling and membrane clogging, resulting in stable and efficient solids separation [35].

2.3 Membrane Selection Criteria

Whether the filtration process on the influent is microfiltration or ultrafiltration, the general guidelines and the membrane selection are based on a comprehensive analysis of the specific membrane pore sizes, wastewater characteristics, and treatment goals. Conducting pilot studies or seeking guidance from membrane manufacturers can help determine the best membrane for a given wastewater treatment. However, optimization by artificial neural networks (ANNs) as an artificial intelligence (AI) technique has been used to forecast the energy requirement and the output at the MBR plant of Ulu-Pandan in Singapore with proof for the successful large-scale application [36].

3 Control and Monitoring of Membrane Fouling

Process operators and engineers need basic instruments that provide immediate information regarding biological safety and the ability to filter mixed liquor and threats by membrane fouling. The membrane bioreactors must be continuously monitored to control clogging due to fouling. Furthermore, proper screening and pre-treatment processes are essential to prevent large suspended debris membrane damage.

The practical tools for monitoring MBR are recommended that the analyses of Capillary Suction Time (CST) and Volatile Solids and Total Suspended (TSS/VSS) should be used regularly total carbon and nitrogen studies for comprehensive characterization [37]. The most effective method to evaluate extracellular polymer concentrations is to measure the Turbidity following centrifugation. Adenosine-triphosphate (ATP) 2nd generation test apparatus are the most reliable and repeatable methods for evaluating the health of microbial biomass.

An amalgamated modeling criteria of 2D-fluorescence statistics and Projection-to-Latent Structures (PLS) modeling predicting an MBR's performance parameters and create a real-time monitoring plan have been suggested [38]. The principal component analysis (PCA) determining potential link within the parameters was performed. The model successfully tracked the fluctuations of the total chemical oxygen demand (COD) in untreated and treated water. The model has limitations in forecasting other performance metrics, indicating that the model's spectroscopy can only partially take the place of traditional MBR monitoring. Data mining techniques in bringing out pertinent particulars from examining a set of data have been implemented [39]. This study employed multivariate statistical process control (MSPC) and PCA to examine relationships between various analytical and parametric quantities for treatment of oil containing wastewater using MBR. The multivariate control charts were used as decision-supporting aids for fouling control.

Discussions on the most current MBR developments in the ultrasonography systems have been carried out and advancing concepts arise from hybrid online and online ultrasound for MBR membrane fouling management [40]. Micro-jets, micro streams, and shock waves are some of the significant physical phenomena produced by ultrasonic irradiation, which can be used to clear the clogged membrane. It should be emphasized that the MBR system uses ultrasonic technology for more than just membrane cleaning. Before the MBR system, wastewater can undergo ultrasonic irradiation for pre-treatment. This can interrupt membrane fouling and lower the organic load on the wastewater.

4 Advancing Membrane Technologies for Water Treatment

As discussed above, membrane technology is a promising approach involving the use of semi-permeable membranes to separate impurities and contaminants from water and produce high-quality water while minimizing energy consumption and waste generation. Some notable developments in up surging membrane technology for sustainable water treatment are discussed herewith.

4.1 Graphene-Based Membranes

This technique has been reviewed with a good recommendation for prospects in several filtration processes, such as atmospheric carbon dioxide gas separation [41]. Graphene, a two-dimensional carbon nanomaterial, has shown remarkable potential in membrane technology. Graphene-based membranes offer excellent water permeability, high selectivity, and exceptional mechanical strength. Researchers are exploring using graphene oxide and nanopores to create advanced membranes for efficient water treatment and desalination [42].

4.2 Forward Osmosis (FO)

This emerging technique utilizes semi-permeable membrane to filter water while removing contaminants. It operates on the principle of osmosis, where water actuates to a high solute concentration from a region of low solute concentration. FO is advantageous as it can operate with lower energy requirements than conventional reverse osmosis (RO) processes. This technique has recently gained attention and is widely studied, mainly due to reduced energy requirement [43, 44]

4.3 Membrane Distillation

Membrane distillation (MD) is a distinctively advancing hybrid method. It uses hydrophobic membranes to separate pure water vapor from saline or contaminated water. It maintains a difference in vapor pressure across the membrane, allowing water particles to traverse through while rejecting contaminants. MD is promising for desalination and wastewater treatment, especially in scenarios where low-grade or waste heat is available [45].

4.4 Nanocomposite Membranes

Nanocomposite membranes combine conventional membrane materials with nanoparticles to enhance their separation efficiency. By incorporating metal–organic frameworks (MOFs), carbon nanotubes or zeolites as nanoparticles, the performance and selectivity of the membranes can be significantly improved. Nanocomposite membranes show promise in several uses like purification of water, heavy metal removal, and organic pollutant dismemberment [46–48].

4.5 Photocatalytic Membranes

Photocatalytic membranes combine membrane separation with photocatalysis to degrade organic pollutants and microorganisms in water. These membranes are coated with a photocatalytic material, such as titanium dioxide (TiO_2) [49], which activates under UV light to generate reactive oxygen species. An amalgamation of photocatalysis and membrane separation enables the removal of contaminants efficiently, offering the potential for sustainable water treatment applications [50].

These emerging membrane technologies offer improved water quality, reduced energy consumption, and enhanced resource efficiency compared to conventional treatment methods, thus holding great promise for sustainable water treatment. Ongoing development and current research efforts focus on utilizing these technologies for large-scale implementation and address challenges related to membrane fouling, scalability, and cost-effectiveness.

5 Conclusion and Perspective

Notably, membrane technology has high treatment efficiency alongside high-quality effluent containment, making it suitable for treating most wastewater sources. The relevance of MBRs in wastewater treatment lies in their ability to provide high-quality treatment, process efficiency, flexibility, resource recovery, and water reuse opportunities. With ongoing advancements in membrane technology and research, MBRs continue to evolve as promising solutions for addressing the challenges of modern wastewater treatment. Therefore, the MBRs and membrane technologies have revolutionized wastewater treatment and water purification, offering innovative solutions to address the challenges of water scarcity, environmental protection, and sustainable development. Their impact on the industry has been remarkable, and their potential for advancement is promising with the emerging membrane technologies in place.

6 Recommendations

Going forward, it is essential to concentrate on a few major areas in order to advance membrane-based water treatment technologies. First, in order to improve the efficiency of MBR systems, optimization of mixed matrix membranes need to be the top priority for future research projects. Furthermore, research into new membrane materials and combinations is required to tackle particular issues like cost-effectiveness and fouling. For well-informed decision-making, comparative studies assessing the effectiveness of developing membrane technologies in various water treatment applications are crucial. In order to create regulations that encourage the use of sustainable membrane technologies, policy planners and decision-makers should work with industry stakeholders to consider aspects like environmental effect and energy efficiency. In order to hasten the integration of cutting-edge membrane technology into practical applications, stakeholder engagement initiatives should promote information sharing. Prioritizing efforts is necessary to solve issues with membrane fouling, scalability, and cost-effectiveness. New technologies, including artificial intelligence, may be used to optimize processes and perform predictive maintenance on MBR systems. It can accelerate the implementation of membrane-based water treatment technology, encourage sustainable water management practices, and guarantee future generations' access to clean water resources by putting these innovations into effect.

Acknowledgements The authors delightedly acknowledge Department of Science and Technology, the Science and Engineering Research Board (SERB) and Government of India (SRG/2022/002212), for their assistance.

References

1. Rahman TU, Roy H, Islam MR et al (2023) The advancement in membrane bioreactor (MBR) technology toward sustainable industrial wastewater management. Membranes 13(2):181
2. Chen F, Qian Y, Cheng H, Shen J, Qin Y, Li YY (2023) Recent developments in anammox-based membrane bioreactors: a review. Sci Total Environ 857:159539
3. Zhi-Guo M, Feng-lin Y, Xing-wen Z (2005) MBR focus: do nonwovens offer a cheaper option? Filtr Sep 42(5):28–30. https://doi.org/10.1016/S0015-1882(05)70558-9
4. Qi B, Jiang X, Wang H et al (2021) Resource recovery from liquid digestate of swine wastewater by an ultrafiltration membrane bioreactor (UF-MBR) and reverse osmosis (RO) process. Environ Technol Innov 24:101830. https://doi.org/10.1016/j.eti.2021.101830
5. Nabi M, Liang H, Zhou Q, Cao J, Gao D (2023) In-situ membrane fouling control and performance improvement by adding materials in anaerobic membrane bioreactor: a review. Sci Total Environ 865:161262
6. Zhang J, Xiao K, Huang X (2020) Full-scale MBR applications for leachate treatment in China: practical, technical, and economic features. J Hazard Mater 389:122138. https://doi.org/10.1016/j.jhazmat.2020.122138
7. Leiknes T, Ødegaard H (2007) The development of a biofilm membrane bioreactor. Desalination 202(1–3):135–143. https://doi.org/10.1016/j.desal.2005.12.049
8. Cicek N (2003) A review of membrane bioreactors and their potential application in the treatment of agricultural wastewater. Can Biosyst Eng 45:6–37
9. Ozgun H, Dereli RK, Ersahin ME, Kinaci C, Spanjers H, Van Lier JB (2013) A review of anaerobic membrane bioreactors for municipal wastewater treatment: integration options, limitations and expectations. Sep Purif Technol 118:89–104. https://doi.org/10.1016/j.seppur.2013.06.036
10. Du X, Shi Y, Jegatheesan V, Haq IU (2020) A review on the mechanism, impacts and control methods of membrane fouling in MBR system. Membranes 10(2):24. https://doi.org/10.3390/membranes10020024
11. Sohail N, Riedel R, Dorneanu B, Arellano-Garcia H (2023) Prolonging the life span of membrane in submerged MBR by the application of different anti-biofouling techniques. Membranes 13(2):217. https://doi.org/10.3390/membranes13020217

12. Zhang R, Hao L, Cheng K, Xin B, Sun J, Guo J (2023) Research progress of electrically-enhanced membrane bioreactor (EMBR) in pollutants removal and membrane fouling alleviation. Chemosphere 331:138791. https://doi.org/10.1016/j.chemosphere.2023.138791
13. Raj Deena S, Kumar G, Vickram AS et al (2022) Efficiency of various biofilm carriers and microbial interactions with substrate in moving bed-biofilm reactor for environmental wastewater treatment. Bioresour Technol 359:127421. https://doi.org/10.1016/j.biortech.2022.127421
14. De Vela RJ (2021) A review of the factors affecting the performance of anaerobic membrane bioreactor and strategies to control membrane fouling. Rev Environ Sci Biotechnol 20(3):607–644. https://doi.org/10.1007/s11157-021-09580-2
15. Luo W, Hai FI, Price WE et al (2014) High retention membrane bioreactors: challenges and opportunities. Bioresour Technol 167:539–546. https://doi.org/10.1016/j.biortech.2014.06.016
16. Mutamim NSA, Noor ZZ, Hassan MAA, Olsson G (2012) Application of membrane bioreactor technology in treating high strength industrial wastewater: a performance review. Desalination 305:1–11. https://doi.org/10.1016/j.desal.2012.07.033
17. Ghangrekar MM (2022) Hybrid aerobic wastewater treatment systems. In: Wastewater to water. Springer Nature Singapore, pp 475–506. https://doi.org/10.1007/978-981-19-4048-4_11
18. Rahman TU, Roy H, Islam MdR, et al (2023) The advancement in membrane bioreactor (MBR) technology toward sustainable industrial wastewater management. Membranes 13(2):181. https://doi.org/10.3390/membranes13020181
19. Lin H, Peng W, Zhang M, Chen J, Hong H, Zhang Y (2013) A review on anaerobic membrane bioreactors: applications, membrane fouling and future perspectives. Desalination 314:169–188
20. Judd S, Judd C (2011) The MBR book: principles and applications of membrane bioreactors for water and wastewater treatment, 2nd edn. Elsevier/Butterworth-Heinemann
21. Nguyen LN, Hai FI, Yang S et al (2013) Removal of trace organic contaminants by an MBR comprising a mixed culture of bacteria and white-rot fungi. Bioresour Technol 148:234–241. https://doi.org/10.1016/j.biortech.2013.08.142
22. Bera SP, Godhaniya M, Kothari C (2022) Emerging and advanced membrane technology for wastewater treatment: a review. J Basic Microbiol 62(3–4):245–259. https://doi.org/10.1002/jobm.202100259
23. Sabia G, Ferraris M, Spagni A (2014) Online monitoring of MBR fouling by transmembrane pressure and permeability over a long-term experiment. Sep Purif Technol 122:297–305. https://doi.org/10.1016/j.seppur.2013.11.022
24. Marrot B, Barrios-Martinez A, Moulin P, Roche N (2004) Industrial wastewater treatment in a membrane bioreactor: a review. Environ Prog 23(1):59–68
25. Naghizadeh A, Mahvi AH, Mesdaghinia AR, Alimohammadi M (2011) Application of MBR technology in municipal wastewater treatment. Arab J Sci Eng 36(1):3–10. https://doi.org/10.1007/s13369-010-0007-7
26. Sun F, Wang X, Li X (2013) An innovative membrane bioreactor (MBR) system for simultaneous nitrogen and phosphorus removal. Process Biochem 48(11):1749–1756. https://doi.org/10.1016/j.procbio.2013.08.009
27. Lin H, Gao W, Meng F et al (2012) Membrane bioreactors for industrial wastewater treatment: a critical review. Crit Rev Environ Sci Technol 42(7):677–740. https://doi.org/10.1080/10643389.2010.526494
28. Wei V, Elektorowicz M, Oleszkiewicz JA (2012) Electrically enhanced MBR system for total nutrient removal in remote northern applications. Water Sci Technol 65(4):737–742. https://doi.org/10.2166/wst.2012.908
29. Lee A, Elam JW, Darling SB (2016) Membrane materials for water purification: design, development, and application. Environ Sci Water Res Technol 2(1):17–42
30. Buer T, Cumin J (2010) MBR module design and operation. Desalination 250(3):1073–1077. https://doi.org/10.1016/j.desal.2009.09.111
31. Fang Y, Chen D, Zhang Y, Field R, Wu JJ, Wang B (2022) Wastewater treatment in large-scale novel corrugated-sheet MBR. J Water Process Eng 50:103215. https://doi.org/10.1016/j.jwpe.2022.103215

32. Ruigómez I, Vera L, González E, Rodríguez-Sevilla J (2016) Pilot plant study of a new rotating hollow fibre membrane module for improved performance of an anaerobic submerged MBR. J Membr Sci 514:105–113. https://doi.org/10.1016/j.memsci.2016.04.061
33. Muhamad Ng SN, Idrus S, Ahsan A, Tuan Mohd Marzuki TN, Mahat SB (2021) Treatment of wastewater from a food and beverage industry using conventional wastewater treatment integrated with membrane bioreactor system: a pilot-scale case study. Membranes 11(6):456. https://doi.org/10.3390/membranes11060456
34. Jin Y, Wang D, Zhang W (2019) Newly designed hydrolysis acidification flat-sheet ceramic membrane bioreactor for treating high-strength dyeing wastewater. Int J Environ Res Public Health 16(5):777. https://doi.org/10.3390/ijerph16050777
35. Psoch C, Schiewer S (2006) Anti-fouling application of air sparging and backflushing for MBR. J Membr Sci 283(1–2):273–280. https://doi.org/10.1016/j.memsci.2006.06.042
36. Kamali M, Appels L, Yu X, Aminabhavi TM, Dewil R (2021) Artificial intelligence as a sustainable tool in wastewater treatment using membrane bioreactors. Chem Eng J 417:128070. https://doi.org/10.1016/j.cej.2020.128070
37. Scholes E, Verheyen V, Brook-Carter P (2016) A review of practical tools for rapid monitoring of membrane bioreactors. Water Res 102:252–262
38. Galinha CF, Carvalho G, Portugal CAM et al (2011) Real-time monitoring of membrane bioreactors with 2D-fluorescence data and statistically based models. Water Sci Technol 63(7):1381–1388
39. Santos AV, Lin ARA, Amaral MCS, Oliveira SMAC (2021) Improving control of membrane fouling on membrane bioreactors: a data-driven approach. Chem Eng J 426:131291
40. Arefi-Oskoui S, Khataee A, Safarpour M, Orooji Y, Vatanpour V (2019) A review on the applications of ultrasonic technology in membrane bioreactors. Ultrason Sonochem 58:104633
41. Sajjad M, Thebo KH, Pakistan S (2018) Graphene-based membranes for CO_2 separation. Published online 2018
42. Yu W, Sisi L, Haiyan Y, Jie L (2020) Progress in the functional modification of graphene/ graphene oxide: a review. RSC Adv 10(26):15328–15345
43. Singh SK, Sharma C, Maiti A (2023) Modeling and experimental validation of forward osmosis process: parameters selection, permeate flux prediction, and process optimization. J Membr Sci 672:121439
44. Wu X, Zhang X, Wang H, Xie Z (2023) Smart utilisation of reverse solute diffusion in forward osmosis for water treatment: a mini review. Sci Total Environ 873:162430
45. Yang S, Jasim SA, Bokov D, Chupradit S, Nakhjiri AT, El-Shafay AS (2022) Membrane distillation technology for molecular separation: a review on the fouling, wetting and transport phenomena. J Mol Liq 349:118115
46. Wang C, Park MJ, Yu H, Matsuyama H, Drioli E, Shon HK. Recent advances of nanocomposite membranes using layer-by-layer assembly. J Membr Sci (2022) Published online 2022:120926
47. Ursino C, Castro-Muñoz R, Drioli E, Gzara L, Albeirutty MH, Figoli A (2018) Progress of nanocomposite membranes for water treatment. Membranes 8(2):18
48. Mohamed A, Yousef S, Tonkonogovas A, Stankevičius A, Baltušnikas A (2023) Fabrication of high-strength graphene oxide/carbon fiber nanocomposite membranes for hydrogen separation applications. Process Saf Environ Prot 172:941–949
49. Liu S, Véron E, Lotfi S, Fischer K, Schulze A, Schäfer AI (2023) Poly (vinylidene fluoride) membrane with immobilized TiO_2 for degradation of steroid hormone micropollutants in a photocatalytic membrane reactor. J Hazard Mater. Published online 130832
50. Mozia S (2010) Photocatalytic membrane reactors (PMRs) in water and wastewater treatment. A Rev Sep Purif Technol 73:71–91. Published online 2010

Membrane Bioreactors: Design, Operation, and Maintenance

Sandeep Singh, Shweta Yadav, Jyoti Singh, Rajesh Singh, Vinay Kumar Tyagi, Sujata Kashyap, Kaptan Singh, and Sandeep K. Malyan

Abstract Population explosion, urbanization, and industrialization are the responsible factors for the huge wastewater generation. Globally, around 380 billion m^3 of wastewater is generated annually and in India, about 35% of the population lives in the urban area and generates around 72 million litres per day. Out of the total generated wastewater, only 28% is treated and the rest is discharged untreated which results in the deterioration of the freshwater resources and also impacts different components of the environment. Wastewater treatment is vital for the protection of the environment and human health, and therefore United Nations emphasized it under sustainable development goals (SDGs). The treated water can be a useful resource if used for secondary applications and offers resource and financial savings. Membrane bioreactor (MBR) is a reliable, robust, and flexible technology in the area of wastewater treatment, and through its application, SDGs can be achieved. In MBR technology, hyphenation of activated sludge process and membrane filtration is practiced for the treatment of organic-laden wastewater. MBR technology has frequently been used to treat industrial and municipal wastewater, where a small footprint, stringent discharge standards, or water reuse is essential. The MBR technology outcompetes with conventional wastewater treatment approach but still has some drawbacks such as fouling of membranes which surges operational costs. The

S. Singh · R. Singh (✉) · V. K. Tyagi · K. Singh
Environmental Hydrology Division, National Institute of Hydrology, Roorkee 247667, India
e-mail: rsingh.nih@gmail.com; rsingh.nihr@gov.in

S. Yadav · J. Singh
Technical Cell, National Institute of Hydrology, Roorkee 247667, India

S. Kashyap
Axa Parenteral Limited, Roorkee 247667, India

K. Singh
Department of Civil Engineering, Madan Mohan Malaviya University of Technology, Gorakhpur 273010, India

S. K. Malyan
Department of Environmental Studies, Dyal Singh Evening College, University of Delhi, New Delhi 110003, India

M. M. Ghangrekar et al. (eds.), *Biological and Hybrid Wastewater Treatment Technology*, Earth and Environmental Sciences Library,
https://doi.org/10.1007/978-3-031-63046-0_5

drawbacks can be overcome through MBR designing and membrane selections. In this study, we cover the different MBR designs, membrane types, and operations for wastewater treatment. The recent development in the research area of MBR is reported in this chapter.

Keywords Wastewater · Environmental pollution · Membrane bioreactors · Fouling · Sustainability · Sustainable development goals

1 Introduction

Human civilization completely depends on water availability and this civilization after utilization, discharges this water as wastewater. Therefore, the volume of wastewater continuously increases and the European Investment Bank reported that around 380 billion m^3 of wastewater is generated globally yearly. However, it is estimated that the volume of wastewater generated will increase by 51% by 2050 and its sustainable management is essential. Freshwater resources are limited and they are getting deleted due to industrialization, urbanization, modernization of agriculture, climate change, etc. Water resources can be protected by reducing the dependence on freshwater through enhanced reuse of treated wastewater. The significance of water treatment and reuse is also covered in The Sustainable Development Goals (SDGs) Report 2023 under goal 6: Clean Water and Sanitation.

Wastewater consists of 99% water and the rest nutrients, microorganisms, micropollutants, heavy metals, suspended, and dissolved particulate matter. Industrial wastewater has heterogeneous pollutants, while domestic municipal wastewater has high microbial contamination. The pulp and paper industries, meat and poultry industries, beverage industries, etc., generate a high volume of wastewater [1–3]. Pulp and paper mill wastewater usually consists of suspended solids, dissolved solids, heavy metals, organic load, inorganic load, etc. [4–7]. The characteristics of pulp and paper industries vary and it depend upon the raw material [4, 5, 7] The meat-processing industry is rapidly growing with the population explosion and it generates a high volume of wastewater. Slaughterhouse wastewater contains a high load of fats, proteins, microbes, total superseded solid (TSS), biological oxygen demand (BOD), chemical oxygen demands (COD), colour, total coliform, etc. [8–11]. Beverage industries wastewater is generally characterized by high COD, BOD, TN, phosphate, and ammonia [12–15]. Discharge of this highly polluted wastewater without treatment into the environment results in toxicity. As per the WHO, report [16], over 2 billion population in the world live in water-stressed countries and at least 1.7 billion people use contaminated water for drinking. Poor sanitation and contaminated water are associated with the transmission of several diseases such as diarrhoea, typhoid, cholera, dysentery, and hepatitis A [16]. Therefore, managing this wastewater is vital for achieving SDGs at global and national levels. Wastewater treatment is not a single-step and easy process. It is the combination of biological, physical, and chemical processes through which different contaminants can be removed from

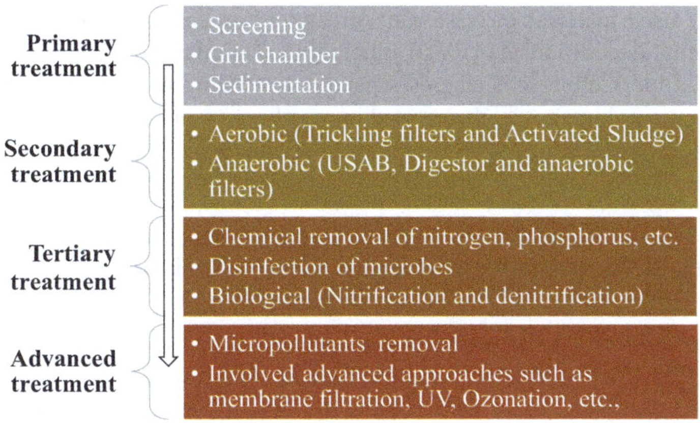

Fig. 1 Steps involved in wastewater treatment processes

wastewater [17–20]. Typically, wastewater treatment processes have several steps which are present in Fig. 1. These steps are involved in different wastewater treatment technologies such as constructed wetlands, membrane filtration, advanced oxidation, electrocoagulation, etc. In this chapter, the wastewater treatment through membrane bioreactors is covered for achieving the SDGs 6.

2 Membrane Material

The ability of the membrane depends on the types of materials, size of pores, solubility, types of wastewater to be treated, and retention time. Material such as polymers, ceramics, and, stainless steel can be used for making the membrane. Based on the type of materials, Asif and Zhang [21] categorised the membranes into: (i) ceramic; (ii) liquid; (iii) polymeric; and (iv) ion exchange. Ceramic membrane uses inorganic material that is resistant to chemical, inert and has high filtration capacity and fouling resistance [22–25]. These properties make it a better choice as compared to polymer for industrial wastewater treatment [24, 25]. Ceramic membrane also has higher hydrophilic ability due to the water contact angle. Ceramic materials such as silicon carbide (SiC), alumina (Al_2O_3), zirconia (ZrO_2), titanium dioxide (TiO_2), zirconium dioxide (ZrO_2), silicon dioxide (SiO_2), etc. [23]. The biggest limitations of ceramic membranes are fragile and their fabrication is economically not feasible [24, 26]. Therefore, the development of low-cost ceramic membranes from natural materials, agro-waste, industrial waste, and animal bones are emerging areas [27, 28]. Luján-Facundo et al. [27] recently fabricated a low-cost ceramic membrane using INKJET technology for municipal wastewater treatment. In this low-cost membrane, they used clay, calcium carbonate, chamotte, and potato starch. The removal efficiency of this membrane is more than 98% for COD [27]. Polymer-based membranes are the widely used material in wastewater treatment processes. Polysulfones (PSFs) and polyvinylidenedifluoride (PVDF) membranes are the most popular one due to its long lifetime. Cellulose acetate (CA) and polytetrafluoroethylene (PTFE) are the most common polymers which are used in membrane fabrication [29]. Usually, UF and MF membranes are used in membrane bioreactors (MBR).

3 Types of Membrane and Their Characteristics

Membrane is used for the production of potable water, desalination, wastewater treatment, recycling, and recycling [30]. Commercialization of the first membrane filters was done by the Sartorius Company in Germany using the Zigmondy process in 1927. Development of membranes takes place depending on the requirements and commercial membrane development process is presented in Fig. 2.

(i) **Nanofiltration**

Nanofiltration (NF) membranes also known as "loose RO" is a type of filtration technology that falls between ultrafiltration (UF) and reverse osmosis (RO) membranes with a pore size of 1 nm which corresponds to molecular weight cut-off (MWCO) of 200–1000 Da [31, 32]. Some of the specifications of commercial NF are given in Table 1. Based on pore size and electrostatic effect, the NF membrane efficiently removed small ions and organic molecules from wastewater of textile effluent [33–36], food [12, 37–41], pharmaceutical [38, 41, 42], and electroplating effluent [20, 43, 44], etc. Important distinguishing characteristics of NF membranes are present in Fig. 3. The NF membrane has a high rejection of divalent ions, low rejection of monovalent ions, and higher flux than RO membranes [32]. Rejection from NF membranes may be attributed to a combination of steric, Donnan, dielectric, and transport effects.

(b) **Ultrafiltration**

Ultrafiltration (UF) membrane is a type of filtration technology that uses a semipermeable membrane to separate, bacteria, proteins, colloids, suspended solids bacteria, and high molecular weight solutes from wastewater. UF membrane pore size ranges

Fig. 2 Development of membrane processes. *Source* Singh [30]

Table 1 Specification of nanofiltration membrane

Membrane	Polymer	Manufacturer	Salt rejection (%)		pH range	Feed	MWCO (Da)	References
			$MgSO_4$	NaCl				
NF245	Polypropylene	Dow Filmtec	1.0	–	3–11	Food/dairy	< 300	[45]
NFX	Polyamide-TFC	Dow Filmtec	99	40	3–10.5	Food/industries/waste	150–300	[45]
Duracid NF	Polyamide-TFC	Dow Filmtec	98	–	0–9	Industrial/commercial	150–200	[45]
NFG	Polyamide-TFC	Dow Filmtec	50	10	4–10	Food/industries/waste	600–800	[45]
NFW	Polyamide-TFC	Dow Filmtec	97	20	4–10	Food/industries/waste	300–500	[45]
NF300	–	Pure Aqua	98	–	3–11	Dairy	–	[46]

Note MWCO molecular weight cut-off

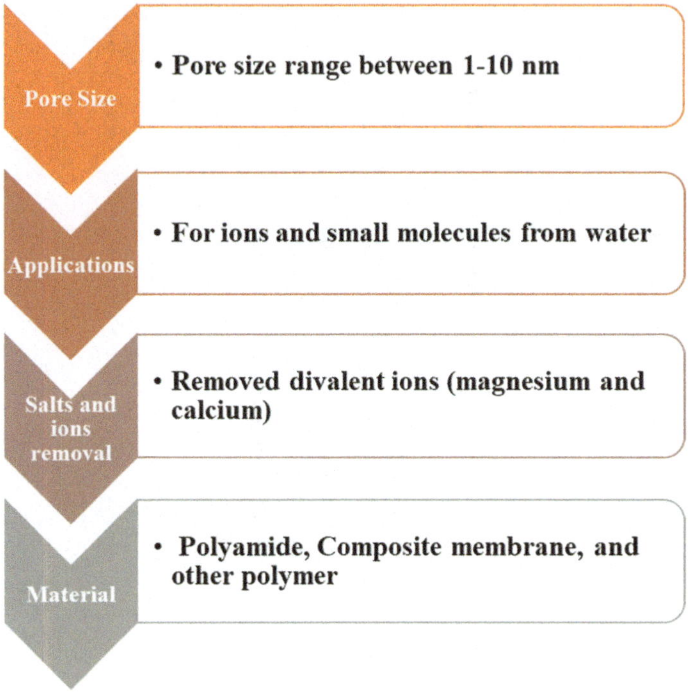

Fig. 3 Some of the characteristics of nanofiltration membrane

from 0.01 to 0.1 μm and this pore size is smaller than microfiltration (MF) but larger than NF and RO [47]. UF membranes typically operate in the pressure range of 3.4–8.3 bar (50–120 PSI) [48]. The product produced after UF is free of any pathogenic and highly pure. Xing et al. [49] investigated the water reclamation of urban wastewater through UF under 162-days pilot-scaled experiment and averaged 97% of COD removal, 100% SS removal, and 96.2% ammonia removal. Zhang et al. [50] also investigated the PVC-UF membrane efficacy for wastewater treatment and found the removal of 88.83% of SS, 50% of colour, and 96.41% of turbidity. Recently, Benkhaya et al. [51] investigated the effect of TiO_2 loading on the removal efficacy of UF membrane for azo dyes and observed up to 97.07% rejection.

(iii) **Microfiltration**

Microfiltration (MF) membrane systems wastewater treatment is becoming widely popular due to its commercial availability, feasibility, process reliability, and relative insensitivity [52–55]. The pore size of typical MF ranges from 0.1 to 10 μm and plays a significant role in removing contaminants. MF is generally fabricated with polymers (polyvinylidene fluoride, polypropylene, polyethylene, etc.) and ceramics as well as with metals. Ceramic MF membrane can be operated in extreme conditions which enhances its employment in the treatment of wastewater. The fouling process is one of the main limitations of the MF. Microbial agents and chemicals including pathogens, microorganisms, organic matter particulates, SS, dyes, and heavy metals mainly contribute to the phenomenon of fouling [54, 56]. The problem of fouling can be potentially overcome in MF membranes through the integration of coagulation with membrane treatments [20].

4 Application of Membrane Bioreactor Technology

Membrane bioreactors have industrial applications and it is exponentially growing with wastewater problems. Wastewater contamination such as BOD, COD, TSS, colour, pathogens, etc., have been removed through MBR technology (Table 2). MBT technology can completely remove the wastewater colour of textiles, landfill leachate, municipal, etc. (Table 2). Xiao et al. [57] quoted that the first super-large scale industrial application of MBR was in the Beijing Wenyu River plant built in 2007 with a 100,000 m^3/d treatment system capacity. Since then, the number of super-large MBRs has continued to increase in the United States, China, India, Europe, and globally. The Sanand automotive wastewater treatment plant of Sanand Gujrat, India has 20 KLD treatment capacity. The Imemflo Hollow fibre membrane made of PVDF having pore size around 0.25 is used for the filtration processes (data from online link http://www.thembrsite.com/). In 2013, a 300 KLD MBR plant was also installed in Zydus Hospital in Ahmedabad, India (Table 3). This MBR treatment plant has a polymeric material membrane and is used to remove biodegradable contaminants, pathogens, COD, TSS, etc. (data from online link http://www.thembrsite.com/). MBR plants are successfully working in the area of wastewater treatment at the industrial level in different regions such as India, China, USA, Belgium, Austria, Italy, South Korea, etc. of the world (Table 3).

5 Monitoring and Control of the Membrane Bioreactor Process

The operational process of the MBR is complex and therefore its monitoring is required for the management of operational issues. The users of MBR have several concerns such as fouling, biosolid production, treatment efficacy, pre-screen device selection, energy consumption, and loss of membrane modules [22]. Membrane fouling is the primary concern of the MBR user and it occurs due to the accumulation of microorganisms, particles, and organic matter on the surface of the membrane. Membrane fouling reduced the removal efficiency and enhanced the operational cost of the treatment. The MBR removal efficiency is directly proportional to the monitoring of the membrane fouling problem and its management. The processes of membrane fouling cannot be completed stop but it can be controlled significantly by

Table 2 Applications of membrane bioreactors for wastewater treatment

References	WW type	Reactor volume (L)	Membrane type	Pore size (μm)	MS area (m^2)	HRT (days)	COD removal (%)	TSS removal (%)	TN removal (%)	Colour removal (%)
Tan et al. [58]	LL	–	RO	0.001	–		98.68	–	–	99.53
Al Bazedi et al. [59]	Pharmaceutical	42.5	NF	–	–	0.5	99.8	–	–	–
Yang et al. [60]	Textile	4	UF-PVDF	0.04	0.05	1–2	91	99.4	–	80
Couto et al. [61]	Textile	–	MF-NF	0.4	–	–	65.44	–	–	100
Friha et al. [62]	Textile	60	UF	–	0.39	2	98	100	–	100
Katsou et al. [63]	Municipal	210	HF, PVDF	0.04	0.93	10.3	508			
Yigit et al. [64]	Textile	230	HF	0.04	–	0.58	97	99	78	98
Viero et al. [65]	Refinery	4.4	HF	0.15	0.0028	10	67	–	–	–
Yuniarto et al. [66]	Palm oil	20	FS	0.4	0.1	0.8	94	–	–	–
Brik et al. [67]	Textile	20	UF	–	0.28	0.7–4	90	–	–	98
Acharya et al. [68]	Food	20	UF	0.04	0.047	2.9	97	–	–	–
Badani et al. [69]	Textile	500	UF	0.025	–	2	97	–	–	–

Note MS membrane surface area; *WW* wastewater; *TSS* total suspended solid; *HRT* hydraulic retention time; *TN* total nitrogen; *LL* landfill leachate

Table 3 Details of MRB applications at industrial level globally (data from online link http://www.thembrsite.com/)

Location	Membrane material	Application	Capacity range (MLD)	Supplier
Sint-Lievens-Houtem, Belgium	Polymeric	Beverage, food and dairy	< 1.0	Blue Foot Membranes NV and Panterein
Sanand, Gujarat, India	Polymeric	Heavy industry (oil, gas and automotive)	< 1.0	Imemflo and Shubham Inc
Belgium	Polymeric	Bespoke solutions	< 1.0	Blue Foot Membranes NV
Dumai, Indonesia	Polymeric	Bespoke solutions	1.0–10	Blue Foot Membranes NV and Aquakimia
Blackfoot, Idaho, USA	Polymeric	Beverage, food and dairy	1.1–10	Veolia Water Technologies & Solutions
Tijuana, Mexico	Polymeric	Beverage, food and dairy	< 1.0	SFC Umwelttechnik GmbH
Austria	Ceramic	Landfill leachate	< 1.0	EnviCare® Engineering GmbH
Brescia, Italy	Polymeric	Municipal	> 100	Veolia Water Technologies & Solutions
Busan, South Korea	Polymeric	Municipal	> 100	Veolia Water Technologies & Solutions and GS Engineering
Wheatland, California, USA	Polymeric	Food, dairy, beverage, Packages systems	< 1.0	Integrated Water Services, Inc. and EPD Consultants
Romania	Polymeric	Slaughterhouse wastewater	< 1.0	MICRODYN-NADIR
Cleveland Bay, Townsville, Queensland, Australia	Polymeric	Municipal	10.1–100	Veolia Water Technologies & Solutions
Eemshaven, The Netherlands	Polymeric	Food, dairy and Beverage	1.1–10	Berghof Membrane Technology GmbH
Kingdom of Saudi Arabia	Polymeric	Municipal	< 1.0	ACWA Services Limited/ACWA Emirates
Bodrum, Turkey	Polymeric	Municipal	< 1.0	MICRODYN-NADIR
Northamptonshire, England	Polymeric	Landfill leachate	< 1.0	Supertec Group

(continued)

Table 3 (continued)

Location	Membrane material	Application	Capacity range (MLD)	Supplier
Vientiane, Laos	Polymeric	Municipal	< 1.0	SFC Umwelttechnik GmbH
Mumbai, India	Polymeric	Municipal	< 1.0	Blue Foot Membranes NV and Ion Exchange
Ireland	Polymeric	Pharmaceutical and fine chemical	1.1–10	Veolia Water Technologies & Solutions
Ahmedabad, Gujarat, India	Polymeric	Municipal and Commercial facilities	< 1.0	Shubham Inc.

controlling the responsible factors. For controlling membrane fouling its mechanism should be clear. The membrane fouling process is affected by several factors such as the material of the membrane, pore size of the membrane, dissolved matter, and configuration, etc. (Fig. 4). The Fouling process has four stages (Fig. 5) and it is a fact that the distinguishing this difference stages of membrane fouling is not easy [22]. Internal pore size of the membrane have a direct correlation with fouling. A larger size particulate of influent will block the pore of the membrane and result in low premature flux. Generally, in industries hydrophobic membrane is used to limit the fouling process. The surface of MBR is also modified with hydrophilic substance and this helps in mitigating the fouling. The feed quality (such as pH, solid content, and ionic strength) also affects the fouling of the membrane.

Monitoring of membrane fouling can be done through tracking transmembrane pressure (TMP). Another conventional method of monitoring membrane fouling be done through tracking variations in flux over filtration time [22]. The limitation of these conventional monitoring processes is that they fail to provide a specific site and stage of membrane fouling. Latest monitoring approaches such as Raman spectroscopy, Synchrotron infrared mapping, and Micro-X-ray overcome these limitations. These advanced monitoring approaches provide the time real-time and on-site results which significantly improve the economic feasibility of the MBR. Mitigation of membrane fouling can be achieved by removal of fouling (physical cleaning, and chemical cleaning), and limitation of fouling (optimization of membrane, optimization of operating conditions, and modification of biomass characteristics) [70]. MBR fouling can be reduced by using an antifouling membrane manufactured with materials such as polyethylene, polyvinylidene fluoride, polyether sulfone, polyacrylonitrile, etc. Membranes made of polyacrylonitrile have high resistance against the fouling process, however, the production cost of this membrane is another limitation.

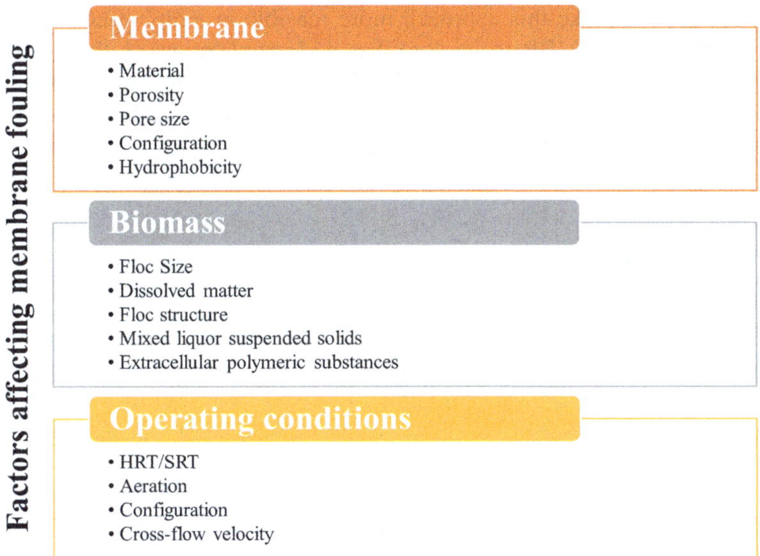

Fig. 4 Pictorial representation of factors affecting membrane fouling in MBR

Fig. 5 Different stages of membrane fouling of the membrane bioreactors

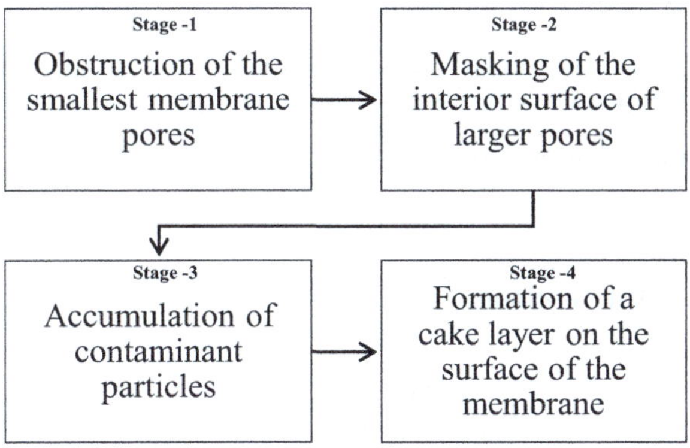

6 Future Prospects and Recommendations

The demand for treated wastewater will gain popularity and become a necessity in the modern era. Under such circumstances, the demand for MBR technology rises and therefore more reliability of the treatment system will be expected from the manufacturing end. Based on the above discussion, it can be reported that the treatment efficacy of MBR has no doubt but the operational cost is still a challenge for the researcher. Therefore, is recommended that more extensive research is required to reduce the expenses of this MBR. Manufacturing of low-cost and high-efficiency membranes is required to achieve this task. Optimization of the operational cost is also required to make this approach more feasible in terms of economic feasibility. At present, the MBR technology has wide applications in the treatment of heterogeneous wastewater. However, MBR's future development in terms of cost-effectiveness will make this technology more popular and reliable for the treatment of industrial wastewater at the global level.

7 Conclusions

Membrane bioreactor (MBR) technology has been gaining popularity for wastewater treatment for the last three decades. This study summarizes the progress and application of MBR technology for achieving sustainable development goals. The efficiencies of MBR technology for pollutants, chemical oxygen demand, biological oxygen demand, total suspended, colour, pathogens, and emerging pollutants, heavy metals are reviewed. Membrane material and types are also reviewed. The industrial-scale application of MBR technology is also discussed. The challenges such as fouling processes in MBR technology have also been covered. MBRs are one the most economically feasible at industrial level wastewater treatment technique and will help achieve the sustainable development goals through the circular water concept.

Acknowledgements All the authors are grateful to the heads of their respective institutes for providing all the necessary facilities.

Conflict of Interest The authors declare no conflict of interests.

Abbreviations

WW	Wastewater
MBR	Membrane bioreactors
NF	Nanofiltration
RO	Reverse osmosis
MWCO	Molecular weight cut-off
UF	Ultrafiltration
MF	Microfiltration
MSA	Membrane surface area
HRT	Hydraulic retention time
TN	Total nitrogen
LL	Landfill leachate
SDGs	Sustainable development goals
TSS	Total suspended solid
BOD	Biological oxygen demand

COD	Chemical oxygen demands
WHO	World Health Organization
PSFs	Polysulfones
PVDF	Polyvinylidenedifluoride
CA	Cellulose acetate
PTEE	Polytetrafluoroethylene
TMP	Transmembrane pressure

References

1. Silva JA (2023) Wastewater treatment and reuse for sustainable water resources management: a systematic literature review. Sustainability 15(14):10940
2. Dutta D, Arya S, Kumar S (2021) Industrial wastewater treatment: current trends, bottlenecks, and best practices. Chemosphere 285:131245
3. Muralikrishna IV, Manickam V (2017) Industrial wastewater treatment technologies, recycling, and reuse. Environ Manag 295–336
4. Shakil MSZ, Mostafa MG (2023) Characterization of paper mill effluent and its impacts on the environment. J Chem Environ 109–122
5. Kumar D, Sharma C (2022) Paper industry wastewater treatment by electrocoagulation and aspect of sludge management. J Clean Prod 360:131970
6. Kumar D, Gaurav VK, Sharma C (2018) Ecofriendly remediation of pulp and paper industry wastewater by electrocoagulation and its application in agriculture. Am J Plant Sci 9(12):2462
7. Devi NL, Yadav IC, Shihua QI, Singh S, Belagali SL (2011) Physicochemical characteristics of paper industry effluents—a case study of South India Paper Mill (SIPM). Environ Monit Assess 177:23–33
8. Sandoval MA, Coreño O, García V, Salazar-González R (2024) Enhancing industrial swine slaughterhouse wastewater treatment: optimization of electrocoagulation technique and operating mode. J Environ Manage 349:119556
9. Mkilima T, Meiramkulova K, Kydyrbekova A, Bazarbayeva T, Gulnur D, Aknur Z, Shegenbayev A, Nurbolat D, Oshanova G, Gulzhakhan K (2023) Biofilm-enhanced natural zeolite material in purification performance for slaughterhouse wastewater. Water 15(19):3501
10. Ng M, Dalhatou S, Wilson J, Kamdem BP, Temitope MB, Paumo HK, Djelal H, Assadi AA, Nguyen-Tri P, Kane A (2022) Characterization of slaughterhouse wastewater and development of treatment techniques: a review. Processes 10(7):1300
11. Bethi CMS, Narayan B, Martin A, Kudre TG (2020) Recovery, physicochemical and functional characteristics of proteins from different meat processing wastewater streams. Environ Sci Pollut Res 27:25119–25131
12. Morker H, Saini B, Dey A (2023) Role of membrane technology in food industry effluent treatment. Mater Today Proc 77:314–321
13. Shroti GK, Tomer P, Hazra S (2023) Valorization of wastewater from the beverage industry for polyhydroxyalkanoate production with its improved applicability to agricultural use. Water Air Soil Pollut 234(4):269
14. Nweke EN, Okechukwu VU, Omokpariola DO, Umeh TC, Oze NR (2022) Pollution evaluation of industrial effluents from consolidated breweries: a case study from Benue State, Nigeria. In: River Basin management-under a changing climate. IntechOpen
15. Abrha BH, Chen Y (2017) Analysis of physico-chemical characteristics of effluents from beverage industry in Ethiopia. J Geosci Environ Protect 5(6):172–182
16. WHO (2023) World Health Organization report-drinking water. Access on: 06 Jan 2024 online link https://www.who.int/news-room/fact-sheets/detail/drinking-water#:~:text=Key%20facts,water%20source%20contaminated%20with%20faeces

17. Dubey M, Vellanki BP, Kazmi AA (2023) Removal of emerging contaminants in conventional and advanced biological wastewater treatment plants in India-a comparison of treatment technologies. Environ Res 218:115012
18. Singh S, Panwar N, Kumar SS, Singh R, Anand G, Kumar A, Malyan SK (2023) Aromatic oils from medicinal plants and their role in nanoparticles synthesis, characterization, and applications. In: Secondary metabolites from medicinal plants. CRC Press, pp 305–314
19. Malyan SK, Yadav S, Sonkar V, Goyal VC, Singh O, Singh R (2021) Mechanistic understanding of the pollutant removal and transformation processes in the constructed wetland system. Water Environ Res 93(10):1882–1909
20. Kumar J, Joshi H, Malyan SK (2021) Removal of copper, nickel, and zinc ions from an aqueous solution through electrochemical and nanofiltration membrane processes. Appl Sci 12(1):280
21. Asif MB, Zhang Z (2021) Ceramic membrane technology for water and wastewater treatment: a critical review of performance, full-scale applications, membrane fouling and prospects. Chem Eng J 418:129481
22. Rahman TU, Roy H, Islam MR, Tahmid M, Fariha A, Mazumder A, Tasnim N et al (2023) The advancement in membrane bioreactor (MBR) technology toward sustainable industrial wastewater management. Membranes 13(2):181
23. Qiu M, Chen X, Fan Y, Xing W (2017) 1.11 ceramic membranes. In: Drioli E, Giorno L, Fontananova E (eds) Comprehensive membrane science and engineering, 2nd edn, pp 270–297
24. Mutamim NSA, Noor ZZ, Hassan MAA, Olsson G (2012) Application of membrane bioreactor technology in treating high strength industrial wastewater: a performance review. Desalination 305:1–11
25. Ciora RJ, Liu PKT (2003) Ceramic membranes for environmental related applications. Fluid Part Sep J 15(1):51–60
26. Hofs B, Ogier J, Vries D, Beerendonk EF, Cornelissen ER (2011) Comparison of ceramic and polymeric membrane permeability and fouling using surface water. Sep Purif Technol 79(3):365–374
27. Luján-Facundo MJ, Mendoza-Roca JA, Bes-Piá A, Zuriaga-Agustí E, Mestre S, Palacios M-D (2023) Low-cost ceramic membranes manufacture using INKJET technology for active layer deposition and validation on membrane bioreactors. Process Saf Environ Protect
28. Hubadillah SK, Jamalludin MR, Othman MHD, Iwamoto Y (2022) Recent progress on low-cost ceramic membrane for water and wastewater treatment. Ceram Int 48(17):24157–24191
29. Al-Asheh S, Bagheri M, Aidan A (2021) Membrane bioreactor for wastewater treatment: a review. Case Stud Chem Environ Eng 4:100109
30. Singh R (2015) Introduction to membrane Technology. In: Membrane technology and engineering for water purification: application, systems design and operation. Elsevier, ISBN. 978-0-444-63362-0, pp 1–80
31. Zhang Z, Fan K, Liu Y, Xia S (2023) A review on polyester and polyester-amide thin film composite nanofiltration membranes: synthesis, characteristics and applications. Sci Total Environ 858:159922
32. Mohammad AW, Teow YH, Ang WL, Chung YT, Oatley-Radcliffe DL, Hilal N (2015) Nanofiltration membranes review: recent advances and future prospects. Desalination 356:226–254
33. Maheshwari K, Agarwal M, Ghosh A (2023) Investigating the tolerance of nano-filtration for the treatment of wastewater. Indian Chem Eng 65(2):213–220
34. Adak MK, Kumar K, Das A, Dhak D (2023) Nano-filtration application in the textile industry for wastewater treatment. In: Nanofiltration membrane for water purification. Springer Nature Singapore, Singapore, pp 69–82
35. Keskin B, Korkut S, Ormancı-Acar T, Turken T, Tas CE, Menceloglu YZ, Koyuncu I et al (2023) Pilot scale nanofiltration membrane fabrication containing ionic co-monomers and halloysite nanotubes for textile dye filtration. Water Sci Technol 87(6):1529–1541
36. Sun Z, Zhu X, Tan F, Zhou W, Zhang Y, Luo X, Cheng X et al (2023) Poly (vinyl alcohol)-based highly permeable TFC nanofiltration membranes for selective dye/salt separation. Desalination 553:116479

37. Rajbongshi A, Gogoi SB (2023) Microfiltration, ultrafiltration and nanofiltration as a post-treatment of biological treatment process with references to oil field produced water of Moran oilfield of Assam. Petrol Res
38. Mallakpour S, Azadi E (2022) Nanofiltration membranes for food and pharmaceutical industries. Emergent Mater 5(5):1329–1343
39. Yadav D, Karki S, Ingole PG (2022) Nanofiltration (NF) membrane processing in the food industry. Food Eng Rev 14(4):579–595
40. Iaquinta M, Stoller M, Merli C (2009) Optimization of a nanofiltration membrane process for tomato industry wastewater effluent treatment. Desalination 245(1–3):314–320
41. Gogoi M, Goswami R, Hazarika S (2023) Membrane-based treatment of wastewater generated in pharmaceutical and textile industries for a sustainable environment. In: Advanced materials from recycled waste. Elsevier, pp 87–109
42. Gulamhussein MA, Saini B, Dey A (2023) Removal of pharmaceutical contaminants through membrane bioreactor. Mater Today Proc 77:260–268
43. Saravanan R, Sathish T, Sharma K, Rao AV, Sathyamurthy R, Panchal H, Zahra MMA (2023) Sustainable wastewater treatment by RO and hybrid organic polyamide membrane nanofiltration system for clean environment. Chemosphere 139336
44. Boricha AG, Murthy ZVP (2009) Preparation, characterization and performance of nanofiltration membranes for the treatment of electroplating industry effluent. Sep Purif Technol 65(3):282–289
45. Sterlitech, Online source. Access on 31 Dec 2023. https://www.sterlitech.com/nanofiltration-nf-membrane.html
46. Pura Aqua, Online source. Access on 31 Dec 2023. https://pureaqua.com/nanofiltration-nf-systems/
47. Hussain CM, Paulraj MS, Nuzhat S (2021) Source reduction and waste minimization. Elsevier. https://doi.org/10.1016/B978-0-12-824320-6.00002-2
48. Khan A, Ali J, Jamil SUU, Zahra N, Tayaba TB, Iqbal MJ, Waseem H (2022) Removal of micropollutants. In: Environmental micropollutants. Elsevier, pp 443–461. https://doi.org/10.1016/B978-0-323-90555-8.00012-X
49. Xing C-H, Tardieu E, Qian Y, Wen X-H (2000) Ultrafiltration membrane bioreactor for urban wastewater reclamation. J Membr Sci 177(1–2):73–82
50. Zhang X, Chen Y, Konsowa AH, Zhu X, Crittenden JC (2009) Evaluation of an innovative polyvinyl chloride (PVC) ultrafiltration membrane for wastewater treatment. Sep Purif Technol 70(1):71–78
51. Benkhaya S, Lgaz H, Tang H, Altaee A, Haida S, Vatanpour V, Xiao Y (2023) Investigating the effects of polypropylene-TiO_2 loading on the performance of polysulfone/polyetherimide ultrafiltration membranes for azo dye removal: experimental and molecular dynamics simulation. J Water Process Eng 56:104317
52. Gryta M, Woźniak P (2024) Application of polypropylene microfiltration membranes for separation of wastewater from car wash. Sep Purif Technol 331:125707
53. Niavarani Z, Breite D, Yasir M, Sedlarik V, Prager A, Schönherr N, Abel B, Gläser R, Schulze A (2024) Removal of endocrine disrupting chemicals from water through urethane functionalization of microfiltration membranes via electron beam irradiation. Front Environ Sci Eng 18(4):45
54. Hakami MW, Alkhudhiri A, Al-Batty S, Zacharof M-P, Maddy J, Hilal N (2020) Ceramic microfiltration membranes in wastewater treatment: filtration behavior, fouling and prevention. Membranes 10(9):248
55. Daels N, De Vrieze S, Sampers I, Decostere B, Westbroek P, Dumoulin A, Dejans P, De Clerck K, Van Hulle SWH (2011) Potential of a functionalised nanofibre microfiltration membrane as an antibacterial water filter. Desalination 275(1–3):285–290
56. AlSawaftah N, Abuwatfa W, Darwish N, Husseini G (2021) A comprehensive review on membrane fouling: mathematical modelling, prediction, diagnosis, and mitigation. Water 13(9):1327

57. Xiao K, Liang S, Wang X, Chen C, Huang X (2019) Current state and challenges of full-scale membrane bioreactor applications: a critical review. Biores Technol 271:473–481
58. Tan YK, Lau WJ, Nawi NSM, Roslan R, Ng PS (2024) Assessing membrane performance for landfill leachate treatment in accordance with local regulatory requirements. J Membr Sci Res 10(1)
59. Al Bazedi G, Abdallah H, Ali SS, Shalaby M, Sabry RM, Ali HM, Gadallah H (2024) Economic evaluation for anoxic biological system integrated with external membrane for pharmaceutical wastewater treatment. Appl Water Sci 14(1):10
60. Yang X, López-Grimau V, Vilaseca M, Crespi M (2020) Treatment of textile wastewater by CAS, MBR, and MBBR: a comparative study from technical, economic, and environmental perspectives. Water 12(5):1306
61. Couto CF, Moravia WG, Amaral MCS (2017) Integration of microfiltration and nanofiltration to promote textile effluent reuse. Clean Technol Environ Policy 19:2057–2073
62. Friha I, Bradai M, Johnson D, Hilal N, Loukil S, Amor FB, Feki F, Han J, Isoda H, Sayadi S (2015) Treatment of textile wastewater by submerged membrane bioreactor: in vitro bioassays for the assessment of stress response elicited by raw and reclaimed wastewater. J Environ Manag 160:184–192
63. Katsou E, Malamis S, Loizidou M (2011) Performance of a membrane bioreactor used for the treatment of wastewater contaminated with heavy metals. Biores Technol 102(6):4325–4332
64. Yigit NO, Uzal N, Koseoglu H, Harman I, Yukseler H, Yetis U, Civelekoglu G, Kitis M (2009) Treatment of a denim producing textile industry wastewater using pilot-scale membrane bioreactor. Desalination 240(1–3):143–150
65. Viero AF, de Melo TM, Torres APR, Ferreira NR, Sant'Anna GL Jr, Borges CP, Santiago VMJ (2008) The effects of long-term feeding of high organic loading in a submerged membrane bioreactor treating oil refinery wastewater. J Membr Sci 319(1–2):223–230
66. Yuniarto A, Ujang Z, Noor ZZ (2008) Performance of bio-fouling reducers in aerobic submerged membrane bioreactor for palm oil mill effluent treatment. J Teknol 2008(49):555–566
67. Brik M, Schoeberl P, Chamam B, Braun R, Fuchs W (2006) Advanced treatment of textile wastewater towards reuse using a membrane bioreactor. Process Biochem 41(8):1751–1757
68. Acharya C, Nakhla G, Bassi A (2006) Operational optimization and mass balances in a two-stage MBR treating high strength pet food wastewater. J Environ Eng 132(7):810–817
69. Badani Z, Ait-Amar H, Si-Salah A, Brik M, Fuchs W (2005) Treatment of textile waste water by membrane bioreactor and reuse. Desalination 185(1–3):411–417
70. Le-Clech P, Chen P, Fane TAG (2006) Fouling in membrane bioreactors used in wastewater treatment. J Membr Sci 284(1–2):17–53

Revolutionizing Wastewater Treatment: Cutting-Edge Technologies for Biological Nutrient Removal

G. Anjali and P. C. Sabumon

Abstract Nutrient removal is crucial in wastewater treatment to counter the negative environmental impacts of excessive nitrogen and phosphorus. From various sources, such as domestic, industrial, and agricultural activities, nutrient pollution poses a significant threat to aquatic systems, necessitating removing nutrients from wastewater. Several pioneering processes have revolutionised wastewater treatment, and this chapter provides a comprehensive overview of advanced technologies for biologically removing nitrogen and phosphorus. It discusses conventional nutrient removal systems based on nitrification and denitrification, followed by advanced nitrogen removal technologies, including simultaneous nitrification, denitrification, phosphorus removal (SNDPR), shortcut nitrogen removal, and post-denitrification technologies. The subsequent sections discuss high-rate, full-scale systems integrating various treatment processes for efficient nutrient removal. Examples include granular biomass systems and biofilm-based treatment systems. These systems provide enhanced nutrient removal capabilities and offer advantages like compactness, high treatment efficiency, and flexibility. The chapter also highlights microalgae-based nutrient removal, emphasising its potential for sustainable nutrient removal and bioenergy production while addressing associated benefits and challenges. Finally, the challenges related to nutrient removal technologies are listed to advance the research and strategies to overcome it. Overall, these pioneering nutrient removal processes have greatly enhanced wastewater treatment, minimising nutrient discharge, preserving water quality and ecosystems, and promoting sustainable development in wastewater management. Further innovation and research in nutrient removal technologies will enhance their effectiveness, contributing to a cleaner and healthier environment.

G. Anjali · P. C. Sabumon (✉)
School of Civil Engineering, Vellore Institute of Technology (VIT), Chennai, Tamil Nadu 600127, India
e-mail: pcsabumon@vit.ac.in

G. Anjali
e-mail: anjaligopakumar@vit.ac.in

M. M. Ghangrekar et al. (eds.), *Biological and Hybrid Wastewater Treatment Technology*, Earth and Environmental Sciences Library,
https://doi.org/10.1007/978-3-031-63046-0_6

Keywords Biological nutrient removal · Granular biomass · Sequencing batch reactor · Microalgae · Circular economy · Wastewater

1 Introduction

Effective wastewater treatment is essential for environmental sustainability and public health, especially as global urbanisation and industrial activities unprecedentedly increase wastewater generation [1]. This chapter focuses on innovative technologies for removing nutrients, particularly nitrogen and phosphorus, from wastewater using microorganisms. Traditional methods often fail to remove these nutrients effectively, resulting in ecological issues like eutrophication. In contrast, biological nutrient removal (BNR) technologies offer sustainable, efficient solutions that improve nutrient removal rates and treatment efficiency. This chapter examines advanced BNR technologies, such as membrane bioreactors and sequencing batch reactors, highlighting their benefits, such as reduced energy consumption, chemical use, and lower carbon emissions. By adopting these technologies, wastewater treatment can shift from a linear to a circular approach, emphasising resource recovery and sustainability by enabling recycling and reusing treated water. The chapter aims to guide researchers, engineers, and policymakers in adopting these advanced methods to enhance wastewater treatment, thereby contributing to the preservation of the ecosystem, improvement in public health, and a sustainable future.

2 Nutrient Removal Systems

Nutrients are essential for organisms to thrive, progress, and perform many significant functions [2]. Carbon, nitrogen, phosphorus, oxygen, and hydrogen are fundamental elements that constitute the structural and functional components of all living organisms [3]. Among these, carbon is intricately woven into the fabric of all living entities, participating in diverse molecular interactions and metabolic activities. On the other hand, both nitrogen and phosphorus have a unique ecological and evolutionary significance. They are considered the macronutrients, supporting the growth and vitality of the entire biosphere, from the tiniest microbes to vast animal kingdoms [2, 4, 5].

Nitrogen is the fundamental element in forming key biological structures like cell walls, proteins, and genetic material [6, 7]. In water bodies, the manifestation of nitrogen is dual-natured. It can be found in simple inorganic states, such as ammonia (NH_3) or ammonium ion (NH_4^+), nitrite (NO_2^-), nitrate (NO_3^-), and the elemental di-nitrogen gas (N_2), and as more intricate organic structures, encompassing compounds like amino sugars and amino acids [8]. Phosphorus in aquatic systems is found in two distinct states: dissolved and particulate. The dissolved phosphorus includes both organic and inorganic forms. In contrast, particulate phosphorus is associated with particles, including bound phosphorus, phosphorus in precipitate form, and its amorphous variety. Inorganic phosphorus primarily consists of orthophosphates and polyphosphates, with orthophosphates being notably stable across various pH levels, including phosphates, hydrogen phosphates, dihydrogen phosphates, and phosphoric acid [9]. These orthophosphates are the readily reactive and accessible form of phosphorus for aquatic organisms. Organic phosphorus is more varied, combining phosphates with carbon-rich molecules, such as inositol phosphate, phosphamides, phosphoproteins, amino phosphoric acids, and sugar phosphates, among other compounds [10].

Though essential nutrients are essential to support aquatic life forms, the excessive release of nutrients, especially nitrogen (N) and phosphorus (P), into water bodies can lead to eutrophication. Eutrophication harms aquatic ecosystems by depleting dissolved oxygen and imposes toxic conditions on several aquatic species [11]. The realisation of the magnitude of the ecological alterations and environmental hazards made World Health Organization (WHO) and United States Environmental Protection Agency (USEPA) set stringent discharge standards for the nutrient release. The discharge limits for total nitrogen (TN) are < 10 mg/L and even 3 mg/L, total phosphorous (TP) are ≤ 0.3 mg/L, levels of NH_4–N are ≤ 1.5 mg/L, nitrates are < 10 mg/L, and nitrite < 0.1 mg/L [12, 13]. To meet these stringent discharge standards, implementing appropriate state-of-the-art treatment technologies in all the relevant wastewater treatment facilities is the need of the hour to protect the aquatic ecosystem from further degradation. The following sections describe the evolution of treatment schemes for biological nutrient removal.

2.1 Natural Nutrient Removal Processes

Natural nutrient removal is a multi-dimensional process involving biological, chemical, and physical interactions. These processes are crucial for maintaining the ecological balance of water bodies and preventing issues like eutrophication. Understanding and preserving these natural mechanisms are key to sustainable water management and environmental conservation. Natural processes are mediated by mainly using renewable energy sources such as solar, wind, water currents etc. as the potential energy for removing the pollutant by oxidation. The simple process of reducing the discharge effect of contaminants in lakes and rivers is through dilution, sedimentation, filtration, precipitation, adsorption, ion exchange etc. Environmental factors like pH, alkalinity, temperature, oxygen and concentration of nutrients plays a major role during the cleaning activity [11].

The rate of pollution is increasing so rapidly because of modern lifestyles and the natural cleansing rate is not able to sustain in completely removing the pollutants from being a hazard or toxic. Human activities have significantly disrupted many natural pathways, particularly in the context of nitrogen fixation. The amount of nitrogen fixed through human activities (210 Tg/year) now matches that of natural processes

(203 Tg/year). This substantial increase in anthropogenic nitrogen flows and their emissions into the environment is drastically altering the global nitrogen cycle [14]. The environmental impact of reactive nitrogen species is one of the most critical. It was found that these species would exceed Earth's carrying capacity by approximately threefold. The key contributors to this issue include using fertilisers, livestock waste, and domestic wastewater. A significant portion of the nitrogen entering the biosphere is ammonium, primarily produced by the Haber–Bosch process used as the primary method of producing ammonia from nitrogen and hydrogen [14, 15]. This process is the principal industrial method, responsible for about one-third of global nitrogen production and consuming 1–2% of the world's energy.

Constructed wetlands utilize natural biogeochemical and physical processes for nutrient removal and are implemented in over 50 countries as a sustainable approach to wastewater treatment. These systems treat wastewater and provide ecosystem services and recreational opportunities. However, their effectiveness can vary based on several factors including local climate, the nature of the wastewater, and how they are managed. Hybrid-constructed wetlands, recognized for their superior water-quality enhancement and greenhouse gas reduction, are influenced by choice of plant species, substrate type, and both environmental and hydraulic conditions. The key factors affecting their performance include temperature, the time wastewater spends in the system, and the rate at which pollutants are loaded into the system [15].

Natural processes are crucial for ecological balance; their limitations and variability under changing environmental conditions necessitate the integration of engineered systems. Engineered solutions provide controlled, efficient nutrient removal and adaptability to ecological and operational challenges. Combining natural and engineered approaches can achieve more sustainable and effective nutrient management, protecting ecosystems and ensuring ecological sustainability.

2.2 Conventional Nutrient Removal Processes

Most developing and underdeveloped countries focus on the preliminary, primary and secondary wastewater treatment. The treated water is then discharged into water bodies after disinfecting considering that the water quality devoid of environmentally altering pollutants. It is understood that the current wastewater characteristics are so complex that a tertiary treatment process is imperative. The tertiary wastewater treatment processes are identified and developed to remove the nutrients efficiently, heavy metals, refractory organics, etc. [2].

The conventional nutrient removal process primarily involves physical, chemical and biological methods. The high cost, inefficiency in operation and the danger in storing and handling the chemicals make the physical and chemical methods inferior to biological methods [16]. The biological processes control the metabolic capabilities of microorganisms to assimilate and transform nitrogen and phosphorus compounds. The traditional route for biological treatment of reactive nitrogen species (NH_4^+, NO_2^-, NO_3^-) involves nitrification and denitrification processes. The NH_4^+

is first oxidized to nitrate (NO_3^-) in a nitrification process. This process consists of two stages: ammonia oxidation to nitrite and nitrite to nitrate. The oxidation process is facilitated by different bacterial groups, with ammonium oxidizing bacteria (AOB) and nitrite oxidizing bacteria (NOB) handling the separate stages. In each stage, NH_4^+ is an energy source for various wastewater bacteria. The *Nitrosomonas* species predominantly carries out ammonium oxidation. The *Nitrosococcus, Nitrospira,* and *Nitrosovibrio* genera also undergo ammonium oxidation. For nitrite oxidation, bacterial genera such as *Nitrospira, Nitrospina, Nitrococcus,* and *Nitrocystis* play a crucial role [17]. Equations (1) and (2) denote the two stages of nitrification process. Subsequently, in an anoxic environment, nitrate is reduced to nitrogen gas (N_2) in the presence of organic carbon by heterotrophic denitrifying bacteria (HDB), allowing the gas to be released harmlessly into the atmosphere [2, 18]. Species such as *Alcaligenes, Paracoccus, and Thiobacillus* categorized as proteobacteria, are known to facilitate the process of denitrification. Equations (3) and (4) denote the denitrification process using nitrate and nitrite respectively. Figure 1 represents the process description followed in conventional nitrogen removal processes.

$$NH_4^+ + 1.5O_2 \xrightarrow{AOB} NO_2^- + H_2O + 2H^+; \quad \Delta G^\circ = -275 \text{ kJ/M} \tag{1}$$

$$NO_2^- + 0.5O_2 \xrightarrow{NOB} NO_3^- \tag{2}$$

$$2NO_3^- + 1.25CH_3COOH \xrightarrow{HDB} 2.5CO_2 + N_2 + 1.5H_2O + 2OH^-; \quad \Delta G^\circ = -946.3 \text{ kJ/M} \tag{3}$$

$$NO_2^- + 0.375CH_3COO^- + 0.125H_2O \xrightarrow{HDB} 0.5N_2 + 0.75CO_2 + 1.375OH^- \tag{4}$$

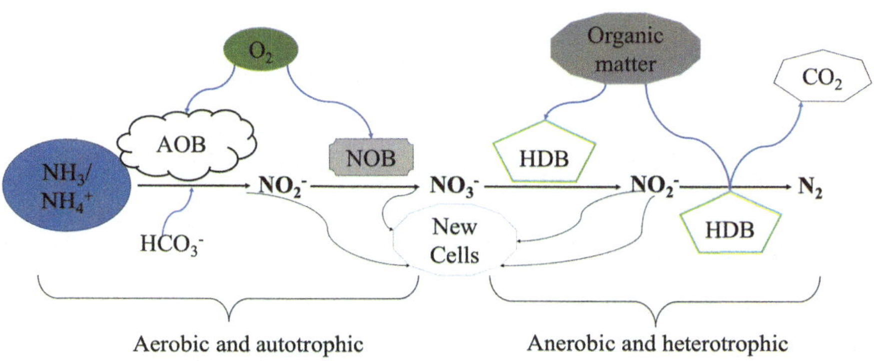

Fig. 1 Process description of conventional biological nitrogen removal. Adapted from Rout et al. [2]

The conventional route for phosphorus remediation includes Enhanced Biological Phosphorus Removal (EBPR). Several bacterial species, known as polyphosphate-accumulating organisms (PAOs), can absorb phosphorus beyond their metabolic requirements when subjected to alternating anaerobic and aerobic conditions. This unique characteristic is used in EBPR to extract phosphorus from wastewater biologically [19].

The major microbial populations mediating EBPR process belong to beta-2 subclass of *Proteobacteria.* PAOs store phosphorus intracellularly as polyphosphate (poly-P). In anaerobic conditions, PAOs consume volatile fatty acids (VFA) using energy from poly-P, converting them into polyhydroxyalkanoates (PHA), which are retained in the cells. In aerobic conditions, PAOs oxidize PHA for energy, facilitating glycogen formation, phosphorus uptake, and cell growth [20]. This way, when excess sludge is discharged, effective phosphorous removal occurs in the treatment system biologically. Figure 2 schematically explains the process of biological phosphorous removal.

The primary contributors of phosphorous in water bodies are sewage and agricultural runoff. Phosphorous removal can be achieved through physical–chemical methods, biological processes, or a combination [11]. The effectiveness of phosphorous elimination and recovery is contingent on the qualities of the influent. Consequently, various high-rate treatment systems have been developed to meet the specified limits for effluent discharge into open water bodies. The chemical pathways for phosphorous removal happen through chemical precipitation using salts such as aluminium, iron, or calcium or Ion exchange process or through struvite precipitation [11]. The typical chemicals employed during chemical precipitation are alum (aluminium sulfate), ferric chloride, and lime (calcium hydroxide). These agents react with soluble phosphates to produce insoluble precipitates that can be separated from the water [21].

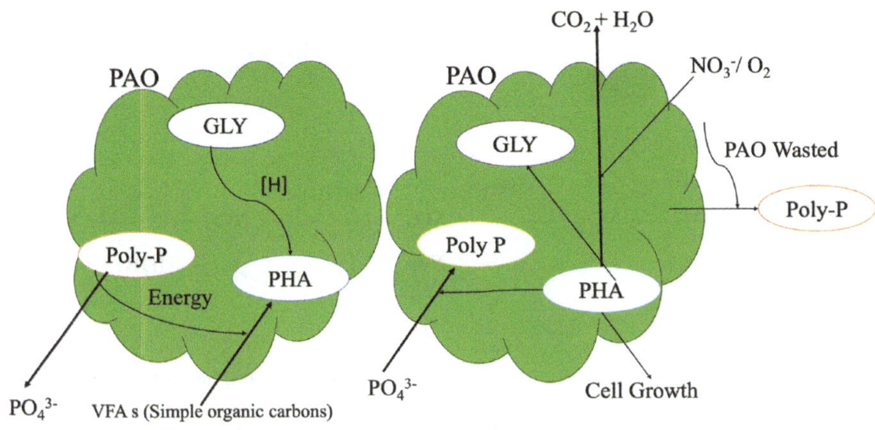

Fig. 2 Process description of biological phosphorous removal. Adapted from Rout et al. [2]

As the stringent regulations for nutrient discharge have been implemented, biological treatment methods have evolved in response. In recent years, various reactor configurations have emerged, designed to provide microorganisms with the optimal conditions for metabolizing and removing nitrogen and phosphorus, sometimes targeting both simultaneously [11, 22]. It includes the A2O (Anaerobic/Anoxic/oxic) process, UCT (University of Cape Town) process, MUCT (Modified UCT) Process, VIP (Virginia Initiative Process), 5-stage Bardenpho process, Westbank process, Johannesburg Process, CNG (Charlotte North Carolina) Process. A detailed study of their process descriptions, advantages and disadvantages are mentioned in Rout et al. [2]. Most of these processes are characterized as complex or complicated operations, require significant footprint for the reactors and effluent characteristics of the recycled effluent were very poor leading to inability for effective nutrient removal. Figure 3 comprehends the various nutrient removal processes available in the literature [11].

The engineered systems for nutrient removal, like BNR technologies, were introduced in the 1970s and have undergone significant scrutiny. Challenges include high energy demand, reliance on external chemicals, and economic and environmental sustainability concerns. To address these issues, innovations have been explored, such as efficient management of influent carbon, simultaneous nitrification and denitrification, and post-anoxic denitrification processes. These approaches aim to reduce energy and chemical usage while adhering to nutrient standards. Efficient nutrient removal requires continuous monitoring and adjustment because variables like pH, temperature, alkalinity, dissolved oxygen (DO) levels and external carbon source requirements can influence the efficiency of the process [23].

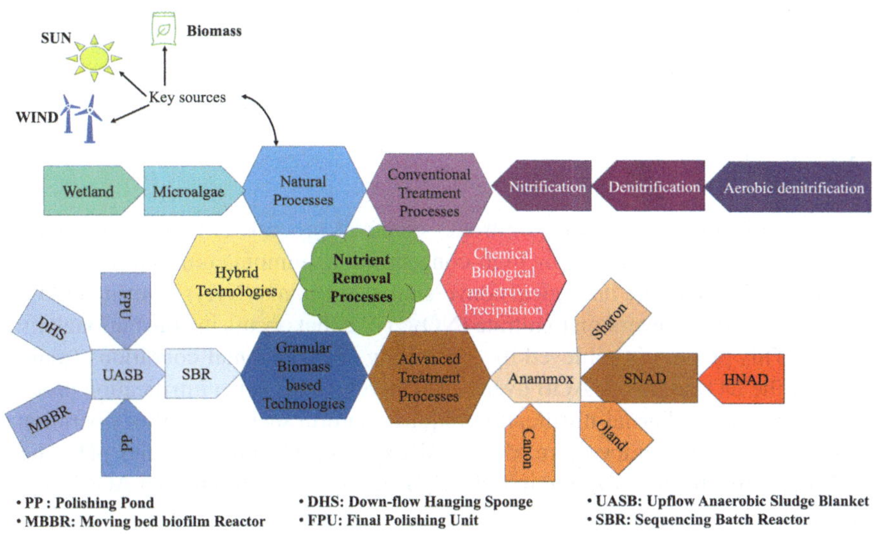

Fig. 3 Various nutrient removal processes. *Source* Hasan et al. [11]

Furthermore, sludge production necessitates sustainable disposal methods, particularly in chemical nutrient removal. Also, the conventional biological nitrogen removal processes are ineffective in controlling the emission of greenhouse gases like CO_2 and N_2O. On the other hand, the advanced approaches aim to reduce energy and chemical usage while adhering to nutrient standards. It is important to note that the research is continually evolving by utilizing the versatile microbial consortium adapting to the pollutant concentrations and enabling an efficient treatment system. Such emerging treatment results in sustainable removal and recovery of nutrients and production of less sludge thereby reducing the disposal load and improving overall efficacy of nutrient removal technologies.

2.3 Advanced Biological Nitrogen Removal Systems

While conventional BNR methods have been widely used in wastewater treatment, advancements in research and technology have paved the way for more innovative and efficient approaches. These advanced methods typically achieve superior nutrient removal efficiencies, adapt better to varying wastewater characteristics, and minimize energy and chemical input required for the treatment.

The major processes developed for reactive nitrogen species removal include partial nitrification (aerobic/microaerobic), simultaneous nitrification–denitrification (SND); anaerobic ammonium oxidation (ANAMMOX) and its modifications like SHARON, CANON, OLAND, etc.; Heterotrophic nitrification and aerobic denitrification (HNAD), NO_X process etc. These processes have reported numerous advantages over conventional processes like less oxygen and energy requirement for operation, no additional carbon requirement, less sludge production, etc. [24]. A few selected process descriptions are given in the following sections.

2.3.1 Partial Nitrification

Partial nitrification, an advanced nitrogen removal process, is increasingly recognized for its efficiency in treating high-concentration ammonia nitrogen wastewater. The process involves the biological conversion of ammonia (NH_3) to nitrite (NO_2^-), bypassing the full conversion to nitrate (NO_3^-) like that occurs in conventional nitrification. This process is achieved by controlling environmental conditions (mainly reducing the DO) to favour AOB over NOB, thereby halting the nitrification process at the nitrite stage. Partial nitrification requires careful management of operational parameters such as temperature, dissolved oxygen concentration, and pH. These conditions must be optimized to inhibit NOB activity while promoting AOB growth selectively.

This approach is more energy-efficient than complete nitrification followed by denitrification [25]. In traditional methods, converting ammonia to nitrate and then to nitrogen gas (N_2) requires significant energy input. Partial nitrification reduces this energy demand by shortening the nitrogen conversion pathway. The process also results in lowering sludge production compared to traditional methods. This reduction in sludge volume can significantly decrease the costs associated with sludge handling and disposal.

This process is particularly beneficial for treating high-strength ammonia wastewater from various Industries. It also applies to mainstream municipal wastewater treatment, especially when integrated with other advanced nitrogen removal processes like the ANAMMOX (anaerobic ammonium oxidation) process [25]. Partial nitritation coupled with denitritation offers a more energy-efficient pathway for nitrogen removal. Recent studies emphasize the control of operational parameters to stabilize nitritation-denitritation processes, reducing the need for external carbon sources [26]. Despite its advantages, partial nitrification can be sensitive to operational fluctuations and may require careful monitoring and control. Ongoing research is focused on enhancing the stability of the process and expanding its applicability to a broader range of wastewater types.

2.3.2 Anammox

Since the process discovery in the year 1995, the anaerobic ammonium oxidation (anammox) process has changed the regular pattern of ammonium oxidation. It was reported by [27] that the process involves the direct conversion of ammonium to nitrogen gas under anaerobic conditions and known as short-cut nitrogen removal. This process can be significantly more energy-efficient than conventional nitrification–denitrification because it requires no aeration, no carbon source, less footprint and low CO_2 emissions. The bacteria responsible for anammox belong to the *Planctomycetes* group and convert ammonium and nitrite directly to nitrogen gas. The process description of ammonia removal using Anammox is shown in Fig. 4.

Anammox is a process that relies on autotrophic anammox bacteria to remove nitrogen. These bacteria do not require organic carbon for their metabolism. The reaction effectively transforms ammonium (NH_4^+) and nitrogen dioxide (NO_2) into

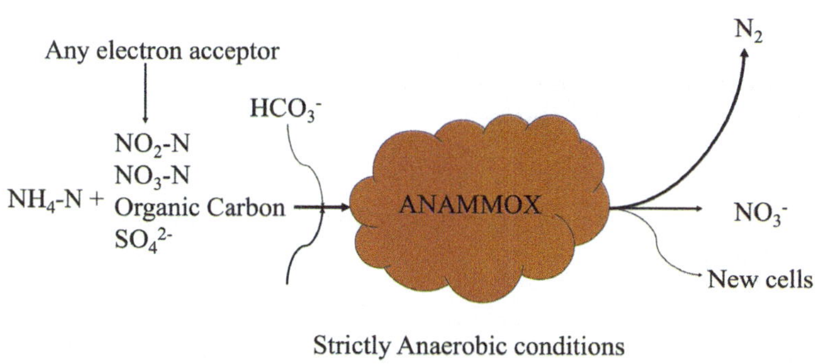

Fig. 4 Process description of anammox process

nitrogen gas (N_2) and water (H_2O). However, it is important to note that this process unavoidably produces some nitrate residuals (approximately 11%), which is a factor to consider in the overall treatment efficiency. Several modifications have been made to the Anammox process to improve its efficiency, adaptability, and application range. The various modifications like (i) the oxidation of ammonia nitrogen to nitrite ion under aerobic conditions (SHARON process) followed by the conversion of the remaining ammonia and nitrite to nitrogen gas by anammox bacteria in anaerobic conditions; (ii) A combination of partial nitration and ANAMMOX processes functioning collaboratively in a single reactor (CANON Process).

In this process, the AOB first oxidize ammonia to nitrite, which is then converted along with residual ammonia to nitrogen gas by anammox bacteria; (iii) **Denitrifying Ammonium Oxidation (Deamox) process** where, nitrate is reduced to nitrite by denitrifying bacteria, which is then used by anammox bacteria along with ammonium to produce nitrogen gas; (iv) **Expanded Granular Sludge Bed Reactor (EGSBR) for Anammox** is a granular sludge method where the Anammox bacteria are cultivated in an up-flow anaerobic sludge blanket reactor, leading to improved biomass retention and process stability; (v) **Oxygen-Limited Autotrophic Nitrification/Denitrification (OLAND)** process, which operates under limited oxygen conditions, allowing both partial nitrification and Anammox to take place in a single reactor, etc. These modifications aim to optimize the anammox process for different operational conditions, wastewater characteristics, and specific treatment goals. They reflect the ongoing innovation in wastewater treatment technology, seeking to make processes more efficient, sustainable, and adaptable to varying treatment needs.

Recent advancements focus on optimizing conditions for anammox processes in mainstream wastewater treatment [28]. Anammox process is diverse and versatile in coexisting with other nitrogen removal species. Hence, an integrated process with anammox is very feasible and efficient. Few integrated processes discovered are partial nitrification/anammox (PNA), Simutaneous Nitrification Anammox and Denitrification (SNAD) process, Simutaneous anammox/denitrification (SAD) process. Reactors involving granular sludge systems, biofilm systems, moving bed biofilm reactors (MBBR), sequencing biofilm batch reactor (SBBR), membrane reactors etc., are designed for enhanced nitrogen removal, especially in low-temperature conditions or in systems with fluctuating ammonia loads. Combining anammox with other biological or chemical processes (like conventional activated sludge) can enhance overall nitrogen removal efficiency, especially in complex wastewater matrices with varying composition.

2.3.3 Nitrogen Removal by Aerobic and Anaerobic Ammonium Oxidizers

Recent advancements in nitrogen removal processes have seen the emergence of microbial consortia such as ammonium oxidizing archaea (AOA). Like aerobic ammonia-oxidizing bacteria (AOB), AOA possess *ammonium mono oxygenase* enzymes, enabling them to actively catalyze ammonia oxidation [29]. Additionally, the complete ammonia oxidation (COMAMMOX) process, catalyzed by the single microorganism *Nitrospira* sp., has shown better adaptation to low dissolved oxygen (DO) conditions [30]. Several studies have also explored the established anammox process with different electron acceptors, including NO_2–N/NO_3–N and ferric oxides (FEAMMOX) [31], sulfates, and organic compounds (using *Candidatus fulgidae*), highlighting its applicability in treating complex wastewaters with low C/N ratios.

Hybrid processes like SNAD, Simultaneous Partial Nitrification and Anammox (SNAP), and SAD, involving the coexistence of different nitrogen removal species, have been developed for wastewater with C/N $\leq$ 2. These processes are underpinned by various pathways mediated by functional enzymes essential for nitrogen transformation in wastewater treatment, as shown in Eqs. (5) to (11). The key enzymes involved in nitrogen transformation include *Ammonia mono oxygenase* (AMO), *hydroxylamine oxidoreductases* (HAO), *nitrite oxidoreductases* (NOR), *Nitrite Reductase* (NIR), *Nitrate reductases* (NXR), *Nitrous Oxide reductases* (NOS), *Cytochrome c Nitrate reductases* (ccNIR), *NCyA* (an enzyme converting NO to NO_2^-), and *Hydrazine hydroxylase* (HH).

(i) Autotrophic nitrification mediated by ammonia oxidizing bacteria (AOB)

$$NH_3 \xrightarrow{\text{AMO}} NH_2OH \xrightarrow{\text{HAO}} NO \xrightarrow{\text{NCyA}} NO_2^- \xrightarrow{\text{NXR(NOB)}} NO_3^- \quad (5)$$

(ii) Reciprocal feeding pathway mediated by Urease positive *Nitrospira moscoviensis (NOB)*

$$\text{Urease} \xrightarrow{\text{NOB}} NH_3 \xrightarrow[]{\substack{\text{AOB} \\ (\text{Urease} \ -\text{ve}, \ \textit{Nitrosomonas europea})}} NO_2^- \xrightarrow{\text{NOB}} NO_3^- \quad (6)$$

(iii) Comammox process mediated by *Nitrospira* sp.

$$NH_4^+ \xrightarrow{\substack{\textit{Nitrospira}\ \text{Sp.} \\ \text{AMO}}} NH_2 \xrightarrow{\text{HAO}} NO_2^- \xrightarrow{\text{NXR}} NO_3^- \quad (7)$$

(iv) Heterotrophic Denitrification

$$NO_3^- \xrightarrow{\frac{\text{NAR}}{\text{NAP}}} NO_2^- \xrightarrow{\text{NIR}} NO \xrightarrow{\text{NOR}} N_2O \xrightarrow{\text{NOS}} N_2 \quad (8)$$

(v) Simultaneous nitrification and denitrification (SND)

$$NH_4^+ \xrightarrow{\text{AMO}} NH_2OH \xrightarrow{\substack{\text{HAO} \\ (O_2)}} NO \xrightarrow{\text{NCyA}} NO_2^- \xrightarrow{\text{NIR}} NO \xrightarrow{\text{NOR}} N_2O \xrightarrow{\text{NOS}} N_2 \quad (9)$$

(vi) Dissimilatory nitrate reduction to ammonia

$$NO_3^- \xrightarrow{NXR} NO_2^- \xrightarrow{ccNIR} NH_4^+ \quad (10)$$

(vii) Anammox Process mediated by anaerobic anammox bacteria

$$NH_4\text{-}N \searrow \quad NO_2^- \xrightarrow{NIR} NO \xrightarrow{HH} N_2H_2 \xrightarrow{HAO} N_2 \quad (11)$$

When the concentration of organic carbon becomes very prominent in effluents from pharmaceutical industry, tanneries, livestock or slaughterhouse waste processing, potato processing industries, alcohol and starch production, formaldehyde production and palm oil mill effluent, etc., the autotrophic processes or the hybrid processes fail to achieve efficiency due to the process limitations and microbial inhibitions [32]. In the treatment of such wastewaters, we require the nitrification to happen heterotrophically or simultaneous nitrification and denitrification (SND) in a single reactor.

2.3.4 Simultaneous Nitrification, Anammox and Denitrification (SNAD) Process

This is a hybrid process experiencing a coordinated and synergetic interaction between three functionally different species, AOB, anammox bacteria, and denitrifiers, in a single reactor system. Most of the modifications of anammox process work well on low oxygen and no carbon condition. Such processes are mainly suitable for wastewaters characterized by low or negligible carbon-to-nitrogen ratios. Also, there is a profound accumulation of nitrate after the anammox reaction (accounting for about 10–20% of the influent load), resulting in the discharge of nitrate in the effluents. Hence, to meet the stringent nitrogen discharge standards, the inclusion of a denitrification step is inevitable. However, to preserve the benefits of the shortcut nitrogen removal process, this denitrification step should be both cost-effective and efficient in minimizing nitrate accumulation. In this context, the SNAD process has garnered significant interest as a potential solution. The SNAD process is developed by combining partial nitrification, anammox and denitrification in a single reactor treating wastewaters containing higher concentrations of NH_4–N and in the presence of COD. The process description of nitrogen removal through SNAD process is represented in Fig. 5. The process was first invented by [33] under limited oxygen condition (0.4–0.6 mg/L) and at C/N ratio about 1.2. The results reported a high percentage removal for nitrogen (70%) and COD (94%). Since then, many works (around 50 numbers of journal publications) have been carried on nitrogen removal using SNAD process.

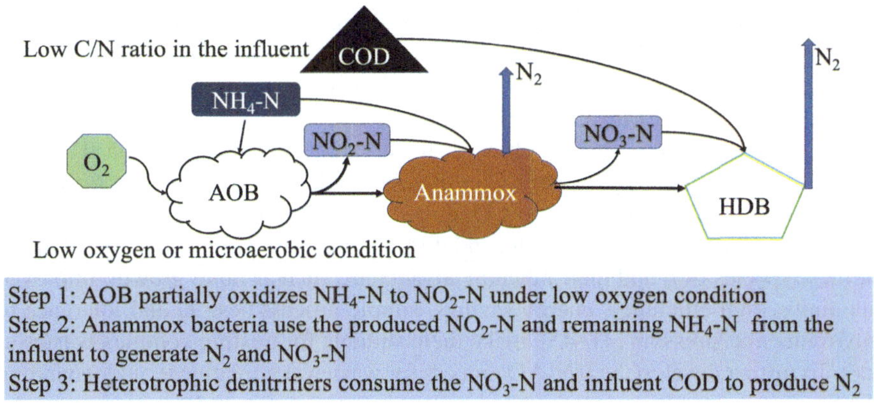

Fig. 5 Process description of SNAD process. Adapted from Chen et al. [34]

The SNAD process demonstrates high-performance efficiency, making it suitable for treating both side streams and domestic wastewater. The critical parameters for optimizing the process are C/N ratio, temperature, pH, DO, inorganic elements, etc. This process integrates three independent methods within a single managed environment, potentially lowering capital expenses. Additionally, its anoxic operation and partial nitrification can cut aeration costs by half. The process is highly effective in minimizing the environmental impact of pollutants like carbon, nitrogen, and phosphorus, subsequently reducing the eutrophication risk from wastewater treatment by 16%. It also lessens greenhouse gas emissions to 1.72 kg of CO_2 equivalent/m^3/d, leading to cost savings of approximately \$ 0.303/m^3/d [34].

2.3.5 Heterotrophic Nitrification and Aerobic Denitrification (HNAD) Process

The heterotrophically oxidizing ammonia to nitrogen in oxygen-rich environments is termed Heterotrophic Nitrification and Aerobic Denitrification (HN-AD). Various pathways have been proposed for this process, reflecting the diversity and specific functional genes or enzyme activities of the involved species. Chen and Ni [35] and Huang et al. [36] have explained the typical pathways followed by HN-AD microbes. In one route, heterotrophic nitrification progresses through the oxidation of ammonium to nitrite, followed by the conversion of nitrite to nitrogen gas, leads to aerobic denitrification ($NH_4^+ \rightarrow NH_2OH \rightarrow NO_2^- \rightarrow N_2O \rightarrow N_2$). Another pathway involves the transformation of hydroxylamine directly to N_2O and/or N_2, bypassing nitrite ($NH_4^+ \rightarrow NH_2OH \rightarrow N_2O \rightarrow N_2$). This process serves as an alternative to conventional BNR methods, favorably enabling simultaneous nitrification and denitrification (SND) in a single reactor [37].

HN-AD species are characterized by the presence of both AMO and periplasmic *nitrate reductase* (NAP) enzymes in a single bacterium, facilitating the conversion of NH_4–N to NO_2–N/NO_3–N and then to N_2. This dual functionality helps avoid competition for substrates and intermediates between heterotrophic denitrifying and autotrophic nitrifying bacteria. HN-AD bacteria also exhibit rapid growth rates, enhancing reactor start-up and stability [38].

The HN-AD process was first observed in the early 1980s by Robertson et al. [39], who identified that *Thiosphaera pantotropha* sp. could perform HN-AD. Since then, numerous HN-AD species have been identified from various environments, including soil, surface water, activated sludge, and different types of wastewater. The ubiquity of these species [40] and their inherent advantages, such as rapid growth, improved alkalinity, high efficiency, energy savings, simple process design, and tolerance to environmental stressors [41–43], make them suitable for treating complex wastewaters. To further develop the HN-AD process for wastewater treatment, understanding the diversity of the biomass, the key biochemical pathways, and the environmental factors regulating these reactions, are crucial.

2.3.6 NO_x Process

This process is characterized by its ability to regulate and enhance the denitrification process carried out by *Nitrosomonas* bacteria. Small amounts of nitrogen oxide (NO_x) can be introduced in to the system to manage the denitrification activity of these bacteria, as demonstrated by Schmidt et al. [44]. The presence of NO_x acts as a regulator, facilitating both nitrification and denitrification processes, even under completely toxic conditions. Equations (12)–(14) illustrate how nitrogen is transformed in a simultaneous nitrification and denitrification process when adequate NO_x is available. This approach has been tested in both laboratory and full-scale settings, revealing that ammonia oxidisers are primarily responsible for reducing nitrogen levels [45].

$$3NH_4^+ + 3O_2 \rightarrow N_2 + NO_2^- + 4H^+ + 4H_2O \quad (12)$$

$$NO_2^- + H^+ + 3[H] \rightarrow 0.5N_2 + 2H_2O \quad (13)$$

$$3NH_4^+ + 3O_2 + 3[H] \rightarrow 1.5N_2^+ + [H] + 6H_2O \quad (14)$$

2.4 *Advanced Biological Phosphorous Removal System*

Recent research highlights the significant role of extracellular polymeric substances (EPS) in phosphorus (P) uptake, contributing to phosphorus removal processes PAOs [46]. However, understanding the specific uptake mechanisms of EPS remains an area requiring further investigation, as emphasized by Diaz et al. [47] to enhance comprehension of the role of EPS in EBPR systems.

The efficiency of biological phosphorus removal can be improved by increasing the presence of PAOs in treatment systems, a strategy still in development. The microbiology of EBPR has evolved from the mid-1970s to the late 1990s, with *Candidatus Accumulibactor phosphatis* emerging as a critical organism [48]. Additionally, Denitrifying PAOs (DPAOs) are increasingly recognized for their role in phosphorus removal. Unlike PAOs, which require oxygen as an electron acceptor under aerobic conditions, DPAOs can utilize nitrate and nitrite in anoxic conditions. Diaz et al. [47] highlights the advantages of DPAOs in EBPR systems, such as reduced carbon and dissolved oxygen needs and lower sludge production.

Operational factors like pH, temperature, salinity, DO, ORP, hydraulic retention time, and solid retention time are crucial for the effective operation of EBPR processes. Diaz et al. [47] reported that the complexity of operational parameters and their interactions with microorganisms make long-term stability and high performance in full-scale wastewater treatment plants challenging. The population of PAOs and the efficiency of EBPR systems are influenced by the Chemical Oxygen Demand to Phosphorus (COD:P) ratio. There are research reports that low COD:P ratios could achieve low effluent phosphorus concentrations and stable removal rates, contrary to earlier findings which suggested higher ratios were necessary [48, 49].

The carbon sources, particularly VFAs and biodegradable organics, significantly impact EBPR processes [50] Recent technological advancements in phosphorus removal using various reactors are outlined in Table 1, focusing on full-scale demonstration and popularising of these new technologies for sustainable wastewater phosphorus removal.

3 High Rate Full Scale Biological Nutrient Removal Systems

High-rate nutrient removal systems are increasingly essential in wastewater treatment to meet stringent effluent standards and protect water bodies from nutrient pollution. These systems are designed to remove nitrogen and phosphorus effectively while minimizing space requirements and operational costs. The recent advancements and applications of high-rate full-scale nutrient removal systems include reactors like granular sludge reactors, Moving Bed Biofilm Reactors (MBBRs), Integrated fixed film reactors, etc. The MBBRs combine the advantages of activated sludge and biofilm systems, to achieving high-rate nitrogen removal with a smaller footprint. Recent studies focus on optimizing biofilm carriers and aeration strategies for improved nitrogen removal efficiency [52]. The Integrated Fixed-Film Activated Sludge systems have gained popularity for high-rate nitrogen removal. They incorporate biofilm carriers into activated sludge, providing additional biomass without

Table 1 Novel P-removal technologies and relative performance at different scales

Technology	Description	TP-removal rates (%)	Phase of development
SBBR	Sequencing batch reactor with fixed biofilm	70–90	Pilot-scale
Sequencing batch moving bed membrane bioreactor	Sequencing batch reactor with integrated membrane, moving-bed Carriers	84	Bench-scale
SBBR	Sequencing batch reactor with fixed biofilm	90	Bench-scale
MBR-UCT	Membrane bioreactor integrated into a continuous-flow EBPR	Up to 88	Full scale
MB-SBBR	Moving bed sequencing batch biofilm reactor	97	Bench-scale
Granular sludge	Advanced activated sludge process operated as SBR	87	Full-scale
MABR-SBR hybrid	Membrane aerated biofilm reactor operating in combination with a sequencing batch reactor	90	Bench-scale
AnoxAn	Vertical flow anaerobic-anoxic reactor	89	Bench-scale

Source Bunce et al. [51]

increasing reactor size. Recent full-scale implementations demonstrate their effectiveness in nitrogen removal under various operational conditions. In contrast to phosphorus, achieving stable and economical control of nitrogen to a low level through tertiary treatment is challenging, making it the primary concern. Several shortcut nitrogen removal processes including nitritation–denitritation and deammonification, have shown promise in high-rate nitrogen removal. They reduce aeration energy requirements and sludge production.

Full-scale studies reported successful implementation in municipal wastewater treatment plants [28]. Removing nutrients from wastewater is an essential strategy for controlling eutrophication and protecting water resources, as noted by Zhou et al. [13]. In full-scale wastewater treatment plants (WWTPs), the anaerobic/anoxic/aerobic (AAO) process is the most widely adopted BNR technique, owing to its effectiveness in simultaneously eliminating organic substances and nutrients. Most WWTPs are tasked with balancing two contradictory goals: reducing pollution and conserving energy. The significant economical alternatives to expanding or upgrading existing WWTPs are continuous monitoring and employing advanced process control (APC) approaches [53]. The APC systems, integrates real-time monitoring and automated control, are crucial for optimizing high-rate BNR systems. They adjust operational parameters dynamically, enhancing system performance and resilience. Various operational strategies have been developed, focusing on routing inflow organic material to anoxic and anaerobic zones [53]. This approach aids in denitrification and phosphate release, thereby improving BNR. Moreover, the application of alternating aeration/non-aeration (A/N-A) cycles elevates denitrification rates and augments the EBPR processes. Such A/N-A tactics, which offer flexible management of simultaneous nitrification and denitrification processes, are commonly employed in smaller WWTPs studies [53, 54].

In larger WWTPs, serving populations over 100,000, the availability of additional staff and technical resources allows for the implementation of more complex and customized APCs. However, with increased complexity, there is a greater need for sophisticated tools to maintain control quality and effectiveness [53]. Integrating traditional controls with advanced techniques such as knowledge-based models, neural networks, or fuzzy logic tools can significantly improve process control and reliability. This enhancement in control and reliability was reported in many researches [53, 55].

3.1 High-Rate Phosphorus Removal Technologies

High-Rate Enhanced Biological Phosphorus Removal (HR-EBPR) systems are designed to increase the selection pressure for PAOs, enhancing phosphorus removal rates. Recent full-scale studies highlight the importance of controlling operational parameters to optimize PAO activity [56]. In another work, the MBRs have been adapted for high-rate phosphorus removal, often coupled with phosphorus recovery technologies such as struvite precipitation. These systems remove phosphorus efficiently and enable resource recovery [57].

The integrated systems and innovative approaches include Hybrid BNR Systems: Combining different nitrogen and phosphorus removal technologies in a single system can achieve high-rate nutrient removal. Full-scale studies demonstrate the feasibility and efficiency of hybrid systems in various wastewater treatment scenarios. Recent trends emphasize not just nutrient removal but also recovery. Technologies like struvite precipitation and algae-based systems are explored in full-scale implementations for sustainable nutrient management [57].

While high-rate nutrient removal systems offer significant benefits, challenges like process stability, energy consumption, and cost-effectiveness need further addressing. Future research should optimize system design, explore energy-neutral or energy-positive approaches, and integrate advanced technologies like artificial intelligence for smarter process control. High-rate full-scale nutrient removal systems represent a critical advancement in wastewater treatment, addressing the pressing need for efficient nutrient removal. Continuous innovation and research in this field are essential to enhance system performance, sustainability, and adaptability to diverse wastewater treatment requirements.

3.2 Granular Sludge Systems

Aerobic granular sludge (AGS) systems offer the advantage of simultaneous nitrification, denitrification, and phosphorus removal in a single reactor. It is a novel biofilm system with no biocarriers (attachment). The compact granules facilitate the creation of different oxygen zones, enabling different biological processes within the same reactor, and thus offer space reduction (70%), energy savings (50%) and excellent effluent quality [58]. The granular sludge process is practiced for several decades and is now recognized as an environmentally friendly and cost-effective method for large-scale biological wastewater treatment. The excellent settling properties of the sludge, dense microbial composition, ability to withstand shock loadings, high retention of biomass and low sludge production, greater tolerance to toxins, and improved adsorption capacity of toxins and heavy metals, makes the process very popular and appropriately feasible among treatment of complex wastewater [59]. Granular technology presents considerable benefits over conventional treatment approaches, as Sarma and Tay [60] observed. It requires just 25% of the space needed for traditional activated sludge facilities and reduces operational expenses, energy usage, and sludge output by up to 30%. However, one significant drawback of this method is its prolonged time to form stable granules. Several elements affect the process of aerobic granulation, such as the design of the reactor, the type of seed sludge used, the carbon source, hydrodynamic selective pressure, the rate at which organic matter is loaded, the time the water is retained in the system, and the generation of extracellular polymeric substances.

The granule formation mechanism is still being researched for full-scale and commercial applications [61]. The formation of microbial aggregates and biofilms is a complex process significantly influenced by physical, chemical, and biological factors. This process, as summarised by Wilén et al. [62], involves four critical steps: (i) Microbes form aggregates influenced by hydrodynamic, diffusion, gravity, and thermodynamic forces; (ii) Aggregation is initiated through various forces including physical (e.g., van der Waals forces), chemical (e.g., ionic pairing), and biochemical interactions; (iii) The secretion of extracellular polymeric substances (EPS) acts as a 'biological glue', essential for aggregate formation; (iv) Hydrodynamic shear forces play a role in stabilizing the granules' structure [63]. Figure 6 shows the sketch of the granulation of aerobic granular sludge. The success of this granulation process, crucial for the stability of granules, depends on the effective management of microbial

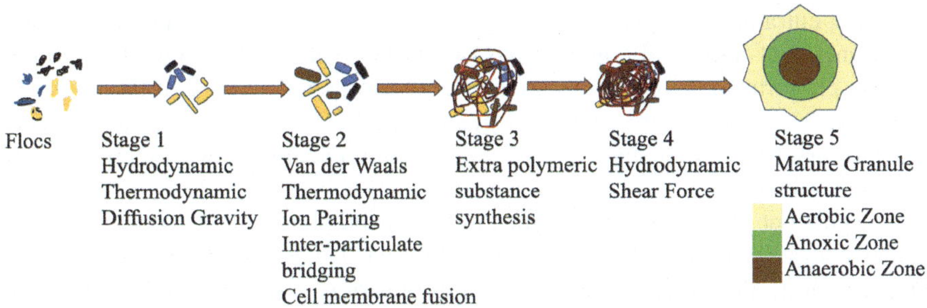

Fig. 6 Schematic of granulation of aerobic granular sludge Adapted from Hamza et al. [58]

activities. This can be achieved by adjusting operational and environmental parameters [64]. Properly cultivated aerobic granules result in stable structures capable of efficiently removing organics, nitrogen, and phosphorus.

The microbial community structure within granular systems is studied by [65]. Understanding the dynamics of microbial structure within granular systems is crucial for optimizing the granulation process and enhancing nutrient removal efficiency [65]. It was reported that diverse microbial populations within granules play a key role in achieving high treatment efficiency.

Even though AGS biotechnology is associated with many advantages, there are few inevitable disadvantages for the process implementation. The key limitation hindering its widespread implementation is the long-term operational instability and the tendency for granule disintegration, as indicated in studies [58]. This decline in granule stability over time significantly affects wastewater treatment efficiency, posing a major challenge for the effective use of AGS in large-scale operations [66]. Furthermore, cultivating granules from municipal wastewater often takes several months [67]. Most AGS biotechnology applications have also been in Sequencing Batch Reactors (SBR). Transitioning to a continuous-flow regime could enable retrofitting existing WWTPs with minimal infrastructural modifications, but AGS in continuous flow system is yet to develop for full scale applications.

3.3 *Biofilm Systems*

Biomass immobilization in biofilm reactors represents a significant advancement in the field of wastewater treatment and bioremediation. This technique involves the confinement of microbial cells within a specific medium, allowing them to form a biofilm. Biofilms are communities of microorganisms that adhere to each other and to surfaces or interfaces, often enclosed within a self-produced matrix of extracellular polymeric substances. This environment provides enhanced stability and protection for the microbes, allowing them to efficiently carry out metabolic processes. The immobilization process can be achieved through various methods, including entrapment in gels or attachment to solid carriers. These methods ensure the retention of microbial communities within the reactor environment, providing a stable platform for biofilm formation. The diversity and activity of the microbial community in a biofilm are crucial for treatment efficiency. These communities can adapt to varying environmental conditions and exhibit resilience against physical and chemical stresses.

The concept of biomass immobilization has been increasingly recognized as an effective strategy to retain slow-growing AOB and Anammox in reactors [68, 69]. This technique also protects these bacteria from various environmental changes [70, 71]. Immobilisation of biomass involves the deliberate enclosure of microbial cells within a certain area for specific metabolic activities. This process is generally divided into two categories: encapsulating biomass in polymeric gel carriers and employing carrier-based biofilm immobilization [71, 72]. The exploration of various immobilization methods for AAOB and the evaluation of key properties of gel carriers, including their solubility and sustainability, are crucial, as discussed by [73]. Materials such as nonwoven fabric and granular activated carbon are commonly used in carrier-based approaches.

In recent years, the immobilization of biomass has become a significant strategy to preserve slow-growing Anammox and AOB in reactors, as discussed by [68, 69]. This technique maintains these bacteria and protects them from environmental changes. Biomass immobilization involves confining microbial cells within a defined space for specific metabolic activities. This can be done in two ways: encapsulating biomass in poly gel carriers or utilizing carrier-based biofilm immobilization [71, 72]. Exploring different methods to immobilize AAOB and identifying essential features of gel carriers, like solubility and sustainability, are crucial, as noted by Wang et al. [73]. Carrier-based biofilm immobilization, using materials like nonwoven fabric and granular activated carbon, significantly influences microbial growth and community structure [74]. Various Anammox immobilization methods and how AAOB behaves in different materials are reported by Ahmad et al. [75].

In biofilm reactors, immobilized biomass is used for various applications, including organic matter degradation, nitrification, denitrification, and phosphorus removal. The reactors can be designed in different configurations, such as moving bed biofilm reactors (MBBR) or membrane aerated biofilm reactors (MABR), to suit specific treatment goals. The MBBR technology, first implemented in 1990 in Lardnal, Norway, represents a fusion of attached growth and suspended growth methods in aerobic wastewater treatment. This technology operates in a two-phase (anoxygenic) or three-phase (oxygenic) system using floating plastic biofilm carriers [76, 77]. These carriers are kept in motion throughout the reactor, either through mechanical mixing or aeration. Critical factors for MBBR's performance include media availability, surface area of the bio-carriers, dissolved oxygen, and organic load.

MBBR merges the advantages of adherent and suspended growth methods, providing an advanced, rapid treatment technique with remarkable efficiency. It is more cost-effective for investment, operation, maintenance, and replacement than traditional approaches, establishing it as a stable and strong alternative [78–80].

Like the activated sludge process, its functioning involves microorganisms attached to plastic carriers within the reactor, as observed by [81]. A primary benefit of MBBR over activated sludge and other biofilm reactors is its elimination of sludge recycle; excess biomass is disposed of without needing circulation. The system adeptly manages fluctuating organic and inorganic loads, achieving removal rates for TKN, COD, and BOD usually in the range of 40–85%, 40–85%, and 75–97%, respectively, according to Barwal and Chaudhary [82].

MBBR also facilitates oxidation of carbon along with nitrification and denitrification, either in a single stage or a combined approach [83]. Despite its effectiveness at both laboratory and large scales, a notable limitation of MBBR is the need for carrier separation to maintain reactor functionality [84]. MABRs have emerged as a promising technology for nitrogen removal, delivering oxygen through gas-permeable membranes to biofilms. This approach has demonstrated increased treatment efficiency and reduced energy consumption compared to conventional methods [85]. This allows for simultaneous nitrification and denitrification as different biofilm layers can be exposed to different oxygen concentrations. MABRs can achieve high nutrient removal efficiencies while using significantly less energy than conventional aeration methods. MABR relies on a spirally wound, air-permeable membrane that is submerged in the wastewater. The biofilm grows on the surface of this membrane utilizing the oxygen diffusing out of the membrane for aerobic treatment processes. The oxygen is supplied directly to the biofilm through the membrane, which allows for more efficient oxygen transfer and significantly reduces the energy required for aeration. MABR typically produces less sludge than conventional systems, including MBBR, making it a more sustainable option.

While biomass immobilization offers numerous advantages, it also presents challenges such as optimizing carrier materials and controlling biofilm thickness. Ongoing research is focused on enhancing the efficiency of these systems, exploring novel materials for immobilization, and understanding the complex interactions within the biofilm. The biomass immobilization in biofilm reactors is a dynamic and evolving field, offering promising solutions for efficient and sustainable wastewater treatment. Its success hinges on a deep understanding of microbial ecology, reactor design, and the interplay between various environmental factors.

3.4 Microalgal Technologies

Treatments based on microalgae offer the dual benefit of wastewater treatment and resource recovery, creating valuable biomass with a broad range of applications. This approach is particularly effective in wastewater treatment as it utilizes microalgae to absorb inorganic nutrients like nitrogen and phosphorus. Simultaneously, microalgae fix atmospheric carbon dioxide (CO_2) in the presence of light [86]. This process not only helps prevent eutrophication but also supports water reclamation efforts. Additionally, microalgae-based treatments stand out for their impressive bioaccumulation efficiency and high rate of biomass production, as highlighted by Díaz et al. [87].

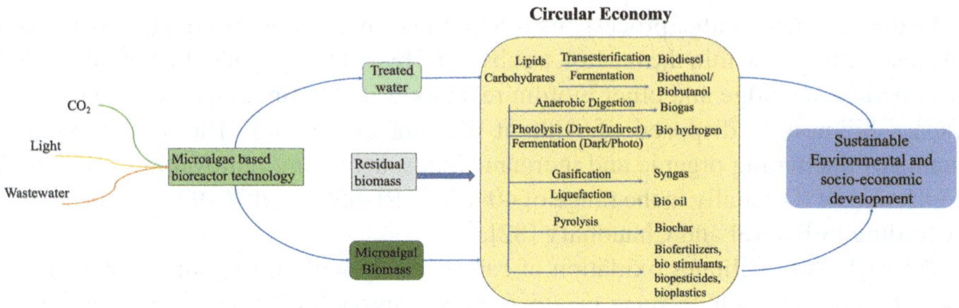

Fig. 7 Circular economy approach in wastewater treatment for resource recovery using microalgae-based bioreactor systems. Adapted from Diaz et al. [87]

This nutrient removal process reduces the concentrations of these pollutants in the treated effluent. Figure 7 depicts circular economy approach in wastewater treatment for resource recovery using microalgae-based bioreactor systems.

Microalgae biomass, derived from wastewater treatment, is notably high in lipids (10–50% of microalgae cells) and carbohydrates. These components are pivotal in producing various biofuels, which serve as energy sources in industries and transportation, contributing to 42% of current biofuels as reported by European Union [88]. Biodiesel, for instance, is produced through the transesterification of lipids, while bioethanol and biobutanol are generated from fermenting carbohydrates like glucose, starch, and cellulose [87, 89]. Moreover, the residual biomass from microalgae post oil extraction is utilized via anaerobic digestion for producing biogas creating biohydrogen by direct and indirect bio-photolysis or fermentation processes, generating syngas via gasification, converting to bio-oil through liquefaction, and forming biochar by pyrolysis. Additionally, the by-products of processing microalgae biomass are transformed into biofertilizers, biostimulants, biopesticides, or serve as sustainable raw materials for producing bioplastics [87, 90].

Significant research has been conducted on algae growth in various types of wastewater, including municipal, agricultural, and industrial [87]. Microalgae can thrive in saline water and arid regions, exhibiting rapid growth. This growth contributes to air quality improvement by reducing greenhouse gases. Microalgae's capability to absorb carbon dioxide and remove nutrients positions it as a viable option for wastewater treatment. The residual biomass from microalgae, post-lipid extraction, can be repurposed as animal feed or crop fertilizer [87].

Microalgae, encompassing cyanobacteria and eukaryotic algae, engage in photosynthesis to produce oxygen, which a microbial consortium utilizes to break down organic matter into simpler inorganic nutrients [17]. These nutrients are then taken up by the microalgae, promoting their growth. Gupta et al. [91] reported how integrating microalgae in Constructed Wetlands (CW) as a secondary treatment can boost nutrient removal. This approach effectively eliminated up to 86% of phosphates, 95% of nitrates, 74% of ammonium, and 87% of COD from concentrated wastewater.

Additionally, improvement in nitrogen removal is observed when combining heterotrophic microbes with microalgae [17]. In such systems, the oxygen generated by microalgae through photosynthesis is consumed by heterotrophic aerobic bacteria to degrade organic substances in wastewater, while the microalgae use the CO_2 released by these bacteria. The coexistence of autotrophic, heterotrophic, mixotrophic, and photoheterotrophic microorganisms alongside microalgae facilitates various applications owing to their diverse metabolic functions. This diversity enables these mixed cultures to adjust to various environmental conditions and nutrient level [17].

The use of microalgae technology in nutrient removal has recently reported the characteristics of absorbing CO_2 during photosynthesis by upgrading biogas and converting harvested microalgae into renewable energy sources like biogas in integrated systems [92]. Microalgae cultivation can occur in either open pond systems or closed photobioreactors, with the choice depending on the application's scale and the desired level of control, according to Li et al. [93].

Microalgae-based BNR systems can be integrated into existing WWTP to complement conventional processes. This integration can potentially reduce energy consumption and operational costs [94]. Effective monitoring and control strategies are essential for optimizing nutrient removal efficiency and preventing issues such as overgrowth of undesirable algae species. Figure 8 represents the formation of microalgal-bacterial biofilm in the presence and the absence of substratum. The microalgal-bacterial biofilms are classified in to three based on their substrate: 2D surface-based, 3D structure-based, and substrate-free (like microalgal-bacterial granular sludge, MBGS). The 2D-based biofilms show enhanced microalgae content and nutrient removal but are impractical for large-scale use due to lengthy processing times. In contrast, MBGS, with lower microalgae content, demonstrate superior pollutant elimination and practical processing times, although their reliance on natural sunlight needs further investigation. The 3D structure-based systems show varied results due to diverse substrate designs. Both 2D and 3D systems exhibit similarities in microbial community structures, whereas MBGS biofilms are richer in EPS-producing microbes, aiding in biofilm formation and denitrification. Further research is needed to understand the specific role of layered EPS in nutrient absorption and condensation. Challenges for microalgal-bacterial biofilm include potential decoupling of microalgal-bacterial synergy in low light, limited control over microbial communities, and increased greenhouse gas emissions from microalgal N_2O production.

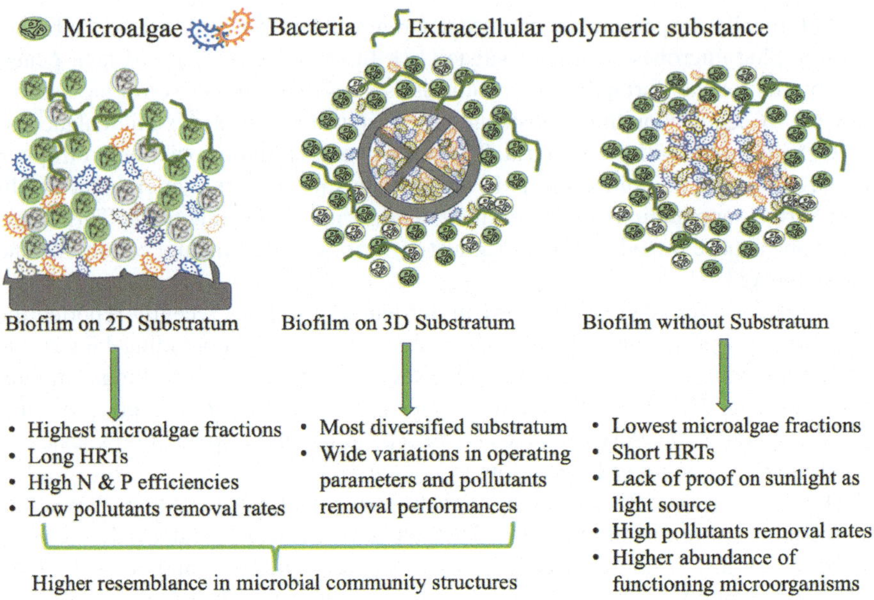

Fig. 8 Microalgal-bacterial biofilm based on substratum configurations. As adapted by Zhang et al. [95]

4 Challenges Associated with Nutrient Removal Technologies

Despite significant advances in understanding and applying nutrient removal techniques in wastewater, there are still considerable obstacles to their widespread realization and in attaining sustainability. The conventional BNR processes are reported with many challenges like (1) energy consumption, (2) operational complexity, (3) sludge production, (4) space requirements, (5) sensitivity to influent variability, (6) chemical usage, (7) greenhouse gas emissions, (8) maintaining microbial health, (9) cost of upgrading and retrofitting and (10) regulatory compliance. To address these challenges many advanced treatment techniques solving one or few challenges was developed. The modifications require careful design, operation, and maintenance of BNR systems and ongoing research and development to improve their efficiency and sustainability.

The performance of the Partial Nitritation (PN) process in treating low-strength municipal wastewater is hindered by challenges in maintaining stability and controlling greenhouse gas (GHG) emissions. Furthermore, the inhibition of NOB during the treatment of wastewater with low carbon-to-nitrogen (C/N) ratios, often aggravated by intermittent aeration strategies, affects the practicality and upkeep of existing BNR processes. Additionally, the effectiveness of BNR processes is compromised at low temperatures and DO levels, owing to the reduced activity of AOB compared to NOB. This limitation restricts the broader application of these processes in mainstream wastewater treatment.

Integration of available microbial consortium suitably and appropriately was found to be more beneficial in terms of improving the efficiency of the treatment. For example: the integration of partial nitritation and anammox for effective nitrogen removal. The integrated process is very promising due to its potential to reduce the energy required for aeration [96]. However, the full operation becomes tedious due to the slow growth of anammox bacteria at lower temperatures, especially in colder climates. The long-term stability of this process requires more research, particularly in maintaining stable nitrite production for the effective functioning of attached partial nitritation and anammox systems.

While the anammox process has several advantages, it also has a major drawback: the tendency to produce high nitrate concentrations in the effluent. Consequently, a post-denitrification step is necessary to meet discharge standards requiring external carbon sources. This additional step leads to increased energy use, more activated sludge, and CO_2 emissions [17]. Other challenges for implementing anammox in mainstream wastewater treatment include a high C/N ratio, lower temperatures, and poor effluent quality. The presence of organic matter can lead to competition between heterotrophic denitrifying bacteria and anammox bacteria, reducing ammonium removal efficiency. Additionally, certain organic chemicals, like methanol, can partially or completely inhibit anammox activity [17].

Considering the current gaps and research issues, future studies on the full-scale implementation of AGS should concentrate on four key areas: (1) Innovations in reactor/system design that facilitate rapid granulation under continuous flow conditions, taking into account factors such as granule selection pressure mechanisms, efficient nutrient and particulate matter removal, the structure of microbial communities, and the influence of operational conditions on key organisms, as well as aspects of system control, equipment, and operational management; (2) Effective management of AGS waste sludge; (3) Exploring the use of AGS in the bioremediation of toxic pollutants, adsorption of heavy metals, and treatment of emerging contaminants; and (4) Developing and scaling up technologies for recovering valuable resources from excess AGS sludge, focusing on maximizing resource production, and improving extraction, characterization, and purification methods, along with assessing the market potential for these resources [11].

While offering efficient wastewater treatment, biofilm reactors face several challenges that can impact their performance and applicability. Controlling the thickness of the biofilm is crucial. Excessively thick biofilms can lead to reduced mass transfer efficiency and uneven distribution of nutrients and oxygen, thereby affecting the reactor's overall performance. Biofilms are susceptible to hydraulic shear stress, which can cause detachment of biofilm from the reactor surfaces. This detachment can lead to a loss of biomass and, consequently, a decrease in treatment efficiency. Over time, biofilm reactors can experience clogging and channelling, especially in systems with small carrier media. This can reduce the flow and mixing efficiency, leading to reduced treatment effectiveness and increased maintenance requirements.

The thickness of the biofilm innately creates issues in the diffusion of substrates and oxygen into the biofilm and in the removal of metabolic products. This leads to reduced metabolic activity and efficiency of the biofilm [85]. Biofilm reactors can be operationally intensive and require regular maintenance to prevent issues like clogging and channelling. This can increase the overall operational costs of the treatment system. Overcoming these challenges requires a deep understanding of biofilm science, careful design and operation of biofilm reactors, and ongoing research and innovation in the field. Furthermore, there are numerous hurdles to overcome for large-scale implementation of microalgae-based methods. This approach requires extensive land for constructing water bodies. The final consideration is the economic viability of the method to ensure consistent nutrient removal along with life cycle assessment of the technology.

The removal of phosphorus from wastewater is critical for preventing eutrophication in water bodies, but it also has several challenges like: (1) chemical demand, (2) sludge production, (3) process control, (4) effluent quality variability, (5) microbial population management, (6) recovery and reuse challenges, (7) energy consumption, (8) limited resource, (9) integration with existing systems and (10) environmental and health risk. Addressing these challenges requires a combination of technological innovation, efficient process management, and ongoing research to develop more sustainable and cost-effective phosphorus removal methods.

5 Future Research Recommendations

Future research in revolutionising wastewater treatment through cutting-edge technologies for biological nutrient removal (BNR) should address several critical avenues to advance the field. Firstly, microbial ecology and genomics present promising directions for exploration. Investigating the microbial communities involved in BNR processes at a molecular level can offer insights into their dynamics and functions. Advanced genomic techniques such as metagenomics, metatranscriptomics, and proteomics can unravel microbial interactions and metabolic pathways [97]. Additionally, there is potential in exploring synthetic biology approaches for engineering microbial consortia with enhanced nutrient removal capabilities [98].

Secondly, artificial intelligence (AI) and machine learning offer opportunities for optimising BNR processes in real-time. Developing AI-driven models considering dynamic influent characteristics and changing environmental conditions can enhance process efficiency [98]. Integration of machine learning algorithms with sensor technologies holds promise for predictive maintenance and fault detection in wastewater treatment plants. Moreover, exploring reinforcement learning techniques for adaptive control and optimization of BNR systems can further improve performance [99].

Thirdly, investigating advanced treatment technologies such as electrochemical processes, photocatalysis, and nanotechnology could enhance nutrient removal efficiency [100, 101]. Novel reactor configurations and hybrid systems integrating multiple treatment processes may achieve synergistic effects. Evaluating these technologies' scalability and practical feasibility for large-scale implementation in wastewater treatment plants is essential.

Fourthly, focusing on resource recovery and circular economy principles is imperative. Innovative approaches for recovering nutrients from wastewater streams for reuse in agriculture and other applications should be explored [102]. Assessing the environmental and economic benefits of circular economy principles in wastewater treatment, including valorization of by-products and waste-to-energy conversion, is crucial. Investigating decentralized wastewater treatment systems for resource recovery and decentralised nutrient management presents promising avenues [103].

Lastly, addressing climate change resilience and sustainability is paramount. Assessing the vulnerability of wastewater treatment infrastructure to climate change impacts and developing adaptive strategies to enhance resilience are necessary steps. Exploring the potential of nature-based solutions, such as constructed wetlands and green infrastructure for climate-resilient wastewater treatment, is essential [104]. Conducting comprehensive sustainability assessments, considering environmental, social, and economic dimensions, can guide decision-making in wastewater treatment planning and design [105].

These recommendations and the latest research references provide a roadmap for future investigations to advance the state-of-the-art in revolutionising wastewater treatment through cutting-edge technologies for biological nutrient removal.

6 Conclusions

In summary, this chapter's exploration of cutting-edge technologies for biological nutrient removal highlights their pivotal role in revolutionising wastewater treatment. From the foundations of natural and conventional processes to the advancements in granular biomass and biofilm-based systems and the innovative use of microalgae, each method presents unique benefits and addresses the critical need to remove excessive nutrients from wastewater. These technologies enhance the efficiency of nitrogen and phosphorus removal and contribute to the sustainability of water management systems.

Despite the significant progress made, the chapter also acknowledges the ongoing challenges these technologies face, including operational complexity, cost implications, and environmental concerns. It highlights the necessity for continued research and development to optimize these systems, making them more accessible, efficient, and environmentally friendly. The incorporation of advanced nutrient removal technologies represents a significant step towards minimizing nutrient discharge into aquatic ecosystems, thus preserving water quality and promoting ecological balance.

The future of wastewater treatment lies in harnessing these innovative technologies, continually refining them, and integrating them seamlessly into existing infrastructure. As the field progresses, these systems are anticipated to become more robust, cost-effective, and adaptable to varying wastewater compositions and volumes.

This evolution is crucial for achieving sustainable development goals in wastewater management and for protecting our water resources for future generations. Key advancements in biological nutrient removal discussed in this chapter represent an achievement in environmental engineering and a commitment to a cleaner, healthier environment. The journey towards optimizing these technologies continues, driven by scientific inquiry and a dedication to take care of the environment.

References

1. Ikram M, Naeem M, Zahoor M, Hanafiah M M, Oyekanmi A A, Ullah R, Al Farraj DA, Elshikh MS, Zekker I, Gulfam N (2022) Biological degradation of the azo dye basic orange 2 by *Escherichia coli*: a sustainable and eco-friendly approach for treating textile wastewater. Water 14(13):2063. https://www.mdpi.com/2073-4441/14/19/2969
2. Rout PR, Shahid MK, Dash RR, Bhunia P, Liu D, Varjani S, Zhang TC, Surampalli R (2021) Nutrient removal from domestic wastewater: a comprehensive review on conventional and advanced technologies. J Environ Manage 296:113246. https://doi.org/10.1016/j.jenvman.2021.113246
3. Varjani S, Upasani VN (2019) Influence of abiotic factors, natural attenuation, bioaugmentation and nutrient supplementation on bioremediation of petroleum crude contaminated agricultural soil. J Environ Manage 245:358–366. https://doi.org/10.1016/j.jenvman.2019.05.070
4. Varjani SJ (2017) Review: microbial degradation of petroleum hydrocarbons. Biores Technol 223:277–286. https://doi.org/10.1016/j.biortech.2016.10.037
5. Mishra B, Varjani S, Agarwal DC, Mandal SK, Ngo HH, Taherzadeh MJ, Chang JS, You S, Guo W (2020) Engineering biocatalytic material for the remediation of pollutants: a comprehensive review. Environ Technol Innov 20:101063. https://doi.org/10.1016/j.eti.2020.101063
6. Saravanan A, Kumar PS, Varjani S, Jeevanantham S, Yaashikaa PR, Thamarai P, Abirami B, George CS (2021) A review on algal-bacterial symbiotic system for effective treatment of wastewater. Chemosphere 129540. https://doi.org/10.1016/j.chemosphere.2021.129540
7. Varjani S, Upasani VN (2021) Bioaugmentation of *Pseudomonas aeruginosa* NCIM 5514—a novel oily waste degrader for treatment of petroleum hydrocarbons. Biores Technol 319:124240. https://doi.org/10.1016/j.biortech.2020.124240
8. Varjani SJ, Upasani VN (2017) A new look on factors affecting microbial degradation of petroleum hydrocarbon pollutants. Int Biodeterior Biodegradation 120:71–83. https://doi.org/10.1016/j.ibiod.2017.02.006
9. Wong A (2015) Investigating the enhancement of biological filtration with capping material designs and nutrient amendments. UWSpace. http://hdl.handle.net/10012/9758
10. Khanal SK, Varjani S, Lin CSK, Awasthi MK (2020) Waste-to-resources: opportunities and challenges. Biores Technol 317:123987. https://doi.org/10.1016/j.biortech.2020.123987
11. Hasan MN, Altaf MM, Khan NA, Khan AH, Khan AA, Ahmed S, Kumar PS, Naushad M, Rajapaksha AU, Iqbal J, Tirth V, Islam S (2021) Recent technologies for nutrient removal and recovery from wastewaters: a review. Chemosphere 277:130328. https://doi.org/10.1016/j.chemosphere.2021.130328
12. EPA U (2009) The US Environmental Protection Agency's strategic plan for evaluating the toxicity of chemicals. In: EPA 100/K-09/001, March 2009. Office of the Science Advisor. Science Policy Council
13. Zhou Z, Wang K, Qiang J, Pang H, Yuan Y, An Y, Zhou C, Ye J, Wu Z (2021) Mainstream nitrogen separation and side-stream removal to reduce discharge and footprint of wastewater treatment plants. Water Res 188:116527. https://doi.org/10.1016/j.watres.2020.116527

14. Spiller M, Moretti M, Paepe DJ, Vlaeminck SE (2022) Environmental and economic sustainability of the nitrogen recovery paradigm: evidence from a structured literature review. Resour Conserv Recycl 184:106406. https://doi.org/10.1016/j.resconrec.2022.106406
15. Qin Y, Wang K, Zhou Z, Yu S, Wang L, Xia Q, Zhao X, Zhou C, Ye J, Wu Z (2023) Nitrogen recovery from wastewater as nitrate by coupling mainstream ammonium separation with side stream cyclic up-concentration and targeted conversion. Chem Eng J 455:140337. https://doi.org/10.1016/j.cej.2022.140337
16. Shahid MK, Kashif A, Rout PR, Aslam M, Fuwad A, Choi Y, Park JH, Kumar G (2020) A brief review of anaerobic membrane bioreactors emphasizing recent advancements, fouling issues and future perspectives. J Environ Manage 270:110909. https://doi.org/10.1016/j.jenvman.2020.110909
17. Rahimi S, Modin O, Mijakovic I (2020) Technologies for biological removal and recovery of nitrogen from wastewater. Biotechnol Adv, 107570. https://doi.org/10.1016/j.biotechadv.2020.107570
18. Winkler MKH, Bassin JP, Kleerebezem R (2018) Unraveling the reasons for disproportion in the ratio of AOB and NOB in aerobic granular sludge. ISME J 12(10):2353–2366. https://link.springer.com/article/10.1007/s00253-012-4126-9
19. Zhou Y, Pijuan M, Zeng RJ, Yuan Z (2020) Free nitrous acid serving as a terminal electron acceptor for enhancing biological phosphorus removal. Environ Sci Technol 54(5):3166–3176
20. Oehmen A, Yuan Z, Blackall LL, Keller J (2005) Comparison of acetate and propionate uptake by polyphosphate accumulating organisms and glycogen accumulating organisms. Biotechnol Bioeng 91(2):162–168. https://doi.org/10.1002/bit.20500
21. Xie W, Ni BJ, Seviour T, Sheng G, Yuan Z (2019) A review of the effect of salts on the microbial metabolism in anaerobic fermentation processes of waste/wastewater. Biores Technol 275:596–605
22. Chew KW, Chia SR, Show PL, Yap YJ, Ling TC, Chang JS (2018) Effects of water culture medium, cultivation systems and growth modes for microalgae cultivation: a review. J Taiwan Inst Chem Eng 91:332–344. https://doi.org/10.1016/j.jtice.2018.05.039
23. Mishra S, Singh V, Cheng L, Hussain A, Ormeci B (2022) Nitrogen removal from wastewater: a comprehensive review of biological nitrogen removal processes, critical operation parameters and bioreactor design. J Environ Chem Eng 10:107387. https://doi.org/10.1016/j.jece.2022.107387
24. Daigger GT (2014) Oxygen and carbon requirements for biological nitrogen removal processes accomplishing nitrification, nitritation, and anammox. Water Environ Res 86:204–209. https://doi.org/10.2175/106143013X13807328849459
25. Ma B, Liang Y, Zhang Y, Wei Y (2023) Achieving advanced nitrogen removal from low-carbon municipal wastewater using partial-nitrification/anammox and endogenous partial-denitrification/anammox. Biores Technol 383:129227. https://doi.org/10.1016/j.biortech.2023.129227
26. Zhou Y, Zhu Y, Zhu J, Li C, Chen G (2023) A comprehensive review on wastewater nitrogen removal and its recovery processes. Int J Environ Res Public Health 20(4):3429. https://www.mdpi.com/1660-4601/20/4/3429
27. Omar A, Almomani F, Qiblawey H, Rasool K (2024) Advances in nitrogen-rich wastewater treatment: a comprehensive review of modern technologies. Sustainability 16(5):2112. https://doi.org/10.3390/su16052112
28. Zhang X, Zhang X, Chen J, Wu P, Yang Z, Zhou L, Zhu Z, Wu Z, Zhang K, Wang Y, Ruth G (2023) A critical review of improving mainstream anammox systems: based on macroscopic process regulation and microscopic enhancement mechanisms. Environ Res 236(Part 1):116770. https://doi.org/10.1016/j.envres.2023.116770
29. Islam GM, Vi P, Gilbride KA (2019) Functional relationship between ammonia-oxidizing bacteria and ammonia-oxidizing archaea populations in the secondary treatment system of a full-scale municipal wastewater treatment plant. J Environ Sci 86:120–130. https://doi.org/10.1016/j.jes.2019.04.031

30. Cotto I, Dai Z, Huo L, Anderson CL, Vilardi KJ, Ijaz U, Khunjar W, Wilson C, De Clippeleir H, Gilmore K, Bailey E (2020) Long solids retention times and attached growth phase favor prevalence of comammox bacteria in nitrogen removal systems. Water Res 169:115268. https://doi.org/10.1016/j.watres.2019.115268
31. Yin S, Li J, Dong H, Qiang Z (2019) Enhanced nitrogen removal through marine anammox bacteria (MAB) treating nitrogen-rich saline wastewater with Fe (III) addition: nitrogen shock loading and community structure. Biores Technol 287:121405. https://doi.org/10.1016/j.biortech.2019.121405
32. Wang XX, Wang SY, Xue TL, Li B, Dai X, Peng Y (2015) Treating low carbon/nitrogen (C/N) waste-water in simultaneous nitrification-endogenous denitrification and phosphorous removal (SNDPR) systems by strengthening anaerobic intracellular carbon storage. Water Res 77:191–200. https://doi.org/10.1016/j.watres.2015.03.019
33. Chen H, Liu S, Yang F, Xue Y, Wang T (2009) The development of simultaneous partial nitrification, anammox and denitrification (SNAD) process in a single reactor for nitrogen removal. Biores Technol 100(4):1548–1554. https://doi.org/10.1016/j.biortech.2008.09.003. 18977657
34. Singh V, Ormeci B, Mishra S, Hussain A (2022) Simultaneous partial Nitrification, ANAMMOX and denitrification (SNAD): a review of critical operating parameters and reactor configurations. Chem Eng J 433:133677. https://doi.org/10.1016/j.cej.2021.133677
35. Chen Q, Ni J (2012) Ammonium removal by *agrobacterium* sp. LAD9 capable of heterotrophic nitrification–aerobic denitrification. J Biosci Bioeng 113(5):619–623. https://doi.org/10.1016/j.jbiosc.2011.12.012
36. Huang X, Li W, Zhang D, Qin W (2013) Ammonium removal by a novel oligotrophic *acinetobacter* sp. Y16 capable of heterotrophic nitrification–aerobic denitrification at low temperature. Biores Technol 146:44–50. https://doi.org/10.1016/j.biortech.2013.07.046
37. Anjali G, Sabumon PC (2023) Heterotrophic nitrification and aerobic denitrification process (Chap. 6). In: Shah M (ed) Ammonia oxidizing bacteria: applications in industrial wastewater treatment. Royal Society of Chemistry Publication. ISBN-13: 9781839166877. https://doi.org/10.1039/BK9781837671960-00103
38. Jia Y, Zhou M, Chen Y, Hu Y, Luo J (2020) Insight into short-cut of simultaneous nitrification and denitrification T process in moving bed biofilm reactor: effects of carbon to nitrogen ratio. Chem Eng J 400:125905. https://doi.org/10.1016/j.cej.2020.125905
39. Robertson LA, Niel EW, Van Torremans RA (1988) Simultaneous nitrification and denitrification in aerobic chemostat cultures of *Thiosphaera pantotropha*. Appl Environ Microbiol 54:2812–2818. https://doi.org/10.1128/aem.54.11.2812-2818.1988
40. Xi H, Zhou X, Arslan M, Luo Z, Wei J, Wu Z, El-Din MG (2022) Heterotrophic nitrification and aerobic denitrification process: promising but a long way to go in the wastewater treatment. Sci Total Environ 805:150212. https://doi.org/10.1016/j.scitotenv.2021.150212
41. Sun Z, Lv Y, Liu Y et al (2016) Removal of nitrogen by heterotrophic nitrification-aerobic denitrification of a novel metal resistant bacterium *cupriavidus* sp. S1. Biores Technol 220:142–150. https://doi.org/10.1016/j.biortech.2016.07.110
42. Huang F, Pan L, Lv N, Tang X (2017) Characterization of novel bacillus strain N31 from mariculture water capable of halophilic heterotrophic nitrification–aerobic denitrification. J Biosci Bioeng 124:564–571. https://doi.org/10.1016/j.jbiosc.2017.06.008
43. Wang H, Li J, Wang B et al (2020) Deciphering pollutants removal mechanisms and genetic responses to ampicillin stress in simultaneous heterotrophic nitrification and aerobic denitrification (SHNAD) process treating seawater-based wastewater. Bioresour Technol 315:123827. https://doi.org/10.1016/j.biortech.2020.123827
44. Schmidt I, Hermelink C, Van de Pas-Schoonen K, Strous M, Op den Camp HJ, Kuenen JG, Jetten MSM (2002) Anaerobic ammonia oxidation in the presence of nitrogen oxides (NO_x) by two different lithotrophs. Appl Environ Microbiol 68:5351–5357. https://doi.org/10.1128/AEM.68.11.5351-5357.2002
45. Ahn YH (2006) Sustainable nitrogen elimination biotechnologies: a review. Process Biochem 41:1709–1721. https://doi.org/10.1016/j.procbio.2006.03.033

46. Li WW, Zhang HL, Sheng GP, Yu HQ (2015) Roles of extracellular polymeric substances in enhanced biological phosphorus removal process. Water Res 86:85–95. https://doi.org/10.1016/j.watres.2015.06.034
47. Diaz R, Mackey B, Chadalavada S, Kainthola J, Heck P, Goel R (2022) Enhanced Bio-P removal: past, present, and future—a comprehensive review. Chemosphere 309(2):136518. https://doi.org/10.1016/j.chemosphere.2022.136518
48. Welles L, Lopez-Vazquez CM, Hooijmans CM, van Loosdrecht MCM, Brdjanovic D (2016) Prevalence of "*Candidatus Accumulibacter phosphatis*" type II under phosphate limiting conditions. AMB Express 6:44. https://amb-express.springeropen.com/articles/10.1186/s13568-016-0214-z
49. Randall CW, Barnard JL, Stensel HD (eds) (1998) Design and retrofit of wastewater treatment plants for biological nutritient removal, vol 5. CRC Press
50. Xie T, Mo C, Li X, Zhang J, An H, Yang Q et al (2017) Effects of different ratios of glucose to acetate on phosphorus removal and microbial community of enhanced biological phosphorus removal (EBPR) system. Environ Sci Pollut Control Ser 24(5):4494–4505
51. Bunce JT, Ndam E, Ofiteru ID, Moore A, Graham DW (2018) A review of phosphorus removal technologies and their applicability to small-scale domestic wastewater treatment systems. Front Environ Sci 6:8. https://doi.org/10.3389/fenvs.2018.00008
52. Le LT, Nghiem LD, Bui XT, Jahng D (2023) Improve nitrogen removal of the biofilm single-stage PN/A process by optimizing the intermittent aeration strategy. Environ Technol Innov 30:103078. https://www.sciencedirect.com/journal/environmental-technology-and-innovation/vol/30/suppl/C. https://doi.org/10.1016/j.eti.2023.103078
53. Palatsi J, Ripoll F, Benzal A, Pijuan M, Romero-Güiza MS (2021) Enhancement of biological nutrient removal process with advanced process control tools in full-scale wastewater treatment plant. Water Res 200:117212. https://doi.org/10.1016/j.watres.2021.117212
54. Santos JMM, Martins A, Barreto S, Rieger L, Reis M, Oehmen A (2020) Long-term simulation of a full-scale EBPR plant with a novel metabolic-ASM model and its use as a diagnostic tool. Water Res 187:116398. https://doi.org/10.1016/j.watres.2020.116398
55. Bernardelli A, Marsili-Libelli S, Manzini A, Stancari S, Tardini G, Montanari D, Anceschi G, Gelli P, Venier S (2020) Real time model predictive control of wastewater treatment plant based on machine learning. Water Sci Technol 81(11):2391–2400. https://doi.org/10.2166/wst.2020.298
56. Qiu G, Law Y, Montanez RZ, Deng X, Lu Y, Roy S, Thi SS, Hoon HY, Nguyen TQN, Eganathan K, Liu NPH, Williams RBH, Wuertz S (2022) Global warming readiness: feasibility of enhanced biological phosphorus removal at 35 °C. Water Res 216:118301. https://doi.org/10.1016/j.watres.2022.118301
57. Sheik M, Harami HR, Rezakazemi M, Cortina JL, Aminabhavi TM, Valderrama C (2023) Towards a sustainable transformation of municipal wastewater treatment plants into biofactories using advanced NH_3–N recovery technologies: a review. Sci Total Environ 904:166077. https://doi.org/10.1016/j.scitotenv.2023.166077
58. Hamza R, Rabii A, Ezzahraoui F-Z, Morgan G, Iorhemen OT (2022) A review of the state of development of aerobic granular sludge technology over the last 20 years: full-scale applications and resource recovery. Case Stud Chem Environ Eng 5:100173. https://doi.org/10.1016/j.cscee.2021.100173
59. Wei D, Si W, Zhang Y, Qiao Z, Yao Z, Zhao W, Zhao J, Chen G, Wei Q, Du B (2012) Aerobic granulation and nitrogen removal with the effluent of internal circulation reactor in start-up of a pilot-scale sequencing batch reactor. Bioprocess Biosyst Eng 35:1489–1496. https://link.springer.com/article/10.1007/s00449-012-0738-9
60. Sarma SJ, Tay JH (2018) Aerobic granulation for future wastewater treatment technology: challenges ahead. Environ Sci Water Res Technol 4:9–15. https://doi.org/10.1039/C7EW00148G
61. Kang AJ, Yuan Q (2017) Long-term stability and nutrient removal efficiency of aerobic granules at low organic loads. Biores Technol 234:336–342. https://doi.org/10.1016/j.biortech.2017.03.057

62. Wilén BM, Liébana R, Persson F, Modin O, Hermansson M (2018) The mechanisms of granulation of activated sludge in wastewater treatment, its optimization, and impact on effluent quality. Appl Microbiol Biotechnol 102:5005–5020. https://link.springer.com/article/10.1007/s00253-018-8990-9
63. De Kreuk MK, Kishida N, van Loosdrecht MCM (2007) Aerobic granular sludge—state of the art. Water Sci Technol 55(8–9):75–81
64. Weissbrodt DG, Neu TR, Kuhlicke U, Rappaz Y, Holliger C (2013) Assessment of bacterial and structural dynamics in aerobic granular biofilms. Front Microbiol 4:175. https://doi.org/10.3389/fmicb.2013.00175
65. Zhao Z, Jiang Y, Yan L, Huang Y, Guo J, Li Y (2021) Microbial community structure within granular systems: implications for optimizing the granulation process and enhancing nutrient removal efficiency. Front Microbiol 12:650285. https://doi.org/10.3389/fmicb.2021.650285
66. Show K-Y, Lee D-Y, Tay J-H (2012) Aerobic granulation: advances and challenges. Appl Biochem Biotechnol 167(6):1622–1640. https://doi.org/10.1007/s12010-012-9609-8
67. Nancharaiah YN, Reddy GKK (2018) Aerobic granular sludge technology: mechanisms of granulation and biotechnological applications. Biores Technol 247:1128–1143. https://doi.org/10.1016/j.biortech.2017.09.131
68. Yan J, Hu YY (2009) Partial nitrification to nitrite for treating ammonium-rich organic wastewater by immobilized biomass system. Biores Technol 100(8):2341–2347. https://doi.org/10.1016/j.biortech.2008.11.038
69. Ali M, Oshiki M, Rathnayake L, Ishii S, Satoh H, Okabe S (2015) Rapid and successful start-up of anammox process by immobilizing the minimal quantity of biomass in PVA-SA gel beads. Water Res 79:147–157. https://doi.org/10.1016/j.watres.2015.04.024
70. Reino C, Ojeda MES, Perez J, Carrera J (2018) Stable long-term operation of an upflow anammox sludge bed reactor at mainstream conditions. Water Res 128:331–340. https://doi.org/10.1016/j.watres.2017.10.058
71. Kumwimba MN, Lotti T, Şenel E, Li X, Suanon F (2020) Anammox-based processes: how far have we come and what work remains? A review by bibliometric analysis. Chemosphere 238:124627. https://doi.org/10.1016/j.chemosphere.2019.124627
72. Manonmani U, Joseph K (2018) Research advances and challenges in anammox immobilization for autotrophic nitrogen removal. J Chem Technol Biotechnol 93(9):2486–2497. https://doi.org/10.1002/jctb.5632
73. Wang J, Liang J, Ning D, Zhang T, Wang M (2022) A review of biomass immobilization in anammox and partial nitrification anammox systems: advances, issues, and future perspectives. Sci Total Environ 821:152792. https://doi.org/10.1016/j.scitotenv.2021.152792
74. Adams M, Xie J, Xie J, Chang Y, Guo M, Chen C, Zhang TC (2020) The effect of carrier addition on anammox start-up and microbial community: a review. Rev Environ Sci Biotechnol 19(2):355–368. https://doi.org/10.1016/j.scitotenv.2023.166398
75. Ahmad HA, Ni S, Ahmad S, Zhang J, Ali M, Ngo HH, Guo W, Tan Z, Wang Q (2020) Gel immobilization: a strategy to improve the performance of anaerobic ammonium oxidation (anammox) bacteria for nitrogen-rich wastewater treatment. Biores Technol 313:123642. https://doi.org/10.1016/j.biortech.2020.123642
76. Weiss JS, Alvarez M, Tang C, Horvath RW, Stahl JF (2005) Evaluation of moving bed biofilm reactor technology. Environ Times, 2085–2102. https://doi.org/10.2175/193864705783867035
77. Odegaard H (2006) Innovations in wastewater treatment: the moving bed biofilm process. Water Sci Technol 53:17–33. https://doi.org/10.2166/wst.2006.284
78. Leyva-Díaz JC, Monteoliva-García A, Martín-Pascual J, Munio MM, García-Mesa JJ, Poyatos JM (2020) Moving bed biofilm reactor as an alternative wastewater treatment process for nutrient removal and recovery in the circular economy model. Biores Technol 299:122631. https://doi.org/10.1016/j.biortech.2019.122631
79. Khan AH, Khan NA, Sirajuddin A, Aastha D, Chandra PS, Khan SU, Ali AM, Changani F, Yousefi M, Alam S, Vambol S, Vambol V, Khursheed A, Ali I (2020) Application of advanced oxidation processes followed by different treatment technologies for hospital wastewater treatment. J Clean Prod 269:122411. https://doi.org/10.1016/j.jclepro.2020.122411

80. Khan AH, Hamidi AA, Khan NA, Sirajuddin A, Mehtab S, Vambol S, Vambol V, Changani F, Islam S (2020) Pharmaceuticals of emerging concern in hospital wastewater: removal of Ibuprofen and Ofloxacin drugs using MBBR method. Int J Environ Analyt Chem. https://doi.org/10.1080/03067319.2020.1855333
81. Qiqi Y, Qiang H, Ibrahim HT (2012) Review on moving bed biofilm processes. Pak J Nutr 11:706. https://doi.org/10.3923/pjn.2012.804.811
82. Barwal A, Chaudhary R (2014) To study the performance of biocarriers in moving bed biofilm reactor (MBBR) technology and kinetics of biofilm for retrofitting the existing aerobic treatment systems: a review. Rev Environ Sci Biotechnol 13:285–299. https://doi.org/10.1007/s11157-014-9333-7
83. Malovanyy A, Yang J, Trela J, Plaza E (2015) Combination of upflow anaerobic sludge blanket (UASB) reactor and partial nitritation/anammox moving bed biofilm reactor (MBBR) for municipal wastewater treatment. Biores Technol 180:144–153. https://doi.org/10.1016/j.biortech.2014.12.101
84. Asri M, Elabed S, Ibnsouda Koraichi S, ElGhachtouli N, (2018) Biofilm-based systems for industrial wastewater treatment. In: Hussain CM (ed) Handbook of environmental materials management. Springer, Cham, pp 1767–1787
85. Li J, Feng M, Zheng S, Zhao W, Xu X, Yu X (2023) The membrane aerated biofilm reactor for nitrogen removal of wastewater treatment: principles, performances, and nitrous oxide emissions. Chem Eng J 460:141693. https://doi.org/10.1016/j.cej.2023.141693
86. Arora K, Kaur P, Kumar P, Singh A, Patel SKS, Li X, Yang YH, Bhatia SK, Kulshrestha S (2021) Valorization of wastewater resources into biofuel and value-added products using microalgal system. Front Energy Res 9:119. https://doi.org/10.3389/FENRG.2021.646571/BIBTEX
87. Díaz V, Leyva-Díaz JC, Almécija MC, Poyatos JM, Munío MM, Martín-Pascual J (2022) Microalgae bioreactor for nutrient removal and resource recovery from wastewater in the paradigm of circular economy. Bioresour Technol 363:127968. https://doi.org/10.1016/j.biortech.2022.127968
88. European Union (2018) Blue bioeconomy, situation report and perspectives. https://www.eumofa.eu
89. Kumar A (2021) Current and future perspective of microalgae for simultaneous wastewater treatment and feedstock for biofuels production. Chem Afr 4(2):249–275. https://doi.org/10.1007/S42250-020-00221-9
90. Acién Fernández FG, Reis A, Wijffels RH, Barbosa M, Verdelho V, Llamas B (2021) The role of microalgae in the bioeconomy. New Biotechnol 61:99–107. https://doi.org/10.1016/J.NBT.2020.11.011
91. Gupta S, Srivastava P, Yadav AK (2020) Simultaneous removal of organic matters and nutrients from high-strength wastewater in constructed wetlands followed by entrapped algal systems. Environ Sci Pollut Res 27:1112–1117. https://doi.org/10.1007/s11356-019-06896-z
92. Nguyen TDP, Le TVA, Show PL, Nguyen TT, Tran MH, Tran TNT, Lee SY (2019) Bioflocculation formation of microalgae-bacteria in enhancing microalgae harvesting and nutrient removal from wastewater effluent. Biores Technol 272:34–39. https://doi.org/10.1016/j.biortech.2018.09.146
93. Li Y, Horsman M, Wu N et al (2008) Biofuels from microalgae. Biotechnol Prog 24(4):815–820
94. Chen M, Tang H, Ma H (2019) Strategies for achieving high-density cultivation of microalgae for sustainable biofuel production. Biores Technol 293:122053
95. Zhang J-T, Wang J-X, Liu Y, Zhang Y, Wang Y-H, Chi Z-Y, Kong F-T (2024) Microalgal-bacterial biofilms for wastewater treatment: operations, performances, mechanisms, and uncertainties. Sci Total Environ 907:167974
96. Zhang Y, Deng J, Xiao X, Li YY, Liu Y (2023) Insights on pretreatment technologies for partial nitrification/anammox processes: a critical review and future perspectives. Biores Technol 384:129351

97. Choubey J, Choudhari JK, Verma MK, Chatterjee T, Sahariah P (2022) Metagenomics and metatranscriptomic analysis of wastewater. In: Microbial community studies in industrial wastewater treatment, 1st edn. CRC Press, p 11. eBook ISBN 9781003354147
98. Ugwuanyi D, Sikhakhane Q, Dada NM, Majemite MT, Obaigbena A (2024), Review of emerging technologies for nutrient removal in wastewater treatment. World J Adv Res Rev 21(2):1737–1749. https://doi.org/10.30574/wjarr.2024.21.2.0520; Li Y, Zhou Z, Zhang Y (2024) Application of artificial intelligence and machine learning in wastewater treatment: a review. Water Res 211:116789
99. Hernández-del-Olmo F, Gaudioso E, Duro N, Dormido R, Gorrotxategi M (2023) Advanced control by reinforcement learning for wastewater treatment plants: a comparison with traditional approaches. Appl Sci 13:4752. https://doi.org/10.3390/app13084752
100. Sravan JS, Matsakas L, Sarkar O (2024) Advances in biological wastewater treatment processes: focus on low-carbon energy and resource recovery in biorefinery context. Bioengineering 11:281. https://doi.org/10.3390/bioengineering11030281
101. Govindaraj D, Pisharody L, Karim AV, Nidheesh PV (2022) Integration of photocatalytic and biological processes for treatment of complex effluent: recent developments, trends, and advances. In: Integrated environmental technologies for wastewater treatment and sustainable development. https://doi.org/10.1016/B978-0-323-91180-1.00024-7
102. Khan NA, Ahmed S, Khan AH, Iqbal J (2021) Recent technologies for nutrient removal and recovery from wastewaters: a review. Chemosphere 130328(1):1–12. https://doi.org/10.1016/j.chemosphere.2021.130328
103. Li Z, Zheng L, Koottatep T, Vinnerås B (2023) Editorial: Decentralized wastewater treatment technologies. Front Environ Sci 11. https://doi.org/10.3389/fenvs.2023.1199552
104. Pamidimukkala A, Kermanshachi S, Adepu N, Safapour E (2021) Resilience in water infrastructures: a review of challenges and adoption strategies. Sustainability 13:12986. https://doi.org/10.3390/su132312986
105. Starkl M, Brunner N, Das S, Singh A (2022) Sustainability assessment for wastewater treatment systems in developing countries. Water 14(2). https://doi.org/10.3390/w14020241

Nutrient Retrieval Techniques in Wastewater Treatment

Divyesh Parde, Rahul Ghosh, Praveen Rajpurohit, Soumyadeep Bhaduri, and Manaswini Behera

Abstract Nitrogen and phosphorus are essential for the growth and well-being of living organisms, making them important nutrients. However, the release of wastewater containing high levels of nutrients (nitrogen and phosphorus) is a significant contributor to water pollution, leading to eutrophication and reduced levels of dissolved oxygen, which in turn pose a threat to aquatic ecosystems. The task of removing nutrients from sewage poses a considerable challenge. To meet stringent standards for nutrient discharge, it is crucial to remove nutrients from wastewater. Additionally, recovering nutrients from waste streams is vital for establishing a circular economy and avoiding the depletion of finite resources. In this comprehensive chapter, a thorough discussion is done on the available technologies for nitrogen and phosphorus removal from wastewater. The focus is on understanding their operational mechanisms, advantages, and limitations. The working principles of each technology are meticulously analyzed, shedding light on how they effectively extract nitrogen and phosphorus from wastewater. The advantages of these technologies encompass factors such as treatment efficiency, cost-effectiveness, and environmental sustainability. However, their implementation is not devoid of challenges. Obstacles such as high operational costs, technical complexities, and energy consumption are identified and discussed. Furthermore, this chapter serves as a platform to propose potential solutions for overcoming these obstacles and enhancing

D. Parde · R. Ghosh · P. Rajpurohit · S. Bhaduri · M. Behera (✉)
School of Infrastructure, Indian Institute of Technology, Bhubaneswar, Jatni, Argul, Odisha 752050, India
e-mail: manaswini@iitbbs.ac.in

D. Parde
e-mail: divyesh.parde@gmail.com

R. Ghosh
e-mail: s22ce09002@iitbbs.ac.in

P. Rajpurohit
e-mail: a22ce09001@iitbbs.ac.in

S. Bhaduri
e-mail: s21ce09003@iitbbs.ac.in

M. M. Ghangrekar et al. (eds.), *Biological and Hybrid Wastewater Treatment Technology*, Earth and Environmental Sciences Library,
https://doi.org/10.1007/978-3-031-63046-0_7

the successful integration of nutrient removal technologies. The goal is to optimize nutrient recovery and make strides towards a sustainable wastewater treatment paradigm. By recognizing the impediments and devising strategies to address them, this chapter aims to pave the way for more effective and widespread implementation of nutrient removal technologies, fostering a cleaner and healthier environment.

Keywords Wastewater treatment · Nitrogen · Phosphorus · Nutrient · Recovery · Cost-effective · Environmental sustainability

1 Introduction

The direct discharge of contaminated water from these sources poses a significant environmental risk, raising increasing concerns due to its diverse array of pollutants. Wastewater (WW) is a complicated mixture of many substances, some of which are present at levels that are extremely harmful to living things. The environment is also exposed to substantial levels of both organic and inorganic nutrients, which raises the biological oxygen demands (BOD). Eutrophication causes an excessive amount of phosphorus and nitrogen to enter aquatic habitats, which causes a number of environmental problems, such as the production of solid waste and unpleasant air pollutants. In addition to endangering aquatic life and degrading drinking water quality, this excess nutrient load also promotes the proliferation of unfavorable bacteria, raising health issues in areas close to the discharge sites [1, 2].

Carbon, nitrogen, and phosphorus (CNP), which are all essential for maintaining a variety of life forms, constitute the majority of the nutrients found in waste streams. DNA, amino acids, and chlorophyll are all made up of important elements like nitrogen and phosphorus. Phosphorus acts as the "energy currency" in eukaryotic cells in the form of adenosine triphosphate (ATP). Both phosphorus and nitrogen are essential for plant growth and the production of food. N_2 gas, which is a relatively stable and nonreactive type of nitrogen, is abundant (78%) in the atmosphere, but its availability in soils is constrained. Therefore, nitrogen is transformed into reactive forms such as amino acids, nitrate, and ammonia to make it useful and improve its accessibility in soils [3]

Biological fixation accounts for the vast majority of naturally produced reactive nitrogen (6.5 Tg (teragram) N/year in the US [2] and 100–300 Tg N/year globally [4], while lightning only contributes 2%. However, natural nitrogen fixation is insufficient to satisfy the rising food and energy needs of the expanding world population. As a result, human-caused reactive nitrogen generation is increasing. In contrast, since phosphorus is a finite resource, depletion is the problem with this element in the environment. Mining is the main technique for obtaining Earth's most common type of phosphorus, which is buried in igneous and sedimentary deposits. Within the next 45–100 years, these easily accessible phosphorus supplies will run out at the current rates of production and use. The quality of the remaining phosphorus will also deteriorate as these sources run out, resulting in decreased phosphorus concentrations and increased contaminants. With a bigger contribution from developing countries, the usage of synthetic fertiliser is showing an upward trend of 15 Tg/year [4]. Nitrogenous and phosphorus-based fertilisers have significantly improved agriculture, supplying an adequate diet for close to one-third of humanity [1]. It is obvious that using fertilisers with nitrogen and phosphorus is essential to reducing world hunger. However, only 4 Tg of the 170 Tg of reactive nitrogen applied each year is actually used by crops; the remaining portion is lost through air and water pathways [5], having an adverse effect on air and water quality, biodiversity, and human health. Two percent of the reactive nitrogen that is added to the atmosphere globally each year escapes as N_2O, which affects the ozone layer and causes global warming, while 12% escapes as NO_x and NH_3, which have different effects on the atmosphere. Furthermore, by subjecting crops to elevated O_3 levels, flooding, intense precipitation, drought, and heat, the consequent climate change and nitrogen interactions may have an impact on agricultural production [6]. Additionally, phosphorus discharge at levels as low as 0.02 mg/L can cause eutrophication in lakes and rivers, making them unusable for agriculture, recreation, or freshwater supplies. It's possible that increasing the output of large-scale fertilizers won't be enough to supply the demand for food unless nitrogen and phosphorus are regulated carefully. Given that a sizeable amount of the production of synthetic fertilizer escapes into the atmosphere, recovering lost nutrients has the potential to help mitigate this issue [7]. It functions according to the 3Rs principle: reduce, reuse, and recycle. Recovering nutrients makes it easier to produce reactive nitrogen and phosphorus, which results in less of these elements entering the environment. The ecosystem would be less negatively impacted, and nutrients would be used more effectively as a result. By reintroducing the recovered nitrogen and phosphorus as fertiliser, nutrient recovery would lower the price of that fertiliser and, in turn, the cost of food as a whole. Aside from the obvious economic advantages, it would also lessen the amount of nitrogen and phosphorus in the atmosphere, making the ecosystem cleaner and more sustainable [8].

Wastewater rich in nitrogen and phosphorus necessitates the application of diverse removal technologies, as detailed in this book chapter. The sources of these nutrients and the consequences if they are not effectively removed are elucidated in the subsequent section. Additionally, the recovery of nitrogen and phosphorus is a topic of discussion. In the treatment of wastewater, effectively removing excess nitrogen and phosphorus is crucial. These nutrients, when present in high concentrations in wastewater, can have detrimental effects on the environment and human health. Nitrogen, primarily in the form of ammonium, nitrate, and nitrite, can cause eutrophication of water bodies, leading to algal blooms and oxygen depletion. This disrupts aquatic ecosystems and can render water unfit for consumption. Similarly, phosphorus, a key component of fertilizers and detergents, can stimulate excessive plant and algae growth in rivers and lakes, resulting in harmful algal blooms and disrupting the balance of aquatic ecosystems. To mitigate these problems, advanced wastewater treatment technologies are employed. Biological nutrient removal (BNR) processes, such as the activated sludge process, utilize microorganisms to convert nitrogen and phosphorus into gas, which can then be released harmlessly into the atmosphere. Enhanced biological phosphorus removal (EBPR) is a notable BNR process that selectively removes phosphorus using specialized bacteria. Additionally, chemical and physical processes like coagulation, flocculation, and ion exchange can also be utilized to effectively remove these nutrients from wastewater. Furthermore, the recovery of nitrogen and phosphorus from wastewater presents an opportunity to recycle these valuable nutrients. Techniques like struvite precipitation can recover phosphorus from wastewater in the form of a useful fertilizer. Similarly, nitrogen recovery technologies like ammonia stripping and membrane-based processes enable the extraction of nitrogen for potential reuse in agricultural or industrial settings. Managing nitrogen and phosphorus in wastewater is vital to prevent environmental pollution. Implementing effective removal and recovery technologies is essential for the sustainable management of these nutrients, minimizing their adverse impacts on the environment and maximizing their potential for reuse and recycling.

2 Sources of Nutrients

Aquatic ecosystems receive nutrients from a variety of point and non-point sources. Direct wastewater treatment plant discharges, combined sewer and septic overflows, municipal wastewater discharges, storm water discharges, and other industrial outflows such as thermal power stations, mineral processing facilities, and detergent manufacturing facilities are among the major point sources. In addition, urban runoff, manure spreading, aquaculture, livestock farming, atmospheric deposition, erosion, and natural processes like rock weathering and release from sediments all contribute to nutrient influx [9]. Urine, faeces, and greywater from domestic and commercial activities all include nutrients that eventually make their way into wastewater systems. Figure 1 illustrates the sources and impacts of nutrients. These wastewater streams have the potential to significantly increase nutrient contamination in natural water bodies like rivers, lakes, and lagoons if they are not effectively treated to meet strict criteria. The expansion of algae and other aquatic plants can be sparked by the release of too many nutrients into these water bodies as a result of insufficient removal or recovery at wastewater treatment facilities (WWTPs), which eventually results in a condition known as eutrophication [10].

Worldwide, sewage has been recognized as a major point source of nutrient pollution, although residential sources are a major non-point source of concern. According to regional variables like climate, socioeconomic conditions, household technology, and more, the nutrient concentration in domestic wastewater varies. Domestic wastewater that hasn't been treated often has TN concentrations between 35 and 100 mg/L and TP concentrations between 18 and 29 mg/L [11].

According to [12], the natural diet makes up 40% of the total domestic nutrient loads, followed by food additives at 29%, dishwashing at 9%, domestic laundry at 14%, personal care products at 1%, food waste disposed down the drain at 1%, and other unspecified sources at 6%. Total kjeldahl nitrogen (TKN) and total phosphorus

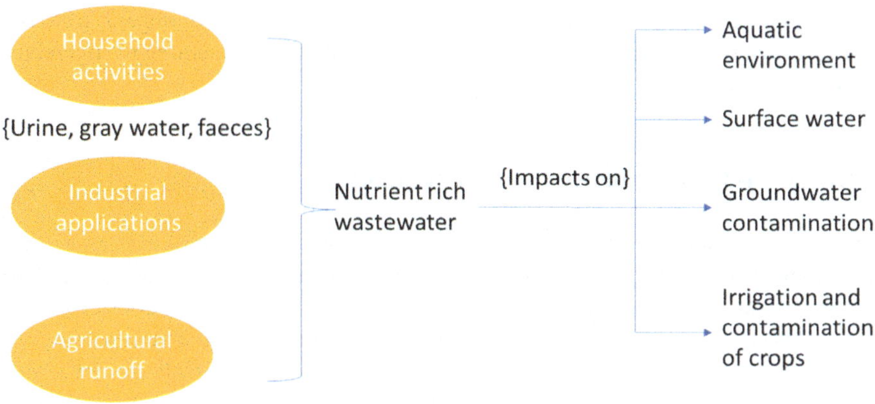

Fig. 1 Sources and impacts of nutrient

(TP) concentrations in cleaning and personal care products differ, offering more information about how they contribute to nutrient loads.

3 Nutrient Removal Technologies

3.1 Nitrogen Removal Technology

Nitrogen is vital for all living organisms, but the excessive release of both organic and inorganic nitrogen compounds can result in severe environmental issues. According to Sharma [13], ammonia is a crucial contaminant that contributes significantly to the oxygen demand in water bodies and is known to promote eutrophication in aquatic ecosystems. Nitrate pollution in drinking water has been linked to increased risks of non-Hodgkin's lymphoma [14], methemoglobinemia [15], as well as bladder, ovarian [16] or gastric cancers [17] in humans. Nitrate loads in surface water are a typical cause of eutrophication and declining water quality. The combination of ground and surface water resources can cause elevated nitrogenous chemicals in surface water to contaminate groundwater as well [18]. Animal waste, septic systems, industrial activities, atmospheric deposition from nitrogen oxide emissions, irrigation, and storm runoff from farmlands are the main causes of nitrogen pollution [19]. Around the world, this problem is widespread and has an impact on the overall value of aquatic ecosystems, ecological systems, and drinking water quality. As an illustration, nitrogen pollution has caused moderate to severe degradation in nearly two-thirds of the country's coastal rivers and bays [20]. The nitrogen pollution has also caused eutrophication and water quality decline in over 85% of China's lakes and around 82% of its 532 major rivers. In response to the negative impacts of excessive nitrogen release, stronger emission rules for pollutants containing nitrogen have been put in place [18].

3.1.1 Iron-Based Nitrogen Removal Technology

The chemical removal of nitrogen-containing pollutants from a range of sources, such as sewage, groundwater, and industrial effluent, has found widespread use in a number of iron-based chemicals [21]. The iron's valence states span from oxides of Fe (III) to ferrates (IV, V, and VI), Fe (II), and Fe0, thus the effluent's characteristics are very diverse. Using iron with various valences, nitrogen can be chemically removed by a variety of oxidation–reduction reactions. N_2, N_2O, NH_4, and NO are the byproducts of the reduction of NO_3 and NO_2 by Fe0 and Fe_2 [22, 23]. As opposed to ammonia, which ferrate (VI) can immediately oxidise to form N_2 or NO_3 [24].

It is possible to speed up electron transmission in these processes by modifying the structural and physical characteristics of iron-based materials. The process of removing pollutants that contain nitrogen is complex, requiring numerous chemical reactions that produce a variety of chemicals, some of which are undesirable. When nitrate is reduced by nFe0, for example, a sizeable amount of it is unfortunately transformed into undesirable ammonia (NH_3-N) instead of the non-toxic gas N_2 [25]. The iron (Fe) can act as a potential support material by providing positions that are appropriate for the deposition of bi- or tri-metals. Due to their selectivity towards non-toxic compounds, nFe0 bimetallic systems [26] and coated nanoparticle systems [27] are receiving more attention as methods for nitrogen removal. The premagnetization Fe0/ (Fe/Cu) system, where larger nitrate concentrations are attracted by the gradient force at the surface of Fe0 or catalysts, is of particular interest. The interaction of a gradient force and a Lorentz force in this configuration promotes N–N recombination [28], which improves selectivity for nitrogen gas [21].

Numerous studies have examined different biological or biochemical mechanisms and activities that involve iron and nitrogen elimination. These studies cover both well-known denitrification techniques and recently developed sewage denitrification procedures. Researchers confirmed the existence of competition between iron nanoparticles and bacteria in integrated denitrification systems, especially in the early stages of denitrification. According to research by An et al. [27], Fe0 has been shown to have a bio stimulatory effect linked to the H_2 created during anaerobic corrosion and an inhibitory effect on bacteria because of competition for nitrate. In addition, hydrogen trophic denitrifying bacteria have been used to successfully remove nitrate and reduce ammonium production.

3.1.2 Conventional Nitrification and Denitrification Method

Biological nitrogen fixation in wastewater involves certain bacteria capable of converting atmospheric nitrogen (N_2) into usable forms like ammonia (NH_3)

or ammonium (NH_4). These bacteria can be harnessed in wastewater treatment processes to enhance nitrogen removal. Biological nitrogen fixation helps improve wastewater treatment and reduce nitrogen pollution in the environment by facilitating the conversion of nitrogen gas into molecules that are simple to remove or use by organisms. During nitrification, chemolithoautotrophic bacteria help turn ammonium into nitrate in an aerobic environment. In comparison to heterotrophic bacteria, chemoautotrophic nitrifiers grow more slowly [8]. Nitrogen conversion process is shown in Fig. 2. Ammonium is first oxidised to nitrite, which is then further oxidised to nitrate in order to complete this conversion. As an energy source and electron acceptor, several species of wastewater bacteria use ammonium or nitrite at each stage. Although other genera, including *nitrosococcus, nitrospira,* and *nitrosovibrio,* can also oxidise ammonium, *nitrosomonas* is most frequently used to do so. *Nitrobacter* is the most common nitrite oxidizer, with *nitrospira, nitrospina, nitrococcus,* and *nitrocystis* also contributing to the process [7]. The yield and oxygen consumption for *nitrosomonas* and *nitrobacter* are given in Eqs. (1) and (2).

$$55NH_4^+ + 76O_2 + 109HCO_3^- \rightarrow C_5H_7O_2N + 54NO_2^- + 57H_2O + 104H_2CO_3 \quad (1)$$

$$400NO_2^- + NH_4^+ + 4H_2CO_3 + HCO_3^- + 195O_2 \rightarrow C_5H_7O_2N + 400NO_3^- + 34H_2O \quad (2)$$

The combined result of Eqs. (1) and (2) is Eq. (3), showing that nitrification is an entirely oxidative process.

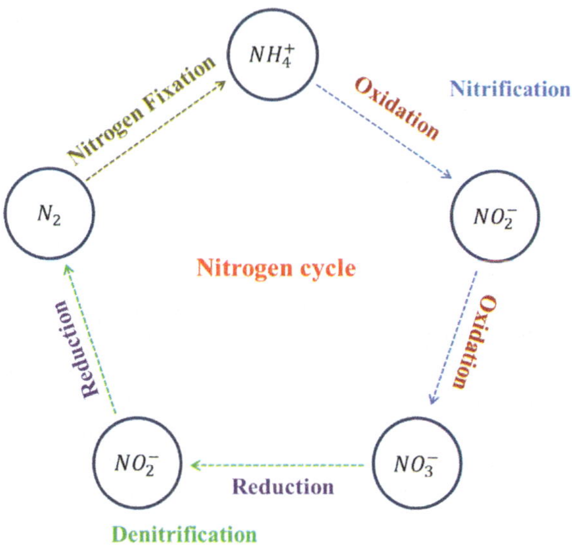

Fig. 2 Nitrogen cycle

$$NH_4^+ + 1.83O_2 + 1.98HCO_3^- \rightarrow 0.021C_5H_7O_2N + 0.98NO_3^- + 1.041H_2O + 1.88H_2CO_3 \quad (3)$$

Due to the formation of acid and consumption of alkalinity during the nitrification process, a decrease in pH was seen. For every mg of ammonium oxidised by the traditional nitrification process, 7.14 mg of alkalinity as $CaCO_3$ were used. The nitrification process is interrupted if the pH is outside the ideal range (7.5–8.0) and there is insufficient alkalinity. To maintain the required pH (7.5–8.0) and suitable alkalinity, additional chemicals such as lime, sodium carbonate/bicarbonate, and magnesium hydroxides are added [29]. Each chemical has a unique set of benefits and disadvantages. For instance, lime produces faster neutralisation but excessive use might impair the biological environment, causing pH values to surpass 10.0. Magnesium hydroxide dosage has a maximum pH limit of 9.0 and a slower rate of neutralization. The increased generation of extra sludge is a noteworthy drawback of chemical addition.

The nitrification process in a medium with free ammonia (FA) entails raising the metabolic rate of AOB while lowering NOB, which results in an increase in the amount of nitrite in the effluent. NOB activity can be decreased by adjusting operational factors such as pH, temperature, solid retention time (SRT), dissolved oxygen (DO), and pH. NOBs in particular are more DO-level sensitive than AOBs. Since AOB has a lower DO value for the coefficient of half saturation (0.2–0.4 mg/L) than NOB (0.7–2.0 mg/L), maintaining a low DO level (0.5 mg/L) can promote their growth. Nitrite accumulation is encouraged because AOB outcompetes NOB formation in an extremely low DO environment. However, AOBs also deteriorate more quickly than NOBs in low-DO environments. When both impacts are considered, a low DO environment generates more NOBs than AOBs, which causes a lower ammonia oxidation rate than nitrite oxidation rate. The results showed that prolonged operation at low DO levels promoted the enrichment of the bacteria AOBs and NOBs in the activated sludge, enhancing the sludge's nitrifying capacity. Low DO concentrations had no negative effects on long-term nitrification performance because the nitrifying community grew and changed over time. As a result, even with low DO procedures, complete nitrification is still feasible.

The substrate used, the ammonium concentration, and the reactor's dimensions all impact the ammonium uptake rate (AUR). AUR was the subject of a comparative analysis by Ge at al. [29] in a number of processes, including nitrification/denitrification, nitritation, and aerobic denitrification. Heterotrophic microorganisms in an anaerobic environment use a process called heterotrophic bioconversion to help denitrification, which takes place in the next stage. Nitrite and nitrate are transformed into gaseous dinitrogen by these microbes. The denitrification process is carried out by specific genera of proteobacteria, including *Alcaligenes, Paracoccus, and Thiobacillus* [30].

In addition, the denitrification process converts the produced nitrite ions (NO_2) into nitrogen gas (N_2) in an anoxic environment. For instance, in a study by Guo et al. [31], the partial nitritation (PN) process in a sequencing batch reactor (SBR) was examined for the removal of nitrogen from domestic wastewater. Over a prolonged operation period of 135 days at temperatures as high as 25 °C, the wastewater treatment system showed a remarkable 90% nitrite accumulation. Ammonium-oxidising bacteria (AOB), which made up about 8.3% of the total bacterial population in the bioreactor and had a very low abundance of nitrite-oxidising bacteria (NOB), were the main nitrifying bacteria in the PN process. In another study, Ge et al. [29], used a continuous plug-flow step-feed method to use the PN process to treat municipal wastewater and get rid of all the nitrogen. The outstanding nitrite accumulation in the effluent, which was reduced to zero under alternating anoxic and oxic conditions, was around 81%. In the anoxic zones, the step-feed method has allowed prompt denitrification. This method produced an effluent with a total nitrogen content of 7.23 ± 2.31 mg/L and a maximum total nitrogen removal of 86%. The effectiveness of the PN procedure in a sequencing batch biofilm reactor (SBBR) was examined by Wei et al. [32]. After 90 days of operation, the system showed an excellent NH_4^+-N removal effectiveness of 98.2% and an average nitrite buildup rate in the effluent of 87.6%. Notably, throughout the operational phase, the protein and polysaccharide contents of the extracellular polymeric substances that were both loosely and tightly bound dramatically decreased. The rate of N_2O emission was determined to be 16.5 mg per cycle, or around 4.79% of the total nitrogen eliminated. Comparatively, the CO_2 emission rate, which was around 67.7 mg per cycle in the same system, was higher than this reported N_2O emission rate.

3.1.3 ANAMMOX Process

The ANAMMOX process, recognized for its energy efficiency is an effective method for the concurrent removal of NH_4-N and NO_2-N from wastewater. It plays a crucial role in biological nutrient removal (BNR) through two distinct steps: firstly, ammonium-oxidizing bacteria (AOB) facilitates the conversion of ammonium to nitrite via partial nitrification (PN), and subsequently, the remaining NH_4-N and NO_2-N are transformed into di-nitrogen (N_2). ANAMMOX-based systems can be categorized into single-stage and two-stage approaches based on their treatment methods. We find systems such as OLAND and CANON within the single-stage category, while the SHARON–ANAMMOX system falls under the two-stage approach [33]. In the SHARON process, AOB is responsible for converting 50% of NH_4-N into NO_2-N, while ANAMMOX takes charge of converting the remaining NH_4-N and NO_2-N into N_2. In contrast, the CANON and OLAND systems operate with low dissolved oxygen (DO) levels, allowing for the simultaneous occurrence of both PN and ANAMMOX processes ([34]. The following chemical reaction (Eq. 4) briefly describes the one-step ANAMMOX method.

$$NH_4^+ + 1.31NO_2^- + 0.066HCO_3^- + 0.13H^+ \rightarrow 1.02N_2 + 0.26NO_3^- + 0.066CH_2O_{0.5}N_{0.15} + 2.03H_2O \quad (4)$$

About 90% of the time, the PN/ANAMMOX process transforms NH_4-N to N_2 gas, and 10% of the time, it produces nitrate. For the ANAMMOX method to be effective, the wastewater must include both nitrite and NH_4-N. In order to pre-treat wastewater, 50% of ammonium must be converted to nitrite using the PN method [32]. In a sequencing batch reactor (SBR), [35] investigated the efficiency of PN granular sludge as prior to treatment for the ANAMMOX process. The effluent's nitrite accumulation rate was reported to be 87.8% after 250 days, with values of NH_4-N and NO_2-N being 277.5 and 280.5 mg/L, respectively. This effluent mixture was considered ideal for the ANAMMOX process. The majority of the bacterial species in the sludge belonged to the Nitrosomonas genus, which represented 24.1% of all bacterial species. It's important to note that the bacteria linked PN/ANAMMOX process required less dissolved oxygen (DO), produced less secondary sludge, and didn't require a higher substrate carbon loading. In a different investigation, Winkler et al. [36] investigated the efficiency of BNR and Chemical Oxygen Demand (COD) reduction in an interconnected aerobic granular sludge reactor with ANAMMOX at a temperature of 18.3 °C over 390 days. The ANAMMOX bacteria took control of the reactor when treating synthetic influent, eliminating 900 gN_2-N/m^3.day and 600 gCOD/m^3.day. The ANAMMOX bacteria can compete for COD and remove nitrate and some acetate at normal temperatures, as evidenced by the fact that there was significantly less sludge than anticipated. Numerous recent studies for the treatment of low-strength wastewater have demonstrated the potential of the PN/ANAMMOX method, demonstrating its ability to effectively remove nitrogen compounds while minimizing sludge production and operating costs. These results highlight how useful and effective the PN/ANAMMOX technique is for treating wastewater. The system maintained an average biological nutrient removal (BNR) rate of 182–46 mgN/L/day while achieving ammonium conversion rates of 315–33 mgN/L.day. The significant 90% reduction in BOD levels in the wastewater was a striking result. The system was successfully run in moderate circumstances, including dissolved oxygen (DO) concentrations of 1–2 mg/L, hydraulic retention times (HRT) of 1.5–2 h, and solid retention times (SRT) of 30–120 days, to further improve specific ammonium conversion. Using a sequencing batch reactor (SBR), [37] evaluated the effectiveness of the PN/ANAMMOX process for treating high-strength ammonia effluent. The PN/ANAMMOX-SBR system demonstrated an amazing 81% BNR rate during a 400-day period, as well as a 68% reduction in greenhouse gas (GHG) emissions. The 39% reduction seen in traditional nitrification and denitrification systems was noticeably inferior to this. The system's BNR efficiency was greatly improved by operating under ideal circumstances, which included maintaining a pH range of 7.0–8.0 and maintaining DO levels at 0.1–0.3 mg/L in the anoxic phase and 0.5–0.6 mg/L in the oxic phase. The addition of ferric chloride to the sewage wastewater and the addition of Fe (0.01 mol/L) to the effluent digester both considerably improved the BNR performance of the system.

3.1.4 Denitrification Process

The biological denitrification process is a key part of the biological cycle. It changes NO_3- or NO_2-nitrogen atoms into gaseous N_2 molecules through a series of biological reductions. The chemical reaction (5) illustrating the denitrification process can be expressed as follows:

$$2NO_3^- + 1.25CH_3COOH \rightarrow N_2 + 2.5CO_2 + 1.5H_2O + 2OH^- \quad (5)$$

Four bio-catalytic nitrogen reductase enzymes—nitrate reductase, nitrite reductase, nitric-oxide reductase, and nitrous-oxide reductase—act in succession during this process. Denitrification is mostly done by facultatively anaerobic bacteria, especially nitrate-oxidizing bacteria, which use gaseous and ionic nitrogen oxide molecules as electron acceptors to make energy when there is no oxygen. Chemolitho-autotrophic nitrification and chemoorgano-heterotrophic denitrification are the two key stages of the bacterially-mediated ANAMMOX process. The denitrification occurs in conventional biological nutrient removal (BNR) when there is little to no dissolved oxygen (DO) (0.2 mg/L) present in the medium. Both the tertiary-fixed-film and denitrifying-activated-sludge processes require the addition of organic carbon as a nutrient and electron donor to the influent in order to increase denitrification rates. The chemical reaction (6), which is best described as follows, serves as an illustration of the combined ANAMMOX-denitrification process:

$$NH_4^+ + 1.06NO_2^- + 0.44CH_4O \rightarrow 1.02N_2 + 0.44CO_2 + 2.908H_2O \quad (6)$$

Numerous studies in the literature highlight the effectiveness of the denitrification process when stimulated by various nutritional substrates. A study by [38] looked at how denitrification enzymes compete for electrons when methanol is used as a carbon source during wastewater treatment. In order to understand how the rate of carbon loading affects the denitrification process, electron competition was examined in both limited and abundant carbon environments. N_2O accumulated during denitrification because the electron flow favoring nitrite reduction was greater than that favoring N_2O reduction. Along with methanol, denitrifying bacteria can also use volatile fatty acids generated during the acidogenic fermentation of sewage sludge as a carbon source. By employing the volatile fatty acid liquid produced by the fermentation of $FeCl_3$-treated sludge as the carbon source for denitrification, [7] showed that nitrate could be removed to a level of 99% without the addition of any additional carbon.

The effectiveness of an integrated endogenous partial denitrification and phosphorus removal system called ANAMMOX was examined in a two-stage sequencing batch reactor (SBR) in a different study by [2]. The biological nutrient removal (BNR) and phosphorus removal efficiencies of the treatment system were 95% and 99%, respectively. Without adding carbon or oxygen, the effluent quality was substantial, with about 6 mg N/L and 0.2 mg P/L. In the reactor system, NO_3-N, NO_2-N with NH_4-N, and phosphorus were largely removed by *Candidatus competibacter, Candidatus kuenenia*, and *Dechloromonas sp.*, according to microbial analyses.

The effectiveness of a two-stage partial denitrification (PD)/ANAMMOX process reactor for the treatment of highly enriched wastewater was examined in a study by [39]. Water with COD, nitrate, and NH_4-N concentrations of 182.5 mg/L, 1000 mg/L, and 58.3 mg/L, respectively, was added to the PD reactor throughout this sequential treatment. The COD in the effluent was decreased by the PD reactor to a range of 76.5–98.6 mg/L before being fed into the ANAMMOX reactor. *Candidatus Brocadia* bacteria were added to the ANAMMOX reactor, which demonstrated nitrogen removal efficiencies of 95.8, 92.8, and 78.9% during the course of the reactor's 215-day operation. The Moving Bed Biofilm Reactor (MBBR) works to treat high-strength wastewater with a lot of ammonium by using a sequence of heterotrophic nitrification and anaerobic denitrification [25]. The reactor showed that *acinetobacter, pseudomonas*, and *paracoccus* were the most common types of bacteria. This led to a 15-day reduction in startup time and an 80.7% removal efficiency of total nitrogen (TN). Bacteria were transported using polyvinyl alcohol gel, which offered a stable environment for effective operation under ammonium stress. In a different study, [40] investigated the simultaneous nitrification and denitrification (SND) process as a means of nitrogen removal from synthetic wastewater (with a C/N ratio of roughly 7.5). *Thauera* spp. dominated the treatment medium which used a heterotrophic ammonium oxidation pathway to convert ammonium into nitrogen gas via intermediate hydroxylamine. On average, ammonium, nitrate, and nitrite removal rates were 2.85, 2.42, and 1.98 mg/L, respectively. This system demonstrates itself to be a technically appealing method for treating wastewater with low carbon content and high NO_3-N enrichment. The performance of biological nitrogen removal technologies are shown in Table 1.

3.1.5 Electrochemical Method

Municipal wastewater, which typically contain roughly 50 mg/L of nitrogen on average, can be efficiently treated. The situation is different for industrial wastewaters that are highly nitrogen-rich and frequently have nitrogen contents that exceed 200 mg/L [44]. While the denitrification process is noticeably slower than nitrification, the former demonstrates a quick reaction rate. As a result, the reaction time is prolonged in situations where nitrate concentrations are high, such as in industrial wastewater. Several methods have been tried to speed up the reaction, such as changing the way the reactors are set up or combining biological nitrification–denitrification processes with other good methods. One such strategy is known as the "bioelectrochemical method", which combines the electrochemical method and the biological denitrification process. Electrochemical systems have benefits that include the selective removal of particular pollutant species, the absence of secondary waste formation, and the simplicity of integration with other approaches [44].

In the bio-electrochemical technique, bacteria that can transmit electrons extracellularly are used as electrochemically active mediators to transfer electrons to the anode, where specific pollutants are oxidised [45]. Microorganisms act as biocatalysts (bioanode or biocathode) in this process. The microbial electrolysis cell is

Table 1 Performance of nitrogen removal technologies

Technology	Wastewater type	Operating conditions	Results	References
Conventional nitrification	Pilot-scale municipal	Five phase treatment (SBR, 7.35, 20 °C, 1.5–2.0 mg/L, 254 ± 116 mg/L, 25 mg/L, 8 h, 3.25 g/L)	Effluent NO_2^--N: 81.5 ± 9.2%	[29]
	Lab-scale synthetic	SBBR, 8–8.5, 25–27 °C, 4 mg/L, 600 mg/L, 200 mg/L, 16 h, 3 g/L)	Effluent NO_2^--N: 87.6%	[35]
	Lab-scale domestic	SBR, 7.5 ± 0.1, 22–26 °C, < 1 mg/L, 160.4–224.8 mg/L, 50.0–72.9 mg/L, 3000 ± 300 mg/L	Effluent NO_2^--N: 97.2%	[10]
ANAMMOX	Lab-scale	Municipal PN/A SBR, 8.3, 30 °C , 4 mg/L, 140 mg/L, 740 mg/L, –, 4 g/L	TN removal: 95%	[34]
	Pilot-scale municipal	Plug-flow granular sludge based reactor, 7–7.5, 19 °C , 2 mg/L, 62 mg/L, 26 mg/L, 1.5–2 h, 4 g/L	TN removal: 182 ± 46 mg/L/day, BOD removal 90%	[41]
	Lab-scale synthetic	Anaerobic baffled reactor, 7.5, 30 °C, 0.8 mg/L, 35 mg/L, 350 mg/L, 7.9 g/L, 3 h Additional 4 g/L phosphorus	TN removal: 86–92%	[42]
Denitrification process	Lab-scale synthetic	Up-flow biofilm reactor, 7, 25 °C, 40 mg/L, 208.7 mg/L, 0.2, 0.5–0.6 h, –	TN removal: 81%	[23]
	Lab-scale municipal	Three phase treatment (SBR and UASB, 7, 18.5–28.2 °C, < 1 mg/L, 164.2 mg/L, 52.1 mg/L, 2, 3–6 h, 3000–3500 mg/L TN: 70 mg/L	TN removal: 97.1%	[43]
	Lab-scale Domestic sewage	SBR, 7.4, 15.4 °C, 0.02 mg/L, 224.6 ± 21.2 mg/L, 67.7 ± 5.1 mg/L, 1.7, 3.3–10 h, –	TN removal: 91.2%	

finished with the addition of a counter electrode. Equation (7), which illustrates a bio-electrochemical process for nitrate reduction

$$2NO_3^- + H_2O \rightarrow N_2 + 2.5O_2 + 2OH^- \quad (7)$$

Microbial Electrolysis Cells (MEC) have a lot of potential for use in industrial settings. MEC can be used successfully in a closed-loop system to treat industrial wastewater, which allows the recovery of valuable resources like hydrogen, methane, and other high-value compounds [46]. It treats difficult wastewater such as molasses, biodiesel, textile, and seafood sector effluents with significant efficiency [47]. The advantages and disadvantages of nitrogen removal technologies are shown in Table 2.

3.2 Phosphorus Removal Technologies

Phosphorus is a crucial component used in agriculture and is frequently obtained through mining [48]. Each year, fossil phosphate reserves yield about 22 million tonnes of phosphorus that are used as fertilisers. However, this phosphorus ultimately enters water bodies through city wastewater and agricultural runoff, with the majority of its source coming from sewage and farm runoff [49]. Physical, chemical, and biological methods can be used separately or in combination to reduce phosphorus pollution [50]. The constituents of the influent determine how well phosphorus removal and recovery processes work. Numerous high-rate treatment systems have been implemented to fulfil the required effluent discharge limits into open water bodies, providing efficient phosphorus removal and recovery.

3.2.1 Precipitation Method

The most commonly used technique for chemically removing phosphorus is phosphorus precipitation. Utilising this method requires adding an appropriate salt to the solution to cause the precipitation of soluble phosphorus. Consequently, the solid phosphorus is separated using physical procedures like gravity separation or filtering [51].

Chemical Precipitation

In the 1950s, Switzerland implemented chemical precipitation techniques on a large scale to mitigate eutrophication by effectively eliminating phosphorus from wastewater. On a global scale, numerous nations have extensively implemented this strategy. The process of chemical precipitation entails the introduction of divalent or trivalent metal salts, such as iron or aluminum in chloride or sulphate compositions, into wastewater. The removal of precipitated phosphorus can be efficiently achieved by gravity settling or filtration. Phosphorus removal can occur in several phases, encompassing the original sludge generation and subsequent secondary sludge treatment processes. The aforementioned objective is accomplished through the utilisation of chemical agents prior to the primary sedimentation stage and inside the aerobic reactor component of an activated sludge procedure. Nevertheless, the removal of phosphorus during tertiary treatment is not recommended because of the exorbitant treatment expenses and the need for extra management of tertiary sludge. The resulting chemical precipitate generally produces a sludge high in phosphorus in the

Table 2 Advantages and disadvantages of nitrogen removal technologies

Nitrogen removal technology	Working principle	Advantages	Disadvantages
Conventional nitrification	Oxidation of ammonium into nitrite, followed by oxidation of nitrite into nitrate, under aerobic conditions [30]	• Effectively removes ammonium and nitrite from wastewater • Well-established and widely used in wastewater treatment	• Requires sufficient oxygen levels • Can lead to excess sludge production • pH and alkalinity conditions need to be maintained for optimal performance
Conventional denitrification	Reduction of nitrate or nitrite into nitrogen gas under anaerobic conditions [30]	• Efficiently reduces nitrate or nitrite to harmless nitrogen gas • Commonly applied in wastewater treatment	• Requires anaerobic conditions • Needs a suitable carbon source for denitrifying bacteria
ANAMMOX process	Anaerobic ammonium oxidation by the bacteria *Candidatus brocadia*, converting ammonium and nitrite into nitrogen gas [34]	• High nitrogen removal efficiency • Reduced energy and carbon footprint • Lower sludge production compared to conventional methods	• Needs specific conditions (e.g., suitable pH, temperature) for optimal performance • Long start-up period and sensitivity to environmental factors
Iron precipitation	Addition of iron salts (e.g., $FeCl_3$) to wastewater resulting in the formation of iron hydroxide precipitates that adsorb and remove nitrogen compounds [21]	• Effective removal of nitrogen compounds • Relatively simple and cost-effective process • Reduces phosphate levels as well	• Requires regular addition of chemicals • High operational and maintenance costs • Sludge disposal and handling challenges
Electrochemical method	Integration of an electrochemical system with biological denitrification, utilizing electrochemically active microbes for enhanced nitrogen removal [44]	• Selective removal of specific pollutant species • Potential for resource recovery (e.g., hydrogen, methane) • No secondary waste generation	• Requires expertise in electrochemistry and microbial electrochemistry • Initial setup costs can be high—complexity in integrating biological and electrochemical processes

form of metal salts. The process of chemically isolating phosphorous and effectively recovering it from the resulting precipitate present's notable difficulties [46].

Electrocoagulation Precipitation

The process of electrocoagulation (EC) involves the formation of hydrolyzed coagulant metal cations in situ and, simultaneously, the application of electric current to sacrificial metal anodes. This process results in the formation of phosphorus (P) precipitates. Aluminium (Al) and iron (Fe) are frequently selected as sacrificial electrodes owing to their cost-effectiveness, widespread accessibility, and exceptional performance and compatibility as coagulants. In the field of electrochemistry, it has been observed that both aluminum (Al) and iron (Fe) anodes exhibit a notable capacity for removing phosphorus (P) with a high efficiency of over 90% over a wide range of pH values. The Al/Fe hybrid electrode demonstrated superior performance, as reported by Kuokkanen et al. [52]. The efficiency of phosphorus removal in an electrocoagulation (EC) system is commonly affected by the molar ratio of aluminum (Al) or iron (Fe) ions to phosphorus (P), as well as the concentration of phosphorus. A study reported that an impressive P removal effectiveness of nearly 100% was attained when a constant current density of 20 A/m^2 was applied for a treatment duration of 80 min [46]. The energy consumption for this process was measured to be 3.422 kWh/m^3. The migrating electric-field-assisted electrocoagulation technique is frequently employed in order to improve the efficiency of the EC system. The implementation of an air breathing cathode and an inert anode, such as graphite, charged through an electric field, has been found to enhance the removal of phosphorus and conserve energy [53]. For example, a research investigation utilized a titanium plate instead of a graphite sheet to serve as the inert electrode positioned between the sacrificial anode and air cathode. The experimental configuration achieved a 98% removal of phosphorus (P) from residential wastewater. This was accomplished by employing an iron electrode and implementing a treatment duration of 5 min, which consisted of five 10-s intervals of inert electrode charging [54]. The fixed current density used in the process was 1 mA/cm^2, resulting in a reduced energy consumption of 0.039 kWh/m^3.

Crystallization

Both chemical precipitation and electrocoagulation (EC) techniques are commonly used for phosphorus (P) removal. However, the crystallisation method specifically emphasises the recovery of P from wastewater and its subsequent transformation into minerals with added value, such as struvite. An economic analysis reveals that the recovery of phosphorus into valuable minerals, such as struvite, is deemed economically feasible when the initial phosphorus concentration ranges from approximately 50–60 mg/L [55]. Hence, it is imperative to employ effective P precipitation methods, such as crystallisation, in order to minimise expenses associated with treatment and optimise the potential for resource recovery and economic gain.

The technique of struvite crystallisation, which has been extensively studied, is commonly employed for the precipitation of phosphorus (P). This method is widely used to simultaneously recover phosphate (PO_4-P) and ammonium (NH_4-N) from wastewater. The struvite crystal formation process often requires an extra supply of magnesium (Mg) in a well-designed reactor, with careful management of pH, molar ratios, and mixing intensity. This is done in order to achieve the simultaneous recovery of phosphorus (P) and nitrogen (N) in the form of $MgNH4PO_4 \cdot 6H_2O$ [51]. The expense associated with struvite crystallisation is closely linked to the concentration of phosphorus in the wastewater, thus requiring precise regulation of the molar ratio of magnesium, ammonium, and phosphate ions (1:1:1) as well as pH adjustment within the range of 8 to 9. While the process of struvite crystallisation has been well recognised as an effective means of phosphorus (P) recovery from wastewater, it is worth noting that the bioavailability of P in the resultant struvite may not be ideal for direct application in agriculture [56]. Furthermore, the use of recovered struvite as a commercial fertiliser may pose possible environmental and public health hazards as it may contain contaminants such as heavy metals, pathogens, and antimicrobial resistance genes [57].

3.2.2 Biological Phosphorus Removal Method

Biological methods are vital in wastewater treatment since they provide the efficient removal of phosphorus from urban and industrial wastewater discharges. In recent times, there has been a notable increase in the acceptance and utilization of biological treatments to effectively and economically reduce phosphorus while also being environmentally sustainable. The utilization of enhanced biological phosphorus removal (EBPR) has been identified as a feasible alternative to conventional approaches [58]. Phosphorus-accumulating organisms (PAOs) have the ability to efficiently ingest and accumulate substantial quantities of phosphorus by undergoing transitions between anaerobic, anoxic, and aerobic environments. During the anaerobic phase, carbon sources are accumulated as polyhydroxyalkanoates (PHAs), whereas the main energy supply is derived from the hydrolysis of polyphosphate stored within the intracellular components. In both anoxic and aerobic settings, polyphosphate-accumulating organisms (PAOs) employ surplus phosphorus to synthesize internal polyphosphate. The effective elimination of phosphorus is eventually accomplished through the utilization of activated sludge, which serves as a suitable substrate for the creation of biochar. Multiple species of bacteria play a role in this process; however, acinetobacter was the initial bacterium identified as the primary contributor to phosphorus removal in enhanced biological phosphorus removal (EBPR) [59]. PAOs assume a pivotal role in the Enhanced Biological Phosphorus Removal (EBPR) process. The environmental conditions present in their habitat have a significant impact on PAOs, just like other living things. These circumstances encompass factors such as oxygen levels, temperature, pH, agitation, and the presence of other waste elements.

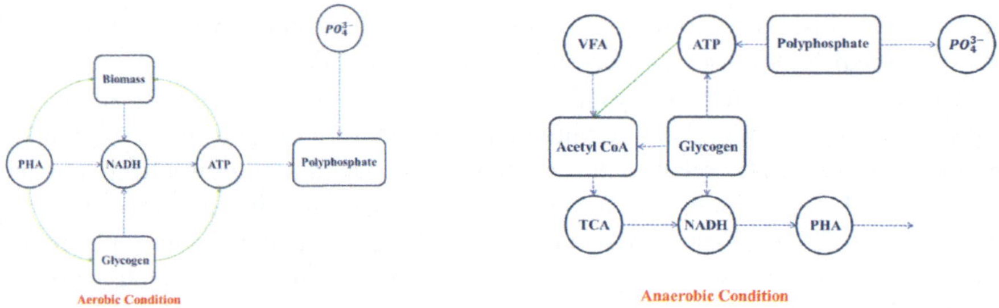

Fig. 3 Metabolic pathway of PAOs under aerobic and anaerobic conditions

The growth of polyphosphate-accumulating organisms (PAOs) can be enhanced by optimizing operational conditions, which involves implementing cycling anaerobic and aerobic phases. Under anaerobic conditions, polyphosphate-accumulating organisms (PAOs) have the ability to absorb volatile fatty acids (VFAs) and subsequently store them as polyhydroxyalkanoates (PHAs). The energy necessary for this particular operation is obtained from adenosine triphosphate (ATP) molecules. The synthesis of polyhydroxyalkanoates (PHA) requires the utilisation of a reduction equivalent to a molecule of nicotinamide adenine dinucleotide (NADH), which can be obtained through either the tricarboxylic acid cycle (TCA) or glycolysis. The process of polyphosphate degradation results in the release of orthophosphate into the surrounding liquid medium [60]. Phosphorus reduction takes place under aerobic conditions through the processes of phosphate absorption and intracellular polyphosphate synthesis [25]. The intracellular PHA is subject to oxidation by an electron acceptor, resulting in the release of energy that is then utilized for the processes of phosphorus intake and glycogen formation [58]. As a result, a considerably greater quantity of phosphorus is deposited in comparison to the anaerobic phase. Metabolic pathway of PAOs under aerobic and anaerobic condition is shown in Fig. 3.

The inadequate effectiveness of phosphorus (P) removal poses a widespread and notable obstacle to numerous full-scale wastewater treatment plants (WWTPs) that utilise enhanced biological phosphorus removal (EBPR). The effectiveness of phosphorus removal is determined by various aspects, such as temperature, pH, salinity levels, organic loading rate, the composition of the wastewater substrate, and the regulation of aeration [61]. In order to enhance the removal of phosphorus (P), it is necessary to utilise specific unit operations and controlling parameters as anticipated by models, with a particular emphasis on rotating between anaerobic and aerobic conditions [33]. The aerobic phase of enhanced biological phosphorus removal (EBPR) could be greatly affected by sudden changes in the concentration of chemical oxygen demand (COD) in wastewater. A negative association between the removal of phosphorus (P) and the concentration of chemical oxygen demand (COD) suggests that excessive loading of COD can result in the release of phosphorus under aerobic conditions [62]. A notable limitation in the process of enhanced biological phosphorus removal (EBPR) is the presence of glycogen-accumulating organisms

(GAOs), which serve as inhibitory factors. The competition between generic activated sludge organisms (GAOs) and polyphosphate-accumulating organisms (PAOs) for substrates and food sources is a common cause of system failures. In municipal wastewater treatment plants (WWTPs) that strive to achieve the concurrent elimination of carbon (C), nitrogen (N), and phosphorus (P), the occurrence of surplus nitrite, nitrate, and ammonia under anaerobic conditions typically results in the failure of enhanced biological phosphorus removal (EBPR). This failure can be attributed to the competitive disadvantage of polyphosphate-accumulating organisms (PAOs) compared to glycogen-accumulating organisms (GAOs) [11]. A range of environmental pollutants have been discovered to have inhibitory effects on polyphosphate-accumulating organisms (PAOs). Various heavy metals, including copper (Cu), silver (Ag), and chromium (Cr), have the potential to impede the activity of polyphosphate-accumulating organisms (PAOs) to differing extents due to their toxic properties. The presence of an Ag^+ concentration of 1 ppm has been found to have a detrimental effect on the effectiveness of phosphorus (P) removal. Similarly, a concentration of 5 ppm of Cr (VI) has been observed to entirely impede P removal in the process of enhanced biological phosphorus removal (EBPR). Engineered nanoparticles, such as nano-metal oxides, when added to the process of enhanced biological phosphorus removal (EBPR), have the potential to release metal ions or elevate levels of reactive oxygen species, hence exerting suppressive effects on polyphosphate-accumulating organisms (PAOs) [27]

Aerobic Granulation Sludge (AGS) technology, which emerged in the 1990s, significantly transformed the conventional activated sludge system. This innovative approach offered a sustainable method for effectively removing phosphorus (P) from wastewater [63]. The Enhanced Biological Phosphorus Removal (EBPR) process is facilitated by this simplification, allowing for the concurrent elimination of carbon, nitrogen, phosphorus, and other elements within sequencing batch reactors. The formation of AGS is assisted by the process of microbial aggregation, which involves the coming together of microorganisms. This aggregation is further supported by the attachment of extracellular polysaccharides (EPS) produced by the microorganisms. Additionally, the stability of AGS is maintained through hydrodynamic shear forces, which exert mechanical pressure on the structure. The AGS system is composed of three distinct zones: aerobic, anoxic, and anaerobic. These zones are arranged in a hierarchical structure, with the surface being the outermost layer and the inner core being the innermost layer [64]. When compared to the traditional activated sludge process, the aerobic granular sludge (AGS) process stands out with distinct features, including larger sludge particles (> 0.2 mm), increased settling velocity (ranging from 30 to 90 m/h), and heightened resistance to toxicity. These characteristics lead to substantial reductions in land area requirements (approximately 75%), energy consumption (ranging from 30 to 50%), and sludge formation (30%) [63].

In the context of enhanced biological phosphorus removal (EBPR), the Moving Bed Biofilm Reactor (MBBR) was introduced in the early 1990s as an innovative aerobic treatment system that combines both suspended and attached growth methods. Within the Moving Bed Biofilm Reactor (MBBR) system, biofilms develop on carriers composed of materials like plastic and sponge-like porous substances.

This development is facilitated by continuous high-rate aeration or mechanical stirring and mixing. The Moving Bed Biofilm Reactor (MBBR) principal aim is to reduce reactor size and eliminate the need for sludge recirculation. Nevertheless, achieving efficient denitrification and phosphorus removal in Moving Bed Biofilm Reactors (MBBR) presents a notable challenge, primarily due to the absence of an anaerobic fermentation phase and prolonged sludge retention time [45].

The application of microfiltration or ultrafiltration membranes in place of clarifiers for sludge separation characterizes membrane bioreactor (MBR) technology. Several anaerobic MBR topologies, including side-stream membrane, submerged membrane, and external membrane modules, have been developed to improve the effectiveness of biological processes in organic removal and nutrient recovery. These configurations achieve this by facilitating optimal sludge separation and maintaining high concentrations of microorganisms. A conventional anaerobic/anoxic/aerobic membrane bioreactor effectively accomplishes concurrent nitrogen (N) and phosphorus (P) removal, achieving removal efficiencies of 81% and 97%, respectively, when operated under conditions of limited aeration [39]. Nonetheless, the issue of membrane fouling, which is attributed to the proliferation of bacteria, extracellular polymeric substances (EPS), and various other by-products on the surface of the membrane, continues to pose a significant problem and remains a prominent area of investigation in the field of membrane bioreactor (MBR) technology. However, it should be noted that membrane bioreactors (MBRs) present an improvement over conventional enhanced biological phosphorus removal (EBPR) systems. This is due to their ability to reduce sludge generation and enhance the possibility of phosphorus recovery. The capital cost of membrane modules in full-scale MBR processes at wastewater treatment plants (WWTPs) is comparatively costly, and the occurrence of membrane fouling can result in escalated energy consumption as a consequence of heightened flux pressure and aeration rates [2].

3.2.3 Adsorption

Adsorption-based phosphorus (P) removal has advanced into a promising approach for wastewater treatment since it was introduced in the 1950s. 5–20 mg/L of P is frequently found in municipal wastewater and secondary treated wastewater (STWW). Because of stricter regulations on P discharge [46], it is often necessary to use more than one treatment method to get the right P concentration. Adsorption-based P removal is considered to be an appropriate treatment method after enhanced biological phosphorus removal (EBPR) processes. Due to adsorption's reversible nature, several desorption-based strategies have been created, allowing for the recovery of P from wastewater and its removal, improving sustainability [1]. Adsorption is more practical and efficient at recovering P from wastewater, even at relatively low levels, in comparison to other chemical and biological procedures that necessitate strict control over treating low-strength waste streams. As a result, adsorption has emerged as a practical method for treating vast quantities of water, including lake water with low P levels and municipal effluent. Additionally, adsorption has many benefits, such as maximising P accumulation prior to recovery, minimising sludge management, and having excellent compatibility with current operations.

Phosphorus (P) adsorption has displayed considerable potential when utilizing metal oxides, hydrooxides, and their composites. Oxyhydroxides based on metals like Al, Ca, Fe, and Mg have been favored as effective P adsorbents for various applications due to their strong affinity for P and widespread availability [46]. Rare earth element oxyhydroxides, including those from lanthanum (La), cerium (Ce), and zirconium (Zr), have also exhibited significant P affinity in water. However, their usage may face challenges due to their cost and accessibility when compared to more common metals.

In general, the adsorption of P onto metal oxyhydroxides is governed by three primary mechanisms: precipitation, inner-sphere complexation due to ligand exchange, and electrostatic attraction facilitated by hydrogen bonding [60]. Within binary oxides, layered double hydroxides (LDH), and layered double oxides, interactions between two metals play a crucial role in P adherence to water. For instance, a study noted that Zr-Fe binary oxides exhibited superior P adsorption capabilities compared to Zr or Fe oxides individually [3]. This improvement was attributed to increased pHZPC (zero point charge), enhanced surface area, and increased surface hydroxyl groups. In a comprehensive investigation of P adsorption on LDH, ZnFeZr and CaZnFeZr as top performers in terms of P removal and recovery, primarily achieved by replacing interlayer OH- with anionic P.

Metal oxyhydroxides have been added to porous adsorbents like zeolites, porous silica, and clay minerals to make them better at absorbing phosphorus (P) [12]. Usually, this process starts with a synthesis method that uses acid and heat, then metal impregnation to make ordered mesoporous or microporous structures. This increases the specific surface area and exposes additional P sorption sites [65]. In large-scale magnetic separation, a magnetic field is used to recover adsorbents with magnetic properties and a high P content. Electrically generated magnetic wires are used to get adsorbents back from the flow and suspension of water. Magnetite, zirconium ferrate, carbonyl iron, and iron oxides are some of the magnetic adsorbents that have been studied for their ability to adsorb and recover P [11]. These P adsorbents are often subjected to multiple application cycles to assess their P recovery and regeneration potential. P recovery through adsorption techniques offers significant sustainability and economic advantages. Several approaches, including saline solutions, complexing agents, acids, alkalis, and elution (a reverse chemical desorption process), can be employed during the regeneration process. After the regeneration process, the P-enriched fluid can undergo P precipitation and be reclaimed as struvite, calcium phosphate products, or hydroxyapatite. But metal oxides, hydroxides, and the adsorbents made from them might not be able to regenerate or be used again as well. For example, adsorbents containing Ca and Mg can lead to surface precipitation, resulting in less P desorption [66].

A lot of research has been done on how different carbon-based adsorbents, like activated carbon (AC), biochar, and biochar-based adsorbents, can treat and recover phosphorus (P) from wastewater. AC is often used to clean water of different environmental pollutants, but because its surface is naturally negatively charged, it can't really soak up P anions [67]. Biochar, on the other hand, is a commercially viable carbonaceous material that is rich in minerals and can better absorb P than AC. Biochar also exhibits potential as a source of fertiliser for P recovery from wastewater [68]. Biochar is usually made by heating up different agricultural and biomass waste types and turning them into charcoal under controlled oxygen conditions [53]. An important distinction is that biochar is generated at lower temperatures than AC, making it approximately five times more cost-effective [69]. Biochar can be made from a number of different lignocellulosic biomass and waste residues, such as hickory wood, municipal waste, peanut shells, bamboo waste, maize residue, sugarcane bagasse, sawdust, and rice husks [25]. The minerals in biochar, especially the amount of ash it has, greatly affect how much P it can hold. Several studies have shown that biochars made from mineral-rich biomass sources, like peanut shells, crawfish shells, and microalgae, often have a high surface potential and large surface areas. This makes them better at absorbing P in water [70].

3.2.4 Ion Exchange

Ion exchange is a simple adsorption method used to separate ions in water based on how they react with other ions. Because it can be done both ways, anion exchange holds great potential for getting rid of and replacing phosphorus (P). The ion exchange resins that are sold in stores are mostly made up of cross-linked polymers like polystyrene, polyvinyl alcohol, sodium alginate, and chitosan [71].

Recent advancements have led to the development of hybrid anion exchange (HAIX) resins. These HAIX resins are created by blending nanoparticles of hydrated metal oxides with polymers to enhance the selectivity and ion retention capacity of anion exchange resins [72]. Commercial HAIX resins like BioPhreeVR, LayneRT, PuroliteVR, and DuoliteVR have demonstrated effectiveness in removing and recovering P from various sources, including municipal wastewater, urine, and surface waters with naturally occurring organic matter [73]. These HAIX resins are typically created by impregnating porous resins with well-dispersed hydrated ferric oxide nanoparticles To achieve synergistic P removal. The macroporous anion exchange resin incorporating nanoscale zirconium (Zr) molybdate exhibited exceptional P capacity and significantly improved bed volume treatment compared to pure resin [74]. A lot of research has shown that anion exchange columns can be used to get rid of P from municipal wastewater. Also, P recovery can be achieved by running these columns through many cycles of adsorption and desorption. However, it is worth noting that ion exchange is still considered a relatively expensive P treatment technology from an economic standpoint [75, 76].

Many research has been done in getting P out of used adsorbents, especially anion exchange resins, but it's still hard to get P directly out of wastewater. 1 mol/L NaOH solution could be used to reuse strong-based anionic exchangers and commercial HAIX resins. An electro-assisted ion exchange approach for P recovery from secondary treated wastewater has also been used. This system consecutively utilized anion and cation exchangers to separate and recover P and Ca. Electro-assisted water splitting made catholytes at pH 11 and anolytes at pH 3. This made it easier for the ion exchanger to regenerate and for Ca-P to be added through precipitation. In a very important way, making acid and base on-site with electricity was a lot cheaper than bringing in bulk chemicals for ion exchange regeneration. The ion exchange removal and regeneration process was found that to increase the amount of P five times. It was also possible to recover over 95% of the P as hydroxyapatite by mixing the regenerates from the catholyte and anolyte (Table 3) [77].

Table 3 Advantages and disadvantages of phosphorus removal technologies

Phosphorus removal technology	Working principle	Advantages	Disadvantages
Precipitation	Addition of chemicals (e.g., alum, ferric chloride) to form insoluble precipitates with phosphorus ions [52]	• Reliable and well established • Removal efficiency 90–95% of P • Commercially available • High efficiency on • P recovery	• High operational cost on chemicals • Difficult recovery and reuse of Fe/Al precipitates • pH control may be needed • External source of Mg_2 and NH_4 required
Biological phosphorus removal (BPR)	Specific microorganisms accumulate phosphorus in their cells, which can be removed during sludge handling [58]	• High phosphorus removal efficiency. Reduced chemical usage • Low operating costs • Simultaneous removal for organics, N, P and other pollutants	• Sensitivity to operational • Conditions and population changes • Requires skilled operation and monitoring • Low effectiveness at trace level • P removal
Adsorption	Use of adsorbents (e.g., activated alumina, iron-based materials) to capture phosphorus from wastewater [71]	• High removal efficiency • Can be regenerated and reused • Easy operation and low costs	• Cost of adsorbent material • Limited adsorption capacity • Regeneration process may be energy-intensive • Influence of competing ions
Ion exchange	Phosphorus ions are exchanged with other ions (e.g., sulfate, chloride) on the surface of ion exchange resins [78]	• Reusable ion exchange resins	• High operating and regeneration costs • Limited resin capacity and susceptibility to fouling

4 Integrated Nitrogen-Phosphorus Removal Technology

4.1 Integrated Up-Flow Anaerobic Sludge Blanket (UASB) and Short Aerobic System

Many sewage Treatment Plants (STPs) were built using the up-flow anaerobic sludge blanket (UASB) reactor known for its rapid speed and anaerobic functions. This method offers several advantages: small energy consumption, less capital expenditure, minimal sludge generation, increased biogas generation, and a smaller footprint. Additionally, it results in lower costs for both operation and upkeep. This method can be utilized for the treatment of domestic sewage and high concentration effluent from industries. The UASB reactor is comprised of a gas-solids separator, which serves the purpose of retaining the anaerobic sludge within the reactor. This separator facilitates the settling of the sludge into a blanket, allowing for the biogas to be released from the upper portion of the reactor. The implementation of effluent recycling to fluidize the sludge bed may not be deemed necessary, given the existing level of interaction between sewage and sludge. A compact sludge bed formed at the base of the reactor, serving as the site for all biological processes. Under ideal circumstances, anaerobic bacteria underwent a process of aggregation, resulting in the formation of thick flocs and granules. The granules said before exhibited enhanced settling characteristics and did not experience any washout from the reactor while running. The interaction between biomass and wastewater was facilitated through the generation of agitation resulting from the motion of wastewater and the production of biogas. The performance of the UASB reactor is contingent upon several key factors, namely pH, temperature, startup duration, and organic input rate. The integration of the UASB system with the anoxic aerobic ultra-filtration membrane bioreactor resulted in a 97% removal efficiency, accompanied by the creation of CH_4. Despite its various advantages, the UASB reactor did not adhere to the rules of wastewater disposal [42].

There exists a necessity to improve the UASB system due to a fundamental constraint of the anaerobic method, which is its inability to effectively eliminate nutrients [79]. The current treatment technologies, including the implementation of short aeration systems, polishing ponds (PP), final polishing units (FPU), or a combination thereof, fail to meet effluent regulations as a result of insufficient retention time. The need for more land for FPU/PP in order to increase detention time poses financial challenges for nations with limited resources. Since the implementation of stricter effluent requirements in 2015, a significant number of WWTPs have become ineffective or necessitate upgrading in order to meet the required effluent standards. In consideration of upgrading, the UASB is transformed into an integrated treatment system, which is then succeeded by a high-rate treatment unit [45].

4.2 Biofilm-Based Treatment System

Moving Bed Biofilm Reactor (MBBR) is a wastewater treatment system that operates at a high rate and utilizes aerobic conditions for effective treatment. The approach exhibits advantages stemming from both attached and growth strategies. The initial implementation of the system took place in 1990 at Lardnal, Norway [80]. MBBR has garnered significant attention due to its versatile nature, efficient use of space, and consistent operational performance. This technology possesses significant capabilities in effectively addressing refractory wastewater from domestic, industrial, and various other sources. MBBR is a system that operates in either an oxygenic or anoxygenic configuration. It utilizes a suspended free-flowing polymer biofilm carrier to facilitate the movement of microorganisms throughout the reactor. This process requires mechanical agitation or aeration to distribute the carriers evenly inside the system. The performance of an MBBR is influenced by several critical elements, including the availability of media within the reactor, surface area of carrier, dissolved oxygen (DO) levels, and organic matter concentration. The MBBR encompasses advantages that are equally provided by both the attached and suspended growth processes. The aforementioned technique is a sophisticated and advanced approach to wastewater treatment, characterized by its rapid and efficient removal capabilities. It offers several advantages, including low initial investment, operational costs, and maintenance expenses. Additionally, it employs a single, reliable, and robust operational methodology. The utilization of MBBR technology necessitates a substantially smaller volume of reactor and eliminates the need for sludge recycling, in contrast with traditional technologies [81]. The functioning of the MBBR can be comparable to that of the activated sludge process (ASP). The dispersion of microbes attached to polymer-carriers can be seen over the entirety of the reactor. These carriers are safeguarded against flowing out by the presence of a protective barrier at the reactor outlet. One additional benefit of this reactor, in comparison to the activated sludge process and other biofilm reactors, is that it is free of sludge recycling. This is due to the absence of sludge recirculation and the separation of excess biomass. The efficacy of this treatment technique has been demonstrated in effectively treating both inorganic as well as organic wastes of different magnitudes [82]. In addition to biochemical oxygen demand (BOD) and chemical oxygen demand (COD), MBBR also facilitates the simultaneous nitrification, and denitrification processes either in a single stage or through a combination [83]. One significant limitation of these approaches is the requirement for subsequent separation of the bio-carrier.

4.3 Microalgae

The efficient removal of nitrogen, phosphorous, and hazardous metals from a diverse array of wastewaters through the utilization of microalgae represents a highly effective approach, particularly when implemented within carefully regulated environmental conditions [11]. A comprehensive investigation was conducted to examine the growth and development of algae in various types of wastewater, including municipal, agricultural, and industrial wastewater. Microalgae possess the capacity to thrive in saline water and infertile regions, exhibiting a notable rate of development in terms of dry weight. The reduction of greenhouse gases leads to an enhancement in air quality. The capacity to sequester CO_2 and facilitate elimination of nutrients presents a viable alternative for the treatment of wastewater [45]. The residual biomass of algae obtained after the separation of lipids from microalgae by-products has the potential to be exploited as animal feed or as fertilizer for agricultural crops [84].

Cyanobacteria and eukaryotic algae, engage in photosynthesis to produce oxygen (O_2). This oxygen is subsequently employed by microbial communities to facilitate the decomposition of organic materials into inorganic elemental nutrients. Moreover, the microalgae themselves rely on these inorganic nutrients for their development, as they play a crucial role in their metabolic processes. An alternative option involves the utilization of a mixed culture comprising heterotrophic bacteria and microalgal cultures to facilitate the process of nitrogen removal [30]. Within a mixed cultural framework, heterotrophic aerobic microorganisms effectively employ the oxygen generated by microalgae through the process of photosynthesis to decompose organic substances found in wastewater. On the other hand, microalgae effectively employ the carbon dioxide generated by heterotrophic microorganisms. In addition, the co-existence of heterotrophic, autotrophic, mixotrophic, and photoheterotrophic microorganisms alongside microalgae offers a wide range of potential uses due to the many metabolic functions they possess [66]. Currently, there is ongoing research into the application of microalgae technology for the purpose of eliminating nutrients from wastewater and biogas production and the utilization of collected microalgae for green energy production, specifically biogas.

4.4 Aerobic Granulation

When wastewater is treated in sequencing batch reactors (SBRs), biofilms are created from self-immobilized cells. These biofilms are referred to as aerobic granules. Aerobic granules have a higher density and weight than standard activated sludge, allowing for speedier settling without the need for flocculating agents [85]. This trait shortens the time spent in the settling phase of wastewater treatment. The majority of current wastewater treatment facilities use conventional activated sludge, which necessitates separate tanks for settling and gravity-based separation of sludge from effluent. However, the use of aerobic granular sludge enables the dual use of a reactor as a reaction and clarifier tank. In addition to minimizing sludge generation and total energy consumption, this dual functionality helps conserve space [86]. It also lowers capital expenses by up to 75%.

The main things that need to be done to make aerobic granular sludge are to use a "feast-and-famine" system and to use short settling times [87]. In cohabiting species, the feast-famine cycle promotes the buildup of polyhydroxyalkanoates (PHA). Organic materials are stored as PHA within microbial cells during the feast phase, and PHAs are used for biomass development during the starving phase Extracellular polymeric substances (EPS) are crucial for the development of aerobic granules. Derivatives of numerous organic substances, including proteins, polysaccharides, lipids, glycoproteins, and nucleic acids, create EPS in granules. They operate as a "glue" to encourage the production of biofilms by bacteria. Alginate-like exopolysaccharides and granulan have both been found as different forms of exopolysaccharides [88].

When *Candidatus accumulibacter phosphatis* is present, aerobic granular sludge can effectively remove phosphorus. Temperature, pH, salt content, organic loading rate, wastewater substrate type, and aeration technique are some of the variables that affect phosphorus removal in aerobic granular systems. *Accumulibacter* and other polyphosphate-accumulating organisms (PAOs) can gradually decline as salt concentrations rise. In contrast to PAOs, glycogen-accumulating organisms (GAOs) are activated at high salt levels (33 g/L NaCl) [89]. Competition between GAOs and PAOs for the available organic substrate results in selective enrichment of PAOs relative to GAOs between pH 6.5 and 7.5 and temperatures between 15 and 30 °C. The capacity of aerobic granules to absorb phosphorus from wastewater and transform it into polyphosphorus inside the cells is essential for effective phosphorus removal. Recent studies have emphasised how aerobic granules' extracellular polymeric substance (EPS) matrix affects phosphorus intake. Along with cations like K^+, Mg^{2+}, and Ca^{2+}, the EPS matrix affects the quantity of phosphorus that can be found there [40].

The use of aerobic granular sludge for biological nutrient removal in a variety of wastewater types, including dairy, textile, farming, and livestock wastewaters, has been demonstrated in numerous studies. Cassidy and Belia reported in 2005 that they had successfully removed nitrogen and phosphorus from abattoir effluent, with high removal percentages of total nitrogen (97.4%) and phosphorus (98.2%). This was done despite the high COD levels in the input water (7685 mg/L) [85]. The use of aerobic granular sludge for biological nutrient removal gained popularity over time and was expanded to include the treatment of difficult wastewater types, like palm oil mill effluent (POME). A study found that aerobic granular sludge had a total nitrogen removal efficiency of 97.6% in POME, which has a high total nitrogen concentration of 650 mg/L [90]. Application of this method on wastewater from the processing of tapioca and rubber was shown to have excellent removal rates of over 80% for both total nitrogen and phosphorus [91].

4.5 Wetlands

Constructed wetlands (CWs) are man-made wetlands that are developed and maintained to improve sewage treatment with the aim of enhancing water quality). Physical, chemical, and biological processes carried out in CWs by substrates, plants, and microorganisms effectively remove a variety of contaminants. A combination of filtration, sedimentation, adsorption, bioconversion, nitrification/denitrification, and absorption are used to remove nutrients. These processes are aided by microbes and wetland macrophytes. In wetlands, phosphorus has been found to accumulate largely in the soil compartment. Sedimentation and ammonium adsorption are two methods for removing nitrogen [92]. Three basic forms of constructed wetlands can be distinguished: hybrid, subsurface flow, and free water surface (FWS). SSF can be further divided into vertical subsurface flow and horizontal subsurface flow systems based on flow direction. These wetlands are mostly covered in marsh, submerged, and floating vegetation. Pebbles, gravel, and sand make up the medium over which waste water is flowing as it comes into contact with plant stems. Maintaining an adequate water basin depth enables sunlight to reach the systems subsurface, which promotes algae development and photosynthetic processes. Adsorption and precipitation processes involving calcium, aluminium, and ferrous compounds promote effective phosphorus removal in gravel-based systems. Plant absorption also significantly contributes to the elimination of phosphorus and occurs uniformly throughout the media. In comparison to the removal of organic matter, wetlands use a variety of macrophyte species, such as *Alocasia macrorrhiza, Alocasia puber, Azolla micrphylla, Eichhornia crassipes, Phragmites australis, Polygonum punctatum, Typha domingensis, and Typha latifolia, to help with nitrogen and nutrient removal* [93].

Macrophytes have demonstrated notable potential in the removal of nutrients, with typical ranges of 30–150 kg of phosphorus (P) per hectare per year and 200–2500 kg of nitrogen (N) per hectare per year. In a recent investigation conducted by Grebenshchykova et al. (2020), specific woody species, including *Salix, Sambucus, Myrica, and Acer* sp., were identified as highly suitable vegetation for use in constructed wetlands (CWs). These species exhibited exceptional adaptation capabilities, resulting in high removal rates of organic matter (ranging from 76 to 88%) and substantial nutrient accumulation. Furthermore, they displayed a notable capacity for evapotranspiration. Less explored species such as *Potamogeton crispus* (known for its enhanced bacterial proportion, robust oxygenation capacity, and efficient chemical oxygen demand removal), *Scripus validus* (notable for its high production of roots and stems, significant radial O_2 release, non-invasive nature, remarkable nutrient tolerance, and ease of establishment), *Acorus calamus* (with a high O_2 release from its roots and effectiveness in removing multiple contaminants, as well as the ability to tolerate high-strength wastewater), *Myriophyllum aquaticum* (efficient in nitrogen removal, substantial release of organic exudates, tolerance to high-strength wastewater and metals), and *Acorus calamus* (with a strong capacity for bacterial enrichment, thriving under high nitrogen loads, and exceptional root aeration capacity) have exhibited favorable characteristics for incorporation as species within CWs. The choice of species suitability may depend on their geographic distribution and availability, as outlined by [14].

5 Potentials and Limitations

For the best nutrient removal effectiveness, traditional biological nutrient removal (BNR) systems need to alternate between anaerobic and aerobic/anoxic environments. Because BNR processes are complex and dependent on specific microenvironments, consistent experimental parameter monitoring and control are difficult. Biological phosphorus removal (BPR) can be affected by the amount of volatile fatty acids (VFAs) in the influent and the amount of organic material that can ferment into VFAs in the anaerobic zone. This can change the performance of the system as environmental conditions and system operation change. Additionally, glycogen-accumulating organisms (GAOs) may compete with phosphate-accumulating organisms (PAOs) in EBPR systems [94]. Additionally, the majority of biological nutrient removal designs currently in use are either targeted at removing total nitrogen (TN) or total phosphorus (TP). Sustainability is a key concern due to the energy-intensive nature of these processes, with aeration being one of the largest energy-consuming operations at wastewater treatment plants using BNR. However, many onsite nutrient removal methods currently in use are not intended for nutrient removal and instead concentrate largely on the removal of organic pollutants. These systems can be energy-intensive and unreliable because some of them are scaled-down versions of centralized plants. Due to the inherent variability in decentralised wastewater characteristics and the practical difficulties in controlling nutrient removal processes at the level of individual habitation, reliability has decreased [95]. Since the nitrification process depletes alkalinity and nitrifying bacteria are inhibited when pH falls below neutral, alkalinity and the preservation of proper pH levels are also essential treatment considerations.

Despite substantial advances in research and expertise about the removal of nutrients from wastewater, significant obstacles still stand in the way of these techniques being widely successful. It may be possible to reduce the energy needed for aeration by using partial nitritation and anammox for nutrient removal. This strategy encounters difficulties, though, because anammox bacteria grow slowly at lower temperatures, making efficient operation during colder months challenging. More research needs to be done to make sure that the approach is stable in the long run, especially when it comes to making stable nitrite, which is needed for the partial nitritation and anammox systems to keep working. The anammox process has a major disadvantage despite its many benefits: the effluent frequently has high nitrate concentrations. In order to meet emission limits, a post-denitrification process is needed. This process requires the use of an outside carbon source and costs more money because it uses more energy, makes more activated sludge, and makes CO_2. Additionally, using the anammox technique to treat conventional wastewater is difficult due to high C/N ratios, low temperatures, and unfavorable effluent water conditions. Anammox bacteria can compete with heterotrophic denitrifying bacteria when organic matter is present, decreasing the effectiveness of ammonium removal. Furthermore, some organic substances, including methanol, have been shown to completely or partially impede the anammox process.

Various technological approaches have traditionally been utilized for recovering nutrients from waste sludge and associated sidestreams in sludge management. Biological phosphate Removal (BPR) systems can achieve optimal phosphate (P) recovery from the mainstream while complying with legal effluent requirements. However, it is crucial to combine conventional BPR with cutting-edge P recovery systems to reduce problems associated with waste sludge management, such as uncontrolled P precipitation that raises maintenance costs and high concentrations of soluble P in supernatants that return to the conventional process, which ultimately reduces overall P recovery.

The common adsorbent biomaterials that are now being developed for ion exchange processes provide a number of attractive benefits, such as low costs, simplicity of use, high efficiency, and no sludge formation. The existence of competing anions in the liquid being treated, however, prevents their possible applicability. Full-scale ion exchange implementation is difficult due to reversible and irreversible adsorbent fouling (such as capacity loss after several adsorption regeneration cycles), which calls for thorough cost–benefit assessments.

6 Conclusions

Macronutrients, specifically nitrogen and phosphorus, play a fundamental role in supporting life on our planet. However, addressing the removal of these elements from wastewater is of paramount importance in combating the environmental issue of eutrophication in various water bodies. A diverse range of technologies exist to achieve this objective. Presently, the most commonly employed processes for eliminating nitrogen and phosphorus are biological nitrification and denitrification for the former, and chemical precipitation for the latter. However, these methods fall short in terms of facilitating the recovery of nitrogen or phosphorus. As easily accessible phosphorus deposits dwindle, and the energy footprint associated with nitrogenous fertilizer production grows, the recovery of these elements from wastewater takes on increasing significance. This shift aligns with a broader change in perspective, wherein the waste stream is not merely viewed as a cost factor for pollutant removal to meet environmental regulations but also as a carrier of valuable, extractable resources.

Various contemporary phosphorus removal technologies, encompassing adsorption, membrane processes, biological phosphorus treatment, and precipitation, exhibit varying levels of effectiveness and employ distinct technical mechanisms to eliminate phosphorus from wastewater. The crucial element in both phosphorus removal and recovery lies in transferring phosphorus from wastewater into concentrated or solid waste streams, which subsequently become reservoirs for phosphorus recovery. For instance, in processes like adsorption, phosphorus can be extracted from spent adsorbents using regeneration solutions. Membrane processes and hybrid techniques excel at enriching phosphorus in wastewater to form highly concentrated liquid solutions. The phosphorus present in these waste stream solutions can ultimately be reclaimed and converted into valuable phosphorus products.

Conversely, biological technologies, including nitrification, ANAMMOX, and denitrification processes, are extensively applied to remove of nitrogen and its derivatives from wastewater. These bioprocesses are favored for their attractive features, such as energy neutrality, eco-friendliness, cost-effectiveness, low environmental risk, and high efficiency. Nevertheless, each approach for nitrogen and phosphorus removal comes with its own set of advantages and disadvantages. It is crucial that all techniques for phosphorus and nitrogen removal continue to evolve while thoughtfully considering the most appropriate approaches to achieve optimal results.

7 Future Scope

Future research should focus on several key aspects and perspectives to further develop and optimize technologies for nutrient recovery, particularly for phosphorus. This includes exploring innovative methods for transferring phosphorus from wastewater into concentrated or solid waste streams, such as through advanced adsorption techniques or membrane processes. Additionally, future studies should investigate the potential of hybrid approaches that integrate multiple nutrient removal technologies to achieve synergistic effects and enhance overall efficiency. This could involve combining biological processes like nitrification, ANAMMOX, and denitrification with chemical precipitation or membrane filtration for comprehensive nutrient removal and recovery. Moreover, research efforts should address the challenges associated with the implementation of nutrient removal technologies, such as high operational costs, technical complexities, and energy consumption. Strategies for improving the cost-effectiveness and sustainability of these technologies through process optimization, resource recovery, and energy-efficient operation should be explored.

References

1. Cheng D et al (2020) Performance of microbial fuel cell for treating swine wastewater containing sulfonamide antibiotics. Bioresour Technol 311:123588. https://doi.org/10.1016/j.biortech.2020.123588
2. Wang X, Zhao J, Yu D, Du S, Yuan M, Zhen J (2019) Evaluating the potential for sustaining mainstream anammox by endogenous partial denitrification and phosphorus removal for energy-efficient wastewater treatment. Bioresour Technol 284:302–314. https://doi.org/10.1016/j.biortech.2019.03.127

3. Fallahi A, Rezvani F, Asgharnejad H, Khorshidi E, Hajinajaf N, Higgins B (2021) Interactions of microalgae-bacteria consortia for nutrient removal from wastewater: a review. Chemosphere 272:129878. https://doi.org/10.1016/j.chemosphere.2021.129878
4. S. Fields, "Global Nitrogen: Cycling out of Control," *Environ. Health Perspect.*, vol. 112, no. 10, Jul. 2004, https://doi.org/10.1289/ehp.112-a556.
5. Smil V (2002) Nitrogen and food production: proteins for human diets. AMBIO A J Hum Environ 31(2):126–131. https://doi.org/10.1579/0044-7447-31.2.126
6. Robertson GP et al (2013) Nitrogen-climate interactions in US agriculture. Biogeochemistry 114(1–3):41–70. https://doi.org/10.1007/s10533-012-9802-4
7. Ali M, Okabe S (2015) Anammox-based technologies for nitrogen removal: advances in process start-up and remaining issues. Chemosphere 141:144–153. https://doi.org/10.1016/j.chemosphere.2015.06.094
8. Cicekalan B, Kosar S, Cingoz S, Eyit N, Ersahin ME, Ozgun H (2023) Techno-economic and environmental assessment of different municipal wastewater treatment systems. J Water Process Eng 53:103822. https://doi.org/10.1016/j.jwpe.2023.103822
9. Saad R, Belkacemi K, Hamoudi S (2007) Adsorption of phosphate and nitrate anions on ammonium-functionalized MCM-48: effects of experimental conditions. J Colloid Interface Sci 311(2):375–381. https://doi.org/10.1016/j.jcis.2007.03.025
10. Zhao G, Du J, Jia Y, Lv Y, Han G, Tian X (2012) The importance of bacteria in promoting algal growth in eutrophic lakes with limited available phosphorus. Ecol Eng 42:107–111. https://doi.org/10.1016/j.ecoleng.2012.02.007
11. Cai T, Park SY, Li Y (2013) Nutrient recovery from wastewater streams by microalgae: status and prospects. Renew Sustain Energy Rev 19:360–369. https://doi.org/10.1016/j.rser.2012.11.030
12. Tjandraatmadja G, Pollard C, Sheedy C, Gozukara Y (2010) Sources of contaminants in domestic wastewater : nutrients and additional elements from household products. In: CSIRO Water a Healthy Country National Research Flagship, Feb 2010, pp 1–118
13. Sharma VK (2011) Oxidation of inorganic contaminants by ferrates (VI, V, and IV)-kinetics and mechanisms: a review. J Environ Manage 92(4):1051–1073. https://doi.org/10.1016/j.jenvman.2010.11.026
14. Shukla A, Patwa A, Parde D, Vijay R (2023) A review on generation, characterization, containment, transport and treatment of fecal sludge and septage with resource recovery-oriented sanitation. Environ Res 216(P1):114389. https://doi.org/10.1016/j.envres.2022.114389
15. Huang W, Zhang B, Feng C, Li M, Zhang J (2012) Research trends on nitrate removal: a bibliometric analysis. Desalin Water Treat 50(1–3):67–77. https://doi.org/10.1080/19443994.2012.708542
16. Weyer PJ et al (2001) Municipal drinking water nitrate level and cancer risk in older women: the Iowa women's health study. Epidemiology 12(3):327–338. https://doi.org/10.1097/00001648-200105000-00013
17. Liu Y et al (2016) Optimization of enhanced bioelectrical reactor with electricity from microbial fuel cells for groundwater nitrate removal. Environ Technol 37(8):1008–1017. https://doi.org/10.1080/09593330.2015.1096962
18. Novotny V, Wang X, Englande AJ, Bedoya D, Promakasikorn L, Tirado R (2010) Comparative assessment of pollution by the use of industrial agricultural fertilizers in four rapidly developing Asian countries. Environ Dev Sustain 12(4):491–509. https://doi.org/10.1007/s10668-009-9207-2
19. Xu J, Hao Z, Xie C, Lv X, Yang Y, Xu X (2012) Promotion effect of Fe2+ and Fe_3O_4 on nitrate reduction using zero-valent iron. Desalination 284:9–13. https://doi.org/10.1016/j.desal.2011.08.029
20. Strobel CJ et al () National Coastal Assessment Field Operations Manual
21. Ren Y, Zhou J, Lai B, Tang W, Zeng Y (2016) Fe0 and Fe0 fully covered with Cu0 (Fe0 + Fe/Cu) in a fixed bed reactor for nitrate removal. RSC Adv 6(110):108229–108239. https://doi.org/10.1039/c6ra24014c

22. Lee C-S, Gong J, Huong CV, Oh D-S, Chang Y-S (2016) Macroporous alginate substrate-bound growth of Fe0 nanoparticles with high redox activities for nitrate removal from aqueous solutions. Chem Eng J 298:206–213. https://doi.org/10.1016/j.cej.2016.03.113
23. Tai Y-L, Dempsey BA (2009) Nitrite reduction with hydrous ferric oxide and Fe(II): stoichiometry, rate, and mechanism. Water Res 43(2):546–552. https://doi.org/10.1016/j.watres.2008.10.055
24. Sharma VK, Bloom JT, Joshi VN (1998) Oxidation of ammonia by ferrate(VI). J Environ Sci Heal Part A Toxic/Hazard Subst Environ Eng 33(4):635–650. https://doi.org/10.1080/10934529809376752
25. Zhang B et al (2020) Microalgal-bacterial consortia: from interspecies interactions to biotechnological applications. Renew Sustain Energy Rev 118:109563. https://doi.org/10.1016/j.rser.2019.109563
26. Hamid S, Bae S, Lee W, Amin MT, Alazba AA (2015) Catalytic nitrate removal in continuous bimetallic Cu–Pd/nanoscale zerovalent iron system. Ind Eng Chem Res 54(24):6247–6257. https://doi.org/10.1021/acs.iecr.5b01127
27. An Y, Li T, Jin Z, Dong M, Xia H, Wang X (2010) Effect of bimetallic and polymer-coated Fe nanoparticles on biological denitrification. Bioresour Technol 101(24):9825–9828. https://doi.org/10.1016/j.biortech.2010.07.110
28. Jiang X et al (2015) Enhanced paramagnetic Cu2+ ions removal by coupling a weak magnetic field with zero valent iron. J Hazard Mater 283:880–887. https://doi.org/10.1016/j.jhazmat.2014.10.044
29. Ge S, Peng Y, Qiu S, Zhu A, Ren N (2014) Complete nitrogen removal from municipal wastewater via partial nitrification by appropriately alternating anoxic/aerobic conditions in a continuous plug-flow step feed process. Water Res 55:95–105. https://doi.org/10.1016/j.watres.2014.01.058
30. Rahimi S, Modin O, Mijakovic I (2019) Technologies for biological removal and recovery of nitrogen from wastewater. Biotechnol Adv 43:107570. https://doi.org/10.1016/j.biotechadv.2020.107570
31. Guo JH, Peng YZ, Wang SY, Zheng YN, Huang HJ, Ge SJ (2009) Effective and robust partial nitrification to nitrite by real-time aeration duration control in an SBR treating domestic wastewater. Process Biochem 44(9):979–985. https://doi.org/10.1016/j.procbio.2009.04.022
32. Wei D et al (2017) Nitrogen removal via nitrite in a partial nitrification sequencing batch biofilm reactor treating high strength ammonia wastewater and its greenhouse gas emission. Bioresour Technol 230:49–55. https://doi.org/10.1016/j.biortech.2017.01.033
33. Rout PR et al (2021) Nutrient removal from domestic wastewater: A comprehensive review on conventional and advanced technologies. J Environ Manage 296:113246. https://doi.org/10.1016/j.jenvman.2021.113246
34. Daverey A, Hung NT, Dutta K, Lin JG (2013) Ambient temperature SNAD process treating anaerobic digester liquor of swine wastewater. Bioresour Technol 141:191–198. https://doi.org/10.1016/j.biortech.2013.02.045
35. Wei D, Ngo HH, Guo W, Xu W, Du B, Wei Q (2018) Partial nitrification granular sludge reactor as a pretreatment for anaerobic ammonium oxidation (Anammox): achievement, performance and microbial community. Bioresour Technol 269:25–31. https://doi.org/10.1016/j.biortech.2018.08.088
36. Winkler M-KH, Kleerebezem R, van Loosdrecht MCM (2012) Integration of anammox into the aerobic granular sludge process for main stream wastewater treatment at ambient temperatures. Water Res 46(1):136–144. https://doi.org/10.1016/j.watres.2011.10.034
37. Arora AS, Nawaz A, Yun CM, Cho H, Lee M (2019) Ecofriendly anaerobic ammonium oxidation system: optimum operation and inhibition control strategies for enhanced nitrogen removal. Ind Eng Chem Res 58(45):20847–20856. https://doi.org/10.1021/acs.iecr.9b04591
38. Pan Y, Ni BJ, Bond PL, Ye L, Yuan Z (2013) Electron competition among nitrogen oxides reduction during methanol-utilizing denitrification in wastewater treatment. Water Res 47(10):3273–3281. https://doi.org/10.1016/j.watres.2013.02.054

39. Du R, Cao S, Peng Y, Zhang H, Wang S (2018) Combined partial denitrification (PD)-Anammox: a method for high nitrate wastewater treatment. Environ Int 126:707–716. https://doi.org/10.1016/j.envint.2019.03.007
40. Wang Q, He J (2020) Complete nitrogen removal via simultaneous nitrification and denitrification by a novel phosphate accumulating *Thauera* sp. strain SND5. Water Res 185:116300. https://doi.org/10.1016/j.watres.2020.116300
41. Lotti T et al (2015) Pilot-scale evaluation of anammox-based mainstream nitrogen removal from municipal wastewater. Environ Technol 36(9):1167–1177. https://doi.org/10.1080/09593330.2014.982722
42. Chernicharo CAL (2006) Post-treatment options for the anaerobic treatment of domestic wastewater. Rev Environ Sci Biotechnol 5(1):73–92. https://doi.org/10.1007/s11157-005-5683-5
43. Chen M et al (2020) Global nitrogen input on wetland ecosystem: the driving mechanism of soil labile carbon and nitrogen on greenhouse gas emissions. Environ Sci Ecotechnol 4:100063. https://doi.org/10.1016/j.ese.2020.100063
44. Soren KR et al (2015) Onl Onl, vol 36, pp 1063–1070
45. Hasan MN et al (2021) Recent technologies for nutrient removal and recovery from wastewaters: a review. Chemosphere 277:130328. https://doi.org/10.1016/j.chemosphere.2021.130328
46. Omwene PI, Kobya M, Can OT (2018) Phosphorus removal from domestic wastewater in electrocoagulation reactor using aluminium and iron plate hybrid anodes. Ecol Eng 123:65–73. https://doi.org/10.1016/j.ecoleng.2018.08.025
47. Jayashree C et al (2016) Treatment of seafood processing wastewater using upflow microbial fuel cell for power generation and identification of bacterial community in anodic biofilm. J Environ Manage 180:351–358. https://doi.org/10.1016/j.jenvman.2016.05.050
48. Reijnders L (2014) Phosphorus resources, their depletion and conservation, a review. Resour Conserv Recycl 93:32–49. https://doi.org/10.1016/j.resconrec.2014.09.006
49. Bowes MJ et al (2015) Characterising phosphorus and nitrate inputs to a rural river using high-frequency concentration-flow relationships. Sci Total Environ 511:608–620. https://doi.org/10.1016/j.scitotenv.2014.12.086
50. Yeoman S, Stephenson T, Lester J, Perry R (1988) The removal of phosphorus during wastewater treatment: a review. Environ Pollut 49(3):183–233. https://doi.org/10.1016/0269-7491(88)90209-6
51. Peng L, Dai H, Wu Y, Peng Y, Lu X (2018) A comprehensive review of phosphorus recovery from wastewater by crystallization processes. Chemosphere 197:768–781. https://doi.org/10.1016/j.chemosphere.2018.01.098
52. Kuokkanen V, Kuokkanen T, Rämö J, Lassi U, Roininen J (2015) Removal of phosphate from wastewaters for further utilization using electrocoagulation with hybrid electrodes—techno-economic studies. J. Water Process Eng. 8:e50–e57. https://doi.org/10.1016/j.jwpe.2014.11.008
53. Lei Y, Saakes M, van der Weijden RD, Buisman CJN (2020) Electrochemically mediated calcium phosphate precipitation from phosphonates: Implications on phosphorus recovery from non-orthophosphate. Water Res 169:115206. https://doi.org/10.1016/j.watres.2019.115206
54. Tian Y, He W, Liang D, Yang W, Logan BE, Ren N (2018) Effective phosphate removal for advanced water treatment using low energy, migration electric–field assisted electrocoagulation. Water Res 138:129–136. https://doi.org/10.1016/j.watres.2018.03.037
55. Jupp AR, Beijer S, Narain GC, Schipper W, Slootweg JC (2021) Phosphorus recovery and recycling—closing the loop. Chem Soc Rev 50(1):87–101. https://doi.org/10.1039/D0CS01150A
56. Ye Y et al (2017) Insight into chemical phosphate recovery from municipal wastewater. Sci Total Environ 576:159–171. https://doi.org/10.1016/j.scitotenv.2016.10.078
57. Zheng Y et al (2019) Reclaiming phosphorus from secondary treated municipal wastewater with engineered biochar. Chem Eng J 362:460–468. https://doi.org/10.1016/j.cej.2019.01.036
58. Acevedo B, Oehmen A, Carvalho G, Seco A, Borrás L, Barat R (2012) Metabolic shift of polyphosphate-accumulating organisms with different levels of polyphosphate storage. Water Res 46(6):1889–1900. https://doi.org/10.1016/j.watres.2012.01.003

59. Barker PS, Dold PL (1996) Denitrification behaviour in biological excess phosphorus removal activated sludge systems. Water Res 30(4):769–780. https://doi.org/10.1016/0043-1354(95)00217-0
60. Lin H, Gan J, Rajendran A, Reis CER, Hu B (2015) Phosphorus removal and recovery from digestate after biogas production. In: Biofuels—status and perspective. InTech
61. Hamza R, Rabii A, Ezzahraoui F, Morgan G, Iorhemen OT (2022) A review of the state of development of aerobic granular sludge technology over the last 20 years: full-scale applications and resource recovery. Case Stud Chem Environ Eng 5:100173. https://doi.org/10.1016/j.cscee.2021.100173
62. Guisasola A, Pijuan M, Baeza JA, Carrera J, Casas C, Lafuente J (2004) Aerobic phosphorus release linked to acetate uptake in bio-P sludge: process modeling using oxygen uptake rate. Biotechnol Bioeng 85(7):722–733. https://doi.org/10.1002/bit.10868
63. de Amorim de Carvalho C, Ferreira dos Santos A, Tavares Ferreira TJ, Sousa Aguiar Lira VN, Mendes Barros AR, Bezerra dos Santos A (2021) Resource recovery in aerobic granular sludge systems: is it feasible or still a long way to go? Chemosphere 274. https://doi.org/10.1016/j.chemosphere.2021.129881.
64. de Sousa Rollemberg SL, Mendes Barros AR, Milen Firmino PI, Bezerra dos Santos A (2018) Aerobic granular sludge: Cultivation parameters and removal mechanisms. Bioresour Technol 270:678–688. https://doi.org/10.1016/j.biortech.2018.08.130
65. Jiang L et al (2021) Evaluation of nutrient removal performance and resource recovery potential of anaerobic/anoxic/aerobic membrane bioreactor with limited aeration. Bioresour Technol 340:125728. https://doi.org/10.1016/j.biortech.2021.125728
66. Delgadillo-Mirquez L, Lopes F, Taidi B, Pareau D (2016) Nitrogen and phosphate removal from wastewater with a mixed microalgae and bacteria culture. Biotechnol. Reports 11:18–26. https://doi.org/10.1016/j.btre.2016.04.003
67. Patwa A, Parde D, Dohare D, Vijay R, Kumar R (2020) Solid waste characterization and treatment technologies in rural areas: an Indian and international review. Environ Technol Innov 20:101066. https://doi.org/10.1016/j.eti.2020.101066
68. Healy M et al (2015) Resource recovery from sewage sludge, vol 14
69. Zhang K et al (2020) Optimization of nitrogen and carbon removal with simultaneous partial nitrification, anammox and denitrification in membrane bioreactor. R Soc Open Sci 7(9):200584. https://doi.org/10.1098/rsos.200584
70. Jung S, Bae S, Lee W (2014) Development of Pd–Cu/hematite catalyst for selective nitrate reduction. Environ Sci Technol 48(16):9651–9658. https://doi.org/10.1021/es502263p
71. Kumar A, Nayak AK, Pani DR, Das BS (2019) Application of phosphorus, iron, and silicon reduces yield loss in rice exposed to water deficit stress. Agron J 111(3):1488–1497. https://doi.org/10.2134/agronj2018.09.0608
72. Bacelo H, Pintor AMA, Santos SCR, Boaventura RAR, Botelho CMS (2020) Performance and prospects of different adsorbents for phosphorus uptake and recovery from water. Chem Eng J 381:122566. https://doi.org/10.1016/j.cej.2019.122566
73. Pinelli D et al (2022) Regeneration and modelling of a phosphorous removal and recovery hybrid ion exchange resin after long term operation with municipal wastewater. Chemosphere 286:131581. https://doi.org/10.1016/j.chemosphere.2021.131581
74. Bui TH, Hong SP, Yoon J (2018) Development of nanoscale zirconium molybdate embedded anion exchange resin for selective removal of phosphate. Water Res 134:22–31. https://doi.org/10.1016/j.watres.2018.01.061
75. Guida S, Rubertelli G, Jefferson B, Soares A (2021) Demonstration of ion exchange technology for phosphorus removal and recovery from municipal wastewater. Chem Eng J 420:129913. https://doi.org/10.1016/j.cej.2021.129913
76. Sendrowski A, Boyer TH (2013) Phosphate removal from urine using hybrid anion exchange resin. Desalination 322:104–112. https://doi.org/10.1016/j.desal.2013.05.014
77. Dong G, Huang X, Bi Y (2022) Anchoring black phosphorus quantum dots on Fe-doped W 18 O 49 nanowires for efficient photocatalytic nitrogen fixation. Angew Chemie 134(30). https://doi.org/10.1002/ange.202204271

78. Karthikeyan P, Banu HAT, Meenakshi S (2019) Synthesis and characterization of metal loaded chitosan-alginate biopolymeric hybrid beads for the efficient removal of phosphate and nitrate ions from aqueous solution. Int J Biol Macromol 130:407–418. https://doi.org/10.1016/j.ijbiomac.2019.02.059
79. Moawad A, Mahmoud UF, El-Khateeb MA, El-Molla E (2009) Coupling of sequencing batch reactor and UASB reactor for domestic wastewater treatment. Desalination 242(1–3):325–335. https://doi.org/10.1016/j.desal.2008.05.011
80. Weiss JS, Alvarez M, Tang C-C, Horvath RW, Stahl JF (2005) Evaluation of moving bed biofilm reactor technology for enhancing nitrogen removal in a stabilization pond treatment plant. Proc Water Environ Fed 2005(14):2085–2102. https://doi.org/10.2175/193864705783867035
81. Leyva-Díaz JC, Monteoliva-García A, Martín-Pascual J, Munio MM, García-Mesa JJ, Poyatos JM (2020) Moving bed biofilm reactor as an alternative wastewater treatment process for nutrient removal and recovery in the circular economy model. Bioresour Technol 299:122631. https://doi.org/10.1016/j.biortech.2019.122631
82. Kermani M, Bina B, Movahedian H, Amin MM, Nikaein M (2008) Application of moving bed biofilm process for biological organics and nutrients from municipal wastewater. Am J Environ Sci 4(6):675–682. https://doi.org/10.3844/ajessp.2008.675.682
83. Malovanyy A, Yang J, Trela J, Plaza E (2015) Combination of upflow anaerobic sludge blanket (UASB) reactor and partial nitritation/anammox moving bed biofilm reactor (MBBR) for municipal wastewater treatment. Bioresour Technol 180:144–153. https://doi.org/10.1016/j.biortech.2014.12.101
84. Spolaore P, Joannis-Cassan C, Duran E, Isambert A (2006) Commercial applications of microalgae. J Biosci Bioeng 101(2):87–96. https://doi.org/10.1263/jbb.101.87
85. Cassidy DP, Belia E (2005) Nitrogen and phosphorus removal from an abattoir wastewater in a SBR with aerobic granular sludge. Water Res 39(19):4817–4823. https://doi.org/10.1016/j.watres.2005.09.025
86. Sarma SJ, Tay JH (2018) Aerobic granulation for future wastewater treatment technology: challenges ahead. Environ Sci Water Res Technol 4(1):9–15. https://doi.org/10.1039/c7ew00148g
87. Fernández I, Vázquez-Padín JR, Mosquera-Corral A, Campos JL, Méndez R (2008) Biofilm and granular systems to improve Anammox biomass retention. Biochem Eng J 42(3):308–313. https://doi.org/10.1016/j.bej.2008.07.011
88. Seviour TW, Lambert LK, Pijuan M, Yuan Z (2011) Selectively inducing the synthesis of a key structural exopolysaccharide in aerobic granules by enriching for Candidatus 'competibacter phosphatis.' Appl Microbiol Biotechnol 92(6):1297–1305. https://doi.org/10.1007/s00253-011-3385-1
89. Bassin JP, Pronk M, Muyzer G, Kleerebezem R, Dezotti M, van Loosdrecht MCM (2011) Effect of elevated salt concentrations on the aerobic granular sludge process: Linking microbial activity with microbial community structure. Appl Environ Microbiol 77(22):7942–7953. https://doi.org/10.1128/AEM.05016-11
90. Abdullah N, Ujang Z, Yahya A (2011) Aerobic granular sludge formation for high strength agro-based wastewater treatment. Bioresour Technol 102(12):6778–6781. https://doi.org/10.1016/j.biortech.2011.04.009
91. Nguyen PTT, Van Nguyen P, Truong HTB, Bui HM (2016) The formation and stabilization of aerobic granular sludge in a sequencing batch airlift reactor for treating tapioca-processing wastewater. Polish J Environ Stud 25(5):2077–2084. https://doi.org/10.15244/pjoes/62736
92. Vymazal J (2014) Constructed wetlands for treatment of industrial wastewaters: a review. Ecol Eng 73:724–751. https://doi.org/10.1016/j.ecoleng.2014.09.034
93. Parde D, Patwa A, Shukla A, Vijay R, Killedar DJ, Kumar R (2021) A review of constructed wetland on type, treatment and technology of wastewater. Environ Technol Innov 21:101261. https://doi.org/10.1016/j.eti.2020.101261

94. Zeng RJ, Lemaire R, Yuan Z, Keller J (2003) Simultaneous nitrification, denitrification, and phosphorus removal in a lab-scale sequencing batch reactor. Biotechnol Bioeng 84(2):170–178. https://doi.org/10.1002/bit.10744
95. Parde D, Behera M (2023) Challenges of wastewater and wastewater management. In: Sustainable industrial wastewater treatment and pollution control. Springer Nature, Singapore, pp 229–255

Anaerobic Digestion (AD) and Resource Recovery

Nagarjuna Kandagatla, Pilli Sridhar, P. V. Rao, and R. D. Tyagi

Abstract This chapter delves into the captivating realm of Anaerobic Digestion (AD) and its remarkable potential for resource recovery. AD is a sustainable and effective biological process for converting organic waste into energy and fertilizers. A comprehensive overview of the AD process, with a focus on its mechanisms, operating parameters, and key factors affecting its performance. In addition, AD can utilize a variety of feedstocks, including farm residue, food waste, and wastewater sludge. In addition, a critical analysis of the benefits and challenges associated with each feedstock, including their availability, composition, and economic feasibility. The chapter also investigates various technological advancements and process optimization strategies to maximize the potential of AD and resource and nutrients recovery. The chapter highlights emerging technologies such as co-digestion, pre-treatment techniques, and reactor configurations that enhance biogas production and digestate quality. Furthermore, the chapter emphasizes the prospects and challenges in circular economy concepts, elucidating its ability to recover and harness valuable resources from organic waste streams. Finally, the techno-economic feasibility of AD technology is discussed.

Keywords Anaerobic digestion · Resources recovery · Circular economy · Feed stocks

N. Kandagatla · P. Sridhar (✉) · P. V. Rao
Department of Civil Engineering, National Institute of Technology Warangal, Warangal 506004, India
e-mail: srenitw@nitw.ac.in

N. Kandagatla
e-mail: kn21cerer14@student.nitw.ac.in

P. V. Rao
e-mail: pvenku@nitw.ac.in

R. D. Tyagi
Dongguan University of Technology, Dongguan, China

M. M. Ghangrekar et al. (eds.), *Biological and Hybrid Wastewater Treatment Technology*, Earth and Environmental Sciences Library,
https://doi.org/10.1007/978-3-031-63046-0_8

1 Introduction

Global annual production of lignocellulosic biomass derived from agricultural waste amounts to 181.5 million tons. Based on crop analysis, the pricing of lignocellulosic biomass varies between 24 to 120 USD per ton. Energy production from lignocellulose biomass contributes 10% of global energy requirements, which illustrates the significance of waste to energy technologies [1]. Various forms of bioenergy can be produced from organic waste using multiple processing methods. In that sense, anaerobic digestion (AD) with different reactors is one of the highly promising technology for converting solid and liquid wastes because it generates methane-rich biogas and nutritious digestate. Pollution control and energy recovery are two desirable outcomes of AD, a waste treatment method. AD is a best option considering the large concentration of readily biodegradable elements in many industrial and agricultural wastes. Waste management and energy recovery from agricultural, industrial, and municipal waste have been handled with anaerobic digestion techniques for many years. Because of the significant scale-up, AD surpasses aerobic alternatives in many ways. Its benefits include the ability to operate at a substantial organic load rate, easy digestate management, less space, incubator requirements, principles of sustainable energy production, and material recovery in line the concepts of CE [2, 3].

AD is limited in applicability due to inadequate methane generation and process instability. Because inhibitory compounds are abundant in waste, they cause anaerobic digester upset or failure [4]. Researchers have conducted numerous studies to determine the inhibition's mechanism and regulatory elements. Ammonia, sulfide, and metal ions are common inhibitors in anaerobic digesters. Still, AD is a rapidly evolving technique that ensures the successful breakdown of organic compounds and biogas generation [3]. The primary research goal is to improve the bioreactor's potential to operate steadily and adapt to adverse process conditions. Increasing phosphorus and nitrogen removal rates would also improve the adaptability of such systems. The first part of the twentieth century witnessed a rise in the production of biogas, but, due to the low cost and widespread availability of fossil fuels, this trend declined in the 1950s. Following the energy crisis of the 1970s, AD became increasingly popular as a renewable energy source. However, there was a slowdown in AD implementation due to digester failure rates, poor design, and inadequate management, but this has now mostly been resolved [5]. In the early stages, research was primarily concerned with improving the operating conditions and adjusting the bioreactor setup and design. Experimental methods with different feedstocks were investigated at the same time. In response to the abundance of lignocellulosic biomass as feedstock, pretreatment strategies have been researched. The effectiveness of AD faces constraints due to drawbacks associated with using single substrates, including inappropriate carbon–nitrogen (C/N) ratios, inadequate substrate pH, inadequate buffering capacity, and high ammonia concentrations. Because of this, the co-digestion of combination substrates for biogas production has recently attracted increased interest [2, 3]. The literature review highlighted six key areas for biogas enhancement: genetics, pretreatments methods, additives, co-digestion, digester design, and functional parameter optimization [6].

2 Pathways for Anaerobic Digestion

During AD, four sequential stages take place: Hydrolysis, acidogenesis, acetogenesis, and methanogenesis. Microorganisms engaged in fermentation initially break down complex substances like carbohydrates, proteins, and lipids into simpler, soluble organic compounds, such as sugars, fatty acids, and amino acids. A critical step limiting the rate of the entire process occurs when hydrolytic microorganisms adjust the pH to break down cellulosic matter. Techniques for accelerating hydrolysis in anaerobic digesters have received a lot of interest. Various pretreatment strategies for waste are being investigated and employed to maximize hydrolysis, particularly for digesters that process a significant quantity of heavy lignocellulosic biomass [7]. Volatile fatty acids (VFAs) and alcohols are formed in the acidogenic stage from the hydrolysis products. Compared to the other anaerobic digestion phases, acidogenesis usually happens more quickly, with acidogenic bacteria regenerating in less than 36 h. At this stage of the process, amino acids degrade into VFAs, and one product of this degrading process is ammonia, a known anaerobic digestion inhibitor at high concentration. Acetogenic bacteria then converts the VFAs into acetate, CO_2, and H_2. A significant amount of acetic acid is generated, together with less byproducts such as propionic acids and butyric acid. During the last phase of methanogenesis, methanogens metabolize acetate to produce biomethane [1]. It, being strictly anaerobic, has a somewhat more complicated process mechanism than those mentioned above. Methanogenic bacteria make biomethane in this stage. Methanogenic bacteria often thrive in environments with reduced redox potential and higher pH levels than those observed in the initial anaerobic digestion stages. Moreover, the regeneration of methanogens tends to occur more gradually compared to other microbes in anaerobic digestion, sometimes taking up to 16 days. Methanogenesis in batch reactors ends when biogas generation ends, which is possible after 40 days. Recent studies indicate that *Methanosarcina sp* are generally resilient to AD, even though methanogenic species are likely the most sensitive species. They can endure inhibitors of AD and adapt to abrupt shifts in pH at levels that would otherwise harm other species [7]. Every stage and its derivatives during the breakdown of lignocellulosic biomass are depicted in Fig. 1.

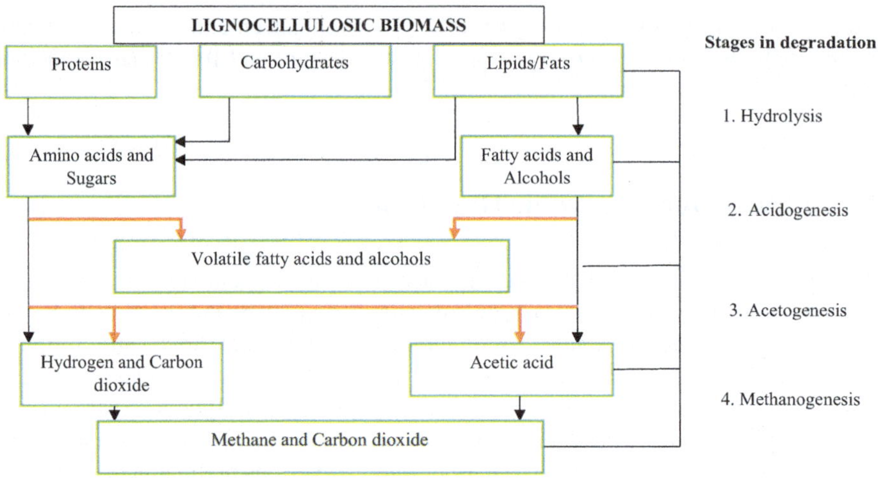

Fig. 1 Stages in anaerobic degradation process

3 Feed Stocks for Anaerobic Digestion

AD traditionally relies on animal manure (from pigs, cattle, and poultry) as its primary feedstock [8]. By the 1970s, AD had expanded its area of use to include industrial and municipal wastes due to environmental awareness and the need for renewable energy sources. Moreover, AD has successfully entered sectors traditionally dominated by aerobic systems, such as processing industrial effluents with low chemical oxygen demand (COD). This is attributed to implementation of high-rate reactor configurations and advanced process control devices. Among the biomass resources, there are three categories: (1) first generation, derived from food crops; (2) second generation, derived from lignocellulosic materials; and (3) third and fourth generation, derived from microalgae [9]. Animal manure, agricultural residue, food waste, and municipal wastewater are important sources of lignocellulosic biomass for bioenergy generation [10]. Table 1 lists the substrate properties (C/N ratio, TS %), methane output and the challenges associated with AD.

Food waste (FW) in landfills can result in notable environmental harm by releasing greenhouse gases, including methane (CH_4) and carbon dioxide (CO_2), contributing to adverse ecological effects. GHG emissions contribute to global warming, and methane is one of the most powerful greenhouse gases, with a greenhouse effect 25 times stronger than that of CO_2 [11]. Utilizing AD for managing household food waste leads to a net-negative emission of greenhouse gases, amounting to − 39 kg CO_2eq/t, and conserves 2 GJ/t of energy. The AD system is confirmed to protect the ozone layer, water, metal, and fossil fuel depletion, as well as possible human and environmental damage [12].

Table 1 Different substrate properties (C/N ratio, TS %), methane output and the challenges associated with anaerobic digestion

Feedstock type	Challenges	C/N ratio	TS %	Methane yield M^3/Kg VS	References
Food waste	Contamination with non-biodegradable materials, variations in waste composition, and the presence of inhibitors can affect the process	11.4–36.4	4.0–41.5	0.46–0.53	[10, 14]
Agriculture residue					
Rice husk	High silica content, which can hinder the digestion process	74.2–99.21	70–95	0.278	[1, 15, 16]
Rice straw	Its high lignocellulosic content, which makes it less readily biodegradable. Effective pretreatment methods are required	43.06–45.52	59.7–75.7	0.302	[1, 15, 16]
Maize straw	Complex structures that can be challenging to break down, Wheat straw may have imbalances in carbon-to-nitrogen, fibrous materials in wheat straw can lead to issues such as fouling, decomposition of maize straw, and wheat straw can be slow	82.88	70–85	0.338	[1, 17]
Wheat straw		158.73	70–85	0.29	[1, 17]

The undeniable reality is that the waste produced from agricultural activities is a major contributor to greenhouse gases (GHGs) emissions, the Intergovernmental Panel on Climate Change (IPCC) in 2007 underscored. Agricultural waste alone is accountable for an annual emission of 405 million metric tons of carbon dioxide equivalent ($MtCO_2eq$). Consequently, converting this waste into energy is a promising and impactful solution [13]. Several forms of lignocellulosic biomass may be classified as either woody or non-woody using just their chemical composition and physical characteristics. Agricultural wastes are the most plentiful and renewable resource on Earth. Straw and other lignocellulosic crop wastes are typically burned in an effort to prepare the soil, however that approach is linked to significant pollution of the environment and nutrient loss [9]. About 50–90% of the total organic matter in agricultural plant wastes is typically made up of lignocellulosic content. The composition of lignocellulosic biomasses of agricultural wastes includes cellulose (45–65%), hemicellulose (25–35%), and lignin (20–40%). There is an indirect relationship between how much lignin is in the starting material of composted organic matter and the breakdown of organic matter.

4 Anaerobic Reactor Conditions with Optimal Performance

4.1 Temperature

As temperature significantly impacts the activity of various organisms, the quality of sludge, hydrolysis rates, and the subsequent production of biogas, it stands as one of the pivotal operational factors in AD [6]. Three temperature ranges are generally suitable for AD microorganisms to operate in: mesophilic, thermophilic and psychrophilic Temperature rises within limited bounds can lead to improved AD performance from the perspective of biochemical processes. It has generally been observed that thermophilic operation performed better at accelerating endergonic metabolic processes, stimulating the formation of biogas, and decreasing ammonia accumulation than mesophilic temperature conditions. However, it has also been mentioned that exergonic metabolic processes may not always benefit from thermophilic AD operation. There is widespread consensus that mesophilic processes are generally considered optimal for achieving efficient degradation of anaerobic organic matter, while thermophilic conditions may be employed in specific circumstances [2]. During thermophilic AD, organic matter degrade faster and pathogens are eliminated more effectively. The methanogenesis phase stands out as the most sensitive level in AD process, serving as the bottleneck where alterations in temperature significantly impact the entire process. Bandgar et al. [18] reported that Ammonia inhibition occurred as temperatures raised from mesophilic to thermophilic environments; however, it is possible to minimize or eliminate this type of inhibition by raising C/N ratio of diverse feedstock to a suitable level. An effective approach to attain the desired digestion performance involves mixing of diverse biomass in anaerobic co-digestion, considering appropriate feed properties like pH, nutrients, and C/N ratio [18].

4.2 Substrate Characteristics

The digestive process is also influenced by the characteristics of the substrate, such as the Microbial Composition, Total Solids, Volatile Solids (VS), C/N Ratio, Trace Elements, and Chemical and Biochemical Oxygen Demands (BOD and COD). Mainly, C/N ratio is indicative of the nutrient composition in a digestion substrate, influencing the sensitivity of digestion systems. A higher C/N ratio leads to a decreased rate of protein dissolution, resulting in lower concentrations of total nitrogen and free ammonia. Optimized C/N value in the AD process can help prevent ammonia inhibition. However, an overly elevated C/N value might offer inadequate nitrogen for sustaining microbial cell growth, leading to rapid nitrogen breakdown and subsequently diminishing biogas production. Conversely, feedstocks characterized by an insufficient C/N ratio carry the potential for ammonia inhibition, posing a threat to methanogens and leading to suboptimal utilization of carbon reservoirs [19]. Hence, various substrates exhibit distinct C/N ratios, with some examples provided in Table 1. The ideal C/N value for better AD generally lies between 20 to 35, and a commonly utilized ratio is 25 [9]. Another crucial parameter is the moisture content (MC) of biomass. Despite these substrates generally having high MC, the levels of moisture can vary among different substrates, and this is a parameter that can influence the process of AD. However, the drawback of high water content poses a limitation on achieving elevated organic input and gas quantity. Therefore, the approach of incorporating more energy-dense biomass (such as crop residues) seems promising for increasing the organic loading rate (OLR) without significantly reducing the hydraulic retention time (HRT).

4.3 pH

The operational consistency and performance of methanogenic and acidogenic bacteria are significantly influenced by pH variations, underscoring pH as a vital factor in the real-time identification and control of the AD process. With the ideal pH values occurring from 7 to 8, methane production occurs within a highly narrow pH range of around 6.5–8.5. The pH must not fall below 6.0 or rise over 8.5 to significantly slow down the process. The pH range of 6 to 8 is generally considered ideal for generating quality biogas. Research has explored pH adjustment in both anaerobic co-digestion and mono-digestion systems to improve biogas production efficiency [6].

4.4 Organic Loading Rate

The OLR denotes the daily input of VS into a reactor in continuous feeding systems. Majority of municipal AD units typically operate with an OLR ranging from 1.2 to 12.0 kg VS/m^3.Day or 2.2 to 33.7 kg COD/m^3.Day. Because of changes in feedstock parameters, operating temperature, HRT, OLRs of different AD processes differed. Inadequate OLR may induce hunger of microorganisms, leading to reduced efficiency in AD. Whereas too much OLR can result in VFA buildup in the digestion system, resulting in suppression of bacteria and potentially leading to operational challenges and system breakdown. In general, when OLR grows, biogas yield increases to some amount. The optimal sustainable OLR value should be computed based on the type of biomass, reactor conditions, and the operating cost [6]. Moreover, thermophilic systems and effluent recirculation can significantly alleviate overloading inhibitions [19].

4.5 Retention Times

Retention times refers to the duration required for the full degradation of substances. It is linked to the pace of microbial growth and is influenced by factors such as process temperature, OLR, and composition of feedstock [19]. This parameter related on two key types of retention time: SRT, denoting average period bacteria reside in a digester, and HRT, determined by the ratio of digester volume to the flow rate over a specific duration. It usually takes5–30 days of HRT for AD process to treat [6]. Excessive HRT results in inadequate feedstock utilization, whereas low HRT results in VFA accumulation, which promotes bacteria inhibition and potentially AD system failure [6]. In essence, adopting a low OLR and an extended HRT proves to be the optimal approach for maintaining consistent and high quantity of methane. Fluctuations in the SRT disrupt and compromise the performance of AD.

5 Resource and Nutrient Recovery from AD

By utilizing biomass containing rich in biodegradable organic matter and nutrients, anaerobic digestion generates biofuels such as VFAs, biohydrogen, and biomethane [20]. Importantly, nitrogen and phosphorus are macronutrients recovered from anaerobic digestate that can offering substitutes for materials derived from non-renewable sources [21]. Currently, one effective waste recycling technique is the valorization of digestate into high-value goods [22]. If sufficiently collected and used, digestate may be a renewable resource. Techniques other than land application are highly relevant from an economic point of view. Methods for recovering nutrients, growing biomass like microalgae, producing charcoal, biofuel, and bio-oil through integrated

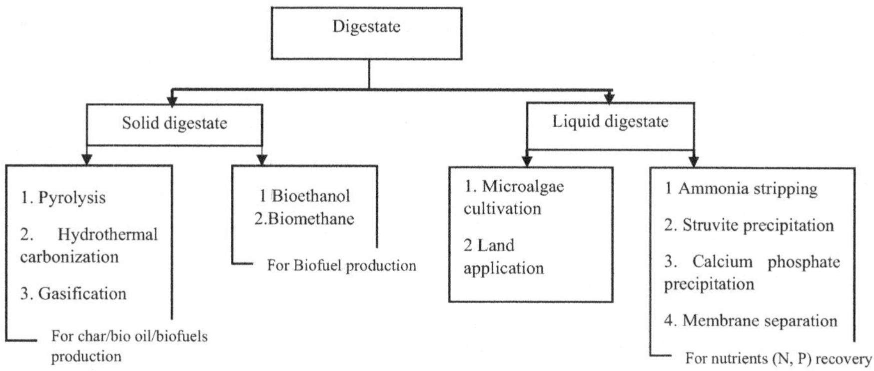

Fig. 2 Various usages and nutrient recovery from digestate

processes have all been outlined in Fig. 2. Recent research has shown that adding pyrochar or hydrochar can increase the generation of biogas in AD [23]. Ambaye et al. [24] studied how adding pyrochar effects on AD and observed that pyrochar inhibited the generation of VFAs with increasing biomethane production. Moreover, recovery of N and P from digestate was assisted by the use of digestate-derived pyrochar, which performed better at recovering P than activated carbon and zeolite. Considering the limited availability of P, research is being conducted on reclaiming P from digestate [23]. The nutrients present in the liquid effluent derived from AD can serve multiple purposes, including supporting algae farming or facilitating the production of struvite, a slow-release fertilizer [5].

5.1 Biohydrogen

By employing specific microorganisms in a bioreactor, biohydrogen can be produced. Typical biohydrogen production approaches compass dark fermentation, photofermentation, direct photolysis, indirect photolysis, microbial electrolysis, and sequential dark and photofermentation [25]. It is becoming more popular to produce biohydrogen via fermentative routes because it consumes less energy than biomass derived from non-petroleum sources. Microbes can exploit different feedstock, such as easily degradable carbohydrates, food leftovers, and industrial discharges such as cheese whey, starch process, and farm residue. As a result, fermentation of biohydrogen has been upgraded from a laboratory to an industrial scale [26]. Research into various aspects of dark fermentation has been conducted in the past four decades [26]. These include determining of operational parameters, selecting suitable inoculums, exploring different types of substrates and their compositions, investigating the role of genetic engineering in suppressing genes encoding uptake hydrogenase, and designing appropriate bioreactors based on the optimal height-to-diameter ratio.

The lack of commercialization of dark fermentation has been due to certain pre-process inhibitors, such as inhibitory H_2-consuming bacteria and metal ions. During the AD process, inhibitors can arise, including the production of undesired phenolic substances and unidentified derivatives during substrate preprocessing. The formation of ammonia is notable during dark fermentation, and a significant issue is the buildup of organic acids and alcohols as by-products [26]. BioH_2 emerges during the acidogenesis and acetogenesis phases of AD. Regarded as a prospective fuel, this gas carries a multitude of societal, financial, and ecological advantages, serving as an alternative to mitigate air pollution and reduce dependency on fossil fuels in the quest for a sustainable energy future [20]. Modern approaches for enhancing biohydrogen production encompass the immobilization of microbial cultures, updating bioreactors, fine-tuning process conditions (such as temperature, pH, OLR, and HRT), selecting and enriching cultures, and metabolic engineering of specialized biohydrogen producers [25].

5.2 Biomethane

By using AD, biomethane is usually produced from lignocellulosic biomass, which can be used as fuel or for generating power. AD biomass is classed depending on treatment process factors such as thermal conditions, MC, and methodology employed for reactor loading. The kind of stage procedure (single, double, multiple stages) also impacts the AD process's performance. AD classification extends to considerations of thermal preferences (thermophilic or mesophilic), the process type (continuous or batch), and the digestion state (dry or wet).Emerging technologies such as anaerobic co-digestion, and reactor configurations two-Stage AD that enhance biogas production and digestate quality [27]. Co-digestion is the process of entails combining different types of substrates that are combined use in AD. Numerous anaerobic digesters employ the mono-digester system, focusing on a single feedstock type. However, this approach comes with notable drawbacks, including fermentation instability, reduced methane production, and restricted access to the feedstock year-round, potentially impacting continuous operational efficiency. In some situations, a quick accumulation of larger concentrations of VFAs, which is harmful to methanogens and has a substantial impact on the digestion. Furthermore, feedstock with a high protein concentration might produce ammonia and hydrogen sulfide, imposing constraints on the overall biogas production [27]. The pre-treatment of biomass before to the AD process is important. This pre-treatment has significance for improving the overall effectiveness of the AD by increasing the anaerobic bacteria to respond. The design of AD reactors aims to increase production by optimizing reactor by feed biomass pre-treatment [1]. List of pretreatment methods for enhancement of bio hydrogen and methane mentioned below in Table 2.

Table 2 Various pretreatment methods for enhancement of AD

Pretreatment methods			
Physical methods	Chemical methods	Biological methods	Hybrid methods
1. Thermal 2. Freezing and thawing 3. Microwave irradiation 4. Ultra sonication 5. Radiation 6. Milling	1. Acid and alkali 2. Ozonation 3. Chloroform 4. Iodopropane 5. Acetylene	1. Enzymatic treatment 2. Fungal treatment	1. Thermal + acid 2. Thermal + alkali 3. Acid + alkali 4. Dilute acid + microwave 5. Dilute alkali + microwave 6. Biological + dilute acid 7. Ultrasonic + ionic 8. Supercritical CO_2 + steam explosion 9. Enzyme hydrolysis + steam explosion

5.3 Volatile Fatty Acids

There is wide use of short-chain VFAs and their derivatives in the food, pharmaceutical, and textile industries. This concept allows the production of VFAs during the acidogenesis of substrates as intermediate products [20]. The production of VFAs through dark fermentation (DF) holds promise, especially when utilizing abundant wastes and wastewater generated on a large scale by industries. Various substrates, including food waste, palm oil effluent, landfill leachate, sewage sludge, and the simultaneous fermentation, have already been investigated for VFAs production through dark fermentation. Blend of these VFAs holds significant applications in facilities for treating wastewater, aiding biological nitrogen and phosphorus removal, biosynthesis of mixed alcohols, and the manufacturing of biodegradable polymers. Furthermore, VFAs may be used to create power in microbial fuel cells while simultaneously acting as a carbon source in biodiesel bioprocessing [28].

5.4 Nutrients from Anaerobic Digestate

The digestate resulting from anaerobic decomposition contain significant amounts of nitrogen and phosphorus. Prior to discharging the effluent into the public water environment, it is essential to eliminate nitrogen and phosphorus to prevent issues related to eutrophication. In the anaerobic conditions, ammonium nitrogen remains in the liquid phase of digestate without experiencing oxidation, allowing for the recuperation of nitrogen during the AD. A common method of treating dewatered sludge from AD processes is to burn, compost, apply to farmland, and landfill it. Biomass waste is stabilized through composting, and pathogens and heavy metals are reduced through composting. Fertilizers made from well-aged compost possessing humic attributes prove beneficial for plant growth [29]. The potential use of AD dewatering digestate as a nutrient recovery resource. Nitrogen recovered as ammonia or ammonia compounds technologies include biological nutrient accumulation via zeolite adsorption, struvite, filtering, reverse osmosis, and NH_3 stripping-absorption. Among all of these methods, the ammonia stripping approach appears to be the most feasible since it creates minimal sludge and is simple to operate [29]. The retrieval of phosphorus in the form of struvite and the novel P recovery method coupled with the granular sludge system. Presently, two predominant approaches exist for utilizing digested liquid ammonia: direct application onto agricultural fields as a fertilizer and recycling ammonia.

5.4.1 Nitrogen (N) and Phosphorus (P) Recovery Methods

A typical phosphorus recovery product from AD is struvite, which is chemically formed from $NH_4MgPO_4 \cdot 6H_2O$. Through struvite production, phosphate and ammonium nitrogen may be extracted from the soluble phase at the same time [29]. The recovery of nitrogen can be achieved through two methods: struvite crystallization and ammonia stripping. Using steam stripping or air, ammonia may be extracted from liquid digestate [30]. A sulfuric acid solution may absorb the ammonia extracted by air stripping to form ammonium sulfate, the most common fertilizer [29]. It is inexpensive, easy to install, and has a high ammonia recovery capability. Temperature and pH are essential process factors. Furthermore, unlike other ammonia recovery processes, stripping permits feed with high solids concentrations, removing the requirement to remove solids from digested liquid. Direct application of side-stream stripping for ammonia can be employed in anaerobic digesters. This approach effectively maintains ammonia levels in the digestate below inhibitory thresholds for methanogenic bacteria. Additionally, it is essential to provide feedstock with elevated nitrogen concentrations through the use of nutrient-rich substrates [30]. The primary sources of the excess phosphorus in the streams are livestock and sewage. Furthermore, phosphorus may be obtained either prior to or during the AD process. Less phosphorus is actually left in the solid fraction than is dissolved in the liquor phase following solid–liquid separation, making up 35–45% of the total digestate [29]. Moreover, phosphorus can be extracted from incineration ash using a thermochemical technique. During the incineration process, all organic matter was eliminated, and the residue had a high potassium and phosphorus content. But, the direct use of sewage sludge ash on farms is restricted due to its limited bioavailability and heavy metal concentration [29]. Frequently, operational problems occur due to crystallization in equipment/piping interference with Struvite crystallization and scaling corrosion due to Ammonia stripping [31]. Economic feasibility analysis Nutrient recovery methods, like as fertilizers, differ not only in efficiency but also in expense. In the case of ammonia stripping, this process of extracting nitrogen from waste effluents will generate financial advantages for farms or organic waste digesters on a sufficient scale [23].

6 Residual Pollutants in Digestate

Digestate serve as crucial recycled fertilizers, imparting beneficial effects on soil parameters. In spite of this, contaminants such as heavy metals and the emissions of greenhouse gases derived from digestate make it imperative to manage properly to ensure its safe discharge [32]. Consequently, the makeup and quality of digestate are highly contingent on the characteristics of the feedstock used. Nonetheless, there is a noticeable lack of knowledge on the presence and concentrations of organic micro pollutants in the residual anaerobic digestion material [33]. Government legislation, along with its associated regulations and quality controls, differs from one country to another. Typically, the optimization of AD processes is geared towards improving biogas yield, often overlooking the quality of digestate. This raises concerns about pollution and the potential transmission of communicable diseases upon digestate disposal. To mitigate contamination risks and prevent disease transmission into the food chain, it is imperative to implement rigorous control measures. This involves establishing appropriate treatment levels and setting standards for the safe disposal of AD digestate [21].

High-quality digestate has the following qualities:

1. **Purity**: The digestate should be free from physical impurities such as plastic, stones, glass, and other non-digestible matter. A high-quality digestate should have a clean and uncontaminated composition.
2. **Hygiene or Sanitization**: High standards of hygiene must be maintained to ensure that the digestate is devoid of any pathogenic and undesirable biological content. This is crucial for preventing disease spread and ensuring agricultural applications' safety.
3. **Safety**: The digestate should be safe for both living organisms and the environment. This involves minimizing or eliminating any harmful substances or contaminants that could have adverse effects on human health, wildlife, or the ecosystem. Ensuring safety is paramount for sustainable and responsible use of digestate in various applications.

6.1 Pollutants

The application of contaminated digestate to soil can pose risk for animal or human health or cause pollution of natural environment. The pollutants in digestate can be divided into three main groups: chemical, biological and physical.

6.1.1 Chemical Pollutants

Digestate, a byproduct of anaerobic digestion, can potentially contain various chemical pollutants. The composition of chemical pollutants in digestate is influenced

Table 3 The permissible levels of heavy metals in quality digestate vary based on regulatory criteria in different countries [35]

Heavy metal	UK PAS 110:2010	Sweden SPCR 120	France NFU 44–051
	mg/kg	mg/kg	mg/kg
Cd	1.5	1	3
Cr	100	100	120
Cu	200	600	300
Hg	1	1	2
Ni	50	50	60
Pd	200	100	180
Zn	400	800	600

by the characteristics of the feedstock used and the anaerobic digestion process. Common contaminants found in digestate include heavy metals, polycyclic aromatic hydrocarbons (PAHs), polychlorinated biphenyls (PCBs), pesticides, and various crop protection products (CPPs) along with disinfectants [34]. Heavy metals such as Cd, Cr, Hg, Pb, Cu, Ni and Zn play an important role in digestate quality and fertiliser applicability. Table 3 presents an overview of quality standards for heavy metals in the utilization of AD digestate, including their respective maximum limits. These standards are taken from quality criteria established in the United Kingdom, Sweden, and France for digestate. In Table 4 Characteristics and heavy metals of different anaerobic digestates mentioned.

6.1.2 Physical and Biological Pollutants

From processing waste material to storing the digested slurry, both pathogenic and nonpathogenic survive in various anaerobic digestion processes. Pathogens such as *Salmonella* and *Mycobacteria paratuberculosis* are mostly destroyed by temperature; however, in the case of Clostridium, the spores were not destroyed by either mesophilic or thermophilic AD. Conversely, it has been documented that C. *perfringens* decay is brought on by high concentrations of organic acids at low pH values (4.5–5.5) during thermophilic AD [21]. Therefore, before applying the digested residue to the soil, they need to be sufficiently treated and shown to be hygienically safe. Although the application of digestate to agricultural land is a straightforward method, it comes with several drawbacks. These include issues such as ammonia volatilization and nutrient leaching, contributing to water bodies' eutrophication. Additionally, there are concerns about heavy metal pollution and biological contamination. The latter can disseminate pathogens and cause physical contamination, such as the presence of stones and plastics, resulting in diminished long-term crop productivity.

Table 4 Characteristics and heavy metals of different anaerobic digestates [21]

Digestate	Characteristics of digestate						Heavy metals (mg/kg)					
	pH	TS%	TC	TN	AN	TP	Zn	Cu	Cd	Cr	Ni	Pb
*Food waste	7.9–8.2	1.4–6.1	311–452	112–157	52–108	6–16	113–474	14–70	0.3–1.0	13–29	25–39	0.1–10
*Agricultural feedstock	7.5–8.4	7.4–24.0	300–400	22–88	6–45	2–66	43–3884	1–1373	0–10	2–55	2–36	0–46
Municipal waste	7.6	–	–	37	–	0.65	401	58.7	1.5	13	6.7	11.7

Total carbon (TC) (g/kg), Total nitrogen (TN) (g/kg), Ammonical nitrogen (NH_4–N) (g/kg), Total phosphorus TP (g/kg),), *Heavy metals Units: g/kg of dry matter

7 Challenges and Prospects of Bio-Based Circular Economy

In terms of sustainability and resource efficiency, the circular economy and bio economy structures complement one other. Globally, the prospect of transitioning to a bio-based circular economy presents a transformative pathway towards sustainable development [36, 37]. The bio-based economy has potential and there are areas that require more work to resolve potential problems. According to the assessed scientific works, the challenges are primarily technological, economic and societal or cultural [37]. Consequently, it is possible to believe that the emergence of nature and knowledge excludes solutions that are not based on economic and technological principles [38]. The integration of a bio-based circular economy contributes to ecological and socio-economic sustainability, yielding positive outcomes for both the environment and society. Numerous studies emphasize these benefits acrossh four key dimensions: resource utilization, environmental impact, economic aspects, and social advantages, including the generation of employment opportunities [39].

7.1 Circular Economy for Economic and Ecological Sustainability

There are many prospects for the transition to a circular economy in terms of both environmental and economic benefit. Moving to more circular economic models promises a much brighter future for the nation's economy, by helping to decouple economic growth from resource use and its impacts, it offers the prospect of sustainable growth [39]. This reduces the possibility of resource use conflict and minimizes the environmental effect of extraction of resources. Furthermore, while the bio-based economy is being built through industrial symbiosis, the cascading use of biomass, and other creative methods of producing bioenergies [38]. The carbon footprint of bio-based processes is often lower than that of traditional processes, contributing to the reduction of greenhouse gas emissions. In a geopolitical context, biomass recovery of nutrients may be significant because most nutrient reserves are found in politically unstable nations. Recovery techniques could be able to decrease energy constraint and wood burning in low-income countries' rural areas. Using biomass might lower the costs of several operations that require carbon-rich materials as fuel [37]. The bio-based economy is driven by the expectation that it would bring about positive changes, such new job possibilities and rural development. Additionally, encouraging the decentralization of biomass processing could help rural communities to function as primary producers as well as processors. Bio-waste refining can be utilized in a bio-based circular economy to produce bioenergy and bioproducts if the biowaste is refined using enormous technology, as outlined in Fig. 3. The European Union (EU) has identified bio-based plastics (or bio-plastics) as a viable alternative

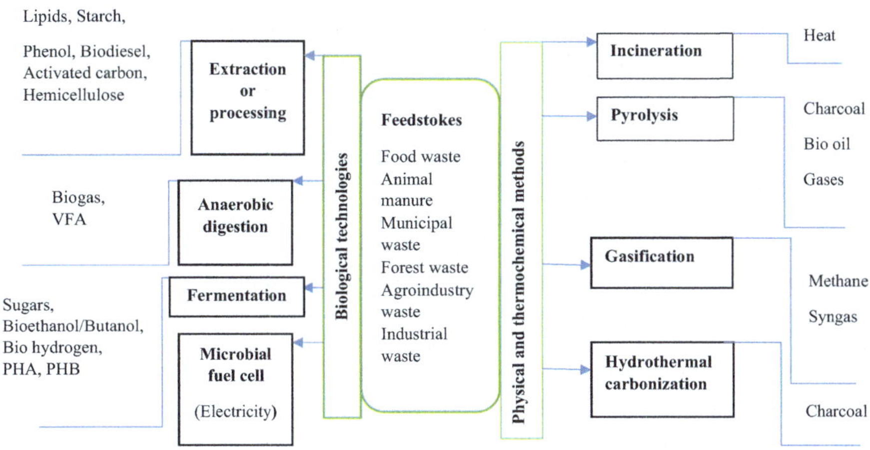

Fig. 3 Bio-based circular economy: sources of bio-waste, byproducts, and bioenergy technologies

to conventional plastics in terms of limited fossil resources, a healthy environment, and climate change. The European plastic strategy intends to connect plastic-related laws with the action plan for the circular economy (CE) [40].

7.2 Challenges

Establishing a bio-based economy faces diverse challenges that depend on factors such as country, geographical region, and the technological requirements of bioenergy production. To ensure the sustainability of these processes, it is essential to conduct a thorough evaluation encompassing both techno-economic analysis and life cycle assessments. The implementation of bioenergy within the context of a circular economy poses various challenges, and some of these are outlined below.

Managing agricultural waste sustainably is essential to advancing the circular economy. The first issue is lack of exact predictions in feedstock to advice end users and regulators on agricultural residue management practices that are both economically and environmentally problematic [41]. The boundaries of the agricultural residue conversion technology available now raise the second challenge. Reusing anaerobic digestate as a renewable fertilizer comes with risks for the environment and public health in addition to storage challenges imposed on by societal perceptions. Moreover, stablishing connections amongst industries, exchanging expertise on residue management, and resolving technical problems, legal in nature and consumer acceptability, throughout implementation are among the major problems [42].

While bio-based sustainable products, such as bioplastics, hold great promise as alternatives to petroleum-based plastics, the introduction of such replacements to the market comes with several challenges and issues. Because waste management laws are not enough, recycling bioplastics raises issues about compromising current recycling facilities [40]. Required legal frameworks worldwide should also better align to enable sustainable alternatives and the bio-based circular economy. Considering the possible advantages, there is still an obstruction to consumer acceptance of items made with bioplastics. Researchers are still in the early phases of analyzing customer preferences, attitudes, choices, and perceptions of bioplastics. Making transition from the conventional petrochemical industry to the circular bio economy for agricultural wastes requires large capital expenditures for new machinery configurations and technological advancements. Recovery technology is emphasized as determining product quality in a bioeconomy context, as well as hygienic challenges, which are crucial. When handling certain substances such as food waste or faecally contaminated waters [37]. Bio-based sectors are expected to benefit from the transition by creating new value chains, products, and goods. However, not all industrial infrastructure and structures are typically flexible enough to facilitate innovation. Moving away from conventional mass manufacturing to more technology- and knowledge-intensive processes can be difficult in traditional sectors because they often have a conservative view, are resistant to change [38]. Regulations, logistics management, geographic spread of businesses, system boundaries and leaks of matter, consumer acceptance, technology development and diffusion, and uncertainty over incentives and investments are the primary challenges that our model would encounter in the context of the circular economy.

8 Techno-Economic Feasibility of AD Technologies

Techno-economic analysis (TEA) and life cycle assessment (LCA) play a critical role in determining a process's feasibility when it comes to bioenergy and products made from biomass. Three areas of the anaerobic digestion process are significant during the techno-economic analysis: (1) unit operations with feed stock collection and transportation stages; (2) providing the method for biogas production; and (3) upgrading the bio-energy for various applications such as electricity and liquid methane for domestic cooking. Scientists and researchers must perform their study on anaerobic digestion at the systems level as well as from the perspectives of techno-economic and environmental advantages, as the global biogas sector is expected to reach $50 billion by 2026 [43]. The implementation of large-scale strategies for waste management, such as AD, requires economic evaluation since the establishment of these approaches largely depends on the technical and geopolitical conditions of the specific location. For this reason, techno-economic evaluations an important role in assessing a project's viability [44]. Integration of process modeling, engineering design, and economic evaluation is common practice in TEA. This means evaluating a process's economic viability, identifying technical feasibility and environmental sustainability cost-risk points throughout the development and implementation stages [28]. Therefore, TEA are essential for determining the feasibility of a projects [44]. Several industries and stakeholders find the production of biogas through anaerobic digestion to be a high-risk investment. This is primarily attributed to challenges related to factors such as year-round availability of feedstock supply, feedstock characteristics, profitability concerns, and the stability of digesters. However, the economic viability of biofuels and other renewable energy fields is very uncertain because inflation occurs in both developed and developing countries [1]. Numerous economic investigations have utilized Excel®, Cost–Benefit Analysis, Net Present Value (NPV), and Aspen Plus® to assess the economic feasibility of AD across diverse feedstocks [45, 46]. Table 5 shows the results of field study investigation of the techno economic feasibility of several locations using various methods.

The economic viability of a waste-to-energy project is significantly influenced by the reactor type and the strategy employed for biogas utilization. In a techno-economic study that compared three biogas utilization strategies and two types of thermophilic anaerobic reactors for generating biogas from vinasse, both the Fixed Bed Reactor (FBR) and Continuous Stirred Tank Reactor (CSTR) were utilized. The estimates for fixed and working capital were based on the Indonesian construction industry database. The investigation, conducted at a 10-day Hydraulic Retention Time (HRT) and an Organic Loading Rate (OLR) of up to 2.4 g/L.day, revealed that the FBR outperforms the CSTR [52]. Despite the potential for significant enhancement in biogas production through pretreatments, the economic viability of anaerobic digestion systems, whether with or without pretreatment, is contingent upon the presence of incentives or subsidies. This conclusion was drawn from a techno-economic investigation of various thermochemical pretreatment techniques (acid, alkaline, sulfite, and alkaline-H_2O_2) applied to the anaerobic digestion of dairy manure [45]. In the event that pretreatment is required, another study substantiated these findings. It disclosed that a mild thermal pretreatment at 60 °C for 1–3 h and 100 °C for 3 h could result in the financial benefit considering the increased biomethane production, the reduced sludge generation, the thermal pretreatment energy cost, and the total capital cost [53]. Advanced technology in anaerobic digestion, such as the Two-stage AD, offers numerous advantages, including increased methane yield and productivity, reduced reactor sizes, optimized processing conditions, enhanced buffer capacity, and the potential for producing value-added products. Recent publications indicate a methane yield increase of 10 to 30% through the implementation of two-stage AD. However, the industrial adoption of this technology remains limited due to reliability and operational issues. Technical challenges associated with two-stage AD include ammonia accumulation in the system due to recirculation and disturbances in the syntrophic relationships of microbes between the two stages. According to techno-economic analysis, the cost of implementing two-stage AD could be approximately 3% higher than that of a single-stage AD [54]. Even though AD technology has several proven benefits and is widely implemented around the world, its adoption in rural households has been limited. There is a lack of well-defined processes or frameworks in the literature for performing TEA, particularly for small-scale farm-based AD. Addressing this gap through a systematic economic evaluation that integrates both technical and economic performance is crucial. Overcoming feasibility challenges is essential for advancing the development of small-scale AD implementation.

Table 5 Case studies of techno economic feasibility with anaerobic digestion

Study area/ substrate	Methods/software followed for analysis	Aim	Most favored choice according to the results	References
Italy/Dairy cheese whey	Territory-oriented approach	• To assess the technical and economic feasibility of implementing anaerobic digestion at the dairy level	• When utilizing a ratio of six for the inoculum to substrate, good methane yields were achieved • Relative to low-frequency ultrasound an increase in ultrasonic energy does not result in a substantial rise in methane production • The results of the study concluded that most of the dairy's energy demand could be covered by anaerobic digestion of cheese whey	[47]
Western Europe/ macro-algae	ECLIPSE process simulation	• An anaerobic digestion of macro-algae combined heat and power system was analyses from a techno-economic perspective	• Several sensitivity analyses addressed the economic uncertainty • Biogas conversion rates and overall cost reductions require more systematic studies to reduce future financial risks	[48]

(continued)

Table 5 (continued)

Study area/ substrate	Methods/software followed for analysis	Aim	Most favored choice according to the results	References
China/ municipal solid waste	NPV, and levelized cost of energy methods	• To evaluate the financial sustainability of anaerobic digestion and landfill gas to energy projects	• Observed that anaerobic digestion is a more viable option than landfill gas to energy, with a higher net present value • Anaerobic digestion technology alone outperformed landfill gas to energy technology in all provinces in terms of its capacity to generate electricity	[49]
West Africa Ghana	TOPSIS	• Find the waste-to-energy technology option that offers the greatest techno-economic return on investment for Ghana	• The order of the most practical WtE technology for investment in Ghana is as follows: gasification, pyrolysis, anaerobic digestion, and plasma arc gasification	[50]
North Atlantic, Barbados	(Microsoft excel® software), waste reduction (WAR) algorithm	• Evaluation of the environmental effects and techno-economic viability of using food waste and brown seaweed as feedstock for bioenergy and fertilizer recovery	• AD and hydrothermal pretreatment bio refinery will benefit Barbados' socioeconomic and environmental conditions in a number of ways • Transfer to energy grid and the 100% exporting of the bio fertilizer to overseas markets can generate a maximum potential income of USD 12.76 million	[51]

(continued)

Table 5 (continued)

Study area/ substrate	Methods/software followed for analysis	Aim	Most favored choice according to the results	References
Cali, Colombia	Super pro designer®	• Technical and economic assessment of anaerobic digestion employing vinasse as feedstock in Colombia	• Proven the economic viability of a vinasse-fed AD plant in Colombia at processing flow rates greater than 25 m^3/h • Found that the solid concentration and the biomethane potential of sugarcane-based vinasse were negatively proportional	[44]

9 Recommendations

Effective communication among supply chain people and stakeholders remains a significant issue. This is strongly related to existing information on agricultural and AD activities. Having sufficient and high-quality feedstock is crucial for smooth plant operations [55]. Different stakeholders should play a key role in decision-making regarding local feedstock management policies, as their site-specific knowledge will improve its performance. Likewise, stakeholders have difficulty dealing with policy uncertainty when it comes to AD. AD can convert faecal matter to biogas, using biochar, and establishing GHG mitigation policies that account for inequity in sanitation access are some of the mitigation strategies for onsite sanitation emissions. Policymakers and regulatory agencies may successfully include climate change issues into sanitation policy. This leads to the establishment of policies targeted at facilitating the transition to a more climate-neutral sanitation [56]. It is possible to improve the performance of industrial-scale anaerobic reactors by optimizing the inoculation process, which would result in a reduction of the lag period [57]. In general, bioH_2 is believed to be a future fuel [26]. This fact is unquestionable because, despite amazing research done in this area so far, the main goal is still to effectively create a biological H_2 generation method that can be used on a large scale.

Anaerobic digestate needs to gain popularity in agriculture as an alternative for or addition to synthetic fertilizers, driven by both agricultural practices and policies aimed at climate change mitigation and soil development [58]. Membrane biological reactors are being used in AD combined with anammox processes, exhibiting significant potential for management waste biomass and phosphorus recovery [59].

10 Conclusions

When it comes to handling lignocellulosic waste, AD offers several advantages. Furthermore, recent findings have indicated the possibility of employing dry AD to extract extra value-added resource from solid waste. These advantages include lower water use, more processing capacity for waste, improved biogas output, and no additional wastewater generation. Suitable pretreatments that reduce health risks and improve biodegradability of lignocellulosic waste are preferred in order to improve AD efficiency and CH_4 yield. Pretreatment of feedstock, co-digestion, and microbial acclimatization can enhance the efficiency of anaerobic reactors in producing methane. More research is needed in anaerobic reactor technology to minimize costs, improve SRT, and simplify process control. Recently, biochar has emerged as a potential enhancer of the AD process, improving digestate quality by retaining nutrients. The cost-effectiveness of biochar production enables AD operators to readily utilize it without the necessity for recovery efforts. However, as biochar wasn't initially tailored for AD, future research should delve into the interplay between biochar and AD microbes.

References

1. Stanley JT, Thanarasu A, Kumar PS, Periyasamy K, Raghunandhakumar S, Periyaraman P, Devaraj K, Dhanasekaran A, Subramanian S (2022) Potential pre-treatment of lignocellulosic biomass for the enhancement of biomethane production through anaerobic digestion—a review. Fuel 318:123593. https://doi.org/10.1016/j.fuel.2022.123593
2. Zieliński M, Kazimierowicz J, Dębowski M (2023) Advantages and limitations of anaerobic wastewater treatment—technological basics, development directions, and technological innovations. Energies 16(1):83. https://doi.org/10.3390/en16010083
3. Wang X, Lu X, Li F, Yang G (2014) Effects of temperature and carbon-nitrogen (C/N) ratio on the performance of anaerobic co-digestion of dairy manure, chicken manure and rice straw: focusing on ammonia inhibition. PloS one 9(5):e97265. https://doi.org/10.1371/journal.pone.0097265
4. Chen Y, Cheng JJ, Creamer KS (2008) Inhibition of anaerobic digestion process: a review. Biores Technol 99(10):4044–4064. https://doi.org/10.1016/j.biortech.2007.01.057
5. Vasco-Correa J, Khanal S, Manandhar A, Shah A (2018) Anaerobic digestion for bioenergy production: global status, environmental and techno-economic implications, and government policies. Biores Technol 247:1015–1026. https://doi.org/10.1016/j.biortech.2017.09.004
6. Zhang L, Loh KC, Zhang J (2019) Enhanced biogas production from anaerobic digestion of solid organic wastes: current status and prospects. Bioresour Technol Rep 5:280–296. https://doi.org/10.1016/j.biteb.2018.07.005
7. Meegoda JN, Li B, Patel K, Wang LB (2018) A review of the processes, parameters, and optimization of anaerobic digestion. Inter J Environ Res Public Health 10:2224. https://doi.org/10.3390/ijerph15102224
8. Steffen R, Szolar O, Braun R (1998) Feedstocks for anaerobic digestion. Institute of Agrobiotechnology Tulin, University of Agricultural Sciences, Vienna 8
9. Manyi-Loh CE, Lues R (2023) Anaerobic digestion of lignocellulosic biomass: substrate characteristics (Challenge) and innovation. Fermentation 9(8):755. https://doi.org/10.3390/fermentation9080755

10. Xu F, Li Y, Ge X, Yang L, Li Y (2017) Anaerobic digestion of food waste–challenges and opportunities. Bioresour Technol 247:1047–1058. https://doi.org/10.1016/j.biortech.2017.09.020
11. Pramanik SK, Suja FB, Zain SM, Pramanik BK (2019) The anaerobic digestion process of biogas production from food waste: prospects and constraints. Bioresour Technol Rep 8:100310. https://doi.org/10.1016/j.biteb.2019.100310
12. Slorach PC, Jeswani HK, Cuéllar-Franca R, Azapagic A (2019) Environmental sustainability of anaerobic digestion of household food waste. J Environ Manage 236:798–814. https://doi.org/10.1016/j.jenvman.2019.02.001
13. Bacenetti J, Negri M, Fiala M, González-García S (2013) Anaerobic digestion of different feedstocks: impact on energetic and environmental balances of biogas process. Sci Total Environ 463:541–551. https://doi.org/10.1016/j.scitotenv.2013.06.058
14. Li L, Peng X, Wang X, Wu D (2017) Anaerobic digestion of food waste: a review focusing on process stability. Bioresour Technol 248:20–28. https://doi.org/10.1016/j.biortech.2017.07.012
15. Contreras LM, Schelle H, Sebrango CR, Pereda I (2012) Methane potential and biodegradability of rice straw, rice husk and rice residues from the drying process. Water Sci Technol 65(6):1142–1159. https://doi.org/10.2166/wst.2012.951
16. Lantasi AI (2020) Rice husk as renewable energy for biogas production from biomass: prospect and challenges. E3S Web Conf 202:06024. https://doi.org/10.1051/e3sconf/202020206024
17. Elsayed M, Andres Y, Blel W, Gad A, Ahmed A (2016) Effect of VS organic loads and buckwheat husk on methane production by anaerobic co-digestion of primary sludge and wheat straw. Energy Convers Manage 117:538–547. https://doi.org/10.1016/j.enconman.2016.03.064
18. Bandgar PS, Jain S, Panwar NL (2021) Current status, opportunities and challenges in anaerobic digestion in Indian context: an overview. Bioresour Technol Rep 16:100830. https://doi.org/10.1016/j.biteb.2021.100830
19. Mao C, Feng Y, Wang X, Ren G (2015) Review on research achievements of biogas from anaerobic digestion. Renew Sustain Energy Rev 45:540–555. https://doi.org/10.1016/j.rser.2015.02.032
20. Silva AF, Brasil YL, Koch K, Amaral MC (2021) Resource recovery from sugarcane vinasse by anaerobic digestion—a review. J Environ Manage 295:113137. https://doi.org/10.1016/j.jenvman.2021.113137
21. Selvaraj PS, Periasamy K, Suganya K, Ramadass K, Muthusamy S, Ramesh P, Bush R, Vincent SG, Palanisami T (2022) Novel resources recovery from anaerobic digestates: current trends and future perspectives. Crit Rev Environ Sci Technol 52(11):1915–1999. https://doi.org/10.1080/10643389.2020.1864957
22. Wang W, Chang JS, Lee DJ (2023) Anaerobic digestate valorization beyond agricultural application: current status and prospects. Biores Technol 373:128742. https://doi.org/10.1016/j.biortech.2023.128742
23. Wang W, Lee DJ (2020) Valorization of anaerobic digestion digestate: a prospect review. Bioresour Technol 323:124626. https://doi.org/10.1016/j.biortech.2020.124626
24. Ambaye TG, Rene ER, Dupont C, Wongrod S, Van Hullebusch ED (2020) Anaerobic digestion of fruit waste mixed with sewage sludge digestate biochar: influence on biomethane production. Front Energy Res 8:31. https://doi.org/10.3389/fenrg.2020.00031
25. Arimi MM, Knodel J, Kiprop A, Namango SS, Zhang Y, Geißen SU (2015) Strategies for improvement of biohydrogen production from organic-rich wastewater: a review. Biomass Bioenerg 75:101–118. https://doi.org/10.1016/j.biombioe.2015.02.011
26. Rao R, Basak N (2021) Development of novel strategies for higher fermentative biohydrogen recovery along with novel metabolites from organic wastes: the present state of the art. Biotechnol Appl Biochem 68(3):421–444. https://doi.org/10.1002/bab.1964
27. Deena SR, Vickram AS, Manikandan S, Subbaiya R, Karmegam N, Ravindran B, Chang SW, Awasthi MK (2022) Enhanced biogas production from food waste and activated sludge using advanced techniques—a review. Bioresour Technol 355:127234. https://doi.org/10.1016/j.biortech.2022.127234

28. Wainaina S, Awasthi MK, Sarsaiya S, Chen H, Singh E, Kumar A, Ravindran B, Awasthi SK, Liu T, Duan Y, Kumar S (2020) Resource recovery and circular economy from organic solid waste using aerobic and anaerobic digestion technologies. Bioresour Technol 301:122778. https://doi.org/10.1016/j.biortech.2020.122778
29. Ma H, Guo Y, Qin Y, Li YY (2018) Nutrient recovery technologies integrated with energy recovery by waste biomass anaerobic digestion. Bioresour Technol 269:520–531. https://doi.org/10.1016/j.biortech.2018.08.114
30. Monfet E, Aubry G, Ramirez AA (2017) Nutrient removal and recovery from digestate: a review of the technology. Biofuels 9:247–262. https://doi.org/10.1080/17597269.2017.1336348
31. Hidalgo D, Corona F, Martín-Marroquín JM, del Álamo J, Aguado A (2016) Resource recovery from anaerobic digestate: struvite crystallisation versus ammonia stripping. Desalin Water Treat 57(6):2626–2632. https://doi.org/10.1080/19443994.2014.1001794
32. Diéguez-Santana K, Chicaiza-Ortiz C, Zhang J, Logroño W (2022) Anaerobic digestate: pollutants, ecotoxicology, and legislation. In: Anaerobic digestate management. IWA Publishing, pp 359–384. https://doi.org/10.2166/9781789062755_0359
33. Al Seadi T, Drosg B, Fuchs W, Rutz D, Janssen R (2013) Biogas digestate quality and utilization. Biogas Handb 267–301. https://doi.org/10.1533/9780857097415.2.267
34. Lamolinara B, Pérez-Martínez A, Guardado-Yordi E, Fiallos CG, Diéguez-Santana K, Ruiz-Mercado GJ (2022) Anaerobic digestate management, environmental impacts, and techno-economic challenges. Waste Manage 140:14–30. https://doi.org/10.1016/j.wasman.2021.12.035
35. Guilayn F, Jimenez J, Rouez M, Crest M, Patureau D (2017) Biogas digestate typologies. S Margherita di Pula 1–29. https://doi.org/10.15454/kpf7-z784
36. Kandagatla N, Kunnoth B, Sridhar P, Tyagi V, Rao PV, Tyagi RD (2023) Rice mill wastewater management in the era of circular economy. J Environ Manage 348:119248. https://doi.org/10.1016/j.jenvman.2023.119248
37. Santagata R, Ripa M, Genovese A, Ulgiati S (2021) Food waste recovery pathways: challenges and opportunities for an emerging bio-based circular economy. A systematic review and an assessment. J Cleaner Prod 286:125490. https://doi.org/10.1016/j.jclepro.2020.125490
38. Bennich T, Belyazid S (2017) The route to sustainability—prospects and challenges of the bio-based economy. Sustainability 9(6):887. https://doi.org/10.3390/su9060887
39. Bennich T, Belyazid S (2017) The route to sustainability—prospects and challenges of the bio-based economy. Sustainability 9(6):887. https://doi.org/10.4172/2252-5211.1000262
40. Shevchenko T, Ranjbari M, Shams Esfandabadi Z, Danko Y, Bliumska-Danko K (2022) Promising developments in bio-based products as alternatives to conventional plastics to enable circular economy in Ukraine. Recycling 7(2):20. https://doi.org/10.3390/recycling7020020
41. Borrello M, Lombardi A, Pascucci S, Cembalo L (2016) The seven challenges for transitioning into a bio-based circular economy in the agri-food sector. Recent Pat Food, Nutr Agric 8(1):39–47. https://doi.org/10.2174/2212798408011603041439 39
42. Gontard N, Sonesson U, Birkved M, Majone M, Bolzonella D, Celli A, Angellier-Coussy H, Jang GW, Verniquet A, Broeze J, Schaer B (2018) A research challenge vision regarding management of agricultural waste in a circular bio-based economy. Crit Rev Environ Sci Technol 48:614–654. https://doi.org/10.1080/10643389.2018.1471957
43. Tolessa A, Louw TM, Goosen NJ (2022) Probabilistic techno-economic assessment of anaerobic digestion predicts economic benefits to smallholder farmers with quantifiable certainty. Waste Manage 138:8–18. https://doi.org/10.1016/j.wasman.2021.11.004
44. Canizales L, Rojas F, Pizarro CA, Caicedo-Ortega NH, Villegas-Torres MF (2020) SuperPro designer®, user-oriented software used for analyzing the techno-economic feasibility of electrical energy generation from sugarcane vinasse in Colombia. Processes 8(9):1180. https://doi.org/10.3390/PR8091180
45. Kim JR, Hu Y, Zavala VM, Karthikeyan KG (2022) Techno-economic analysis of pretreatments to dairy manure biomass for enhanced biogas production. Bioresour Technol Rep 20:101275. https://doi.org/10.1016/j.biteb.2022.101275

46. Fuess LT, Klein BC, Chagas MF, Rezende MC, Garcia ML, Bonomi A, Zaiat M (2018) Diversifying the technological strategies for recovering bioenergy from the two-phase anaerobic digestion of sugarcane vinasse: an integrated techno-economic and environmental approach. Renew Energy 122:674–687. https://doi.org/10.1016/j.renene.2018.02.003
47. Mainardis M, Flaibani S, Trigatti M, Goi D (2019) Techno-economic feasibility of anaerobic digestion of cheese whey in small Italian dairies and effect of ultrasound pre-treatment on methane yield. J Environ Manage 246:557–63. https://doi.org/10.1016/j.jenvman.2019.06.014
48. Dave A, Huang Y, Rezvani S, McIlveen-Wright D, Novaes M, Hewitt N (2013) Techno-economic assessment of biofuel development by anaerobic digestion of European marine cold-water seaweeds. Bioresour Technol 135:120–7. https://doi.org/10.1016/j.biortech.2013.01.005
49. Cudjoe D, Han MS, Nandiwardhana AP (2020) Electricity generation using biogas from organic fraction of municipal solid waste generated in provinces of China: techno-economic and environmental impact analysis. Fuel Process Technol 203:106381. https://doi.org/10.1016/j.fuproc.2020.106381
50. Afrane S, Ampah JD, Jin C, Liu H, Aboagye EM (2021) Techno-economic feasibility of waste-to-energy technologies for investment in Ghana: a multicriteria assessment based on fuzzy TOPSIS approach. J Cleaner Prod 318:128515. https://doi.org/10.1016/j.jclepro.2021.128515
51. Thompson TM, Ramin P, Udugama I, Young BR, Gernaey KV, Baroutian S (2021) Techno-economic and environmental impact assessment of biogas production and fertiliser recovery from pelagic Sargassum: a biorefinery concept for Barbados. Energy Convers Manage 245:114605. https://doi.org/10.1016/j.enconman.2021.114605
52. Bayu AI, Mellyanawaty M, Budiman A, Budhijanto W (2021) Techno-economic analysis of reactor types and biogas utilization schemes in thermophilic anaerobic digestion of sugarcane vinasse. Renew Energy 201:864–875. https://doi.org/10.1016/j.renene.2022.10.087
53. Kim GB, Cayetano RD, Park J, Jo Y, Jeong SY, Lee MY, Pandey A, Kim SH (2022) Impact of thermal pretreatment on anaerobic digestion of dewatered sludge from municipal and industrial wastewaters and its economic feasibility. Energy 254:124345. https://doi.org/10.1016/j.energy.2022.124345
54. Rajendran K, Mahapatra D, Venkatraman AV, Muthuswamy S, Pugazhendhi A (2020) Advancing anaerobic digestion through two-stage processes: current developments and future trends. Renew Sustain Energy Rev 123:109746. https://doi.org/10.1016/j.rser.2020.109746
55. Roder M (2016) More than food or fuel. Stakeholder perceptions of anaerobic digestion and land use; a case study from the United Kingdom. Energy Policy 97:73–81. https://doi.org/10.1016/j.enpol.2016.07.003
56. Manga M, Muoghalu CC (2024) Greenhouse gas emissions from on-site sanitation systems: a systematic review and meta-analysis of emission rates, formation pathways and influencing factors. J Environ Manage 357:120736. https://doi.org/10.1016/j.jenvman.2024.120736
57. Hayyat U, Khan MU, Sultan M, Zahid U, Bhat SA, Muzamil M (2024) A review on dry anaerobic digestion: existing technologies, performance factors, challenges, and recommendations. Methane 3(1):33–52. https://doi.org/10.3390/methane3010003
58. Fagbohungbe MO, Herbert BM, Hurst L, Ibeto CN, Li H, Usmani SQ, Semple KT (2017) The challenges of anaerobic digestion and the role of biochar in optimizing anaerobic digestion. Waste Manage 61:236–249. https://doi.org/10.1016/j.wasman.2016.11.028
59. Ma H, Guo Y, Qin Y, Li YY (2018) Nutrient recovery technologies integrated with energy recovery by waste biomass anaerobic digestion. Biores Technol 269:520–531. https://doi.org/10.1016/j.biortech.2018.08.114

Anaerobic Digestion for Treatment and Resource Recovery

Lakshmi Pathi Thulluru, Indrajit Chakraborty, Shamik Chowdhury, and M. M. Ghangrekar

Abstract Anaerobic digestion (AD) is the most primitive form of metabolism. The process typically involves the breakdown of complex organic compounds into fatty acids by bacteria, which work in tandem with methanogens to produce carbon dioxide and methane, collectively known as biogas. Biogas has been widely considered as one of the most promising renewable energy sources for the past several decades. Apart from biogas generation, mineralising organics of higher molecular weight organics is particularly interesting to environmental scientists and engineers. It is possible as AD can be replicated in engineered vessels to treat organic wastes and wastewater. In addition, tailored AD of specific substrates can yield other valuables, such as hydrogen through dark fermentation, as well as nitrogen and phosphorus-rich liquid fertiliser recovery from the digestate supernatant liquid of the anaerobic digester. A revisit to the concepts and the translation of these concepts into resource recovery, solid organic waste stabilisation and wastewater treatment is imperative. This chapter, therefore, collates the fundamental principles of AD, emphasising its potential and limitations in wastewater/waste treatment and resource recovery.

Keywords Acidogenesis · Acetogenesis · Hydrolysis · Methanogenesis · Resource recovery · Syntrophy · Xenobiotic compounds

---

L. P. Thulluru · S. Chowdhury
School of Environmental Science and Engineering, Indian Institute of Technology Kharagpur, Kharagpur 721302, India
e-mail: thullurulakshmipathi@kgpian.iitkgp.ac.in

S. Chowdhury
e-mail: shamikc@iitkgp.ac.in

I. Chakraborty (✉)
Environmental Science and Engineering Department, Indian Institute of Technology Bombay, Mumbai 400076, India
e-mail: indra.esed@iitb.ac.in

M. M. Ghangrekar
Department of Civil Engineering, Indian Institute of Technology Kharagpur, Kharagpur 721302, India
e-mail: ghangrekar@civil.iitkgp.ac.in

M. M. Ghangrekar et al. (eds.), *Biological and Hybrid Wastewater Treatment Technology*, Earth and Environmental Sciences Library,
https://doi.org/10.1007/978-3-031-63046-0_9

1 Introduction

Environmental engineering processes traditionally aim to remove polluting compounds from entering the environment, considering the aspects of hygienization and ecological protection. When handling biodegradable waste, anaerobic digestion (AD) is superior to traditional technologies such as landfills and incineration. It reduces waste sludge generation, kills pathogens, provides essential nutrients, and is less energy-demanding [1]. The biodegradable waste or biomass employed in the technology, including industrial effluents, food waste, agricultural waste, and sewage sludge, is regarded as future renewable resources [2]. The biogas generated here is a sustainable alternative, completing the life cycle of organic waste and favouring the concept of a circular economy [3]. Interestingly, the energy derived from AD can meet about one-fifth of the current global primary energy demand [4].

The AD is a cost-effective, eco-friendly technology that has been successfully demonstrated for biowaste treatment and biofuel generation globally. It constitutes a series of microbial processes, broadly classified into hydrolysis, acidogenesis, acetogenesis, and methanogenesis, carried out through complex intracellular and extracellular biochemical reactions. These reactions lead to the degradation of organic waste to volatile fatty acids (VFAs) and eventually into biofuels, such as methane (CH_4) and hydrogen (H_2). The advantages of AD include: (i) stabilisation and deactivation of pathogens of waste biomass, avoiding secondary pollution [5], (ii) requirement of a low energy input [6], and (iii) production of less biomass compared to aerobic processes [7]. Therefore, it is considered a valuable technology for treating organic waste.

In the past decade, traditional research on recovering valuables like CH_4 has shifted towards H_2, hythane, VFA production, and nutrient recovery. On the other hand, a particular focus on eliminating xenobiotic compounds from wastewater has also grown recently. Understanding the basic principles and how each scope of application is derived from these principles is imperative. Hence, this chapter focuses on the fundamental mechanism of AD, followed by the factors and pathways that affect its operation. Also, emphasis is provided on different interspecies electron transfer and syntrophy that synergetically support the whole process. An example of the application of AD in treating effluents from the brewery and dairy industries, along with its unique ability to treat xenobiotic-laden wastewater is also briefly explained. Finally, recent research integrating resource recovery with AD is also discussed.

2 Fundamentals of AD

2.1 Mechanism

The intricate communities in AD encompass a diverse group of microorganisms, each performing a specialised function of degrading organic matter in two stages. The first stage entails breaking down complex organic compounds, viz., carbohydrates, proteins, and lipids, by different groups of facultative and anaerobic microbes into simpler VFAs as well as gases comprising carbon dioxide (CO_2) and H_2 [8]. Consequently, the second stage involves CH_4 and CO_2 formation through a syntropic interaction between acid-forming microorganisms and methanogenic archaea. In other words, the second stage involves the assimilation of dissociated organic compounds obtained from the first stage.

The VFA concentration in an operating AD reactor is a crucial factor that affects the ecological balance inside the reactor [9]. The biochemical degradation of organic matter leads to the formation of short-chain volatile acids, notably formic (HCOOH), acetic (CH_3COOH), propionic (CH_3CH_2COOH), and butyric ($CH_3(CH_2)_2COOH$) acids. However, valeric and iso-valeric acids are observed in minuscule quantities [10]. The composition of biogas changes based on the existing environmental conditions in the anaerobic digester. The varying composition of biogas is noticeable during different phases, such as at start-up, during stable operation periods, and even at reactor inhibition conditions. The ratio of CH_4/CO_2 in biogas also varies based on the characteristics of the degraded organic matter.

Hydrolysis, the first phase of AD (Fig. 1), is the disintegration and dissolution of complex organic matter using exoenzymes secreted by the hydrolytic fermentative bacteria. The common phyla of these hydrolytic fermentative bacteria reported are Bacteroidetes, Firmicutes, Proteobacteria, Actinobacteria, and Chloroflexi [11]. However, at the genus level that participate in active hydrolysis are species of *Enterobacter* and *Streptococcus*, apart from *Sphingomonas*, *Sporobacterium*, *Bifidobacterium*, *Propionibacterium*, *Bacteroides*, and *Lactobacillus* [12].

The next step facilitates further breakdown of soluble products inside the facultative bacteria (Fig. 1), eventually excreting them. The simpler compounds coming out of the cell consist of VFAs, alcohols, lactic acid, CO_2, H_2, ammonia (NH_3), and hydrogen sulphide (H_2S) [13]. Some of the substrate also contributes to building new bacterial cells. A diverse group of fermentative bacteria, such as Ruminococcus and Paenibacillus, perform acidogenesis. Other families of spore-forming microbes include *Clostridia* sp. and *Bacteroidaceaea* sp., commonly found in the digestive tracts also participate in the acidogenesis [14].

Acetogenic bacteria produce substrate for methanogenic microbes by oxidising the compounds generated during acidogenesis (Fig. 1). The acetogenesis products comprise acetic acid, CO_2, and H_2, causing accumulation of H_2 that significantly lowers pH in the aqueous medium. This phenomenon is counteracted by either directing H_2 and CO_2, resulting in the formation of CH_4 by methanogenesis, or yielding acetic, propanoic, and butyric acid using H_2, CO_2, and acetic acid. About 50% of biodegradable chemical oxygen demand (bCOD) is further decomposed by

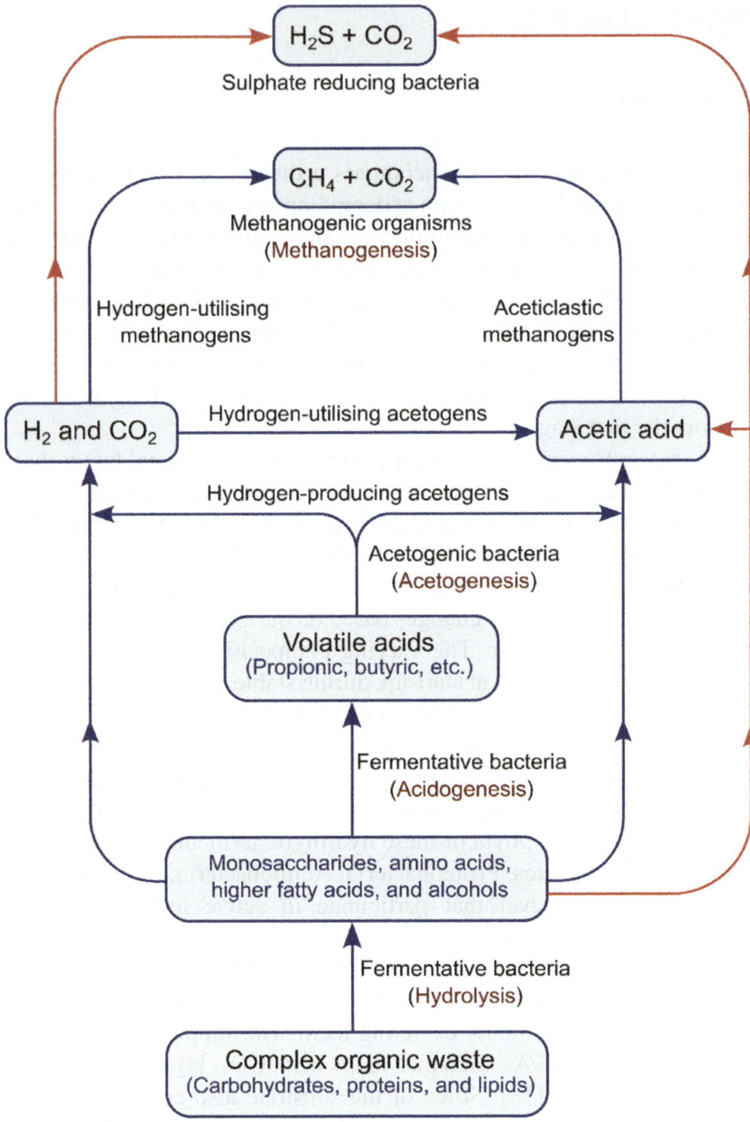

Fig. 1 Metabolic pathways of different microbial groups involved in AD

the acetogens by utilizing acetic, propanoic, and butyric acid as substrates [15]. Two groups of bacteria, the acetogenic bacteria and the syntropic acetate-oxidising bacteria (SAOB), act in a syntropic manner for producing and degrading acetate, respectively. The acetogenic bacteria are responsible for utilising H_2 along with CO_2, to form acetate; whereas, the SAOB performs a reverse reaction, oxidising acetate to CO_2 and H_2 [16].

The final step completely degrades organic matter into CH_4 and CO_2, termed methanogenesis (Fig. 1). The methanogens use only a limited number of substrates, precisely H_2/CO_2, carbon monoxide (CO), methylamines, acetic acid, formic acid, and methanol [17]. Based on the affinity of methanogens with the above mentioned substrates, they are broadly categorised into two groups. The first group of methanogens capable of forming CH_4 from the cleavage of methanol or acetic acid are known as acetoclastic methanogen. The second group are the hydrogenotrophic methanogens that yield CH_4 by reducing CO_2. This is more common in natural environments such as marshlands. The acetoclastic methanogens are only a few species responsible for around 60–70% of CH_4 generation in AD [18].

Acetoclastic methanogens reduce the methyl (CH_3^-) group in acetic acid to CH_4, while CO_2 production occurs by oxidising the carboxylic (− COOH) group (Eq. 1) [19]. The acetoclastic methanogens can be further classified based on their predominance at acetic acid concentrations of less than or greater than 10^{-3} M. *Methanosarcina* sp. can prevail at acetate concentrations above the stated level, whereas *Methanosaeta* sp. operates at low acetate concentrations. *Methanosaeta* sp. is filamentous, exclusively uses acetate, demands higher solid retention time, and it is sensitive to varying pH compared to *Methanosarcina* sp. [20]. *Methanosarcina* sp. are coccus-forming and more versatile than most methanogens, demonstrating a greater growth rate and showcasing the capability to use H_2 and methylamines.

$$CH_3COOH \rightarrow CH_4 + CO_2 \ \Delta G_o = -33.0\,\mathrm{kJ\,mol^{-1}} \tag{1}$$

Hydrogenotrophic methanogens, such as species of *Methanobrevibacter*, *Methanobacterium*, and *Methanospirillum*, utilise H_2/CO_2 for metabolism (Eq. 2). As the levels of CO_2 are always abundant in an anaerobic digester, the CH_4 formation is not rate limited by the concentration of CO_2 [21].

$$CO_2 + 4H_2 \rightarrow CH_4 + 2H_2O \ \Delta G_o = -135.0\,\mathrm{kJ\,mol^{-1}} \tag{2}$$

Another class of bacteria present in AD-treating wastewater due to the presence of sulphate (SO_4^{2-}) or sulphite (SO_3^{2-}) is sulphate-reducing bacteria (SRB). The SO_4^{2-} is an electron acceptor, yielding H_2S during the oxidation of organic matter by the SRB (Fig. 1). The SRB utilises the intermediate compounds produced during AD, competing with the fermentative, acetogenic, and methanogenic communities. This action results in a decreased CH_4 yield for a given amount of organic matter in the influent [22]. The activity of SRB is more prevalent when the relative concentration of SO_4^{2-} is increased with respect to chemical oxygen demand (COD) concentration. Theoretically when $COD/SO_4^{2-} < 0.67$ all electrons flow to SO_4^{2-} and organic matter is primarily treated by SRB. On the other hand, when the ratio exceeds 0.67, the SRB and other bacteria compete for electron donors [23]. Further, the classification of SRB first includes demonstrating incomplete organic matter oxidation to acetate. The common genera belonging to this group include *Desulfobulbus* sp., *Desulfomonas* sp., *Desulfotomaculum* sp., and *Desulfovibrio* sp. The groups that display complete oxidation of organic substrate to CO_2 commonly include

Desulfonema sp., *Desulfococcus* sp., *Desulfobacterium* sp., and *Desulfobacter* sp. [24].

The major pathways for SO_4^{2-} reduction in SRB include assimilated and dissimilatory sulphate reduction (ASR and DSR) [25]. The biomolecular synthesis method of ASR is principally responsible for the reduction of SO_4^{2-} into sulphur (S). The energy provided by methionine production drives ASR [26]. Secondly, the major portion of SO_4^{2-} is reduced into sulphide (S^{2-}) by the DSR pathway [27]. Here, the DSR reaction is a two-step process involving the activation of SO_4^{2-}, then reducing it into S^{2-}. Additionally, from a thermodynamic point of view, SO_4^{2-} is an unfavourable electron acceptor ($E^0(SO_4^{2-}/SO_3^{2-}) = -516$ mV) compared to the intercellular mediator ferrite protein ($E^0 = -398$ mV) that reduces it. Initially, SO_4^{2-} is activated by sulphate adenylyltransferase to form adenosine 5-phosphosulfate (APS), consuming two equivalents of adenosine triphosphate (ATP). The APS is further reduced to HSO_3^{2-} along with adenosine monophosphate (AMP) (Eq. 3 through Eq. 5) [28]. The second step involves HSO_3^{2-} reduction by bisulfite reductase using 12 electrons (Eq. 6). The final reaction is represented by Eq. (7) [29].

$$SO_4^{2-} + ATP^{4-} + H^+ \rightarrow APS^{2-} + HP_2P_7^{3-} \tag{3}$$

$$2APS^{2-} + 4e^- + H^+ \rightarrow HSO_3^- + SO_3^{2-} + 2AMP^{2-} \tag{4}$$

$$HP_2PO_7^{3-} + H_2O \rightarrow H_2PO_4^- + HPO_4^{2-} \tag{5}$$

$$HSO^{3-} + SO_3^{2-} + 12e^- + 14H^+ \rightarrow HS^- + H_2S + 6H_2O \tag{6}$$

$$\begin{aligned} &2SO_4 + 2ATP^{4-} + 16e^- + 17H^+ \rightarrow HS^- + H_2S \\ &+ 2AMP^{2-} + 2HPO_4^{2-} + 4H_2O \end{aligned} \tag{7}$$

The methanogenic community in the anaerobic consortia prevents the accumulation of intermediate acids by converting them into CO_2 and CH_4. These intermediate acids neutralise with the inherently occurring alkalinity in the medium. However, the process is hindered if the utilisation of volatile acids by the methanogenic archaea is less compared to the rate of intermediate acid production by the acidogenic bacteria. It results in complete alkalinity consumption, causing the souring of the anaerobic reactor [30].

About 85% of the total CH_4 generated is from acetic (72%) and propionic acid (13%) while degrading complex substrates such as sludge from sewage treatment plants. The degradation of remaining acids, such as formic and butyric acid, is responsible for the remaining 15% of CH_4 generation [31]. The presence of carbohydrates and proteins in biodegradable waste is majorly accountable for propionic acid accumulation [32]. The final phase should also fulfil micro-environmental conditions like oxidation–reduction potential < 250 mV versus SHE, VFA accumulation < 2500 mg L^{-1}, and ammonia < 1000 mg L^{-1} [33].

The thermodynamic aspect of degrading propionate, butyrate, and ethanol under standard conditions of pH 7.0, pressure 1 atm, and temperature 25 °C is unfavourable as the standard Gibbs free energy (ΔG_o) is greater than 0 in water (Eqs. 8 to 11) [12]. However, the degradation is led in a forward direction if electron acceptors like bicarbonate and sulphate continuously consume the produced H_2 in the presence of H_2-consuming microbial communities (Eqs. to 15) [34]. An optimally performing methanogenic digestor ensures the13 H_2-partial pressure is (usually 10^{-6} atm and) less than 10^{-4} atm [35].

Oxidation reactions:

$$CH_3CH_2COO^- + 3H_2O \rightarrow CH_3COO^- + HCO_3^- + H^+ + 3H_2 \quad \Delta G_o = +76.1\,kJ\,mol^{-1} \tag{8}$$

$$CH_3CH_2CH_2COO^- + 2H_2O \rightarrow 2CH_3COO^- + H^+ + 2H_2 \quad \Delta G_o = +48.1\,kJ\,mol^{-1} \tag{9}$$

$$CH_3CH_2OH + H_2O \rightarrow CH_3COO^- + H^+ + 2H_2 \quad \Delta G_o = +9.6kJ\,mol^{-1} \tag{10}$$

$$CH_3CHOCOO^- + 2H_2O \rightarrow 2CH_3COO^- + HCO_3^- + H^+ + 2H_2 \quad \Delta G_o = +48.1\,kJ\,mol^{-1} \tag{11}$$

$$CH_3CH(OH)COOH + H_2O \rightarrow CH_3COOH + 2H_2 \quad \Delta G_o = -4.2\,kJ\,mol^{-1} \tag{12}$$

Reduction reactions:

$$2HCO_3^- + H^+ + 4H_2 \rightarrow CH_3COO^- + 4H_2O \quad \Delta G_o = -104.6\,kJ\,mol^{-1} \tag{13}$$

$$HCO_3^- + H^+ + 4H_2 \rightarrow CH_4 + 3H_2O \quad \Delta G_o = -135.6\,kJ\,mol^{-1} \tag{14}$$

$$SO_4^{2-} + H^+ + 4H_2 \rightarrow HS^- + 4H_2O \quad \Delta G_o = -151.9\,kJ\,mol^{-1} \tag{15}$$

The proper operation of the anaerobic digester is a function of organic matter type and concentration, the loading rate of organic matter, nutrient concentration and other decisive physio-chemical characteristics of the reactor, such as temperature, pH, inorganic nutrients, alkalinity, and presence of potentially toxic compounds.

2.1.1 Organic Matter

Quantifying the organic matter in the form of biodegradable solids present in the substrate as well as in soluble form can be done using preliminary tests like biochemical oxygen demand (BOD) and COD tests. However, understanding the concentrations of carbohydrates, proteins, and lipids is essential. An appropriate quantification of these compounds can give a clearer picture of the state of the reactor [36]. The presence of complex substrates in an anaerobic reactor results in a lag phase, slowing the growth of anaerobic microorganisms. The COD fraction degraded by the anaerobic reactor includes the amount of influent bCOD converted into CH_4 and the fraction consumed by microorganisms [37]. On the other hand, the bCOD fraction converted into acidified COD, along with biologically recalcitrant COD, remains in the treated effluent. The recalcitrant COD refers to the biologically resistant substrate not easily degraded by the fermentative microorganisms.

The non-biological mechanism of COD removal can also occur through adsorption and precipitation [38]. Usually, fat adsorption onto the surface of bacterial sludge is seen in the case of soluble COD. The insoluble particulate COD is removed non-biologically by retainment in the sludge. Systems such as up-flow anaerobic sludge blanket (UASB) reactors depend on biomass self-immobilisation, resulting in the retention of insoluble COD on the surface of the bacterial sludge. This retention causes a non-bacterial sludge to form over the active biomass, hampering the methanogenic activity. Alternatively, the second way to remove non-biological COD is by changing pH by adding calcium-based alkaline compounds, resulting in precipitation that eventually settles down along with the sludge [39].

Estimating the maximum CH_4 formed for a given chemical composition of wastewater can be determined using the Buswell stoichiometric equation (Eq. 16) [40].

$$C_nH_xO_yN_z + \left(n - \frac{x}{4} - \frac{y}{2} + \frac{3z}{4}\right)H_2O \rightarrow CH_4 + \left(\frac{n}{2} - \frac{x}{8} + \frac{y}{4} + \frac{3z}{8}\right)CO_2 + (z)NH_3 \quad (16)$$

The $C_nH_xO_yN_z$ represents the biodegradable organic matter in an anaerobic digester. The presence of sulphate and nitrate results in decreased CH_4 generation. This reduction in CH_4 generation is due to the reduction reaction of sulphate and nitrate to N_2 and H_2S gas, respectively (Eqs. 17 and 18) [41]. In other words, the sulphate and nitrate act as alternate terminal electron acceptors, diverting the metabolic pathway from methane formation as the end product.

$$6H_2 + 2NO_3^- \rightarrow N_2 + 6H_2O \quad (17)$$

$$4H_2 + SO_4^{2-} \rightarrow H_2S + 2H_2O + 2OH^- \quad (18)$$

The magnitude of H_2S and CH_4 production depends on parameters such as pH, temperature, COD/SO4 $^{2-}$ ratio, electron donor, oxidation/reduction potential (ORP), hydraulic retention time (HRT), concomitant ions, and syntrophic/competitive groups of communities [23]. For instance, in acidic conditions (pH 4.5–5.0), it exists mostly in the form of H_2S (~99%) (Eq. 19), whereas under neutral and slightly alkaline conditions, it presents in the form of HS^- and S^{2-} in the aqueous solution (Eq. 20) [42]. The inhibition of methanogenic activity occurs when the COD/SO_4 $^{2-}$ ratio < 7. It is known that H_2S poses a serious problem of odour, corrosion, and the cost involved in purification if the biogas is intended to be used [43].

$$H_2S \leftrightarrow HS^- + H^+ \quad pK_{a1} = 6.97\,(25\,°C) \tag{19}$$

$$HS^- \leftrightarrow S^{2-} + H^+ \quad pK_{a2} = 12.9\,(25\,°C) \tag{20}$$

2.1.2 Nutrients

The nutrients and physical factors determine the microbial community in the anaerobic environment. The empirical composition of the microbial cell determines its nutritional needs. Nutrients ideal for the methanogenic community include phosphorous, nitrogen, sulphur, cobalt, iron, nickel, riboflavin, vitamin B12, selenium and molybdenum [44]. The reactors treating several types of wastewater can compensate for the required micronutrients. Generally, domestic sewage is considered appropriate for the growth of microorganisms in AD. Industrial effluents have specific compositions and may require the addition of nutrients for optimal degradation of organic matter [45].

2.1.3 Nitrogen, Phosphorus and Sulphur

The primary nitrogen source for microbes in the anaerobic digestor is the fraction of organic nitrogen released during degradation and ammonia. The bacterial growth is observed to be (about six times) greater in the case of wastes containing abundant carbohydrates supporting bacterial growth than in waste containing protein and VFAs, demanding high amounts of nitrogen in the first case [46]. However, the presence of phosphorous should be 1/5 to 1/7 of nitrogen content in wastewater. Microbes can consume inorganic forms of phosphorus like orthophosphate [45]. On the other hand, the source of sulphur in the form of sulphate present in domestic sewage is sufficient for microbial growth. The inorganic sulphate is reduced into sulphide, which is consumed by most methanogens apart from the other sources of sulphur, such as cysteine. However, excess sulphate hinders methanogenesis as the sulphate-reducing bacteria compete for substrates produced during AD, such as hydrogen and acetate.

2.2 *Metabolic Pathways*

Tapping the full potential of AD for capturing biomethane requires decrypting the metabolic pathways, understanding the behaviour of organisms or the synergistic interaction in microbial consortia and balancing the substrate flow between the microbial communities [12]. Nevertheless, there are a few inherent complications that compromise the efficiency in terms of the type of feedstock used, slow mass transfer depending on the type of substrate, accumulation of VFA and presence of lignocellulosic recalcitrant residues, imbalance in carbon to nitrogen ratio, micronutrient deficiency, and ammonia inhibition [47].

The acidogenesis step is most crucial for the metabolic pathways during biomass degradation. A particular metabolic pathway adapted is based on the type of VFA formed. Therefore, understanding the different metabolic pathways and the co-existing syntropic interaction plays an essential role in improving the performance of anaerobic digesters [48]. Apart from the metabolic pathway, formation of the intermediate compound pyruvate during glycolysis is responsible for the formation of VFA and alcohols. As discussed later, the amount of pyruvate distributed in these pathways depends on the substrate, conditions in the digester and the dominance of a given microbial strain [49].

2.2.1 Acetate-Ethanol Forming Pathway

The acetate-ethanol forming metabolic pathway occurs when acetate and ethanol are formed from pyruvate using the acetyl-CoA pathway along with syntropic oxidation of VFAs like propionate and butyrate (Eqs. 21 to 24) [49]. Ethanol formation also occurs through the fermentation of glucose (Eq. 16). It follows a two-step process, where the decarboxylation of pyruvate initially results in acetaldehyde, followed by the reduction, forming ethanol [50]. On the other hand, in some microbial communities, pyruvate to ethanol conversion occurs in three steps: intermediate acetyl-CoA generation, then acetaldehyde formation, followed by its reduction. The acetate and ethanol distribution amount depends on the feedstock type (Eqs. 21 and 23). The feedstock with higher carbohydrates favours the acetate formation [51].

$$C_6H_{12}O_6 + 2H_2O \rightarrow 2CH_3COOH + 2CO_2 + 4H_2 \tag{21}$$

$$CH_3CH_2OH + 2H_2O \rightarrow CH_3COOH + 2H_2 \quad \Delta G_o = +9.6\ \text{kJ mol}^{-1} \tag{22}$$

$$C_6H_{12}O_6 + H_2O \rightarrow CH_3CH_3OH + CH_3COOH + 2H_2 + 2CO_2 \tag{23}$$

$$2CH_3CH_2OH + CO_2 \rightarrow 2CH_3COO^- + 2H^+ + CH_4 \quad \Delta G_o = -151.9\ \text{kJ mol}^{-1} \tag{24}$$

Some bacterial species of *Clostridium* genera (*C. acetobutylicum*, *C. beijerinckii*) have produced acetone ($(CH_3)_2CO$) and butanol ($CH_3(CH_2)_3OH$) using glucose [52]. The conversion process follows acetyl-CoA synthesis from glucose. It then forms acetacetyl-CoA before synthesising acetone. Similarly, butanol is formed from acetoacetyl-CoA followed by 3- hydroxybutyl-CoA, crotonyl-CoA, and butyryl-CoA as intermediates.

2.2.2 Propionate Forming Pathway

Propionate synthesis is achieved via two pathways (Eqs. 25 and 26). The first pathway involves the reduction of pyruvate to lactate, catalysed by lactate dehydrogenase. The lactate is then reduced to propionate in the presence of propionate dehydrogenase [53]. The second pathway follows a transcarboxylase cycle, where propionate is formed following intermediates, such as 3-oxaloacetate, succinate, fumarate, and propionyl CoA. The propionate is greatly depending on environmental conditions like pH (4.0 – 4.5), inoculum, and product inhibition [54].

$$C_6H_{12}O_6 + 2H_2 \rightarrow 2CH_3CH_2COOH + 2H_2 \quad (25)$$

$$3C_6H_{12}O_6 \rightarrow 4CH_3CH_2COOH + 2CH_3COOH + 2CO_2 + 2H_2O \quad (26)$$

2.2.3 Butyrate Forming Pathway

The butyrate-forming pathway involves the formation of butyrate and acetate as the main product and H_2 as a by-product (Eqs. 27 and 28). It is initiated by converting acetyl-CoA to butyryl-CoA and acetoacetyl-CoA. Butyrate synthesis is observed after a series of intermediates like 3-hydroxybutyryl-CoA, crotonyl-CoA, and butyryl-CoA [51]. The final reaction of butyryl-CoA to butyrate is catalysed by butyrate-kinase and phospho-trans-butyrylase. Compared to acetate, reducing equivalents like $NADH_2$ significantly reduces the intermediates [55].

$$C_6H_{12}O_6 \rightarrow CH_3(CH_2)_2COOH + 2CO_2 + 2H_2 \quad (27)$$

$$4C_6H_{12}O_6 \rightarrow 3CH_3(CH_2)_2COOH + 2CH_3COOH + 8CO_2 + 8H_2 \quad (28)$$

2.2.4 Amino Acid Conversion Pathways

Compared to carbohydrates (70%), proteins show a lesser conversion efficiency of about 40%; therefore, they are considered as a rate-limiting step during acidogenic fermentation [56]. About 70% of the total VFA generated during the acetogenesis of protein is acetate, apart from propionate (15%) and butyrate (10%) [57]. On the other hand, amino acids are initially converted into pyruvate and then to VFAs through transamination and deamination reactions (Eqs. 29 to 31) [58]. Amino acids like asparagine and aspartate undergo transamination reactions to form oxaloacetate. However, non-polar amino acids, such as methionine, valine, and isoleucine follow a propionate production pathway, converting to succinyl-CoA and propionyl-CoA.

$$\text{Alanine} + \alpha\text{-keto glutarate} \rightarrow \text{pyruvate} + \text{glutamate} \quad (29)$$

$$\text{Alanine} + H_2O \rightarrow \text{pyruvate} + NH_4^+ \quad (30)$$

$$\text{Serine} \rightarrow \text{pyruvate} + NH_4^+ \quad (31)$$

2.2.5 Fatty Acid Oxidation Pathways

The VFAs generated during acidogenesis should be further decomposed into acetate and formate for further methanogenesis. This conversion is accomplished by the syntrophic fatty acids oxidising bacteria; these include *P. thermopropionicum*, *S. fumaroxidans*, and *Smithella propionica* in case of propionate oxidation (Eq. 32). The menaquinone system facilitates the electron transport in *P. thermopropionicum* and *S. fumaroxidans* [59]. The combination provides an ease in carrying electrons for membrane-linked catalytic reactions using dehydrogenase/hydrogenase enzymes. The *S. fumaroxidans* participates in re-oxidising NADH using Rhodobacter nitrogen fixation complex, following a reverse electron transfer, resulting in the oxidation of propionate [60].

$$CH_3CH_2COOH + 2H_2O \rightarrow CH_3COOH + CO_2 + 3H_2 \quad \Delta G_o = +76.1\ \text{kJ mol}^{-1} \quad (32)$$

Similar syntrophs that help in generating acetate (Eq. 33) and formate from butyrate and valerate through β-oxidation include *Syntrophus aciditrophicus*, *Syntrophothermus lipocalidus* and *Syntrophomonas wolfei* [61]. Apart from the ribosome modulation factor (Rmf)-like complex and menaquinones that facilitate reverse electron transfer, these syntrophs also provide flavoprotein and iron-sulphur oxidoreductase. The adenosine TATP produced during the β-oxidation of butyrate is also utilised during the reverse electron transport mechanism [62]. Under thermodynamically unfavourable conditions, poly-β-hydroxyalkanoate (PHA) produced by *S. wolfei* has been used in ATP synthesis to maintain the reverse electron transfer.

$$CH_3(CH_2)_2COOH + 2H_2O \rightarrow 2CH_3COOH + 2H_2 \quad \Delta G_o = +48.1\ kJ\ mol^{-1} \tag{33}$$

2.2.6 Acetate Oxidation Pathway (Methanogenesis)

Acetate oxidation occurs through the syntrophic interaction between the hydrogenotrophic methanogens and SAOB, wherein 70% of the CH_4 is generated through acetate conversion [63]. The H_2 and CO_2 produced by SAOB are used by hydrogenotrophic methanogens, generating CH_4. Most SAOB oxidise acetate following the Wood-Ljungdahl (WL) pathway; these include *Tepidanaerobacter acetatoxydans, Thermacetogenium phaeum, Clostridium ultunense, Thermotoga lettingae, Syntrophaceticus schinkii*, amongst others [64]. The electrochemical gradient generated by cytoplasmatic protons supports ATP generation, which is supported by ATP synthase and hydrogenases [65]. The acetate oxidation is supported by high temperature; nevertheless, a high concentration of ammonia has been reported to inhibit the acetoclastic methanogens [66]. The syntrophic acetate oxidation depends on the interspecies H_2 and/ or $HCOO^-$ transfer, where the syntrophic partners like hydrogenotrophic methanogen consume the fermentation products. The overall stoichiometry for the reactions in acetoclastic methanogenesis (Eq. 34) and syntrophic acetate oxidation (Eqs. 35 and 36) remains the same, along with a small amount of energy generation during this process (-31.0 kJ mol^{-1}) [67].

$$CH_3COO^- + H_2O \rightarrow CH_4 + HCO_3^- \quad \left(\Delta G^o = -31.0\ kJ\ mol^{-1}\right) \tag{34}$$

$$CH_3COO^- + 4H_2O \rightarrow 2HCO_3^- + 4H_2 + H^+ \quad \left(\Delta G^o = +104.6\ kJ\ mol^{-1}\right) \tag{35}$$

$$4H_2 + HCO_3^- + H^+ \rightarrow CH_4 + 3H_2O \quad \left(\Delta G^o = -135.6\ kJ\ mol^{-1}\right) \tag{36}$$

2.3 *Interspecies Electron Transfer and Syntrophy*

A nutritional situation is when a substrate is not catabolised by a single organism alone, but rather, organisms combine their metabolic capabilities to catabolise it, known as syntrophy [68]. The syntrophic interaction depends on the thermodynamic dependence between the acidogenic bacteria and methanogens achieved through interspecies electron transfer, maintaining a chemical equilibrium [69]. An efficient transformation of organic matter to CH_4 is ensured through the electron transfer mechanism between the electron donor (or acidogenic bacteria) along with acceptors (or methanogens). Therefore, the electron transfer can occur through two routes, namely, mediated transfer using H_2 (Eq. 2) or formate (Eq. 37) and direct electron transfer [70].

$$4HCOOH \rightarrow CH_4 + 3CO_2 + 2H_2O \quad \Delta G^{o} = -130.0 \text{ kJ mol}^{-1} \tag{37}$$

In the case of mediated interspecies electron transfer, mediators like H_2 assist in carrying electrons to support the syntropy between acidogenic microbes and methanogens. Similarly, formate also participates in transferring electrons in the syntrophic metabolic pathway. The presence of these mediators greatly influences the pH [71]. The pH, in turn, impacts the ΔG^{o} for converting formate to H_2, altering the concentration ratio of H_2 to formate in the anaerobic digestor. This is countered by methanogens that consume both the mediators, ensuring the stability in the reactor.

The interspecies electron transfer using the species H_2 (Fig. 2a) and formate (Fig. 2b) is spontaneous in the presence of methanogens in a syntrophic methanogenesis community. The standard redox potential in both cases is similar, with − 414 mV in H^+/H_2 and − 420 mV in CO_2/formate. The interspecies electron carrier is based on hydrogenase and formate dehydrogenase activity. In the case of butyrate-oxidising syntrophic communities, such as *Syntrophomonas wolfei* and *Syntrophobotulus glycolicus*, H_2 acts as an electron carrier. On the other hand, in the case of propionate degradation syntrophic communities, such as *Syntrophobacter fumaroxidans*, both H_2 and formate act as interspecies electron carriers.

Interestingly, methanogens also showcase the ability of transferring electrons primarily through direct interspecies electron transfer (DIET). This way of moving electrons occurs in many ways, such as (i) cell appendages like pili (Fig. 2c), (ii) via c-type cytochrome on the bacterial cell membrane (Fig. 2d), and (iii) conductive materials (CMs) (Fig. 2e). These appendages, such as flagella, pili, and membrane cytochromes, establish physical contact between the syntrophic microbes. Carbon-based materials, viz., carbon cloth, biochar, granulated carbon, and activated carbon, are hypothesised to enhance DIET. Among the carbon materials that may support DIET, biochar also demonstrates additional benefits of mitigating limonene and ammonia inhibition and regulating methyltransferase in AD. Similarly, granulated activated carbon is responsible for improving the *Firmicutes* sp. population and helping the cell metabolism of *Elusimicrobia* sp. Additionally, CMs offer features like adsorption, release of trace elements, and stimulation of enzyme activity [72].

3 AD in Wastewater Treatment

The AD is implemented as a practical and relevant technology in treating effluents from industries, such as breweries, diaries, canaries, and even conventional sewage sludge, helping a transformation from a linear economy to a circular economy. The advantages catered in this context include promoting waste energy recovery, reducing environmental pollution, and providing environmentally friendly products: biogas and bio-fertiliser.

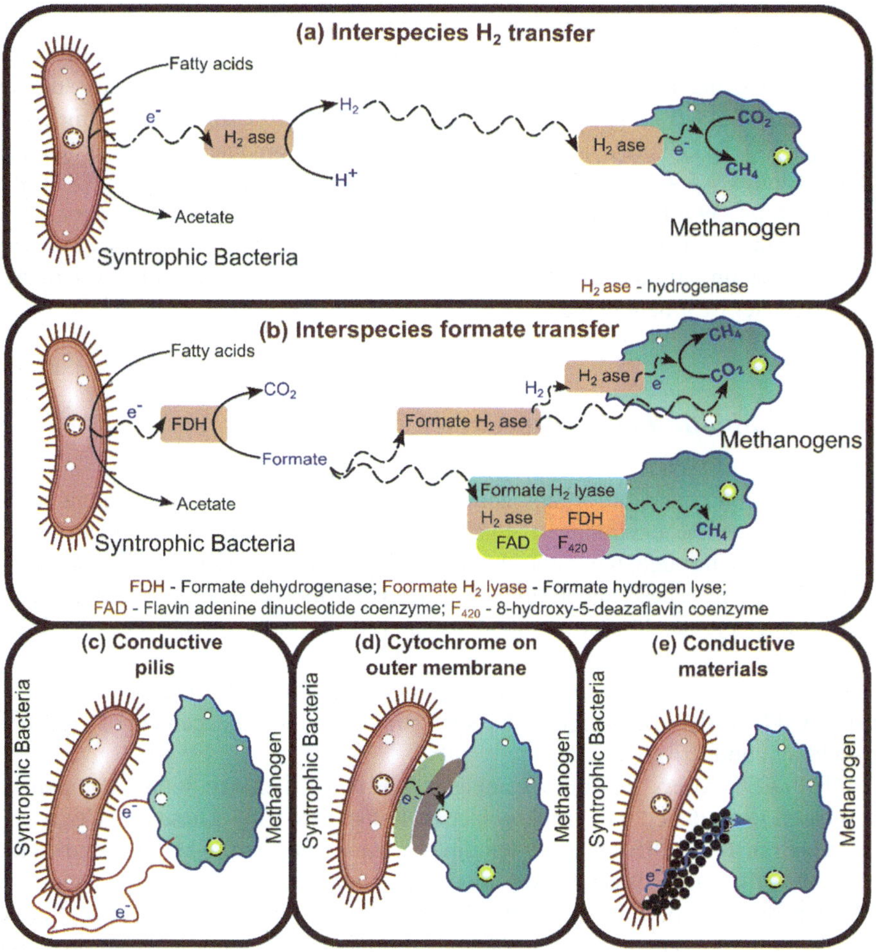

Fig. 2 Interspecies electron transfer mechanisms: **a** H_2 and **b** formate-mediated indirect electron transfer, **c** conductive pilis, **d** cytochrome, and **e** conductive material direct electron transfer

Even though the volume of brewery wastewater generated depends on the specific water usage and production, ~ 3–10 L of brewery wastewater is generated per litre of beer produced in breweries. The characteristics of wastewater include high organic matter content, readily biodegradable, non-toxic, and devoid of heavy metals. The concentration of organic content depends on the wastewater-to-beer ratio. Both nitrogen and phosphorus concentrations range between 30 and 100 mg L^{-1}. The sources of nitrogen include malt, adjuncts, nitric acid used for cleaning, the amount of yeast discharged, and the usage of other cleaning agents. Other characteristics include pH $= 6.5 \pm 0.4$, COD $= 1250 \pm 100$ mg L^{-1}, NH_3-N $= 16 \pm 5$ mg L^{-1}, TN $= 24 \pm 3$ mg L^{-1}, suspended solids $= 500 \pm 50$ mg L^{-1} [3].

The characteristics of dairy wastewater include pH = 3.3–12.0, COD = 0.7–92.8 g L^{-1}, TSS = 0.1–23 g L^{-1}, sodium (Na) = 84–2325 mg L^{-1}, potassium (K) = 8–690 mg L^{-1}, and phosphorus (P) = 29–1099 mg L^{-1}. Metals such as iron (Fe) = 0.03–6.7 mg L^{-1}, cobalt (Co) = 0.001–0.022 mg L^{-1}, and nickel = 0.004–1.000 mg L^{-1} are in low concentration. The high organic matter also includes carbohydrates = 5.4–62.9 g L^{-1}, dairy proteins (casein/whey = 1.3–13.1 g L^{-1}), and lipids = 0.07–39 g L^{-1}. Generally, the risk of failure in an anaerobic reactor while treating dairy wastewater is due to ammonia inhibition, trace metal deficiency along with long-chain fatty acid inhibition [73]. Table 1 highlights the application of AD in wastewater treatment plants.

3.1 Xenobiotic-Laden Wastewater Treatment

Xenobiotic compounds are organic compounds originating from anthropogenic activities that enter the environment. Such xenobiotic compounds accumulate in the trophic levels through the food chain and may severely threaten human health in the long run. These xenobiotic compounds include azo dyes, pharmaceutical compounds, organochlorine compounds, and antibiotic-resistant genes (ARGs). Therefore, treating these pollutants before discharging the effluent into the environment is imperative. Treatment of xenobiotic compounds in AD processes can be expedited by different strategies. One such strategy is the enhancement of the DIET by adding conductive materials.

The DIET offers several advantages while treating xenobiotic compounds compared to mediated interspecies electron transfer (MIET). These include: (i) the electron transfer in DIET is about ten times greater compared to MIET based on the electron-diffusion-electrochemistry model, enhancing the metabolic processes and pollutant degradation efficiency in the microbial community; (ii) DIET is not affected by H_2-partial pressure as it does not depend on any mediators (like H_2); (iii) electron competition and diffusion losses are mitigated as the electrons are transferred directly to methanogens in DIET [82].

The microbial metabolism of a xenobiotic compound can occur through (i) direct metabolism, (ii) co-metabolism, and (iii) syntrophic metabolism (Fig. 3). Direct metabolism is difficult for the microorganism to metabolise as it has to use the contaminant as a nutrient/carbon source directly. In comparison, the microbe uses simple organic matter, such as glucose, ethanol and acetate, as exogenous carbon sources, improving the AD of the contaminant [83]. However, the DIET-dominated syntrophic metabolism involves a complex interchange of compounds, intermediates, and energy flow in the microbial community. It offers an enhanced electron transport mode for contaminant removal and methanogenesis [84]. More research in this domain will unravel the substrate conversion pathways as well as surer proof of DIET through future research.

AD effectively treats organic waste containing emerging contaminants, especially antibiotics widely used in the agriculture and health sectors. Investigations that

Table 1 Application of AD in wastewater treatment

Anaerobic reactor type	Microbial community type	Source of wastewater	Performance	References
I-stage acidogenic reactor containing H_2-producing bacteria with II-stage methanogenic bioreactor containing encapsulated microbes (CH_4E reactor) parallelly an AnMBR	Mesophilic	Brewery wastewater	VFA content increased by 40.4% in the I-stage stage at an influent COD of 1970 ± 853 mg L^{-1}. Additionally, the off-gas contained CH_4 > 60% after 14 d in the CH_4E reactor in the II-stage	[74]
Anaerobic membrane bioreactor (AnMBR)	Mesophilic (Methanosaeta dominant)	Synthetic (oleate) wastewater	Stable operation at a loading rate of 1 g L^{-1} d^{-1}	[34]
Continuous stirred tank reactor (CSTR) and anaerobic membrane bioreactor (AnMBR)	Mesophilic	Ice cream process wastewater	AnMBR operated better compared to CSTR. Lipid concentration of 0.05 g g^{-1} of total solids negatively influenced its stability	[75]
AnMBR	Mesophilic	Synthetic wastewater	Stable operation was observed at SRT = 40 d, LCFA = 0.115 g COD g^{-1} VSS, and lipid loading rate = 0.79 g COD L^{-1} d^{-1}	[76]
Expanded granular sludge bed (EGSB)	Psychrophilic (Methanosaeta dominant)	Synthetic wastewater	CH_4 conversion rate = 44%–51% and LCFA loading rate = 0.67–0.99 g COD L^{-1} d^{-1}. Flotation and granular sludge washout was observed	[77]
UASB reactor	Mesophilic	Brewery wastewater	Average COD removal = 78.97% from an initial effluent COD = 2005.73 ± 1139.85 mg L^{-1}	[78]
UASB reactor	Mesophilic	Brewery wastewater	CH_4 production rate increased from 0.29 to 1.46 L CH_4 g^{-1} COD when the loading rate was increased from 2.0 to 8.26 g COD L^{-1} d^{-1}	[79]

(continued)

Table 1 (continued)

Anaerobic reactor type	Microbial community type	Source of wastewater	Performance	References
Lab-scale biogas-lift reactors	Mesophilic	Brewery wastewater	Inoculated using residual activated sludge and anaerobic granular sludge. 90% COD removal efficiency was observed at an organic loading rate of 8.3 kg COD m^{-3} d^{-1}, resulting in an effluent COD of 240 mg L^{-1}	[80]
Pre-acidification + UASB reactor with recycling	Mesophilic	Cheese whey wastewater	The lipid removal efficiency was 63%–89% at loading rate of 3.5 g L^{-1} d^{-1}, additionally caused biomass flotation	[81]

have demonstrated the removal of antibiotics using AD include tetracycline [85], norfloxacin [86], ampicillin [87] and sulfonamide [88], amongst others. However, antibiotics inhibit methanogens, negatively affecting CH_4 production and VFA accumulation [89]. Additionally, the usage of biochar has demonstrated enhanced CH_4 generation and antibiotic removal [88].

Furthermore, long-term antibiotic exposure produces antibiotic-resistant bacteria and/or genes. These proliferate through vertical and horizontal gene transfer depending on the reproduction along with the transfer of mobile genetic elements (MGEs) between different bacterial species [90]. Therefore, adequate measures are necessary to abate the risk posed by ARGs to human health and the environment. Optimisation of environmental factors and operating conditions like hydrolysis pretreatment, temperature, and usage of auxiliary materials can aid in removing ARGs [91].

Using biochar in AD has been shown to enhance the elimination of ARGs by altering the microbial community structure by restraining the abundance of MGEs [92]. However, there is less understanding of the elimination mechanism of ARGs by biochar as investigations have focused on only a few ARGs. Nevertheless, the rich pore structure adsorption mechanism, mainly through π-π, electrostatic and hydrophobic interaction along with persisting free hydroxyl radical (•OH), essentially damages the structure of ARGs [91]. For instance, these target the phosphodiester bonds to the base stacking structure and phosphate bonds present in the nucleotides [93]. The CMs-led DIET has performed exceptionally in removing xenobiotics, such as medical antibiotics, organochlorine compounds, industrial dyes, and ARGs. Table 2 highlights the participation of different CMs aiding the DIET in removing xenobiotics from wastewater.

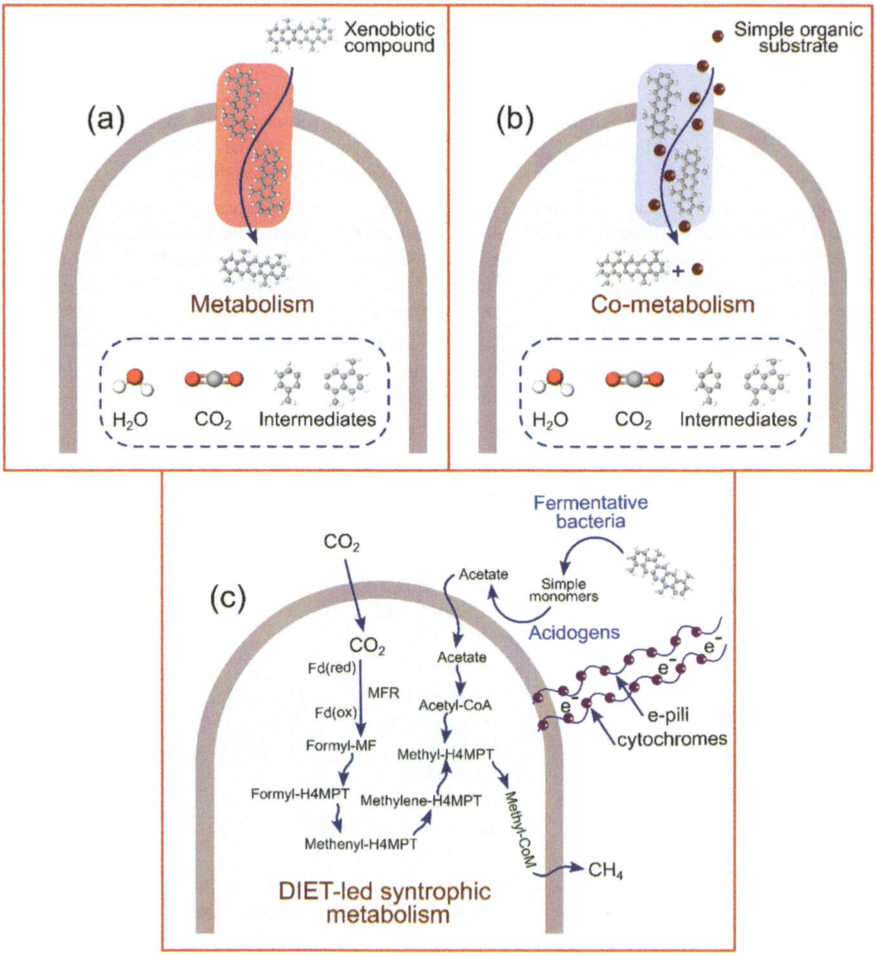

Fig. 3 Different metabolic pathways of xenobiotic compound degradation by microorganisms: **a** direct metabolism, **b** co-metabolism, and **c** syntrophic metabolism

4 Resource Recovery

Simultaneous waste management and generation of biogas, mainly CH_4 using AD, is a lucrative option for low cost, low production of residual waste, and a renewable energy source. The digestate produced is considered as an alternative to fertiliser or soil conditioner.

Table 2 Application of AD in treating xenobiotic-laden wastewater

Xenobiotic compounds	CMs	Mechanism involved	Removal efficiency	References
Phenols/ N-heterocycles	Biochar/GAC/ Magnetite	Enhanced electron transfer activity and sludge conductivity	Improved the removal of aromatic and N-heterocycles	[94]
Red 2 dye, direct yellow 12 dye, acid orange 7 dye, Congo red dye, direct black 38 dye (azo dye)	Nano manganese dioxide	Electrons were majorly used for dye removal and not for methanogenesis	The degradation rate of azo dye enhanced but suppressed the CH_4 production	[95]
Acid orange 7 (azo dye)	Fe_3O_4	Iron-reducing bacteria and methanogens were enriched. It resulted in superior electron transfer capacity of the sludge	The degradation rate of azo dye and sulphate increased	[96]
Methyl orange (azo dye)	Modified lignin	Electron transfer was enhanced using redox mediators like phenols, ketones, and quinones	Decolorisation rate increased from 20 to 95%	[97]
Methyl orange (azo dye)	Magnetic nanoparticles	CMs enriched the iron-reducing bacteria, establishing a stable DIET and accelerated sludge granulation	The dye removal rate enhanced from 70.64 to 96.87%	[98]
2,4,6-Trichlorophenol	Fe_3O_4 supported on water hyacinth biochar	The CMs enhanced microbial enzyme activity and sludge granulation	CMs improved CH_4 generation and 2,4,6-Trichlorophenol removal	[99]
Cationic red X-GRL (azo dye)	Fe_3O_4-sludge carbon	DIET between *Geobacter* and *Methanosaeta*. This enhanced sludge granulation, aniline removal, and CH_4 production	97.6% decolorisation rate	[25]

(continued)

Table 2 (continued)

Xenobiotic compounds	CMs	Mechanism involved	Removal efficiency	References
RR2 (azo dye)	Fe_3O_4	Fe_3O_4 supported in the redox medium that promoted DIET and conductivity	The CMs enhanced the removal rate of azo dye. It also helped in shortening the lag period and improving the CH_4 generation	[100]
AO7 (azo dye)	$Fe(OH)_3$	Iron-reducing bacterium enhanced DIET. $Fe(OH)_3$ increased COD removal and decolorisation efficiency	89.5% COD removal and 99.8% of decolorisation rate	[27]

4.1 CH_4 and H_2 Recovery

The amount of CH_4 recovered largely depends on factors such as reactor configuration, type of organic substrate, external supply of H_2, and operational conditions. A promising strategy to enhance the CH_4 yield is reutilising the CO_2 generated as the acidogenic off-gas in the two-phase anaerobic reactor [101]. The investigation focused on utilising the acidogenic off-gas by introducing it into the methanogenic UASB reactor, which enhanced the CH_4 recovery by 38%. In addition to CO_2, pulsed or regular injection of H_2 (0.3–1.7 L_{H2} L_{sludge}^{-1} d^{-1}) in the reactor can strategically enhance the abundance of hydrogenotrophic methanogen (*Methanobacterium*), enhancing biomethanisation [102]. Employing different packing materials (rashig rings and alumina ceramic sponge) and providing a separate chamber for H_2 injection can also be an alternative to enhance gas–liquid mass transfer for CH_4 generation [103]. Additionally, intensifying the gas transfer of H_2 (0.93–1.44 L d^{-1}) using a hollow fibre membrane (HFM) attached to the reactor also supports the biogas upgrading technique [104]. Nevertheless, the H_2 supply can result in CO_2 depletion, acetate, and other VFA inhibition apart from CH_4 generation. Hence, regulating the H_2 addition rate is crucial by considering the H_2 uptake rate, CO_2 production rate, and growth of hydrogenotrophic methanogens to prevent VFA inhibition due to its accumulation [105]. Though these investigations address the bottleneck related to CH_4 upgradation by externally supplying H_2, the source of H_2 can be a great challenge. A few investigations suggest that as the storage of H_2 is problematic due to safety issues and lack of infrastructure, the H_2 generated from the electrolysis of water that sources its electrical energy from surplus renewable energy sources can be a suitable option [105].

In comparison to CH_4, research has also embarked towards producing H_2 from dark fermentative pathways. Hydrogen provides the benefits of both being compressed as well as a high energy density (141 MJ kg^{-1}) compared to CH_4

(50 MJ kg^{-1}) [106]. It depends on two enzymes, hydrogenase, and nitrogenase, that catalyses a simple reaction represented by Eq. (38). The biological fermentation of carbohydrates generates (about 20 times) higher H_2 than the fermentation of protein or fat-rich organic matter [107]. The H_2 production depends on factors like composition and pretreatment of organic waste and process configuration. Interestingly, the presence of proteins ensures nitrogen source enhancing the activity and growth of H_2-producing bacteria. Strategies to enhance such bio-H_2 production have also been explored. For instance, the co-digestion of food waste and sewage sludge (10:1) has increased the H_2 production rate [108]. Optimising the organic protein in the feedstock is a feasible alternative to intensify the H_2 production further [109]. Complex organic wastes slow down biodegradability; therefore, adapting mechanical or chemical pretreatment can improve the degradation. Other pretreatment adapted in the case of complex organic waste include heat, ultrasonication, acid, base, and their combination thereof [110].

$$2H^+ + 2e^- \leftrightarrow H_2 \tag{38}$$

Hythane is a H_2 and CH_4 mixture with a varying H_2 concentration between 10 and 25% (v/v) [111]. It is advantageous over compressed natural gas (CNG) in terms of flammability range observed as flame speed due to the presence of H_2 [111]. Other advantages include reduced greenhouse gas (GHG) emissions, excellent reducing agents, lower exhaust temperatures, and stimulation of CH_4 combustion [112]. Hythane production occurs when different organic substrates are employed in the sequential anaerobic phase of a dark fermentation for H_2 production, followed by a secondary digestion step for CH_4 production [113]. The two stages improve the H_2/CH_4 ratio, reduce the fermentation time, increase degradation efficiency by enrichment of selective microbes, help in increasing the organic loading rate, and facilitate separate optimisation of H_2 and CH_4 [109]. Though the two-stage AD appears to be a lucrative option for hythane production, problems like maintenance issues and high initial investments have limited the commercialisation of this technology. On the other hand, the H_2 present in hythane demands the usage of steel in the gas pipeline, which is less prone to H_2 embrittlement under pressure in the gas distribution systems [113]. As such, fulfilling these requirements is necessary to adopt hythane as a more sustainable energy source.

4.2 VFAs Recovery

Novel bio-processing routes help transform the economic status of low-value feedstocks like organic waste by converting it to higher value-added products, such as VFAs. Herewith, the VFAs produced in an anaerobic reactor have tremendous applications, such as usage as carbon substrate while denitrification in wastewater, microbial lipid production, acting as precursors in producing higher-value biofuels, bioplastics, and other biochemical generation [94]. Separating these intermediate degradation products from rewired anaerobic digestors is possible by arresting the methanogenesis [114]. Methanogenesis is arrested by inoculum pretreatment like methanogenic inhibitors or by regulating parameters such as retention time, redox potential, temperature, and pH [115].

The product spectrum in terms of VFAs and alcohol found in mixed culture fermentation depends on environmental factors, such as pH, temperature, dilution rate in chemostat conditions, type of carbohydrate substrate, and feeding pattern, among others [116]. For instance, lactate production is dominant at a low pH of 3.9, supporting the enrichment of the *Lactobacillus* strain in the mixed inocula [117]. On the other hand, neutral to alkaline pH of 7.0–8.5 at thermophilic temperatures intensifies ethanol production [118]. The technologies practised for the separation of VFAs from effluent include nanofiltration [119], liquid/liquid extraction [120], anion exchange [121], and secondary conversion processes [122].

Membrane separation, an in-situ technology, has shown a potential for scalability in wastewater treatment. The effluent from AD consists of high concentrations of solids (biomass) and salts that cause membrane fouling, adversely affecting the performance of the separation membrane [123]. Employing pretreatment techniques, such as effluent dilution, sieving, and microfiltration, can help overcome this problem. Further, applying nanofiltration on pretreated effluent at a neutral pH of 7.0 helps achieve a retention ratio of up to 75%, separating the acids in dissociated form [119]. On the other hand, anion-exchange technology can also accomplish the separation of dissociated acids (anions). It has showcased a maximum lactate recovery yield of up to 92% from water at a low pH of 2.0 [121]. Additionally, in the case recovery followed by esterification of succinic acid, the anion-exchange membrane helped to achieve a high yield of 96% dimethyl succinate [124].

Besides directly separating carboxylates, secondary fermentation to biopolymers like PHAs can facilitate product recovery. These are produced as intracellular storage compounds or as internal energy (carbon reserve) that are easily separated by solid–liquid separation [125]. Structurally, it can be compared to petrochemical plastics and participates as a building block in other biochemical reactions [126]. Employing synthetic substrate acetate or lactate has demonstrated accumulation of up to 0.90 gPHA $gVSS^{-1}$ at lab-scale experiments [127]. Additionally, the application of wastewater has also resulted in a PHA content of 0.75 gPHA $gVSS^{-1}$ [128] from fermented molasses and 0.77 gPHA $gVSS^{-1}$ from paper mill [127].

A different approach for valuables recovery would be through further chain elongation of medium-chain fatty acids (MCFAs). These MCFAs are produced through anaerobic fermentation using ethanol and short-chain VFAs. It is advantageous as longer carbon chain molecules are more apolar, facilitating easy extraction from fermentation broth. Continuous extraction of MCFAs can enable continuous production as high concentrations of MCFAs are toxic to the microbial cell [129]. Apart from the sustainable contribution of AD for conventionally producing CH_4-containing biogas, the bioprocesses aiming to produce biopolymers, MCFAs, and other value-added products can replace them. Identifying the niche market for a particular product is often necessary to maximise the waste valorisation from a given organic waste, helping to achieve a value addition within a short value chain [125].

4.3 Nutrient Recovery

During the AD of waste, biomass releases nitrogen and phosphorus in the liquid phase. When discharged into the aquatic ecosystem, it becomes the potential cause for eutrophication. Proteins present in the biomass are the source of nitrogen in AD. The first step of hydrolysis forms amino acids, followed by deamination, resulting in the formation of ammonium nitrogen. Usually, protein degradation produces equivalent moles of bicarbonate alkalinity and ammonium nitrogen. The bicarbonate alkalinity is responsible for buffering action, preventing the pH drop due to VFA accumulation and protecting the methanogens. However, high proteins (C/N > 20) can cause high levels of free ammonia, inhibiting methanogenesis. Ammonium (NH_4^+-N) remains in the liquid digestate without undergoing further oxidation, along with phosphorous and metal salts. The proportion of NH_4^+-N in the digested liquid is greater than in the organic matter used in AD. Its application as liquid fertiliser directly spreading on arable land is an effective solution. Nevertheless, the impact of digested liquid on human health and the environment remains unexplored, and its efficacy as a fertiliser is yet to be investigated. The N-loading required for a particular crop should also be considered prior to applying this liquid fertiliser to avoid excess N_2O emissions or NO_3^- formation.

Ammonia removal through air stripping and absorption by sulphuric acid (H_2SO_4) produces ammonium sulphate ($(NH_4)_2SO_4$), which is a physico-chemical process. The $(NH_4)_2SO_4$ crystals formed act as an agricultural fertiliser. Adjusting pH to an alkaline range and aeration transfers NH_3 from liquid to air. NH_3 stripping is observed in the case of the digestate of diary manure fermentation. About 90% of NH_3 removal was observed when the digestate effluent maintained at a mesophilic temperature of 35 °C and at an optimised pH of 10.3 by dosing with 3.3 g L^{-1} lime [130]. On the other hand, practising vacuum thermal NH_3 stripping coupled with H_2SO_4 absorption results in high-purity $(NH_4)_2SO_4$. It is accomplished by boiling the digestate at a temperature less than the normal boiling point under a vacuum [131]. The pH of the discharged effluent after NH_3 stripping is usually > 9.0. Though adding acid as part of post-treatment controls the pH, it further increases the operational cost. However, passing the biogas through the effluent can simultaneously decrease the pH of the effluent by scrubbing the gases like H_2S and CO_2 present in biogas [132].

The conventional way to recover phosphorus is by absorption, crystallisation into hydroxyapatite (HAp) or struvite ($NH_4MgPO_4{\cdot}6H_2O$), and thermo-chemical method by incinerating digested sludge. The enhanced biological phosphorus removal (EBPR) also removes about 90% phosphorus and precipitation into sludge in a typical wastewater treatment plant [133]. The $NH_4MgPO_4{\cdot}6H_2O$ formation can simultaneously remove phosphate (PO_4^{3-}) and NH_4^+-N from the soluble phase at low pH [134].

Unlike traditional anaerobic digesters, granular sludge processes like UASB reactor and expanded granular sludge bed (EGSB) reactor have the merit of high volumetric loading and microbial retention capacity [135]. These systems support the formation of granules which act as nuclei for microbial colonisation and calcium

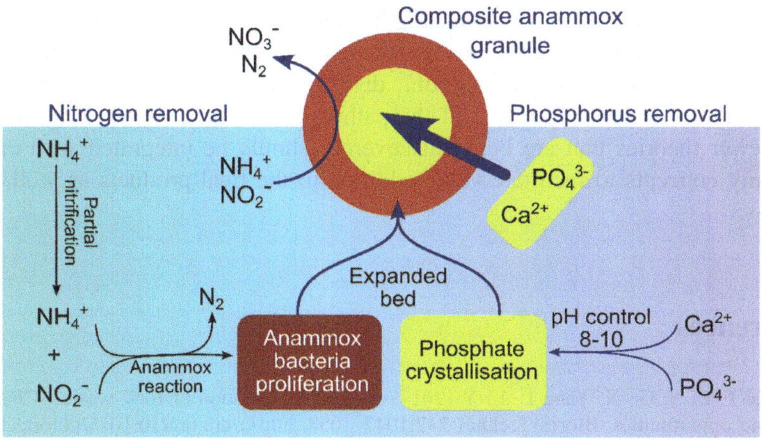

Fig. 4 Mechanism for simultaneous phosphorus recovery and nitrogen removal in an expanded bed reactor. Adapted from Ma et al. [4]

phosphate precipitation. It has been observed that external calcium addition has been shown to precipitate 89% of phosphate present in the black water treated in the UASB reactor [135].

The anaerobic ammonium oxidation (anammox) supports ammonia removal with the advantage of a partial nitrification-anammox process, thus saving about 60% of the cost involved in aeration along with no further addition of carbon source compared to the traditional nitrification and denitrification process [136]. Investigations have explored combining the anammox process with phosphorus recovery, resulted in $Ca_3(PO_4)_2$ precipitation in the anammox granules (Fig. 4) [137]. The quantitative elemental and spectral molecular analysis indicated the presence of phosphorus in the inner part of the composite granule in HAp [135]. The composite granule consists of anammox biofilm bound to the inorganic core. The phosphorus recovered as HAp is suitable as a precursor for the phosphorus industry [138].

5 Conclusion and Future Scope

The AD technology helps degrade organic waste by a cascade of reduction and oxidation reactions by anaerobic microbial consortia to attain final products of the lowest oxidation state, such as CH_4. The intricate phases of breaking down organic molecules involve limiting steps that can hamper the process. These include the type of organic waste, presence of toxic substances, ammonia inhibition, accumulation of VFAs, and inhibition of methanogens. Additionally, nutrient-rich effluent sometimes requires additional treatment before discharge into the environment. Therefore, integration of AD with other technologies based on the nature of organic matter and the available options for further waste valorisation products can benefit the whole process. Such processes should also take consider the environmental risk involved in releasing the final effluent into the environment and ensure that discharge is within the permissible standards. The anaerobic digestion technology is evolving with new insights into the intricate details of the pathways and the cross feeding involved. The newer theories that are being discovered should be integrated with circular economy concepts to facilitate value addition to the final products as well as the process.

References

1. Xu F, Li Y, Ge X, Yang L, Li Y (2017) Anaerobic digestion of food waste—Challenges and opportunities. Biores Technol 247:1047–1058. https://doi.org/10.1016/j.biortech.2017.09.020
2. Raut MP, Pandhal J, Wright PC (2021) Effective pretreatment of lignocellulosic Co-substrates using Barley Straw-adapted microbial consortia to enhanced biomethanation by anaerobic digestion. Biores Technol 321:124437. https://doi.org/10.1016/j.biortech.2021.125235
3. Maria MP, Torres NH, Nascimento VRS, Chagas TSA, Saratale GD, Mulla SI, Bharagava RN, Cavalcanti EB, Romanholo Ferreira LF (2023) Current advances in the Brewery wastewater treatment from anaerobic digestion for biogas production: a systematic review. Environ Adv 13:100394. https://doi.org/10.1016/j.envadv.2023.100394
4. Ma H, Guo Y, Qin Y, Li Y (2018) Nutrient recovery technologies integrated with energy recovery by waste biomass anaerobic digestion. Biores Technol 269:520–531. https://doi.org/10.1016/j.biortech.2018.08.114
5. Zhen G, Lu X, Kato H, Zhao Y, Li Y (2017) Overview of pretreatment strategies for enhancing Sewage Sludge disintegration and subsequent anaerobic digestion: current advances, full-scale application and future perspectives. Renew Sustain Energy Rev 69:559–577. https://doi.org/10.1016/j.rser.2016.11.187
6. Li Y, Park SY, Zhu J (2010) Solid-state anaerobic digestion for methane production from organic waste. Renew Sustain Energy Rev 15(1):821–826. https://doi.org/10.1016/j.rser.2010.07.042
7. Khalid A, Arshad M, Anjum M, Mahmood T, Dawson L (2011) The anaerobic digestion of solid organic waste. Waste Manage 31(8):1737–1744. https://doi.org/10.1016/j.wasman.2011.03.021
8. Wainaina S, Lukitawesa, Kumar Awasthi M, Taherzadeh MJ (2019) Bioengineering of anaerobic digestion for volatile fatty acids, hydrogen or methane production: a critical review. Bioengineered 10(1):437–458. https://doi.org/10.1080/21655979.2019.1673937
9. Mei B, Zhi L, Cheng Q (2019) Evaluation of parameters for monitoring an anaerobic digestion process. In: IOP conference series: earth and environmental science. IOP Publishing, p 227, 022021. https://doi.org/10.1088/1755-1315/227/2/022021
10. Patel A, Mahboubi A, Horvath IS, Taherzadeh MJ, Rova U, Christakopoulos P, Matsakas L (2021) Volatile fatty acids (VFAs) generated by anaerobic digestion serve as feedstock for freshwater and marine oleaginous microorganisms to produce biodiesel and added-value compounds. Front Microbiol 12:614612. 10.3389%2Ffmicb.2021.614612
11. Luo L, Xu S, Jin Y, Han R, Liu H, Lü F (2018) Evaluation of methanogenic microbial electrolysis cells under closed/open circuit operations. Environ Technol 39(6):739–748. https://doi.org/10.1080/09593330.2017.1310934
12. Yadav M, Joshi C, Paritosh K, Thakur J, Pareek N, Masakapalli SK, Vivekanand V (2021) Organic waste conversion through anaerobic digestion: a critical insight into the metabolic pathways and microbial interactions. Metab Eng 69:323–337. https://doi.org/10.1016/j.ymben.2021.11.014

13. Menzel T, Neubauer P, Junne S (2020) Role of microbial hydrolysis in anaerobic digestion. Energies 13(21):5555. https://doi.org/10.3390/en13215555
14. Patakova P, Branska B, Sedlar K, Vasylkivska M, Jureckova K, Kolek J, Provaznik I (2019) Acidogenesis, solventogenesis, metabolic stress response and life cycle changes in *Clostridium beijerinckii* NRRL B-598 at the transcriptomic level. Sci Rep 9(1):1371. https://doi.org/10.1038/s41598-018-37679-0
15. Karekar S, Stefanini R, Ahring B (2022) Homo-acetogens: their metabolism and competitive relationship with hydrogenotrophic methanogens. Microorganisms 10(2):397. https://doi.org/10.3390/microorganisms10020397
16. Westerholm M, Schnürer A (2019) Microbial responses to different operating practices for biogas production systems. In: Rajesh Banu J, Ravi YK (eds), Anaerobic digestion, Intech Open, pp. 1–36. https://doi.org/10.5772/intechopen.82815
17. Buan NR (2018) Methanogens: pushing the boundaries of biology. Emerg Top Life Sci 2(4):629–646. https://doi.org/10.1042/etls20180031
18. Enzmann F, Mayer F, Rother M, Holtmann D (2018) Methanogens: biochemical background and biotechnological applications. AMB Express 8:1–22. https://doi.org/10.1186/s13568-017-0531-x
19. Karakashev D, Batstone DJ, Trably E, Angelidaki I (2006) Acetate oxidation is the dominant methanogenic pathway from acetate in the absence of *Methanosaetaceae*. Appl Environ Microbiol 72(7):5138–5141. https://doi.org/10.1128/aem.00489-06
20. Chen S, He Q (2015) Persistence of *Methanosaeta* populations in anaerobic digestion during process instability. J Ind Microbiol Biotechnol 42(8):1129–1137. https://doi.org/10.1007/s10295-015-1632-7
21. Lackner N, Hintersonnleitner A, Wagner AO, Illmer P (2018) Hydrogenotrophic Methanogenesis and Autotrophic Growth of *Methanosarcina thermophila*. Archaea 2018. https://doi.org/10.1155/2018/4712608
22. Mahesh M, Arivizhivendhan KV, Nivetha K, Swarnalatha S, Sekaran G (2018) Anaerobic digestion of sulphate-rich post-tanning wastewater at different COD/Sulphate and F/M Ratios. 3 Biotech 8:1–15. 10.1007%2Fs13205-018-1154-x
23. Dar SA, Kleerebezem R, Stams AJ, Kuenen JG, Muyzer G (2008) Competition and coexistence of sulfate-reducing bacteria, acetogens and methanogens in a lab-scale anaerobic bioreactor as affected by changing substrate to sulfate ratio. Appl Microbiol Biotechnol 78:1045–1055. https://doi.org/10.1007/s00253-008-1391-8
24. Yilmaz T, Erdirencelebi D, Berktay A (2012) Effect of COD/SO ratio on anaerobic treatment of landfill leachate during the start-up period. Environ Technol 33(3):313–320. https://doi.org/10.1080/09593330.2011.572920
25. Zhang Z, Zhang C, Yang Y, Zhang Z, Tang Y, Su P, Lin Z (2022) A review of sulfate-reducing bacteria: metabolism, influencing factors and application in wastewater treatment. J Clean Prod 376:134109. https://doi.org/10.1016/j.jclepro.2022.134109
26. Rückert C (2016) Sulfate reduction in microorganisms—Recent advances and biotechnological applications. Curr Opin Microbiol 33:140–146. https://doi.org/10.1016/j.mib.2016.07.007
27. Li J, Cai M, Miao Y, Luo G, Li W, Li Y, Li A (2019) Bacterial community structure and predicted function in an acidogenic sulfate-reducing reactor: effect of organic carbon to sulfate ratios. Biores Technol 293:122020. https://doi.org/10.1016/j.biortech.2019.122020
28. Muyzer G, Stams AJ (2008) The ecology and biotechnology of sulphate-reducing bacteria. Nat Rev Microbiol 6(6):441–454. https://doi.org/10.1038/nrmicro1892
29. Keller KL, Wall JD (2011) Genetics and molecular biology of the electron flow for sulfate respiration in *Desulfovibrio*. Front Microbiol 2:9742. https://doi.org/10.3389/fmicb.2011.00135
30. Harirchi S, Wainaina S, Sar T, Nojoumi SA, Parchami M, Parchami M, Taherzadeh MJ (2022) Microbiological insights into anaerobic digestion for biogas, hydrogen or volatile fatty acids (VFAs): a review. Bioengineered 13(3):6521–6557. https://doi.org/10.1080/21655979.2022.2035986

31. Czatzkowska M, Harnisz M, Korzeniewska E, Rusanowska P, Bajkacz S, Felis E, Jastrzębski JP, Paukszto Ł, Koniuszewska I (2021) The impact of antimicrobials on the efficiency of methane fermentation of sewage sludge, changes in microbial biodiversity and the spread of antibiotic resistance. J Hazard Mater 416:125773. https://doi.org/10.1016/j.jhazmat.2021.125773
32. Vidal G, Carvalho A, Méndez R, Lema J (2000) Influence of the content in fats and proteins on the anaerobic biodegradability of dairy wastewaters. Biores Technol 74(3):231–239. https://doi.org/10.1016/j.jhazmat.2021.125773
33. Atasoy M, Owusu-Agyeman I, Plaza E, Cetecioglu Z (2018) Bio-based volatile fatty acid production and recovery from waste streams: current status and future challenges. Biores Technol 268:773–786. https://doi.org/10.1016/j.biortech.2018.07.042
34. Zhang N, Wu C, Zhang J, Han S, Peng Y, Song X (2023) Impacts of lipids on the performance of anaerobic membrane bioreactors for food wastewater treatment. J Membr Sci 666:121104. https://doi.org/10.1016/j.memsci.2022.121104
35. Junicke H, Van Loosdrecht MCM, Kleerebezem R (2016) Kinetic and thermodynamic control of butyrate conversion in non-defined methanogenic communities. Appl Microbiol Biotechnol 100:915–925. https://doi.org/10.1007/s00253-015-6971-9
36. Meegoda JN, Li B, Patel K, Wang LB (2018) A review of the processes, parameters, and optimisation of anaerobic digestion. Int J Environ Res Public Health 15(10):2224. https://doi.org/10.3390/ijerph15102224
37. Botheju D, Bakke R (2011) Oxygen effects in anaerobic digestion—a review. Open Waste Manage J 4:1–19. https://doi.org/10.2174/1876400201104010001
38. Choi YY, Baek SR, Kim JI, Choi JW, Hur J, Lee TU, Lee BJ (2017) Characteristics and biodegradability of wastewater organic matter in municipal wastewater treatment plants collecting domestic wastewater and industrial discharge. Water 9(6):409. https://doi.org/10.3390/w9060409
39. Collivignarelli MC, Abbà A, Caccamo FM, Calatroni S, Torretta V, Katsoyiannis IA, Rada EC (2021) Applications of up-flow anaerobic sludge blanket (UASB) and characteristics of its microbial community: a review of bibliometric trend and recent findings. Int J Environ Res Pub Health 18(19):10326. https://doi.org/10.3390/ijerph181910326
40. Ruiz LM, Fernández M, Genaro A, Martín-Pascual J, Zamorano M (2023) Multi-parametric analysis based on physico-chemical characterisation and biochemical methane potential estimation for the selection of industrial wastes as co-substrates in anaerobic digestion. Energies 16(14):5444. https://doi.org/10.3390/en16145444
41. Conrad R (2020) Methane production in soil environments—Anaerobic biogeochemistry and microbial life between flooding and desiccation. Microorganisms 8(6):881. https://doi.org/10.3390/microorganisms8060881
42. Janyasuthiwong S, Rene ER, Esposito G, Lens PN (2016) Effect of pH on the performance of sulfate and thiosulfate-fed sulfate reducing inverse S-fluidised bed reactors. J Environ Eng 142(9):C4015012. https://doi.org/10.1061/(ASCE)EE.1943-7870.0001004
43. Madden P, Al-Raei AM, Enright AM, Chinalia FA, de Beer D, O'Flaherty V, Collins G (2014) Effect of sulfate on low-temperature anaerobic digestion. Front Microbiol 5:376. https://doi.org/10.3389/fmicb.2014.00376
44. Watanabe F, Bito T (2018) Vitamin B12 sources and microbial interaction. Exp Biol Med (Maywood, N.J.) 243(2):148–158. https://doi.org/10.1177/1535370217746612
45. de Lemos Chernicharo CA (2007) Anaerobic reactors. IWA Publishing. https://doi.org/10.2166/9781780402116
46. Liu C, Yuan X, Zeng G, Li W, Li J (2008) Prediction of methane yield at optimum pH for anaerobic digestion of organic fraction of municipal solid waste. Biores Technol 99(4):882–888. https://doi.org/10.1016/j.biortech.2007.01.013
47. Thanh PM, Ketheesan B, Yan Z, Stuckey D (2016) Trace metal speciation and bioavailability in anaerobic digestion: a review. Biotechnol Adv 34(2):122–136. https://doi.org/10.1016/j.biotechadv.2015.12.006

48. Zhou M, Yan B, Wong JW, Zhang Y (2017) Enhanced volatile fatty acids production from anaerobic fermentation of food waste: a mini-review focusing on acidogenic metabolic pathways. Biores Technol 248:68–78. https://doi.org/10.1016/j.biortech.2017.06.121
49. Chen Y, Luo J, Yan Y, Feng L (2013) Enhanced production of short-chain fatty acid by Co-fermentation of waste activated sludge and kitchen waste under alkaline conditions and its application to microbial fuel cells. Appl Energy 102:1197–1204. https://doi.org/10.1016/j.apenergy.2012.06.056
50. Bensaid S, Ruggeri B, Saracco G (2014) Development of a photosynthetic microbial electrochemical cell (PMEC) reactor coupled with dark fermentation of organic wastes: medium term perspectives. Energies 8(1):399–429. https://doi.org/10.3390/en8010399
51. Chaganti SR, Kim D, Lalman JA (2011) Flux balance analysis of mixed anaerobic microbial communities: effects of linoleic acid (LA) and pH on biohydrogen production. Int J Hydrogen Energy 36(21):14141–14152. https://doi.org/10.1016/j.ijhydene.2011.04.161
52. Yen H, Li R, Ma T (2011) The development process for a continuous acetone–butanol–ethanol (ABE) fermentation by immobilised *Clostridium acetobutylicum*. J Taiwan Inst Chem Eng 42(6):902–907. https://doi.org/10.1016/j.jtice.2011.05.006
53. Lee HS, Salerno MB, Rittmann BE (2008) Thermodynamic evaluation on H_2 production in glucose fermentation. Environ Sci Technol 42(7):2401–2407. https://doi.org/10.1021/es702610v
54. Wang K, Yin J, Shen D, Li N (2014) Anaerobic digestion of food waste for volatile fatty acids (VFAs) production with different types of inoculum: effect of pH. Biores Technol 161:395–401. https://doi.org/10.1016/j.biortech.2014.03.088
55. Vital M, Howe AC, Tiedje JM (2014) Revealing the bacterial butyrate synthesis pathways by analysing (Meta) genomic data. MBio 5(2):10–1128. https://doi.org/10.1128/mbio.00889-14
56. Battista F, Bolzonella D (2018) Some critical aspects of the enzymatic hydrolysis at high dry-matter content: a review. Biofuels, Bioprod Biorefin 12(4):711–723. https://doi.org/10.1002/bbb.1883
57. Strazzera G, Battista F, Garcia NH, Frison N, Bolzonella D (2018) Volatile fatty acids production from food wastes for biorefinery platforms: a review. J Environ Manage 226:278–288. https://doi.org/10.1016/j.jenvman.2018.08.039
58. Newsholme P, Brennan L, Rubi B, Maechler P (2005) New insights into amino acid metabolism, β-cell function and diabetes. Clin Sci 108(3):185–194. https://doi.org/10.1042/cs20040290
59. Kosaka T, Kato S, Shimoyama T, Ishii S, Abe T, Watanabe K (2008) The genome of *Pelotomaculum thermopropionicum* reveals niche-associated evolution in anaerobic microbiota. Genome Res 18(3):442–448. https://doi.org/10.1101/gr.7136508
60. Worm P, Stams AJ, Cheng X, Plugge CM (2011) Growth-and substrate-dependent transcription of formate dehydrogenase and hydrogenase coding genes in *Syntrophobacter fumaroxidans* and *Methanospirillum hungatei*. Microbiology 157(1):280–289. https://doi.org/10.1099/mic.0.043927-0
61. McInerney MJ, Sieber JR, Gunsalus RP (2009) Syntrophy in anaerobic global carbon cycles. Curr Opin Biotechnol 20(6):623–632. https://doi.org/10.1016/j.copbio.2009.10.001
62. Sieber JR, Crable BR, Sheik CS, Hurst GB, Rohlin L, Gunsalus RP, McInerney MJ (2015) Proteomic analysis reveals metabolic and regulatory systems involved in the syntrophic and axenic lifestyle of *Syntrophomonas wolfei*. Front Microbiol 6:115. https://doi.org/10.3389/fmicb.2015.00115
63. Venkiteshwaran K, Bocher B, Maki J, Zitomer D (2015) Relating anaerobic digestion microbial community and process function: supplementary issue: water microbiology. Microbiol Insights 8:MBI-S33593. 10.4137%2FMBI.S33593
64. Balk M, Weijma J, Stams AJ (2002) *Thermotoga lettingae* Sp. Nov., A novel thermophilic, methanol-degrading bacterium isolated from a thermophilic anaerobic reactor. Int J Syst Evol Microbiol 52(4):1361–1368. https://doi.org/10.1099/00207713-52-4-1361
65. Manzoor S, Bongcam-Rudloff E, Schnürer A, Müller B (2016) Genome-guided analysis and whole transcriptome profiling of the mesophilic syntrophic acetate oxidising bacterium

Syntrophaceticus schinkii. PLoS ONE 11(11):e0166520. https://doi.org/10.1371/journal.pone.0166520
66. Westerholm M, Dolfing J, Sherry A, Gray ND, Head IM, Schnürer A (2011) Quantification of syntrophic acetate-oxidising microbial communities in biogas processes. Environ Microbiol Rep 3(4):500–505. https://doi.org/10.1111/j.1758-2229.2011.00249.x
67. Dyksma S, Jansen L, Gallert C (2020) Syntrophic acetate oxidation replaces acetoclastic methanogenesis during thermophilic digestion of biowaste. Microbiome 8(1):105.s https://doi.org/10.1186/s40168-020-00862-5
68. Stams AJ, Plugge CM (2009) Electron transfer in syntrophic communities of anaerobic bacteria and archaea. Nat Rev Microbiol 7(8):568–577. https://doi.org/10.1038/nrmicro2166
69. Zhang Y, Li J, Liu F, Yan H, Li J, Zhang X (2018) Reduction of gibbs free energy and enhancement of *Methanosaeta* by bicarbonate to promote anaerobic syntrophic butyrate oxidation. Biores Technol 267:209–217. https://doi.org/10.1016/j.biortech.2018.06.098
70. Leng L, Yang P, Singh S, Zhuang H, Xu L, Chen W, Dolfing J, Li D, Zhang Y, Zeng H, Chu W, Lee P (2017) A review on the bioenergetics of anaerobic microbial metabolism close to the thermodynamic limits and its implications for digestion applications. Biores Technol 247:1095–1106. https://doi.org/10.1016/j.biortech.2017.09.103
71. Schink B, Montag D, Keller A, Müller N (2017) Hydrogen or formate: alternative key players in methanogenic degradation. Environ Microbiol Rep 9(3):189–202. https://doi.org/10.1111/1758-2229.12524
72. Wang Z, Yin Q, Gu M, He K, Wu G (2018) Enhanced Azo Dye reactive red 2 degradation in anaerobic reactors by dosing conductive material of ferroferric oxide. J Hazard Mater 357:226–234. https://doi.org/10.1016/j.jhazmat.2018.06.005
73. Ye M, Li Y (2023) Methanogenic treatment of dairy wastewater: a review of current obstacles and new technological perspectives. Sci Total Environ 866:161447. https://doi.org/10.1016/j.scitotenv.2023.161447
74. Chen S, Arnold W, Wright N, Zhu K, Ajayi O, Novak P (2022) Encapsulation technology for decentralised brewery wastewater treatment: a small pilot experiment. Biores Technol 347:126435. https://doi.org/10.1016/j.biortech.2021.126435
75. Lutze R, Engelhart M (2020) Comparison of CSTR and AnMBR for anaerobic digestion of WAS and lipid-rich flotation sludge from the dairy industry. Water Resour Ind 23:100122. https://doi.org/10.1016/j.wri.2019.100122
76. Szabo-Corbacho MA, Pacheco-Ruiz S, Míguez D, Hooijmans CM, García HA, Brdjanovic D, van Lier JB (2021) Impact of solids retention time on the biological performance of an AnMBR treating lipid-rich synthetic dairy wastewater. Environ Technol 42(4):597–608. https://doi.org/10.1080/09593330.2019.1639829
77. Singh S, Rinta-Kanto JM, Kettunen R, Tolvanen H, Lens P, Collins G, Kokko M, Rintala J (2019) Anaerobic treatment of LCFA-containing synthetic dairy wastewater at 20 °C: process performance and microbial community dynamics. Sci Total Environ 691:960–968. https://doi.org/10.1016/j.scitotenv.2019.07.136
78. Enitan AM, Kumari S, Odiyo JO, Bux F, Swalaha FM (2018) Principal component analysis and characterisation of methane community in a full-scale bioenergy producing UASB reactor treating brewery wastewater. Phys Chem Earth Parts A/B/C 108:1–8. https://doi.org/10.1016/j.pce.2018.06.006
79. Enitan AM, Adeyemo J, Swalaha FM, Bux F (2015) Anaerobic digestion model to enhance treatment of brewery wastewater for biogas production using UASB reactor. Environ Model Assess 20:673–685. https://doi.org/10.1007/s10666-015-9457-3
80. Xu F, Huang Z, Miao H, Ren H, Zhao M, Ruan W (2013) Identical full-scale biogas-lift reactors (BLRs) with anaerobic granular sludge and residual activated sludge for brewery wastewater treatment and kinetic modeling. J Environ Sci 25(10):2031–2040. https://doi.org/10.1016/S1001-0742(12)60268-X
81. Erdirencelebi D (2011) Treatment of High-fat-containing dairy wastewater in a sequential UASBR system: influence of recycle. J Chem Technol Biotechnol 86(4):525–533. https://doi.org/10.1002/jctb.2546

82. Storck T, Virdis B, Batstone DJ (2016) Modelling extracellular limitations for mediated *versus* direct interspecies electron transfer. ISME J 10(3):621–631. https://doi.org/10.1038/ismej.2015.139
83. Tian X, Jin X, Wang J, Shen Z, Zhou Y, Wang K (2021) Iron foam coupled hydrolysis acidification for trichloroacetaldehyde treatment: strengthening characteristics and mechanism. Biores Technol 342:126047. https://doi.org/10.1016/j.biortech.2021.126047
84. Wu L, Shen Z, Zhou Y, Zuo J (2023) Stimulating anaerobic digestion to degrade recalcitrant organic pollutants: potential role of conductive materials-led direct interspecies electron transfer. J Environ Manage 344:118337. https://doi.org/10.1016/j.jenvman.2023.118337
85. Xiong Y, Harb M, Hong PY (2017) Performance and microbial community variations of anaerobic digesters under increasing tetracycline concentrations. Appl Microbiol Biotechnol 101:5505–5517. https://doi.org/10.1007/s00253-017-8253-1
86. Zhao L, Ji Y, Sun P, Deng J, Wang H, Yang Y (2019) Effects of Individual and combined zinc oxide nanoparticle, norfloxacin, and sulfamethazine contamination on sludge anaerobic digestion. Biores Technol 273:454–461. https://doi.org/10.1016/j.biortech.2018.11.049
87. Zhou P, Su C, Li B, Qian Y (2006) Treatment of high-strength pharmaceutical wastewater and removal of antibiotics in anaerobic and aerobic biological treatment processes. J Environ Eng 132(1):129–136. https://doi.org/10.1061/(ASCE)0733-9372(2006)132:1(129)
88. Cheng D, Ngo HH, Guo W, Chang SW, Nguyen DD, Nguyen QA, Zhang J, Liang S (2020) Improving sulfonamide antibiotics removal from swine wastewater by supplying a new pomelo peel derived biochar in an anaerobic membrane bioreactor. Biores Technol 319:124160. https://doi.org/10.1016/j.biortech.2020.124160
89. Gurmessa B, Pedretti EF, Cocco S, Cardelli V, Corti G (2020) Manure anaerobic digestion effects and the role of pre- and post-treatments on veterinary antibiotics and antibiotic resistance genes removal efficiency. Sci Total Environ 721:137532. https://doi.org/10.1016/j.scitotenv.2020.137532
90. Shi Z, Zhao R, Wan J, Li B, Shen Y, Zhang S, Luo G (2021) Metagenomic analysis reveals the fate of antibiotic rsistance genes in two-stage and one-stage anaerobic digestion of waste activated sludge. J Hazard Mater 406:124595. https://doi.org/10.1016/j.jhazmat.2020.124595
91. Liu B, Yu K, Ahmed I, Gin K, Xi B, Wei Z, He Y, Zhang B (2021) Key factors driving the fate of antibiotic resistance genes and controlling strategies during aerobic composting of animal manure: a review. Sci Total Environ 791:148372. https://doi.org/10.1016/j.biortech.2021.124679
92. Yang S, Chen Z, Wen Q (2021) Impacts of biochar on anaerobic digestion of swine manure: methanogenesis and antibiotic resistance genes dissemination. Biores Technol 324:124679. https://doi.org/10.1016/j.biortech.2021.124679
93. Wu C, Fu L, Li H, Liu X, Wan C (2022) Using biochar to strengthen the removal of antibiotic resistance genes: performance and mechanism. Sci Total Environ 816:151554. https://doi.org/10.1016/j.scitotenv.2021.151554
94. Zhu D, Wang Z, Liu K, Si B, Yang G, Tian C, Zhang Y (2023) Multi-cycle anaerobic digestion of hydrothermal liquefaction aqueous phase: role of carbon and iron based conductive materials in inhibitory compounds degradation, microbial structure shaping, and interspecies electron transfer regulation. Chem Eng J 454:140019. https://doi.org/10.1016/j.cej.2022.140019
95. Che L, Xu H, Wei Z, Wei R, Yang B (2022) Activated carbon modified with nano manganese dioxide triggered electron transport pathway changes for boosted anaerobic treatment of dyeing wastewater. Environ Res 203:111944. https://doi.org/10.1016/j.envres.2021.111944
96. Jin Z, Zhao Z, Liang L, Zhang Y (2021) Effects of ferroferric oxide on azo dye degradation in a sulfate-containing anaerobic reactor: from electron transfer capacity and microbial community. Chemosphere 286:131779. https://doi.org/10.1016/j.chemosphere.2021.131779
97. Chen R, Cai J, Li Q, Yu L, Wei X, Gan C, Liu R, Yong Q (2021) Enhancement on the microbial extracellular electron transfers by modified lignin materials: application on decolorisation of Azo Dye. J Market Res 15:5265–5276. https://doi.org/10.1016/j.jmrt.2021.10.114

98. Qin J, Qian L, Zhang J, Zheng Y, Shi J, Shen J, Ou C (2020) Accelerated anaerobic biodecolorization of sulfonated Azo Dyes by magnetite nanoparticles as potential electron transfer mediators. Chemosphere 263:128048. https://doi.org/10.1016/j.chemosphere.2020.128048
99. Zhuang H, Shi J, Shan S, Ping L, Zhang C (2020) Enhanced anaerobic treatment of Azo Dye wastewater *via* direct interspecies electron transfer with Fe_3O_4/sludge carbon. Int J Hydrogen Energy 45(53):28476–28487. https://doi.org/10.1016/j.ijhydene.2020.07.234
100. Wang C, Liu Y, Gao X, Chen H, Xu X, Zhu L (2018) Role of biochar in the granulation of anaerobic sludge and improvement of electron transfer characteristics. Biores Technol 268:28–35. https://doi.org/10.1016/j.biortech.2018.07.116
101. Yan BH, Selvam A, Wong JW (2016) Innovative method for increased methane recovery from two-phase anaerobic digestion of food waste through reutilization of acidogenic off-gas in methanogenic reactor. Biores Technol 217:3–9. https://doi.org/10.1016/j.biortech.2016.03.116
102. Agneessens LM, Ottosen LDM, Voigt NV, Nielsen JL, De Jonge N, Fischer CH, Kofoed MVW (2017) In-situ biogas upgrading with pulse H_2 additions: the relevance of methanogen adaption and inorganic carbon level. Biores Technol 233:256–263. https://doi.org/10.1016/j.biortech.2017.02.016
103. Bassani I, Kougias PG, Angelidaki I (2016) In-situ biogas upgrading in thermophilic granular UASB reactor: key factors affecting the hydrogen mass transfer rate. Biores Technol 221:485–491. https://doi.org/10.1016/j.biortech.2016.09.083
104. Luo G, Angelidaki I (2013) Hollow fiber membrane based H_2 diffusion for efficient in-situ biogas upgrading in an anaerobic reactor. Appl Microbiol Biotechnol 97:3739–3744. https://doi.org/10.1007/s00253-013-4811-3
105. Mulat DG, Mosbæk F, Ward AJ, Polag D, Greule M, Keppler F, Nielsen JL, Feilberg A (2017) Exogenous addition of H_2 for an In-Situ biogas upgrading through biological reduction of carbon dioxide into methane. Waste Manage 68:146–156. https://doi.org/10.1016/j.wasman.2017.05.054
106. Aathika AR, Kubendran S, Yuvarani D, Thiruselvi M, Amudha D, Karthik TP, Sivanesan S (2018) Enhanced biohydrogen production from leather fleshing waste co-digested with tannery treatment plant sludge using anaerobic hydrogenic batch reactor. Energ Sour Recovery Utilisation Environ Effect 40(5):586–593. https://doi.org/10.1080/15567036.2018.1435754
107. Patel SK, Kumar P, Kalia VC (2012) Enhancing biological hydrogen production through complementary microbial metabolisms. Int J Hydrogen Energy 37(14):10590–10603. https://doi.org/10.1016/j.ijhydene.2012.04.045
108. Kim D, Kim S, Kim H, Kim M, Shin H (2011) Sewage sludge addition to food waste synergistically enhances hydrogen fermentation performance. Biores Technol 102(18):8501–8506. https://doi.org/10.1016/j.biortech.2011.04.089
109. Cavinato C, Giuliano A, Bolzonella D, Pavan P, Cecchi F (2012) Bio-hythane production from food waste by dark fermentation coupled with anaerobic digestion process: a long-term pilot scale experience. Int J Hydrogen Energy 37(15):11549–11555. https://doi.org/10.1016/j.ijhydene.2012.03.065
110. Elbeshbishy E, Hafez H, Dhar BR, Nakhla G (2011) Single and combined effect of various pretreatment methods for biohydrogen production from food waste. Int J Hydrogen Energy 36(17):11379–11387. https://doi.org/10.1016/j.ijhydene.2011.02.067
111. Sun C, Xia A, Liao Q, Fu Q, Huang Y, Zhu X (2019) Life-cycle assessment of biohythane production *via* Two-Stage anaerobic fermentation from microalgae and food waste. Renew Sustain Energy Rev 112:395–410. https://doi.org/10.1016/j.rser.2019.05.061
112. Roy S, Das D (2016) Biohythane production from organic wastes: present state of art. Environ Sci Pollut Res 23:9391–9410. https://doi.org/10.1007/s11356-015-5469-4
113. Rawoof SAA, Kumar PS, Vo DVN, Subramanian S (2021) Sequential production of hydrogen and methane by anaerobic digestion of organic wastes: a review. Environ Chem Lett 19:1043–1063. https://doi.org/10.1007/s10311-020-01122-6
114. Badgett A, Milbrandt A (2019) A summary of standards and practices for wet waste streams used in waste-to-energy technologies in the United States. Renew Sustain Energy Rev 117:109425. https://doi.org/10.1016/j.rser.2019.109425

115. Jiang Y, May HD, Lu L, Liang P, Huang X, Ren ZJ (2019) Carbon dioxide and organic waste valorisation by microbial electrosynthesis and electro-fermentation. Water Res 149:42–55. https://doi.org/10.1016/j.watres.2018.10.092
116. Temudo MF, Mato T, Kleerebezem R, Van Loosdrecht MC (2009) Xylose anaerobic conversion by open-mixed cultures. Appl Microbiol Biotechnol 82:231–239. https://doi.org/10.1007/s00253-008-1749-y
117. Liang S, Gliniewicz K, Mendes-Soares H, Settles ML, Forney LJ, Coats ER, McDonald AG (2015) Comparative analysis of microbial community of novel lactic acid fermentation inoculated with different undefined mixed cultures. Biores Technol 179:268–274. https://doi.org/10.1016/j.biortech.2014.12.032
118. Zhang F, Chen Y, Dai K, Zeng RJ (2014) The chemostat study of metabolic distribution in extreme-thermophilic (70 C) mixed culture fermentation. Appl Microbiol Biotechnol 98:10267–10273. https://doi.org/10.1007/s00253-014-6157-x
119. Zacharof MP, Lovitt RW (2014) Recovery of volatile fatty acids (VFA) from complex waste effluents using membranes. Water Sci Technol 69(3):495–503. https://doi.org/10.2166/wst.2013.717
120. Li Z, Qin W, Dai Y (2002) Liquid−liquid equilibria of acetic, propionic, butyric, and valeric acids with trioctylamine as extractant. J Chem Eng Data 47(4):843–848. https://doi.org/10.1016/S1369-703X(02)00024-4
121. Cao X, Yun HS, Koo Y (2002) Recovery of l-(+)-lactic acid by anion exchange resin amberlite IRA-400. Biochem Eng J 11(2–3):189–196. https://doi.org/10.1016/S1369-703X(02)00024-4
122. Marang L, Jiang Y, Van Loosdrecht MC, Kleerebezem R (2013) Butyrate as preferred substrate for polyhydroxybutyrate production. Biores Technol 142:232–239. https://doi.org/10.1016/j.biortech.2013.05.031
123. Le-Clech P, Chen V, Fane TA (2006) Fouling in membrane bioreactors used in wastewater treatment. J Membrane Sci 284(1–2):17–53.c https://doi.org/10.1016/j.memsci.2006.08.019
124. López-Garzón CS, Van der Wielen LA, Straathof AJ (2013) Green upgrading of succinate using dimethyl carbonate for a better integration with fermentative production. Chem Eng J 235:52–60. https://doi.org/10.1016/j.cej.2013.09.017
125. Kleerebezem R, Joosse B, Rozendal R, Van Loosdrecht MC (2015) Anaerobic digestion without biogas? Rev Environ Sci Bio/Technol 14:787–801. https://doi.org/10.1007/s11157-015-9374-6
126. Chen GQ (2009) A microbial polyhydroxyalkanoates (PHA) based bio-and materials industry. Chem Soc Rev 38(8):2434–2446. https://doi.org/10.1039/b812677c
127. Jiang Y, Marang L, Tamis J, Van Loosdrecht MC, Dijkman H, Kleerebezem R (2012) Waste to resource: converting paper mill wastewater to bioplastic. Water Res 46(17):5517–5530. https://doi.org/10.1016/j.watres.2012.07.028
128. Albuquerque M, Torres C, Reis M (2010) Polyhydroxyalkanoate (PHA) production by a mixed microbial culture using sugar molasses: effect of the influent substrate concentration on culture selection. Water Res 44(11):3419–3433. https://doi.org/10.1039/C0EE00282H
129. Steinbusch KJ, Hamelers HV, Plugge CM, Buisman CJ (2011) Biological formation of caproate and caprylate from acetate: fuel and chemical production from low grade biomass. Energy Environ Sci 4(1):216–224. https://doi.org/10.1039/C0EE00282H
130. Jiang A, Zhang T, Zhao Q, Li X, Chen S, Frear CS (2014) Evaluation of an integrated ammonia stripping, recovery, and biogas scrubbing system for use with anaerobically digested dairy manure. Biosys Eng 119:117–126. https://doi.org/10.1016/j.biosystemseng.2013.10.008
131. Ukwuani AT, Tao W (2016) Developing a vacuum thermal stripping—acid absorption process for ammonia recovery from anaerobic digester effluent. Water Res 106:108–115. https://doi.org/10.1016/j.watres.2016.09.054
132. Serna-Maza A, Heaven S, Banks C (2015) Biogas stripping of ammonia from fresh digestate from a food waste digester. Biores Technol 190:66–75. https://doi.org/10.1016/j.biortech.2015.04.041

133. Yuan Z, Pratt S, Batstone DJ (2012) Phosphorus recovery from wastewater through microbial processes. Curr Opin Biotechnol 23(6):878–883. https://doi.org/10.1016/j.copbio.2012.08.001
134. Le Corre KS, Valsami-Jones E, Hobbs P, Parsons SA (2009) Phosphorus recovery from wastewater by struvite Crystallisation: a review. Crit Rev Environ Sci Technol 39(6):433–477. https://doi.org/10.1080/10643380701640573
135. Cunha JR, Schott C, Van der Weijden RD, Leal LH, Zeeman G, Buisman C (2018) Calcium addition to increase the production of phosphate granules in anaerobic treatment of black water. Water Res 130:333–342. https://doi.org/10.1016/j.watres.2017.12.012
136. Lackner S, Gilbert EM, Vlaeminck SE, Joss A, Horn H, Van Loosdrecht MC (2014) Full-scale partial nitritation/Anammox experiences—An application survey. Water Res 55:292–303. https://doi.org/10.1016/j.watres.2014.02.032
137. Johansson S, Ruscalleda M, Colprim J (2017) Phosphorus recovery through biologically induced precipitation by partial nitritation-anammox granular biomass. Chem Eng J 327:881–888. https://doi.org/10.1016/j.cej.2017.06.129
138. Tervahauta T, Van der Weijden RD, Flemming RL, Hernández Leal L, Zeeman G, Buisman CJ (2013) Calcium phosphate granulation in anaerobic treatment of black water: a new approach to phosphorus recovery. Water Res 48:632–642. https://doi.org/10.1016/j.watres.2013.10.012

Anaerobic Digestion of Wastewater and Resource Recovery

Nupur Kesharwani and Samir Bajpai

Abstract Rapid industrialization, population growth, and excessive resource consumption significantly challenge the world for sustainable resources. It is imperative to use sustainable technologies for resource recovery from the wastewater stream to reduce the burden on natural resources. Anaerobic digestion is a promising resource and nutrient recovery technology for wastewater treatment plants. In anaerobic digesters, bacteria and archaea break down organic compounds present in wastewater, converting them into biogas and digestate. This biological treatment process effectively reduces the organic pollutant load in the wastewater, making it safer for release into the environment or for further treatment. The most significant resource recovered during anaerobic digestion is biogas. By capturing and utilizing biogas, not only can harmful emissions be reduced, but valuable energy can also be generated. Alternatively, the Implications of biogas for clean fuel in vehicles is also the need of the day apart from heat generation and electricity. In addition to biogas, anaerobic digestion can aid in nutrient recovery. The digestate produced during the process has abundant amounts of Nitrogen and Phosphorus, nutrients which are essential for agriculture. By properly treating and managing digestate, these nutrients can be reclaimed and reused as fertilizers, reducing the need for synthetic fertilizers, and minimizing nutrient pollution in water bodies. This chapter delves into the current state of resource and nutrient recovery technologies from wastewater and explores recent advancements in the field addressing challenges, perspective, and techno-economic feasibility.

Keywords Wastewater · Resource recovery · Nutrients · Biogas · Techno-economy

N. Kesharwani
Department of Civil Engineering, Government Engineering College, Koni, Bilaspur, Chhattisgarh, India

S. Bajpai (✉)
Department of Civil Engineering, National Institute of Technology, Raipur, Chhattisgarh, India
e-mail: sbajpai.ce@nitrr.ac.in

M. M. Ghangrekar et al. (eds.), *Biological and Hybrid Wastewater Treatment Technology*, Earth and Environmental Sciences Library,
https://doi.org/10.1007/978-3-031-63046-0_10

1 Introduction

Wastewater, or sewage, is any water adversely affected in quality by anthropogenic influence and includes domestic, industrial, and agricultural effluents. It contains a complex mixture of contaminants, including organic matter, pathogens, nutrients, heavy metals, and emerging pollutants like pharmaceuticals, toxic substances, and chemicals [1, 2]. Untreated or inadequately treated wastewater severely threatens human health and the environment. According to the United Nations, about 4/5th of the total global wastewater is inadequately treated and allowed to get mixed with the receiving streams without any prior treatment [3]. Wastewater treatment is a critical process that safeguards our health, protects our environment, and recycles precious water resources [4].

Traditional wastewater treatment methods typically involve physical, chemical, and biological processes to remove contaminants. However, these methods often require significant energy and generate sludge, a semi-solid byproduct that needs further treatment and disposal [5]. One of the most promising methods of wastewater treatment is anaerobic digestion. Anaerobic digestion offers an alternative solution that overcomes these challenges. A biological process that treats wastewater without the need for oxygen [6]. This method not only cleans the water but also generates valuable biogas that can be used as an energy source. Anaerobic digestion is a series of biological processes in which microorganisms break down biodegradable material without oxygen. It is widely used for treating organic wastes, including wastewater, manure, food waste, and industrial effluents. The anaerobic digestion process is typically carried out in a digester, a closed system that provides the optimal conditions for the microbes. The process involves four key stages: hydrolysis, acidogenesis, acetogenesis, and methanogenesis. One of the key advantages of anaerobic digestion is production of the biogas, a renewable energy source. This gas can be collected, cleaned, and used as a fuel for heat and power generation, replacing fossil fuels. The digested sludge, known as digestate, is also a rich source of nutrients and can be used as a soil conditioner, contributing to the circular economy [7].

Furthermore, anaerobic digestion reduces sludge volume, thus mitigating the challenges associated with sludge management. It also significantly reduces the emission of greenhouse gases compared to conventional treatment methods. The process can also degrade hard-to-remove pollutants and pathogens by improving effluent quality.

Anaerobic digestion also plays a critical role in sustainable environmental management. It aligns with several United Nations Sustainable Development Goals (SDGs), including clean water and sanitation, affordable and clean energy, and climate action [8]. Despite its numerous benefits, there are challenges associated with the implementation of anaerobic digestion for wastewater treatment. These include the need for pre-treatment, the sensitivity of the process to environmental conditions, and the requirement for skilled operation. However, ongoing research and technological advancements are addressing these issues, making the process more efficient and feasible.

2 Fundamentals of Anaerobic Digestion

Anaerobic digestion involves a series of microbial reactions that convert complex organic compounds into simpler molecules (Fig. 1). The process occurs without oxygen and is facilitated by a consortium of microorganisms, including bacteria and archaea. These microbes work together to break down organic matter primarily into methane (CH_4) and carbon dioxide (CO_2) through a multi-step process. The substrates for anaerobic digestion may include a variety of organic materials such as agricultural residues, organic constituents of municipal solid waste, sewage sludge, and organic industrial waste [9, 10]. These materials are rich in complex organic compounds, including carbohydrates, proteins, and lipids, which serve as the microbial community's primary energy sources. Several groups of microorganisms participate in anaerobic digestion. Hydrolytic bacteria are the first to act, breaking down complex polymers into simpler compounds such as sugars, amino acids, and fatty acids. Acidogenic bacteria then convert these intermediates into organic acids. Acetogenic bacteria further metabolize these acids into acetate, hydrogen, and carbon dioxide. Finally, methanogenic archaea convert these simple compounds into methane and carbon dioxide. Several factors influence the efficiency of anaerobic digestion, including temperature, pH, and nutrient availability. The process typically occurs within a mesophilic (30 °C to 38 °C; moderate temperature) or thermophilic (50 °C to 60 °C; higher temperature) range, depending on the microbial consortium involved. The optimal pH for anaerobic digestion is usually between 6.5 and 7.5. Nutrient balance in the form of C/N, particularly nitrogen and phosphorus, is critical for microbial growth and activity. Various reactor designs are employed for anaerobic digestion, each with its advantages and limitations. Common types include batch reactors, continuous stirred-tank reactors (CSTR), plug-flow reactors and high-rate anaerobic bioreactors [2]. The selection of a specific reactor depends on the characteristics of the substrate, desired retention time, and the specific needs of the process.

Anaerobic digestion can be integrated into various systems to enhance its overall efficiency. Combined Heat and Power (CHP) systems utilize the heat generated during biogas combustion for additional energy production. Co-digestion involves the simultaneous digestion of multiple substrates to optimize nutrient balance and enhance biogas yield. While anaerobic digestion offers numerous benefits, challenges such as process instability, substrate variability, and inhibition by certain compounds may arise. Research focuses on addressing these challenges through advanced monitoring and control systems, improving microbial engineering, and reactor designs. Anaerobic digestion plays a crucial role in sustainable waste management, reducing greenhouse gas emissions, and promoting circular economies. By converting organic waste into valuable resources like biogas and digestate, the process contributes to a more sustainable and resource-efficient future.

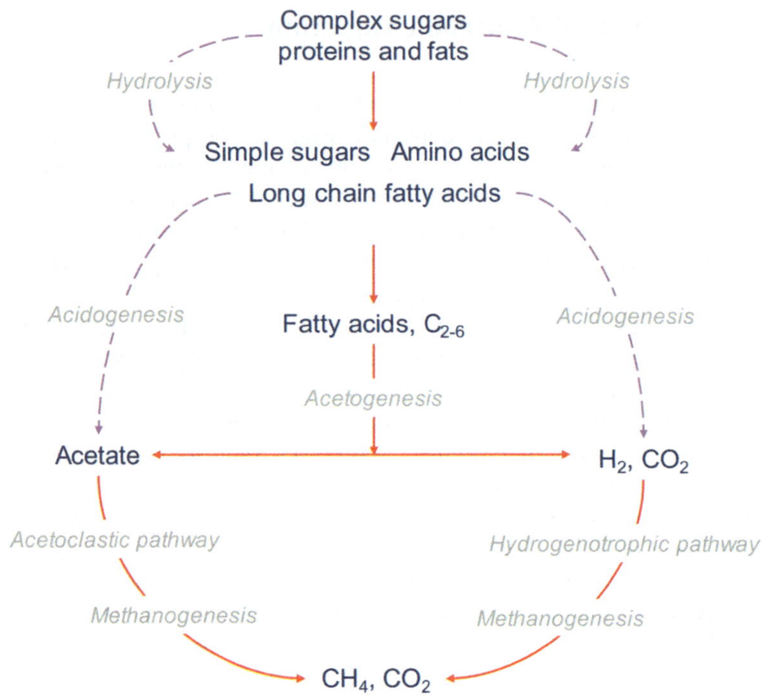

Fig. 1 Anaerobic digestion steps

3 Resource and Nutrient Recovery from Anaerobic Digestion

Anaerobic digestion also produces nutrient-rich byproducts known as digestate. The digestate is a nutrient-rich material, containing nitrogen (N), phosphorus (P), potassium (K), and other micro-nutrients, which are crucial for plant growth and soil fertility. The concept of nutrient recovery from anaerobic digestion of wastewater emerges from the intersection of several environmental and agricultural challenges. As the global population continues to grow, the demand for nutrient-rich fertilizers to sustain agricultural productivity also increases. However, traditional fertilizer production often comes with substantial environmental footprints and relies on finite natural resources. Simultaneously, the improper disposal of nutrient-rich wastewater can lead to eutrophication in water bodies, causing serious harm to aquatic ecosystems. Nutrient recovery from AD of wastewater offers a sustainable solution to these challenges, creating a circular economy, and providing viable solutions by transforming potential waste into valuable resources. Several methodologies for nutrient recovery from digestate can be broadly divided into physical, chemical, and biological methods. Separation methods, such as sedimentation, filtration, and centrifugation, separate solid and liquid fractions of the digestate, with the solid fraction usually containing a higher concentration of nutrients. Chemical methods involve chemical reactions to recover specific nutrients. For example, struvite precipitation is a common method for Phosphorus recovery, which involves the reaction of magnesium, ammonia, and phosphate in the digestate to form struvite crystals. Biological methods, like the use of microalgae and duckweed, exploit the nutrient uptake capabilities of these organisms to recover nutrients from the digestate. While nutrient recovery from AD of wastewater offers significant potential, several challenges need to be overcome to make this approach viable and efficient. The quality and nutrient content of the digestate may switch off and on, greatly based on the type of feedstock/substrate with operational conditions of the AD process, posing challenges for the design and operation of nutrient recovery processes. In addition, the recovered nutrients often require further treatment to meet the quality standards for agricultural use. Furthermore, the economic viability of nutrient recovery processes is a critical factor to consider, as the cost of nutrient recovery could be higher than the market value of the recovered nutrients. Lastly, integrating nutrient recovery into existing wastewater treatment infrastructure requires careful planning and coordination to ensure process compatibility and regulatory compliance. Table 1 summarizes the recent studies on wastewater anaerobic digestion (Fig. 2).

4 Residual Pollutants from AD Treated Effluent

While anaerobic digestion offers several environmental and economic benefits, it has challenges. One of the primary concerns associated with the process is the potential for the production of residual pollutants in the effluent. These are substances that remain in the water even after treatment, and they can pose significant environmental and health risks if not properly managed. The nutrient-laden effluent from anaerobic digestion, particularly nitrogen and phosphorus are of concern. While these nutrients benefit plant growth, their excessive presence in streams may propagate eutrophication, a phenomenon where surplus nutrients may exacerbate the hasty growth of algae and other plant life. This may depreciate the dissolved oxygen content in the receiving streams, leading to the complete end of aquatic organisms. Nitrogen, present predominantly as ammonia in the effluent, can be toxic to aquatic life at high concentrations. Furthermore, it can lead to nitrate formation through nitrification, which can contaminate groundwater and pose health risks to humans if it enters drinking water sources. Phosphorus, on the other hand, is not toxic but can contribute to the eutrophication of surface waters. It is also a finite resource, and its recovery from waste streams is becoming increasingly important to ensure the sustainability of global food production.

Anaerobic digestion has the potential to reduce the number of disease-causing organisms or pathogens in waste. However, some pathogens may still survive the process, particularly if the temperature and retention time are not sufficient for their complete inactivation. These can include bacteria, viruses, and parasitic organisms, all of which can pose risks to human health if the effluent is not properly treated before

Table 1 Recent studies on anaerobic digestion of different types of wastewater

Wastewater	Anaerobic technology	Key remarks on affected parameters for specific resource/nutrient recovery	References
Distillery wastewater	600 mL batch bioreactor	• Retention time mostly influenced the biogas production and methane percentage • The optimal biogas production was obtained at 31.21 °C at a pH of 7.35	[11]
Sewage sludge + expired hand sanitiser	1 L CSTR	• Hindrance in the biogas production was observed when expired sanitiser exceeded volume by 0.3% during co-digestion • The COD removal was 76% while the specific biogas production was 30 L/kg COD	[12]
Slaughterhouse wastewater	Tubular bioreactor with 3.5 L of working volume	• The COD removal for HRT of 6 and 11 days was 70 and 93% respectively • The presence of the baffles in the tubular reactor reduced the short circuit effects	[13]
Protein-rich wastewater	Anaerobic sequential batch reactor	• The anaerobic sequential batch reactor was able to remove 90% of COD • The protein degradation was observed to be 80% • A significant correlation was observed between the efficiency of ammonification and microbiome	[14]
Distillery wastewater	250 mL batch bioreactor	• The addition of distillery wastewater to sewage sludge anaerobic digestion improved its biodegradability and reduced the hydrolysis rate • The mono digestion of distillery wastewater showed 302 L/kg COD of methane yield	[15]
Piggery wastewater	3 L CSTR	• In-situ ammonia extraction improved methane yield by 33% in anaerobic digestion • The COD and volatile solid removal efficiency were improved by 34 and 78% respectively • The maximum volatile fatty acids removal efficiency was observed at a pH of 9	[16]

(continued)

Table 1 (continued)

Wastewater	Anaerobic technology	Key remarks on affected parameters for specific resource/nutrient recovery	References
Textile wastewater + domestic wastewater	1 L batch suspended growth anaerobic reactor	• Anaerobic co-digestion of textile wastewater with domestic wastewater improved COD and colour removal efficiency • Long retention time and higher pH (8–9) favoured anaerobic digestion and colour removal	[17]
Wine vinasse	Single and two-stage batch anaerobic reactor	• Energy recovery was similar in both, single and two-stage anaerobic bioreactor • Singe stage process showed 55 L/kg VS_{added} of methane while two-stage showed 45 L/kg VS_{added7} of methane yield	[18]

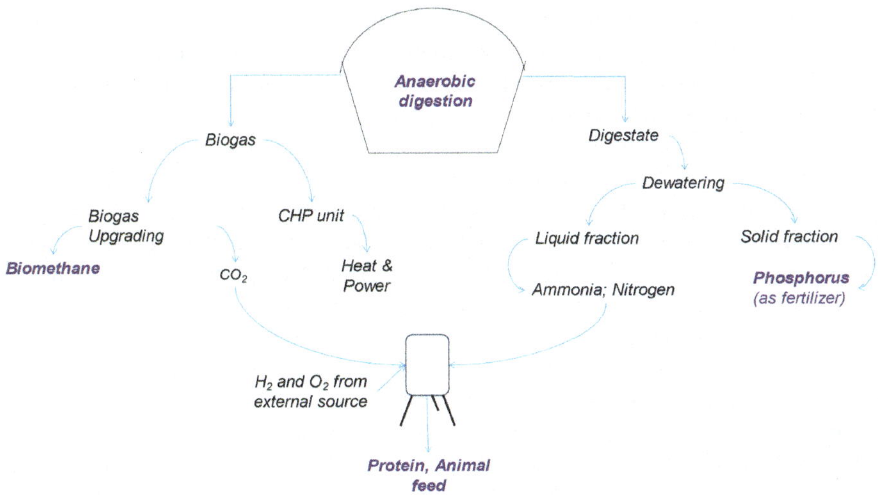

Fig. 2 Resource recovery from anaerobic digestion of wastewater

being discharged or reused. Trace elements are essential for the growth and activity of the microbes intricating the AD process. However, an excess of these elements can be toxic and may accumulate in the effluent. These can include elements like copper, zinc, and nickel. The presence of these elements in the effluent depends on their concentration in the feedstock and the operating conditions of the digester.

Emerging pollutants such as microplastics and pharmaceuticals are of increasing concern in the effluent of anaerobic digestion. Microplastics, tiny fragments of plastic less than 5 mm in size, can originate from a variety of sources, including synthetic clothing, personal care products, and the breakdown of larger plastic items. They can accumulate in the environment and the bodies of organisms, with potential impacts on wildlife and human health. Pharmaceuticals, including antibiotics and hormones, can also be present in the effluent, particularly when the feedstock includes sewage sludge or manure from livestock treated with these substances. These can have ecotoxicological effects and contribute to the development of antibiotic resistance. While the potential for residual pollutants in the effluent of anaerobic digestion is a concern, several measures can be taken to mitigate these risks. These include further treatment of the effluent, such as composting or aerobic treatment, to reduce the level of pollutants. Nutrient recovery technologies can also be employed to recover nitrogen and phosphorus from the effluent, reducing their potential for environmental harm and providing a source of valuable fertilizers.

5 Techno-Economic Feasibility of AD Technologies

The techno-economic feasibility analysis of anaerobic digestion technology includes transportation, pretreatment of the feedstock, digestion process and upgradation of the biogas and digestate. There are vital financial indicators that form the foundation of a techno-economic assessment. These include the net present value, internal rate of return, return on investment, and static payback time. The net present measures the present value of all cashflows, including fixed cost, operational cost, and expected future revenues [19]. A positive net present value indicates the project's financial viability. The internal rate of return is a key financial metric representing a project's return on investment. It is the discount rate at which the sum of discounted cash flows equals the initial investment [20]. The internal rate of return is used to assess project viability and compare different investment options. Return on investment is a financial indicator that shows the efficiency of an investment and aids in comparing different investment options [21]. Payback time is the period required for the initial capital expenditure to equal the total cash inflow.

In a study, authors reported that when the efficiency of anaerobic digestion was between 26 and 35%, the profit would be feasible if the selling price of biomethane lies between 135 and 183 £/MWh [22]. Furthermore, the sensitivity analysis of the process showed that the selling price of electricity was the most sensitive to the net present value and internal rate of return. The techno-economic feasibility of wastewater could be profitable if co-digestion is adopted in anaerobic treatment. Vinardell et al. [23] reported that when sewage sludge was co-digested with food waste, the revenue from the electricity offsets the running cost of the anaerobic bioreactor. Vinardell et al. [23] also reported that the gate fee (fee to accept food waste by WWTP operators in some countries) of food waste at wastewater treatment plants was the most sensitive to the techno-economic indicators. Similar results were also observed in another study where the co-digestion of three substrates, showed higher profits compared to mono digestion, namely sewage sludge, food waste and fish sludge [24].

6 Conclusion and Future Perspective

In summary, anaerobic digestion is a complex biological process with far-reaching implications for waste management, energy production, and nutrient recovery. Understanding its fundamentals, from microbial interactions to reactor design, is essential for optimizing performance and realizing the full potential of this environmentally friendly technology. As global interest in renewable energy and sustainable waste management continues to grow, anaerobic digestion stands out as a key player in the transition towards a sustainable and decarbonized society.

References

1. Li A, Zhang Y, Ge W, Zhang Y, Liu L, Qiu G (2022) Removal of heavy metals from wastewaters with biochar pyrolyzed from MgAl-layered double hydroxide-coated rice husk: Mechanism and application. Bioresour Technol 347. https://doi.org/10.1016/j.biortech.2021.126425
2. Paritosh K, Kesharwani N (2024) Biochar mediated high-rate anaerobic bioreactors: a critical review on high-strength wastewater treatment and management. J Environ Manage 2024(355):120348. https://doi.org/10.1016/j.jenvman.2024.120348
3. Jason Ren Z, Schnoor JL, Pagilla KR (2022) Toward a net zero circular water economy. In: Pathways to water sector decarbonization, Carbon capture and utilization, IWA Publishing, pp 1–13. https://doi.org/10.2166/9781789061796_0001
4. Li X, Lee H-S, Wang Z, Lee J (2021) State-of-the-art management technologies of dissolved methane in anaerobically-treated low-strength wastewaters: a review. Water Res 200:117269. https://doi.org/10.1016/j.watres.2021.117269
5. Lu L, Guest JS, Peters CA, Zhu X, Rau GH, Ren ZJ (2018) Wastewater treatment for carbon capture and utilization. Nat Sustain 1:750–758. https://doi.org/10.1038/s41893-018-0187-9
6. Moerland MJ, Bruning H, Buisman CJN, van Eekert MHA (2021) Advanced modelling to determine free ammonia concentrations during (hyper-)thermophilic anaerobic digestion in high-strength wastewaters. J Environ Chem Eng 9:106724. https://doi.org/10.1016/j.jece.2021.106724
7. Wang W, Chang J-S, Show K-Y, Lee D-J (2022) Anaerobic recalcitrance in wastewater treatment: a review. Bioresour Technol 363:127920. https://doi.org/10.1016/j.biortech.2022.127920
8. Paritosh K, Bose A (2024) Multi-criteria-based decision-making assessment for anaerobic digestion of ammonia-rich distillery wastewater: effect of pyrochar and temperature. Bioresour Technol 397:130493. https://doi.org/10.1016/j.biortech.2024.130493
9. Moerland MJ, Bruning H, Buisman CJN, van Eekert MHA (2021) Advanced modelling to determine free ammonia concentrations during (hyper-)thermophilic anaerobic digestion in high strength wastewaters. J Environ Chem Eng 9. https://doi.org/10.1016/j.jece.2021.106724
10. Wang X, Liu G, Sun W, Cao Z, Liu H, Xiong Y (2023) Removal of toilet paper fibres from residential wastewater: a life cycle assessment. Environ Sci Pollut Res 30:84254–84266. https://doi.org/10.1007/s11356-023-28291-5
11. Weldehans MG (2023) Optimization of distillery-sourced wastewater anaerobic digestion for biogas production. Cleaner Waste Syst 6:100118. https://doi.org/10.1016/j.clwas.2023.100118
12. Vu HP, Cai Z, Tra V-T, Wang Q, Nghiem LD (2023) Anaerobic co-digestion of expired alcohol-based hand sanitizer with synthetic wastewater for biogas production. Environ Technol Innov 32:103319. https://doi.org/10.1016/j.eti.2023.103319
13. Guamán-Marquines CW, Mendoza-Loor RJ, Gómez-Salcedo Y, Baquerizo-Crespo RJ (2023) Assessment of the start-up of tubular reactors on a laboratory scale for the anaerobic digestion of

slaughterhouse wastewater. Int J Thermofluids 19:100378. https://doi.org/10.1016/j.ijft.2023.100378
14. Deng Z, Muñoz Sierra J, Ferreira ALM, Cerqueda-Garcia D, Spanjers H, van Lier JB (2024) Effect of operational parameters on the performance of an anaerobic sequencing batch reactor (AnSBR) treating protein-rich wastewater. Environ Sci Ecotechnol 17:100296. https://doi.org/10.1016/j.ese.2023.100296
15. Ripoll V, Solera R, Perez M (2022) Kinetic modelling of anaerobic co-digestion of sewage sludge and Sherry-wine distillery wastewater: effect of substrate composition in batch bioreactor. Fuel 329:125524. https://doi.org/10.1016/j.fuel.2022.125524
16. Rivera F, Sepúlveda-Muñoz CA, Prádanos P, Hernández A, Palacio L, Muñoz R (2023) Influence of pH on the performance of anaerobic piggery wastewater treatment coupled with membrane-based NH_3 extraction. J Water Process Eng 55:104226. https://doi.org/10.1016/j.jwpe.2023.104226
17. Bidu JM, Njau KN, Rwiza M, Van der Bruggen B (2023) Textile wastewater treatment in the anaerobic reactor: influence of domestic wastewater as co-substrate in colour and COD removal. S Afr J Chem Eng 43:112–121. https://doi.org/10.1016/j.sajce.2022.10.007
18. Auxiliadora López R, Tena M, Solera R, Pérez M (2023) Anaerobic co-digestion of sewage sludge and wine vinasse mixtures in single-stage and sequential-temperature processes. Fuel 348:128531. https://doi.org/10.1016/j.fuel.2023.128531
19. Budzianowski WM, Budzianowska DA (2015) Economic analysis of biomethane and bioelectricity generation from biogas using different support schemes and plant configurations. Energy 88:658–666. https://doi.org/10.1016/j.energy.2015.05.104
20. Tena M, Buller LS, Sganzerla WG, Berni M, Forster-Carneiro T, Solera R (2022) Techno-economic evaluation of bioenergy production from anaerobic digestion of by-products from ethanol flex plants. Fuel 309:122171. https://doi.org/10.1016/j.fuel.2021.122171
21. Cardoso J, Silva V, Eusébio D (2019) Techno-economic analysis of a biomass gasification power plant dealing with forestry residues blends for electricity production in Portugal. J Clean Prod 212:741–753. https://doi.org/10.1016/j.jclepro.2018.12.054
22. Michaels S, Walker M, Moody A, Poggio D, Pourkashanian M (2020) Biomethane production using an integrated anaerobic digestion, gasification and CO_2 biomethanation process in a real wastewater treatment plant: a techno-economic assessment. Energy Convers Manag 209:112663. https://doi.org/10.1016/j.enconman.2020.112663
23. Vinardell S, Astals S, Koch K, Mata-Alvarez J, Dosta J (2021) Co-digestion of sewage sludge and food waste in a wastewater treatment plant based on mainstream anaerobic membrane bioreactor technology: a techno-economic evaluation. Bioresour Technol 330:124978. https://doi.org/10.1016/j.biortech.2021.124978
24. Fernando-Foncillas C, Varrone C (2021) Potential of the sewage sludge valorization in Scandinavia by co-digestion with other organic wastes: a techno-economic assessment. J Clean Prod 324:129239. https://doi.org/10.1016/j.jclepro.2021.129239

Study of Anaerobic Digestion Mediated Micropollutant Degradation by UPLC-MS/MS

Aatiya Shaikh and Srikanth Mutnuri

Abstract Antibiotics have received considerable international interest but with its increased production and abuse and the risk of Antibiotic-Resistant bacteria and Antibiotic Resistance Genes (ARG's) dispersing across several habitats it has become a serious issue. Residues of various organic micro-pollutants occur at low concentrations (ng/L to micro gram/L) in water bodies and possibly influence the environment and living organisms even at those low concentrations. Processes such as oxidation processes (advanced), chlorination, ozonation, electro-chemical as well as biological have been studied for micropollutant degradation. Anaerobic digestion (AD) can prove to be an effective way for biodegradation of this compound as its less time and energy consuming. Biomethane Potential Tests (BMP) were set up using Food, Sludge, Septage in four different ratios with six different loading rates in BMP vials of capacity 120 mL. Working volume of 80 mL was maintained. The digestate from nearby anaerobic reactor was used as an inoculum and was stabilized for a week prior to use. Loading rate of 1 kgVS/m^3 (8:1:1) was used for the purpose of LC/MS as it gave highest biogas production for a 30-day monitoring period. The setup was spiked with Trimethoprim (TMP), Ciprofloxacin (CIP) and Estrone (EST) (Concentration 1 mg/L) followed by monitoring every 24, 48, and 72 h. The experiment aimed to establish a dependable and sensitive UPLC-MS/MS (Ultra Performance Liquid Chromatography Mass Spectrometry) method to study TMP, CIP and EST degradation rates in AD.

Keywords UPLC-MS/MS · Micropollutants · Antibiotics · Anaerobic digestion

A. Shaikh · S. Mutnuri (✉)
BITS Pilani, Goa Campus, Zuarinagar, Sankval 403726, India
e-mail: srikanth@goa.bits-pilani.ac.in

M. M. Ghangrekar et al. (eds.), *Biological and Hybrid Wastewater Treatment Technology*, Earth and Environmental Sciences Library,
https://doi.org/10.1007/978-3-031-63046-0_11

1 Introduction

Antibiotics have received considerable international interest since the advent of Penicillin in 1929 due to their ability to enhance both human and animal health. With its increased production and use the potential for the emergence of antibiotic resistance genes (ARGs) as well as bacteria resistant to antibiotics spreading in different environments have become serious [1]. Mostly these issues occur due to low antibiotic degradation efficiencies in hospital, pharmaceutical, livestock and aquaculture industries and in municipal wastewater treatment. The recent years have seen a significant increase in research on antibiotic fate and resistance.

Residues of various antibiotics occur in water bodies at low quantities (ng/L to μg/L) [2] and possibly influence the environment and living organisms even at those low concentrations. After their application, it enters the environment via industry, agriculture, households, stormwater runoff with septage being one of the most important continuous paths into water bodies. The negative impact of micropollutants is determined by its physicochemical and ecotoxicological properties and its water concentration.

A number of durable indicators have been established to learn more about how wastewater has contaminated environmental compartments such as Carbazepine, Crotamiton, Sucralose and Acesulfame. Detection gives direct info about its contamination caused by wastewaters and the degree of raw wastewater contamination can be accessed using the ratio of conservative to labile pollutants. However, these markers alone don't give any more details concerning the source of the wastewater in particular and calls for the usage of a chemical marker [3].

Immunoassays, bioassays, or detectors compatible with HPLC, such as fluorescence and UV diode-arrays are typically used to determine antibiotic residue. For several quantitative analytical applications, liquid chromatography integrated with mass spectrometry (LC–MS/MS) has now become the technique of choice due to its robustness, high sensitivity, selectivity, and structure elucidation capabilities. The majority of approaches for determining multiple residues that have been published focus on similar compounds, typically those that fall within one category of antibiotics. Recently, multi-class analysis has been the focus of innovative methodologies. However, many residues in human serum were the subject of relatively little research. As a result of this, a more sensitive quantitative LC–MS/MS method was developed using a triple quadrupole mass spectrometer to assess the degree of percentage reduction of micropollutants in the Anaerobic Digestion process [4].

Trimethoprim (TMP) and Sulfamethoxazole (SMX); two antibiotics that are typically taken together as a single medication, are two most commonly consumed antibiotics. Co-trimoxazole has been used since 1968 [5] in SMX:TMP at a 5:1 ratio permits the two compounds to work together synergistically. After Penicillin's, Macrolides, and Fluoroquinolones; Sulphonamides and TMP are currently the fourth most commonly used drugs in the world. Bladder infections are treated with TMP for effective treatment of urinary problems. When acting alone, it halts the spread of a significant number of bacteria. The 2-Aminopyrimidine scaffold is the drug's most distinctive feature along with two rings that allowing for the structural alterations that lead to various forms of resistance. Dihydrofolate Reductase (DHFR) is where it binds and has a 3000 times stronger affinity for bacterial DHFR than human DHFR. The key enzyme DHFR (EC 1.5.1.3) is responsible for reducing Tetrahydrofolate from dihydrofolate and has been the focus of extensive research as an antimicrobial agent target. The selective inhibitors of DHFR include Methotrexate, Trimethoprim, and Pyrimethamine all of which have a 2-Aminopyrimidine moiety. The two rings of TMP result in different lipophilicity and resistance profiles [6].

Regardless of how TMP is used in antibiotics for medical aquaculture and veterinary purposes, humans and animals inadequately metabolize it [7, 8]. In aquaculture its used to prevent and treat infectious fish and shrimp infections using live food. Moreover in veterinary field most countries use it as a feed supplement [9]. Consumers eliminate 80% of this medicine through their urine and feces due to poor metabolization [10].

Two of multiple estrogens, Estrone (sometimes referred to as E1) and Estradiol (17-estradiol or E2), are steroid hormones that have a role in the formation and operation of physiological and anatomical features and processes in females, such as menstrual cycle [11]. Two significant groups whose life quality might be improved by evaluating oestrogen, testosterone, and perhaps progesterone levels below the present limits of detection are perimenopausal or postmenopausal women and geriatric males [11, 12]. Physical (osteoporosis, metabolic syndrome and hot flashes) and psychological (irritability, anxiety, sleeplessness, mood instability, and depression) consequences may accompany women throughout the menopausal transition [13]. Elderly men tend to experience a decline in cognitive function and physical stature which is due to a decrease in testosterone production associated with aging [14].

Fluoroquinolones (FQ) are strong, bactericidal antimicrobials that are effective against both types of Gram-Negative bacteria as well as Gram-Positive. Since their tissue and cerebrospinal fluid penetration is high, they have been employed extensively in adult patients [15]. Although FQ resistance is reportedly on the rise, this class of antibiotics is still effective against 90% of Gram-negative pathogens seen in the general population [15].

One such fluoroquinolone, Ciprofloxacin (CFL), is frequently used to treat or prevent bacterial infections. It can also be used in biological warfare to treat and stop the deliberate spread of severe diseases such as Tularemia, Plague, and Anthrax of the skin or mouth. Given its effectiveness in treating multi-resistant bacterial infections which are occasionally linked to meningitis [16–18] it is also administered to severely unwell newborns. Since CFL is a fluoroquinolone antibiotic most frequently used to treat UTIs, it was selected for the assessment [19].

In the past UV and HPLC-based methods were often used to analyse pharmaceuticals involving biological materials, including plasma, serum, and urine from humans. These detection techniques, however, were only effective at concentrations nearing mg/L. As a result, a technique that combines mass spectrometry (MS) and ultra-performance liquid chromatography (UPLC) has been used to identify the analytes in the anaerobic digestate at relatively low concentrations (ng/L).

Analytical techniques with enough sensitivity and specificity to identify and measure residual levels in sample are required to detect the required residue levels. Several authors [8] have devised analytical techniques to find analyte residues. For instance, various techniques have been developed to identify sulfonamides or trimethoprim with sulfonamides in fishery products. Ultra-high performance liquid chromatography (UHPLC) Q-Orbitrap has been utilised to identify sulfonamides in salmon. Tandem mass spectrometry-Liquid Chromatography has been employed to detect sulfonamides in tilapia fillets. Furthermore, quadrupole time-of-flight mass spectrometry (LC-Q TOF/MS) has been used to determine sulfonamide and trimethoprim residues present in fish tissue.

Antibiotic removal is less effective in conventional WWTP's because they are not intended to remove pharmacological load. Partial elimination of erythromycin (7–59%), sulfamethoxazole (24–49%), and sulfamethazine (41–47%) has been observed [20]. Similar results were observed by Aziz et al. [1] concerning Trimethoprim removal (40%). For emerging compounds, treatment methods such as membrane separation, adsorption, traditional, enhanced oxidation procedures and electrocoagulation are successful but are expensive, energy-intensive and could create hazardous metabolites. Hence, such technologies are recommended as a pre or post-treatment alternative rather than the primary treatment process. The treatment of industrial as well as domestic wastewater with high concentration has been done for many years using Anaerobic Digestion (AD) and proves to be a promising technology. The removal of antibiotics has recently been studied using the technique. The benefits include reduced sludge generation, excellent organic removal, increased loading rates, reduced nutrient requirements, low energy requirements, low energy demand and significant bio-energy generation.

Organic Loading Rate (OLR) influences AD's equilibrium and production significantly. A high OLR results in the enrichment of a range of bacterial species, reduction in the reactor's capacity requirements and the conservation of heating energy. Additionally, the accumulation of ethanol and Volatile Fatty Acids (VFA) might occur, resulting in inadequate heat transfer that may ultimately cause irreparable failure. Therefore, it is important to determine the optimal loading rate in order to facilitate anaerobic digestion [21, 22].

This work seeks to provide a stable and sensitive UPLC-MS/MS technique for examining the rate of degradation of micropollutants such as TMP, EST and CFL in AD. It aims to investigate how biogas production is affected and identify any potential variations in how antibiotics affecting the methanogenic and acetogenic bacteria during anaerobic digestion of various substrates.

2 Methodology

2.1 Materials

Trimethoprim (TMP), Ciprofloxacin (CIP) and Estrone (EST) was purchased from Sigma-Aldrich (Germany). Purchases of LC–MS grade acetonitrile (CAN), methanol (MeOH), formic acid (FA) and ethyl acetate were made from HIMEDIA. The water used was Ultra-Pure water (Millipore, Merck). The stock solutions of TRI were prepared in MeOH. The stock solutions of CIP was prepared in methanol whereas that of EST was prepared 1% Formic Acid in Milli-Q water.

For the process of Biomethane Potential Tests (BMP), 120 mL serum vials were used which were purchased from a local vendor. The inoculum was got from an Anaerobic Digestor (AD) in BITS Pilani K K Birla Goa campus. The food sample was got from college mess in the same campus and septage and sludge samples from the Sequence Batch Reactor (SBR) in the campus itself. Chemicals for the preparation of Macro-Nutrients such as NH_4Cl, KH_2PO_4, $MgCl_2$.6H20, $CaCl_2.2H_2O$ and for Micro-Nutrients like $FeCl_2$.4H2O, $CoCl_2$.6H2O, $MnCl_2$.4H2O, $NiCl_2$.6H2O, $ZnCl_2$, H_3BO_3, Na_2SeO_3, $CuCl_2.2H_20$, $Na_2MoO_4.2H_20$ as well as Ethanol were purchased from HIMEDIA.

Ammonium chloride (NH_4Cl), Potassium Chloride (KCl), Sodium Di-Hydrogen Phosphate (NaH_2PO_4), HIMEDIA was the supplier of Sodium Hydrogen Carbonate ($NaHCO_3$) and sodium acetate (CH_3COONa).

2.2 LC–MS/MS Instrument and Conditions

An Agilent Technologies Series-1290 Infinity-11 HPLC system was used to perform the UPLC-MS/MS analysis of TMP, CIP, and EST. The equipment consists of a thermostat column oven, flexible pump, and autosampler. Within the autosampler, the sample vials were maintained at 4 °C.

A C-8 column (150 mm × 4.6 mm, 5 um) by Agilent Technologies was used to determine the TMP. Mobile phases A—0.1% FA in MeOH (90%) and B—(0.1% FA in Milli-Q water (10%) were used to accomplish the elution. Isocratic mode analysis of EST was conducted using the same column. Phase A comprised of 10 mM ammonium acetate containing 0.01% FA, whereas phase B was composed of 0.01% MeOH. All chromatographic conditions that were used are detailed in Table 1.

EST was analysed using the same column in Isocratic mode. All analytes were detected and quantified using an Agilent Technology, 6460 Triple Quad UPLC-MS/MS mass spectrometer, manufactured in Santa Clara, CA, USA with an ion spray voltage of 4 kV and a positive-mode electrospray interface (ESI). Multiple Reaction Monitoring (MRM) was used to ensure that every analyte was quantified. The suitable parameters such as elemental composition and mass measurements of ions isolated from standard solutions for each drug are summarised in Table 2. An

Table 1 LC/MS conditions used

Compound name	Column used	Mobile phase	Flow rate (min)	Elution type	Retention time (min)	Injection volume (μL)	Linearity range (ng/mL)
TRI	Agilent C8 (150 × 4.6 mm, 5.0 um)	0.1% FA in Methanol (90%) & MilliQ (10%)	0.3 mL/min	Isocratic	3	5	0.78–100
EST	Agilent C8 (150 × 4.6 mm, 5.0 um)	0.01% FA in 1 mM $NH_4CH_3CO_2$ in MeOH (1:1)	0.3 mL/min	Isocratic	4.2	5	0.78–100
CIP	C18 column (100 × 2.1 mm, 1.8 μm)	0.5% FA in MilliQ	0.3 mL/min	Gradient	10	5	0.78–100

Table 2 BMP nutrient media composition (in mL)

	Test 1	Test 2	Positive control	Negative control
Micro-nutrients	1	1	1	1
Macro-nutrients	1	1	1	1
Sodium Bicarbonate buffer	5	5	5	5
D/W	10	10	10	11
Ethanol	0	0	1	0
Digestate	62	62	62	62
Sample	1	1	0	0

Agilent Technologies 6460 triple quad UPLC-MS/MS mass spectrometer was used to identify and quantify all analytes. This instrument had an electrospray (ESI) interface operating in the positive direction and an ion spray voltage of 4 kV. A triple quad system was configured with respect to the resolution of the unit. Mass hunter V.B.08 software was used to ascertain the peak area of all compounds.

2.3 BMP Studies and Spike Experiments

The biodegradability of septage, food waste, and sludge was studied using anaerobic digestion tests, as referenced in [23]. Composite food samples comprised of food waste samples obtained during weekly samplings were subjected to digestion tests.

Representative samples were selected from the resulting mixture to constitute the composite samples, subsequent to the combination of all the food waste samples.

The composite food samples were digested with sewage sludge and septage in the ratios of 6:2:2, 7:1:2, 7:2:1, 8:1:1 in serum vials of 120 mL capacity with six different loading rate of 0.5 kg VS/L/m^3, 1 kg VS/L/m^3, 1.5 kg VS/L/m^3, 2 kg VS/L/m^3, 3 kg VS/L/m^3, 4 kg VS/L/m^3 each in duplicate at mesophilic temperature. The effective volume of the digester was 80 mL. At the start of the digestion tests, 60 mL of bacterial inocula (digestate) was combined with food waste in each digester. The exact amount was established based on the initial VS loading and VS content of the food waste in addition to other substrates. The inoculum was collected from mesophilic digestor. Micro-nutrients, Macro-nutrients and 5% Sodium Bicarbonate buffer was added in amount as given in Table 1.

A screw cap and rubber septum were used to tightly secure the vials. To ensure anaerobic conditions, the head space was purged with Nitrogen gas for 5 min. Two controls were set up for the ratios. The positive control contained Ethanol as substrate and negative control contained purified distilled water. Once daily, each digestor was manually mixed and gas production was monitored for a period of 30 days.

1 kgVS/L/m^3 of 8:1:1 was the loading rate that was selected for spiking it with TMP as it gave the highest biogas production in the previous experiments. Analytes were introduced at a final concentration of 1 mg/L, and LC/MS 'zero hour' analysis was conducted. The analysis was followed with 48 h and 72 h and the reduction percentage was calculated. The efficacy of pharmaceutical doses on BMP was investigated by monitoring the respective transformation products produced during AD using MRM study (Multiple Reaction Monitoring).

Analytical methods—The substrates parameters such as pH, conductivity, Chemical Oxygen Demand (COD), Total Kjeldahl Nitrogen, Volatile and Total Solids were evaluated before commencing the batch assay. For the calculation of daily biogas production, Water Displacement (Gas Counter) technique was used. A biogas sample was obtained from the vials and subjected to composition analysis each week. The biogas composition was analysed using a Gas Chromatogram (GC) from Thermos Fisher equipped with a Thermal Conductivity Detector (TCD) and a TDX-01 packed column containing nitrogen, helium, and zero air as the carrier gases. Injection port, column oven, and detector operative temperatures were at 60 °C, 120 °C and 120 °C respectively. The volume of injection was 1 mL. The aforementioned analyses were performed in duplicates, and the results are presented as the mean ± standard deviation.

2.4 Preparation of Calibration Standards and Quality Control Samples

The working stock solution was prepared with MeOH at concentrations ranging from 0.78 to 100 μg/mL.

The blanks were spiked with a working stock solution of the corresponding concentration in order to prepare calibration standards and Quality Control samples (QC's). The following concentrations were used to develop calibration curve standards: 0.78, 1.56, 3.12, 6.25, 12.5, 25, 50 and 100 ng/mL.

2.5 Extraction and Sample Preparation

The contents of the BMP vials was centrifuged at 10,000 RPM for 10 min post the experiment period such as 0 h, 24 h, 48 h and 72 h. The centrifugate was transferred to a separate beaker and it was subjected to filtration through 0.45μ syringe filter. Each analyte for all respective concentrations was then poured into LCMS vials and given for analysis through the autosampler.

2.6 Method Validation and Quantification Procedure

Parameters such as selectivity, linearity, and the lower limit of detection were assessed and verified for the technique. The peaks detected at the time of retention were used to identify the analytes. From their working stock solutions, calibration standards for the analyte were made in the concentration range of 0.78 ng/mL-100 ng/mL.

In order to study the method's selectivity, chromatograms of un-spiked solution and sample spiked with the drug were compared. The carryover effect was reduced by optimising the run time of the analyte and interference in the control was examined with MRM technique.

3 Result and Discussion

3.1 Characterisation of Substrates and BMP Studies

The average values for substrates such as food, sewage sludge and septage pH, Conductivity, Total Solids, Volatile Solids, COD (Chemical Oxygen Demand), TKN (Total Kjeldahl Nitrogen), VFA (Volatile Fatty Acids), moisture content is as given in Table 3.

The VS/TS was constant and its high value favoured anaerobic conversion.

The Biogas production measured during the digestion of wastes is given in Figs. 1 and 2. The daily and cumulative readings for the 30-day period are mentioned with the standard deviation. OLR 8:1:1 gave a high biogas yield of 800 mL/gVS added. This could be because of high proportion of food waste in comparison to sewage sludge and septage. An average of around 50% methane production was observed

Table 3 Characterisation of substrates

	Food	Sludge	Sewage
pH	6.86	5	5.67
Conductivity	2.1	3	2.8
Total solids	19	66	81
Volatile solids	96	17	12
COD (mg/L)	35,000	50,000	30,000
TKN (%)	1.57	1.35	1.86
VFA (g/L)	2.4 ± 0.1	0.8 ± 0.1	0.3 ± 0.1
Moisture content	74	60	55

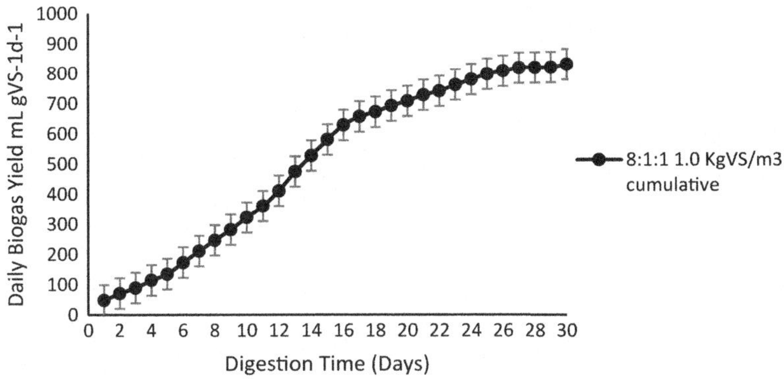

Fig. 1 Biogas production for 8:1:1 ratio (Cumulative)

in all ratios. Immediate degradation of substrate was observed. The digestion was carried out until the gas production rate was negligible.

3.2 *Method Development and Standardization*

The ionisation method used in the process was Electro Spray ionisation (ESI). Hence, the solvent needs to be compatible with this method e.g. water, methanol, acetonitrile, Ethanol and chloroform. Higher organic contents in the solvents improve the ionisation efficiency of ESI [24]. It is critical to avoid the usage of buffers that include inorganic ions such as phosphate and sodium acetate as these cause significant ion suppression. Presence of a small percentage of formic acid improves resolution of the peaks. This study used ammonium acetate and methanol as mobile phases throughout the separation process. C_{18} columns were used to enhance separation efficiency, retention time, and baseline shape [25] (Fig. 3).

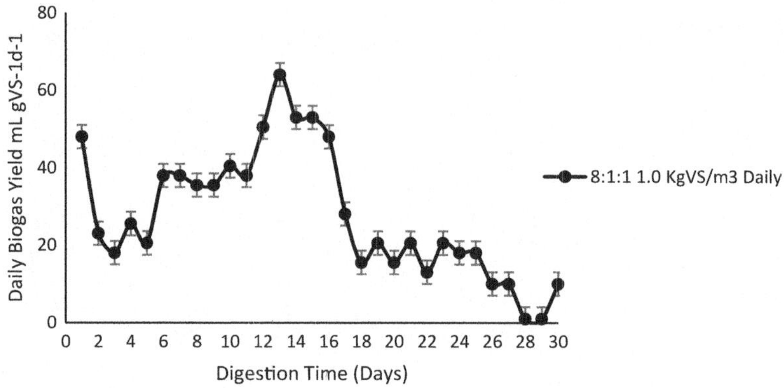

Fig. 2 BMP nutrient media composition

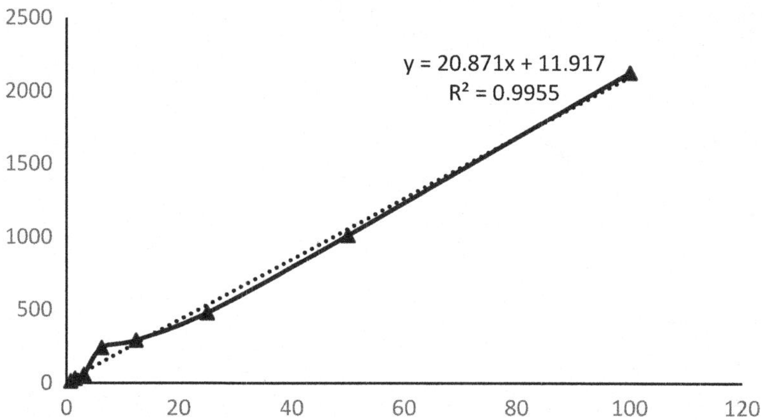

Fig. 3 Standardisation of ciprofloxacin

Optimisation was carried out by repeated injections of individual standard solutions prepared in respective solvents at 1 μg/mL concentrations. The method was validated as per FDA Bioanalytical Technique Validation guidelines [26]. The analytes exhibited a linear response spanning the calibration ranges of 0.78–100 ng/mL with a r^2 value of 0.998. There were no interfering peaks observed in the blank sample of biogas digestate which confirmed specificity. The compounds were effectively separated using all the chromatographic conditions mentioned previously as displayed in Table 1 (Figs. 4 and 5).

Linear regression equation was used to analyze the results obtained. The achieved r^2 values were 0.9955, 0.9985, 0.9971 for Estrone, Trimethoprim and Ciprofloxacin respectively. The LLOQ for all three analytes was found to be 0.78 ng/mL. Values obtained for the intercept, slope, and standard deviation of each parameter was used to validate the method.

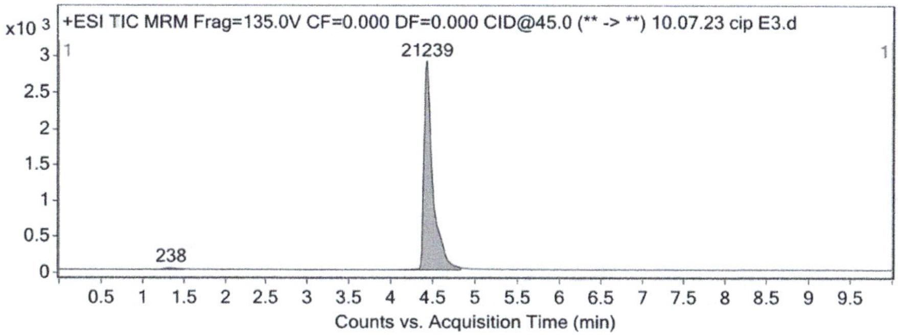

Fig. 4 Chromatogram for ciprofloxacin (Sample)

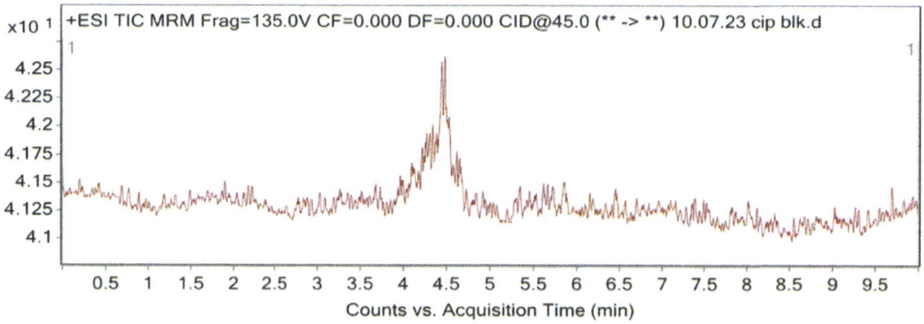

Fig. 5 Chromatogram for ciprofloxacin (Blank)

MRM (Multi-Reaction Monitoring) reduces the interference and enhances the sensitivity of the method hence the same was used for analysis. ESI was operated in positive and negative ways; positive ions were shown to provide optimum sensitivity and selectivity.

3.3 *Percentage Reduction Experiments*

8:1:1 ratio which gave highest cumulative biogas and good proportion of methane was selected for the purpose of spiking with antibiotics. It was spiked with the respective antibiotic concentration at final concentration of 1 mg per litre. Constant monitoring was done with respect to the degradation pattern at the required interval. It was observed that Estrone offered highest percentage reduction over a 72 h duration. This could be because of the role of laccase enzyme in degradation of estrone and is well documented. Laccase belongs to monophenol oxidase group and uses enzyme bound Cu^{+2} to transfer hydrogen to molecular oxygen (Figs. 6, 7, 8, 9, 10 and 11).

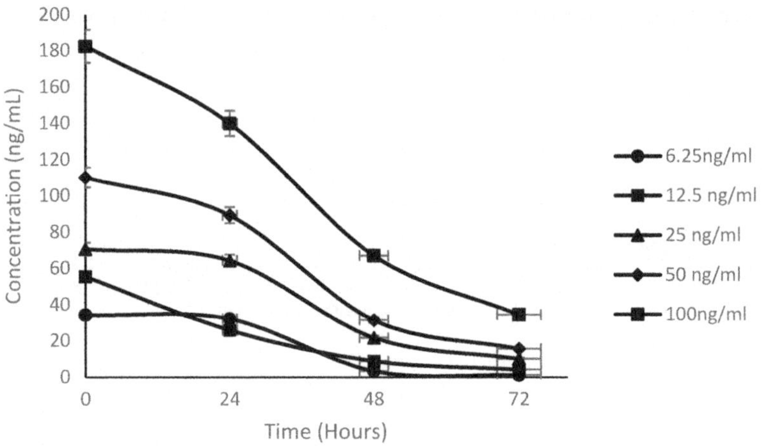

Fig. 6 Reduction of estrone over a 24 h, 48 h, 72 h period

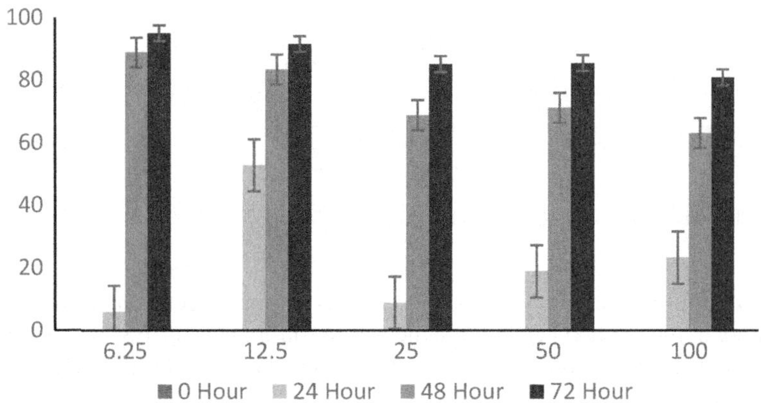

Fig. 7 Percentage reduction efficiency for estrone over a 24 h, 48 h, 72 h period

All the analytes used achieved a good range of percentage reduction across the experimental duration. Trimethoprim showed an average of 60% percentage reduction at the end of 72 h period, whereas Ciprofloxacin showed 75% percentage reduction for the same duration.

In contrast to other antibiotics, Trimethoprim is typically recalcitrant to degradation under anaerobic conditions. Studies imply that heterotrophic bacteria are crucial for degradation of the antibiotic.

Fluoroquinolones like Ciprofloxacin are most effectively transformed by photolysis and by photo-catalysis. Even when the analyte is subjected to photo-transformation, it's not completely eliminated as the quinoline structure remains

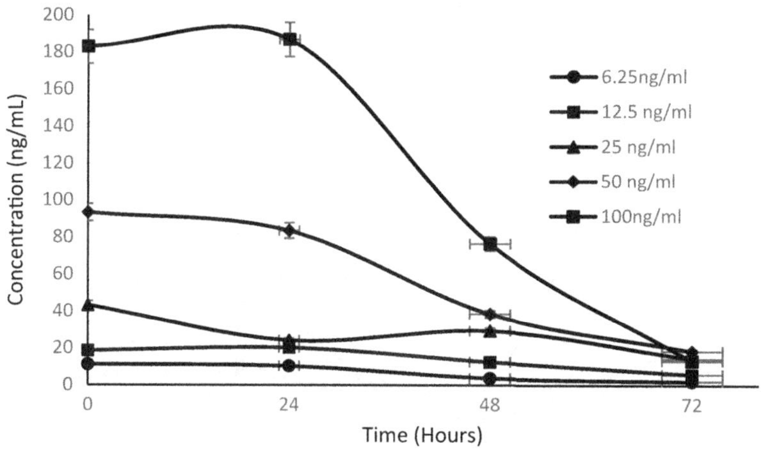

Fig. 8 Reduction of ciprofloxacin over a 24 h, 48 h, 72 h period

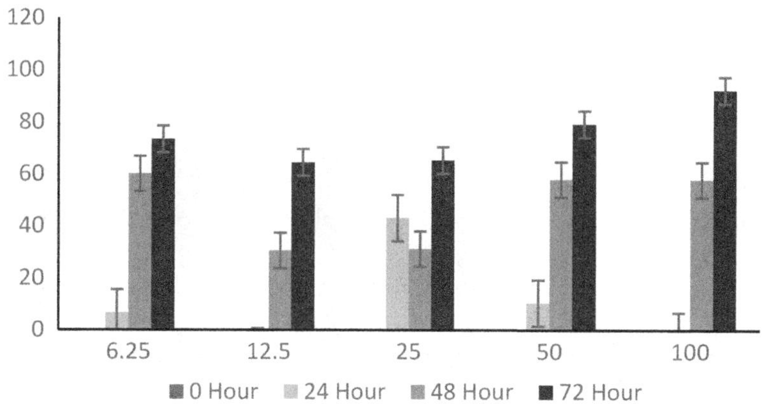

Fig. 9 Percentage reduction efficiency for ciprofloxacin over a 24 h, 48 h, 72 h period

intact. The Anaerobic Digestion method in the study has successfully showed over 50% degradation of CFL.

In literature studies, Estrone has been subject to removal due to adsorption. Here, it has been effectively managed.

3.4 Method Validation

FDA bioanalytical method validation was applied to the above mentioned technique. The method showed a linear response for the calibration curve of 0.78–100 ng/ml

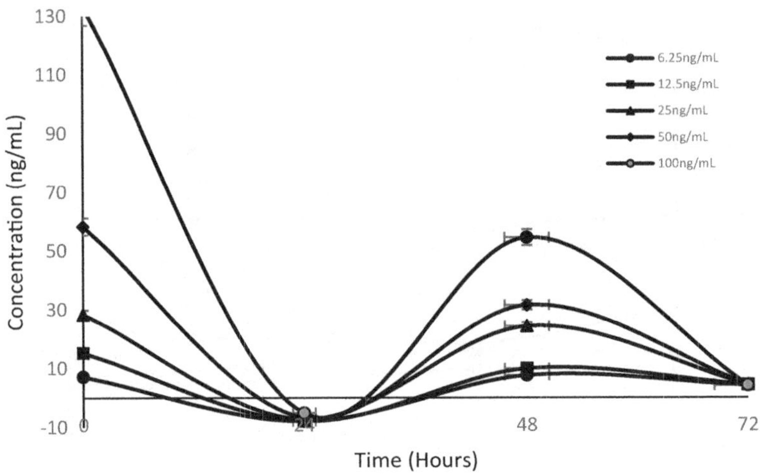

Fig. 10 Reduction of trimethoprim over a 24 h, 48 h, 72 h period

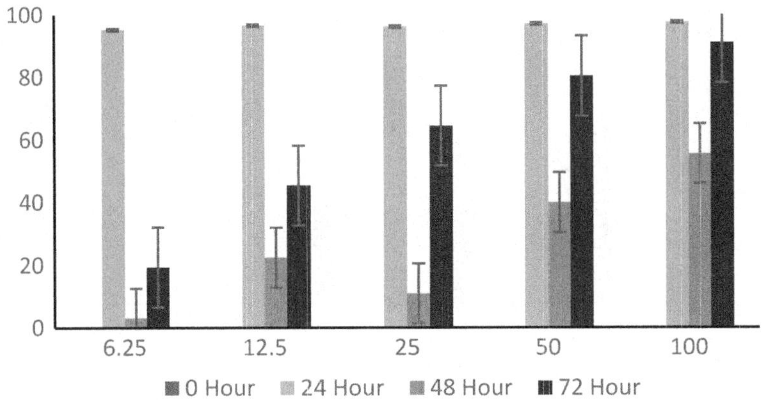

Fig. 11 Percentage reduction efficiency for trimethoprim over a 24 h, 48 h, 72 h period

with an r^2 value of 0.994 for the given analyte. The absence of any interference peaks in the blank sample corresponds to the retention duration of the analyte and thus validates specificity. The MRM spectrum showed excellent separation for all the compounds.

The linear regression equation derived from the calibration curve was used to analyse the obtained r^2 value. The achieved value of 0.994 with $y = 71.266 \times - 50.363$ confirms the linearity of the method.

The value of the intercept and slope together with their corresponding standard deviations were used to validate this method.

4 Conclusion

For BMP experiments 1 kgVS/L/m^3 gave the highest biogas production for 8:1:1 ratio for substrates such as food, sewage sludge and septage.

The peaks derived from the UPLC-MS/MS process were precise and distinct. Upon spiking the BMP setup with Trimethoprim, Ciprofloxacin and Estrone, reduction percentage was monitored and it was above 50% for all the concentrations over a 72-h study. For 24 h and 48 h study it was less which could be attributed to the thing that hydrolysis is a rate-limiting step of AD and breakdown of carbohydrates, fats and proteins take time accordingly.

AD has proven to be an efficient technique for reducing the micro-pollutant load and has paved a way for a better world.

5 Future Scope and Recommendations

Ground water pollution due to micropollutants is a cause of concern in India and globally. Anaerobic digestion can be effectively used to reduce micropollutant load on the environment. Since it's a biological method it carries the added advantage of less to no pollution. Policymakers need to resort to biological methods in contrast to chemical methods such as sonication and other chemical degradation methods. Lifestyle conditions such as Diabetes, Cardiovascular related concerns are on rise which warrants excessive use of pharmaceuticals, the risks of which can be mitigated using the aforementioned techniques. The study could also be conducted in thermophilic AD's which may give a better result due to temperature sensitivity.

Acknowledgements UGC funding to Ms. Aatiya Shaikh for this study is acknowledged by the authors. Authors also acknowledge Department of Science & Technology, Government of India for the funding under BRICS scheme DST/PRAO/TSA/01/2022/498.

Conflict of Interest No conflict of interest is declared by the authors.

References

1. Aziz A, Sengar A, Basheer F, Farooqi IH, Isa MH (2022) Anaerobic digestion in the elimination of antibiotics and antibiotic-resistant genes from the environment—a comprehensive review. J Environ Chem Eng 10(1). https://doi.org/10.1016/j.jece.2021.106423
2. Aziz A, Sengar A, Basheer F, Farooqi IH, Isa MH (2022) Anaerobic digestion in the elimination of antibiotics and antibiotic-resistant genes from the environment—a comprehensive review. J Environ Chem Eng 10(1):106423. https://doi.org/10.1016/J.JECE.2021.106423
3. Goldman JL, Koen YM, Rogers SA, Li K, Leeder JS, Hanzlik RP (2016) Bioactivation of trimethoprim to protein-reactive metabolites in human liver microsomes. Drug Metab Dispos 44(10):1603. https://doi.org/10.1124/DMD.116.072041

4. Cao Z (2019) Development and validation of a LC-MS / MS method for the determination of multi-class antibiotics in human serum and its application to maternal women exposure. Msacl
5. Salter AJ (1982) Overview trimethoprim-sulfamethoxazole: an assessment of more than 12 years of use. Rev Infect Dis 4(2):196–236. https://doi.org/10.1093/CLINIDS/4.2.196
6. Rashid U et al (2016) Design, synthesis, antibacterial activity and docking study of some new trimethoprim derivatives. Bioorganic Med Chem Lett 26(23):5749–5753. https://doi.org/10.1016/j.bmcl.2016.10.051
7. Mpatani FM et al (2021) A review of treatment techniques applied for selective removal of emerging pollutant-trimethoprim from aqueous systems. J Clean Prod 308:127359. https://doi.org/10.1016/J.JCLEPRO.2021.127359
8. Ramirez AJ (2007) Analysis of pharmaceuticals in fish using liquid chromatography-tandem mass spectrometry. Anal Chem 79(8):3155–3163. https://doi.org/10.1021/ac062215i
9. Gleckman R, Blagg N, Joubert DW (1981) Trimethoprim: mechanisms of action, antimicrobial activity, bacterial resistance, pharmacokinetics, adverse reactions, and therapeutic indications. Pharmacotherapy 1(1):14–19. https://doi.org/10.1002/J.1875-9114.1981.TB03548.X
10. Wróbel A, Arciszewska K, Maliszewski D, Drozdowska D (2020) Trimethoprim and other nonclassical antifolates an excellent template for searching modifications of dihydrofolate reductase enzyme inhibitors. J Antibiot (Tokyo) 73(1):5. https://doi.org/10.1038/S41429-019-0240-6
11. Bromberger JT, Kravitz HM (2011) Mood and menopause: findings from the study of women's health across the nation (SWAN) over ten years. Obstet Gynecol Clin North Am 38(3):609. https://doi.org/10.1016/J.OGC.2011.05.011
12. Weismiller DG (2009) Menopause. Prim Care 36(1):199–226. https://doi.org/10.1016/J.POP.2008.10.007
13. Polotsky HN, Polotsky AJ (2010) Metabolic implications of menopause. Semin Reprod Med 28(5):426–434. https://doi.org/10.1055/S-0030-1262902
14. Maggio M et al (2012) The hormonal pathway to cognitive impairment in older men. J Nutr Health Aging 16(1):40–54. https://doi.org/10.1007/S12603-012-0002-7
15. Owens RC, Ambrose PG (2000) Clinical use of the fluoroquinolones. Med Clin North Am 84(6):1447–1469. https://doi.org/10.1016/S0025-7125(05)70297-2
16. Alghasham AA, Nahata MC (2000) Clinical use of fluoroquinolones in children. Ann Pharmacother 34(3):347–359. https://doi.org/10.1345/APH.18146
17. Redmond TM et al (1998) Rpe65 is necessary for production of 11-cis-vitamin A in the retinal visual cycle. Nat Genet 20(4):344–435. https://doi.org/10.1038/3813
18. Smith A, Weber A, Pandher R, Williams-Warren J, Cohen ML, Ramsey B (1997) Utilization of salivary concentrations of ciprofloxacin in subjects with cystic fibrosis. Infection 25(2):106–108. https://doi.org/10.1007/BF02113587/METRICS
19. Matta MK et al (2018) LC-MS/MS based quantitation of ciprofloxacin and its application to antimicrobial resistance study in Balb/c mouse plasma, urine, bladder and kidneys. Anal Methods 10(10):1237–1246. https://doi.org/10.1039/c7ay02923c
20. Han S et al (2021) Simultaneous determination of seven antibiotics and five of their metabolites in municipal wastewater and evaluation of their stability under laboratory conditions. Int J Environ Res Public Health 18(20). https://doi.org/10.3390/ijerph182010640
21. Orhorhoro EK, Ebunilo PO, Sadjere GE (2018) Effect of organic loading rate (OLR) on biogas yield using a single and three-stages continuous anaerobic digestion reactors. Int J Eng Res Afr 39(11):147–155. https://doi.org/10.4028/www.scientific.net/JERA.39.147
22. Liu C, Wang W, Anwar N, Ma Z, Liu G, Zhang R (2017) Effect of organic loading rate on anaerobic digestion of food waste under mesophilic and thermophilic conditions. Energy Fuels 31(3):2976–2984. https://doi.org/10.1021/acs.energyfuels.7b00018
23. Angelidaki I et al (2009) Defining the biomethane potential (BMP) of solid organic wastes and energy crops: a proposed protocol for batch assays. Water Sci Technol 59(5):927–934. https://doi.org/10.2166/wst.2009.040
24. Pitt JJ (2023) Principles and applications of liquid chromatography mass spectrometry in clinical biochemistry. Biochem, Clin Biochem Rev (1):19–34. https://www.ncbi.nlm.nih.gov/pmc/articles/PMC2643089/

25. Dolan JW, Snyder LR (2000) Selectivity for C18 and C8 reversed-phase columns as a function of temperature and gradient steepness: I optimising selectivity and resolution
26. Fda, Cder (2018) Bioanalytical method validation guidance for industry biopharmaceutics bioanalytical method validation guidance for industry biopharmaceutics contains nonbinding recommendations. http://www.fda.gov/Drugs/GuidanceComplianceRegulatoryInformation/Guidances/default.htm and/or http://www.fda.gov/AnimalVeterinary/GuidanceComplianceEnforcement/GuidanceforIndustry/default.htm

Constructed Wetland: Design, Operation, and Maintenance Techniques

Shweta Yadav, Sandeep K. Malyan, Rajesh Singh, Sujata Kashyap, Vinay Kumar Tyagi, Omkar Singh, and Jyoti Singh

Abstract Globally constructed wetlands (CWs) have been adopted as the green solution for wastewater treatment owing to their low economic costs, reduced energy demand, and eco-friendly approach. They are artificially created systems for wastewater treatment utilizing natural processes involving vegetation species, soil, and microbes. The CW involves several contaminant removal strategies such as phytoextraction, phytovolatilization, phytostabilization, and rhizo-filtration. It provides decentralized wastewater treatment to both rural and urban areas and offers multiple ecosystem services such as improved water quality, reuse of treated water, and biodiversity development. However, the poorly understood design criteria, operation, and maintenance of CW limits its widespread adoption and sustainability, thus presenting a setback to the technology as a whole. Poor operation and maintenance of the CW lead to negative environmental outcomes such as excessive odor, and rerelease of the contaminants into the system. Appropriate human-centered design and regular maintenance is required to ensure effective and continuous functioning of CW treatment mechanisms (physical, chemical, and biological). This chapter provides a comprehensive overview of the CW design, operation, and maintenance in the context of developing nations. The study further aims to ensure that the technology is widely accepted, implemented, and appropriately designed for effective wastewater treatment while minimizing negative environmental impacts. The classification of CW, their design criteria, operation, and maintenance are discussed in detail. Furthermore,

S. Yadav · O. Singh · J. Singh
Technical Cell, National Institute of Hydrology, Roorkee 247667, India

S. K. Malyan
Department of Environmental Studies, Dyal Singh Evening College, University of Delhi, New Delhi 110003, India

R. Singh (✉) · V. K. Tyagi
Environmental Hydrology Division, National Institute of Hydrology, Roorkee 247667, India
e-mail: rsingh.nih@gmail.com; rsingh.nihr@gov.in

S. Kashyap
Axa Parenteral Limited, Roorkee 247667, India

M. M. Ghangrekar et al. (eds.), *Biological and Hybrid Wastewater Treatment Technology*, Earth and Environmental Sciences Library,
https://doi.org/10.1007/978-3-031-63046-0_12

appropriate recommendations were suggested for the confronting challenges in the design, operation, and maintenance of the CW system.

Keywords Constructed wetlands · Design criteria · Operation · Maintenance

Abbreviations

ASP	Activated Sludge Process
BOD	Biological Oxygen Demand
CH_4	Methane
COD	Chemical Oxygen Demand
CW	Constructed Wetlands
DO	Dissolved Oxygen
FWS	Free-water Constructed Wetlands Systems
GHG	Greenhouse Gas
HCW	Hybrid Constructed Wetlands
HLR	Hydraulic Loading Rate
HRT	Hydraulic Retention Time
HSFCW	Horizontal Sub-Surface Flow Constructed Wetlands
MBBR	Moving-Bed Biofilm Reactor
MTS	Mechanized Treatment Systems
NTS	Natural Treatment Systems
SBR	Sequencing Bio-Reactors
SDG	Sustainable Development Goal
SFCW	Subsurface Flow Constructed Wetlands
TN	Total Nitrogen
TP	Total Phosphorus
TSS	Total Suspended Solid
UN	United Nations

1 Introduction

Globally, ~ 380 billion m^3 of wastewater is generated annually and only 24% of household wastewater undergoes some kind of treatment on average [1]. The target-6.3 of United Nations (UN) Sustainable Development Goal-6 (SDG-6) emphasizes on improving the quality of water by reducing pollution, and reducing the untreated wastewater proportion to half together with recycling and safe reuse (UN-SDG2030: https://sdgs.un.org/goals). With the increase in population, the wastewater generation volume will increase by 24% in 2030 [2]. There are several conventional and advanced treatment technologies for wastewater treatment across the globe such as activated sludge process (ASP), sequencing bio-reactors (SBR), moving-bed biofilm reactor (MBBR), trickling filters, and Membrane technology. However, these mechanized treatment systems (MTS) need skilled manpower, high energy, and high operation and maintenance costs, which makes them relatively unsustainable for long-term application [3]. Many of these technologies have not considered the sustainable development goal (6.3) by releasing eco-hazardous waste including chemicals in the environment, unsafe waste disposal, and reuse and recycling at high cost in terms of energy. Natural treatment systems (NTS) such as Constructed Wetlands (CW) appear to be an alternative solution to wastewater treatment in place of MTS both economically and environmentally.

Constructed Wetlands (CW) are artificially engineered wetlands consisting of shallow basins (< 1 m), vegetation (i.e., aquatic macrophytes), microorganisms (e.g., bacteria, protozoa, algae, fungi, yeast), and they may or may not have the substrate (e.g., sand or gravel) depending on the type of CW [4, 5]. CW mimics the natural wetlands processes (physical, biological, and chemical) in a more controlled manner [6]. Since 1950, CW has been widely used to treat a variety of wastewater (industrial, agricultural, stormwater, and municipal wastewater), worldwide [7–16]. They are less complex than the conventional treatment system and have high pollutant removal efficiency for longer operational periods [14]. The horizontal sub-surface flow constructed wetlands (HSFCW) planted with *Typha latifolia* and *Phragmites australis* showed the removal of pesticide (terbuthylazine) from agricultural wastewater by 58% and 78%, respectively [17]. In cold climates, a hybrid constructed wetland designed to treat the dairy wastewater exhibited high removal efficiency of 70–96% for chemical oxygen demand (COD), 32–86% for Ammonium (NH_4–N), 39–90% Total Nitrogen (TN), and 70–93% Total Phosphorus (TP). A floating treatment wetland (FTW) with anaerobic biofilter showed high removal efficiency for Total Coliform (99.9%), Turbidity (91%), and COD (71%) from urban wastewater [18].

Nevertheless, due to improper design, lack of operational and maintenance know-how, inadequate standard assessment protocols, and the general assumption that CW remains eco-friendly on its own lead to a significant adverse effect on its performance and subsequent negative impact on the environment. Consequently, effective wastewater treatment and long-term sustainable operation of CW remains a challenge worldwide. Poorly managed CW can potentially contribute to environmental challenges such as greenhouse gas (GHG) emission, odor, and growth of invasive species [14, 19]. Methane (CH_4) is 25 times potent GHG and contribute significantly to global warming [20, 21]. Studies have shown that CW can be a possible source of GHG emissions such as methane (CH_4: 30–450 mg/m^2.h) and nitrous oxide (N_2O: 59–1252 mg/m^2.h) in two year period [22]. The performance and the sustainability of the CW critically depend on several design parameters (i.e., surface area, water depth, hydraulic retention time and loading rate, and wastewater type), its operation, and regular maintenance (i.e., influent feeding mode, performance monitoring, and biomass harvesting). Therefore, in this chapter, an attempt has been made to comprehend the CW design techniques and related influencing factors for appropriate and efficient design. Furthermore, the operation and maintenance techniques were precisely discussed for the sustainable and long-term effective performance of CW.

2 Classification of Constructed Wetlands (CW)

Multiple design configurations of CW are reported and classified such as based on flow pattern (free-surface flow, and subsurface flow), type of vegetation (emergent, free-floating, and submerged), stages (single stage, multi-stage or hybrid system), type of feeding mode (continuous, and intermittent feeding), type of treatment (primary, secondary, and tertiary), and enhancement (artificially aerated, tidal flow, and baffled flow) [4, 23–26]. Principally, typical CW are classified based on wetland hydrology, and flow regime into two broad categories, (i) Surface flow (free surface also known as free-water surface CW), and (ii) Subsurface flow CW as shown in Fig. 1 [27]. Several CWs have been developed by merging the two or more to obtain the desired effluent quality. The combination of Free-water surface CW (FWS) and various forms of Subsurface flow CW (SFCW) forms the third category of wetland, (iii) Hybrid CW (HCW) [3, 28].

The FWS constructed wetland consists of a shallow basin with impervious lining or permeable substrate or both to support the vegetation roots and the water flow control structure. The wastewater flows above the saturated substrate (e.g., gravel, sand). In FWS, the water flow is mainly regulated by aquatic macrophytes, low flow velocity, and water depth [15]. The surface layer of FWS remains aerobic and the deeper water layers are generally anaerobic therefore, they are frequently used to

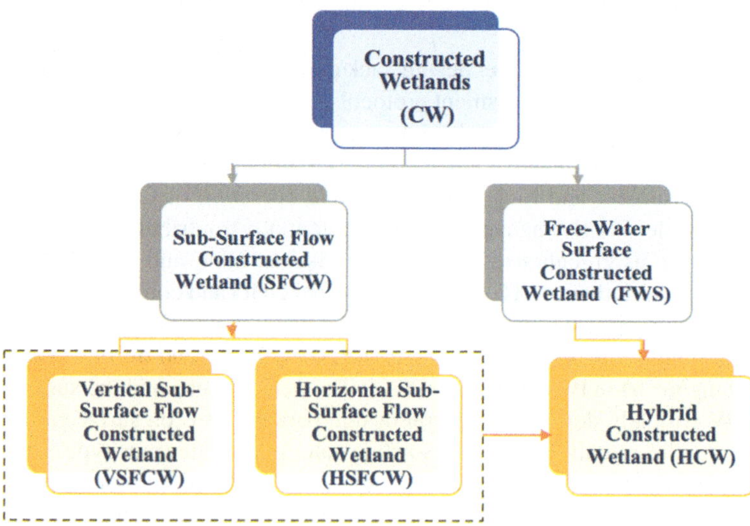

Fig. 1 Classification of constructed wetland based on wetland hydrology and flow regime

treat stormwater, and mine drainage [4]. FWS are low-cost, easy to maintain, and are less prone to clogging than other CWs [18]. However, the decay of emergent vegetation, algal growth, litter, and cold climate often reduced the pollutant removal efficiency of FWS [15, 19].

On the other hand, SFCW consists of an artificially built basin with a seepage resisting material with porous substrate (sand, gravel, rock) and emergent vegetation through which wastewater flows either horizontally or vertically. The water level remains 10–15 cm below the top surface of the SFCW therefore, minimize the pest, and odor problems [26]. The porous substrate acts as the filter media and provides a greater surface area for wastewater treatment contact than FWS thus, being efficient in removing a wide range of pollutants such as organic loads, heavy metals, nutrients, and suspended solids. Subsequently, SFCW shows enhanced assimilation capacity per unit area and is smaller than FWS for the same volume of wastewater. The system also has high tolerance to cold climate [4, 26, 29]. However, SFCW has relatively high per unit construction cost than FWS. Owing to the hydraulic constraint posed by the substrate it can treat wastewater with low solid concentrations and at uniform flow condition. Clogging in SFCW is mainly associated with the solid accumulation due to high retention, microbial biofilm growth, and unmanaged vegetation [24]. The SFCW is further categorized based on the flow direction into, (a) vertical sub-surface flow CW (VSFCW) where the effluents move vertically (up-flow or down-flow) through the substrate, and (b) horizontal sub-surface flow CW (HSFCW) where the effluent moves horizontally through the substrate. In general, HSFCW is used for the removal of BOD, TSS and for enhanced denitrification due to limited oxygen transfer capacity. Conversely, VSFCW has greater oxygen transfer capacity and is used where nitrification is required [4, 19, 30].

Hybrid CW (HCW) is a slightly complex combination of various forms of SFCW and FWS, used to enhance the pollutant removal efficiency, depending on the type of wastewater [14, 31]. Studies have shown that the pollutant removal efficiency increases with an increase in design complexity in the order of HSFCW < VSFCW < HCW or intensified CW, however, difficult to maintain [32]. The most preferred combination used in studies is VSFCW-HSFCW followed by HSFCW-VSFCW. Contaminant removal efficiency increases per unit area of the hybrid treatment system even in the cold climate [33]. The combined system reduces the problem of nitrification and denitrification simultaneously making the HCW more efficient in pollutant removal and transformation [33–35]. Artificial aeration is required to maintain the dissolve oxygen in the CW as in most cases nitrification takes place when DO > 1.5 mg/l and denitrification takes place at DO < 0.5 mg/l [26].

3 Design Techniques of CW

The CW design configuration should be planned to take the advantage of the natural topography. The design must be simple, easy to maintain, utilizing natural energies, and planned for extreme weather and climate conditions (e.g., storms, floods, and droughts) [29]. Principal criteria for CW design include, site selection, water depth, vegetation selection, substrate selection, hydraulic loading rate (HLR), Hydraulic retention time (HRT), wastewater type, and its operation and maintenance (Table 1).

Table 1 Design criteria for FWS and SFCW for the desired effluent quality of BOD < 30 mg/L and TSS < 30 mg/L [4, 18, 23, 24, 26, 29, 30, 36, 37]

Design/sizing parameters	Equations	FWS	SFCW
Design flow	Q_m or Q_{av}	Maximum or average flow	Maximum or average flow
Size of bed or area (A) in (m^2)	L * W	Larger	< 2500
Land requirement (population equivalent) in m^2/PE		–	0.8–3 (VSFCW) 5–10 (HSFCW)
Length (L): width (W) or (aspect ratio)		> 1:1 (plug flow hydraulics); 3:1–5:1	1:1 and 1:2; < 3:1 [L: 61 m; W: 12–30 m (~ 15 m)]
Water depth (m)	D_w	0.1–2; 0.6–0.9 m (fully vegetated zone); 1.2–1.5 m (open water zones); 1 m (inlet settling zone)	0.3–1.5
Bottom slope (%)	dh/dL	< 0.5%	0.5–1%
Hydraulic retention time (HRT) in (days)	V_w/Q_o $[V_w = A * D_w * n]$	2–3 or 5–30 days (climate dependent)	2–5
Hydraulic loading rate (HLR) in (m/day)	$C * V_w/A$	< 0.1	< 0.5
Inlet and outlet structures		Uniform distribution and collection of influent and effluent	Uniform distribution and collection of influent and effluent
Substrate (filter media size)		< 20 mm (porosity 0.3–0.5)	40–80 mm (inlet zone, outlet zone) 20–30 mm (treatment zone) 5–20 mm (top 10 cm of planting media)
Vegetation		Locally available native species	Locally available native species

Note *L* length, *W* width, *dh/dL* hydraulic gradient, V_w useable CW water volume or mean volume, Q_o average outflow flow rate: (inflow + outflow/2), Q_m maximum flow, Q_{in} wastewater inflow, *A* CW surface area, *C* parameter concentration, *n* porosity

3.1 Site Selection for CW

Selection of appropriate site for CW can save significant construction cost and prevent negative environmental impacts [10]. For sustainable and effective functioning of the CW, the critical factors for selecting the suitable site involves topography (slope), land use and soil layers, land availability, natural resources near the site, hydrology, and the climatic conditions [4]. In principle, the site for the CW must be near the wastewater source to avoid lengthy conveyance and a low, flat area with gentle slope is conducive for the gravity flow, reducing the cost of pumping and construction. However, the FWS flat area lowers the cost while the sloped area requires construction of dikes or gradings. Ideally, the CW site consists of the soil layer with low permeability ($< 10^{-6}$ cm/s) which acts as the barrier to minimize contaminant infiltration to groundwater. For soil with high permeability liners such as imported clay, asphalt and synthetic liners (PVC) must be used, which raises the construction cost [29]. Furthermore, it is essential to consider the depth of aquifer, water from streams and other drainages problems before site selection. It is recommended to avoid construction in floodplains or in natural wetlands, unless specific measures are required. Tournebize et al. [38] found that off-stream CW enhance biodiversity with less areal requirement and better performance in pesticide removal than the in-stream CW. The performance of the CW often limits due to inadequate retention time which is limited by the size of the wetland as the CW with larger size often has higher retention time and consequently higher removal rate. Therefore, CW site must have adequate space or land available for construction [19]. Local climate and seasonal variations significantly affect the efficacy of CW.

Owing to the shallow depth, the open to atmosphere CW are greatly, influenced by rainfall, temperature, snowmelt, drought, and spring runoff. Change in temperature largely affects the microbial community in the CW. For instance, denitrification in CW reduces at temperature < 15 °C while stops at 5 °C affecting the ammonia and nitrogen removal. Conversely, microbial activity significantly influences organic matter removal and is less dependent on temperature. Microbial community can function at the temperature < 15 °C [39]. Precipitation dilutes the pollutant, increase water level and reduces HRT whereas, evapotranspiration (ET) concentrates the pollutants in CW and increase the HRT. In general, ET in CW is reported to be 1.5–2 times the pan evaporation [29]. Singh and Chakraborty [40] reported that rainfall noticeably reduced the treatment of acid mine drainage in an unsheltered CW (sulfate: 50%, Fe: 85%, Al: 70%, Zn: 93%, Co: 82%, Ni: 83%, and Cr: 90%) than the sheltered CW (sulfate: 60%, Fe: 90%, Al: 79%, Zn: 94%, Co: 92%, Ni: 91%, and Cr: 96%) due to shortening of HRT and insignificant dilution. The first flow storm often carries the higher pollutants therefore excess runoff need to be diverted from the CW to avoid runoff overflow. Selection of site therefore requires the proper understanding of the climatic and seasonal variation for effective functioning of the CW. The runoff due to precipitation need to be estimated and included in the design flow.

3.2 CW Sizing: Governing Equations

In general design criteria involves areal loading rate (g/m^2.d), and volumetric loading rate (g/m^3.d). Many of the pollutant removal mechanisms in CW depends on the surface area theoretically, such as oxygen transfer, plant growth, and sedimentation therefore, areal loading rate ((influent flow rate * pollutant concentration)/surface area) is frequently used for the design [29]. SFCW can be calculated based on BOD, total nitrogen, and ammonia nitrogen, etc. Typically, P-K-C* first order model has been used to estimate the design area of SFCW [41]. And first order plug flow model used for FWS assuming linear and areal kinetics [5, 9, 42].

EPA suggests the use of Darcy Law for the CW area and width estimation depends on the areal loading rate, water depth, flow, and hydraulic conductivity, assuming the laminar flow through the media [29], as given in Eq. 1.

$$Q = KAS = KWD_w\left(\frac{dh}{dL}\right) \tag{1}$$

where, K is hydraulic conductivity (m/d or m^3/m^2 day); Q is flow rate (m^3/day); A is cross sectional area which is normal to flow given by Width (W) * Water Depth (Dw) in m^2; S is hydraulic gradient given by dh/dL. Hydraulic conductivity in the initial 30% remains 1/4–1/3 of the remaining 70% of the SFCW. While the recommended bottom slope is 0.5–1% for easy maintenance and proper draining from inlet to outlet. The width of the CW system can be calculated using the Darcy law as in Eq. 1.

Several modification has been made in different literatures for estimating the CW area. UN-HABITAT [23] provide the surface area of the wetland used for sizing based on influent and effluent BOD concentration Eq. 2.

$$A = \frac{Q(\ln C_i - \ln C_e)}{K_{BOD}} \tag{2}$$

where, A is surface area of bed (m^2); Q is average flow rate per day; Ci and Ce are the concentration of pollutants (mg/l) at the influent and effluent of the CW and K_{BOD} is the decay rate constant increases 10% at every 1 °C rise in temperature expressed as $K_T dn$ where K_T is given by Eq. 3:

$$K_T = K_{20}\left(1.06^{T-20}\right) \tag{3}$$

where, K_{20} is rate constant (at 20 °C/day), T is system temperature in °C, d and n is column water depth (m) and substrate porosity (%). K_{BOD} increases rapidly in VSFCW than in HSFCW.

Design criteria of CW involves the maximum pollutant load (rate) to meet the discharge standard. While designing CW, the end-use of the treated wastewater need to be considered. For instance, emphasize should not be on excessive nutrient removal when treated wastewater is used for irrigation where nitrogen (N) and phosphorus (P) are beneficial whereas, the same need to be efficiently removed before discharging it to the water bodies [41]. Land requirement for CW is often calculated based on the population equivalent (PE) for wastewater of uniform quality and quantity. Ilyas and Masih [43] reported the land requirement of HSFCW (5–10 m^2/PE) and VSFCW (1–3 m^2/PE). In Europe, typical land requirement for CW is 1–3 m^2/PE (VSFCW) and 5 m^2/PE (HSFCW). For tertiary treatment areal requirement of 1 m^2/PE and for storm water treatment 0.5 m^2/PE is recommended by Environment Protection Agency. UN-HABITAT [23] recommends land requirement of 0.8–1.5 m^2/PE (VSFCW) and 1–2 m^2/PE (HSFCW). However, the small land requirement is often not suitable for nitrate removal particularly in colder regions [26].

3.3 Water Depth

Water depth in CW is a critical factor influencing the system's functioning, vegetation growth, biochemical reaction in the root zones by altering the redox status and dissolved oxygen (DO) [14, 44]. Species richness in the CW are directly linked with water depth.

Studies have suggested that the optimum water depth in SFCW should be equivalent to maximum root depth as the plant root depth is critical for the removal of pollutants [25]. Nevertheless, plants with abundant nutrient supply near surface might not grow their roots deeper in the system. Better BOD removal was observed in the system with 45 cm media depth than in the 18 cm media depth, however unclear about the effect of increased HRT due to increased depth [29]. Singh et al. [41] showed the abundance of bacterial community in the shallow cells of HSFCW (0.8 m) than in the deeper cells (1.5 m) such as aerobic denitrifiers, anoxic denitrifiers, hetrotrophic nitrifiers, and phosphate accumulating bacteria influencing pollutant removal, however, nitrogen fixing bacteria and nitrifiers were found missing in the shallow cells. This suggests the selection of water depth could be based on target contaminant and the different HRT required to remove or transform them. Millot et al. [36] studied the treatment performance of CW at different depths and found that increasing depth beyond 0.27 m does not significantly influence the removal of TSS and COD except ammonium removal which increased (60%) with depth from 0.40 to 1 m. Most of the COD removal occurs at first 0.20 m of depth. Typically the recommended water depth is 0.1–0.7 m for SFCW however, greater depths were suggested in other studies [41]. And for FWS the recommended depth is 0.6–0.9 m (fully vegetated zone), 1.2–1.5 m (open water zone) [29]. In FWS, the water depth needs to be monitored, as over time with the accumulation of detritus, the water depth decreases.

3.4 Vegetation Selection

Macrophytes or wetland plants are one of the important component of CW design as it supports several pollutant removal processes in the system such as uptake of nutrients, filtration in water column (FWS), reduce water velocity, platform for microbial biofilm development, transmit and release oxygen, provide habitation and supports biodiversity, more often prevent clogging of the treatment system [15, 29]. Selection of plants for CW design depends entirely on type of wastewater to be treated, availability and properties of plants. UN-HABITAT [23] recommended plant selection based on availability of locally available plants, species with deep roots, rhizomes and large fibrous roots, good stem density and biomass for translocation of nutrients, and plants with sufficient oxygen release in the root zones. Plants tolerant to waterlogging and anoxic condition; also efficient in hyper-eutrophic situation with greater pollutant absorption capacity are also recommended for CW design [5, 24, 45]. Contrarily, considering the limited contribution of plants in oxygen transfer (< 2 mg/L) and nutrient uptake (12–120 g N/m^2/year) the selection of plants were suggested to be based on aesthetics, impact on CW operation, and feasibility in the given location [29]. Among the emergent, floating, and submerged vegetation, the emergent plants were commonly used in the FWS and SFCW. Studies have indicated that the plant with higher biomass development has higher pollutant uptake potential which could be a criteria for plant selection. Most commonly used plant species in CWs are *Phragmites* sp. and *Typha* sp. such as *Phragmites australis*, *Typha latifolia*, and *Phragmites karka* which has rapid growth, high biomass, and tolerant to variable climatic conditions [23]. Other commonly used vegetation species are *Typha augustifolia*, *Typha domingensis*, *Scirpus validus*, *Scirpus lacustric* L., *Juncus effuses*, *Juncus articulates*, *Eleocharis sphacelate*, *Cyperus papyrus*, *Canna indica* [3, 15].

3.5 Substrate Selection

Substrate selection is crucial for the CW's effective functioning as it provides media for plant growth, surface area for pollutant absorption and adsorption and aids in filtration. The pollutant exchange, precipitation, adsorption and complexation are the main processes through which pollutant removal and transformation occur in the CW system [3]. The availability and the cost of the substrate usually determine the substrate selection. Substrate with low adsorption capacity often hampers the pollutant removal efficiency in CW whereas the substrate with low hydraulic conductivity could lead to system clogging further reducing the CW performance. The substrate with high hydraulic conductivity enhance the phosphorus sorption capacity of the CW and avoid short-circuiting [24]. Therefore, adsorption, and hydraulic permeability are important properties of substrate for its selection. In FWS, the use of loose soil with high humic content, and the use of loam, silt, and sand promote the plants growth, and rapid biomass generation [29]. In general, the other substrate selection criteria involves physical factors such as particle size, shape, surface area, porosity, hydraulic conductivity; chemical factors such as toxicity, and chemical stability; and biological factors such as electron acceptor and electron donor. Particle size of the substrate influence its porosity, and hydraulic conductivity. For instance, fine sized substrate often enhances the pollutant removal efficiency, however, due to filtration and accumulation of the pollutants they are highly susceptible to clogging. UN-HABITAT [23] recommends the use of rounded substrate over the sharp edged substrate as the round shape provides loose packing and not damaging the plant roots. In SFCW, the recommended size of the substrate at the inlet/outlet zone is 40–80 mm while at the treatment zone it is 20–30 mm [4]. Essentially, substrate must have a long life, should not lead to any secondary pollution such as leaching of toxic heavy metals causing adverse environmental impact, and must be disposed safely whenever required. Commonly used natural substrate which are easily available, low cost and has effective pollutant removal efficiency are soil, sand, gravel, and rock. Singh et al. [44] used gravel substrate in the HSFCW for the treatment of domestic wastewater and obtained 89–91% of organic load removal, 67–72% of nutrient removal, and 53% of phosphorus removal. Phosphorus removal can be enhanced by the substrate with calcium (Ca), Iron (Fe), and aluminum (Al) content such as limestone, alum sludge and slag. Other substrate such as compost, rice husk and peat can be used to removal the nitrogen and organic load in CW. Mixed substrate such as strate gravel, and vermiculite provide reactive surface for microbial growth and aid in removing water with low nutrients [24]. Biochar formed by the pyrolysis process (biomass) has high surface area enhance the pollutant removal in CW [3, 46]. Furthermore, the use biochar, zeolite, and gravel reduces the N_2O emission from CW [14].

3.6 Hydraulic Loading Rate and Retention Time

The hydraulic capacity of the CW is principally indicated by the hydraulic loading rate (HLR) and hydraulic retention time (HRT). Hydraulic capacity is crucial for designing, and effective functioning of the CW. HLR signifies the maximum loading of the pollutants in the CW per unit area for a specific time (days), and governs the planning as well as designing of the treatment system [47]. While HRT is the average time the wastewater retained or remained within the CW also indicating the contact time with the substrate, microbial biofilm, and the plants. At higher HRT the contact time of the wastewater with the microbial community increases which further increases the removal efficiency of the contaminants or parameters needing higher HRT [26, 48, 49]. On the other hand, increased HLR reduces the contact time through quick flow of wastewater within the system [23, 45, 47, 50, 51]. García et al. [47] found that with increase in HLR concentration of COD and BOD increases in the effluent from the HSFCW. Studies have shown that nitrate removal enhanced with an increase in HLR upto 1.5 m/day and at an HRT of 2 days, however further increase in HLR reduces the nitrate removal efficiency [52]. In general, optimum HRT largely influenced by the optimum HLR for the treatment system. At HLR (75–100 L/day) and high HRT (> 3 days) leads to increased plant density in the CW compared to lower HRT (1–2 days) and HLR (150–300 L/day) [24, 36]. In some studies, different HLR was suggested for variety of wastewater and different treatment system such as for the treatment of domestic wastewater the optimum HLR remains 0.1 m/day (FWS), 0.36 m/day (VSFCW), 0.46 m/day (HSFCW) and 0.3 m/day (HCW), for industrial wastewater the HLR remains 0.1 m/day (FWS), 0.57 m/day (VSFCW), and 0.59 m/day (HCW) [26, 53]. Longer HRT reduces the peak of the pesticide concentration in the vegetated cell by 91% at HRT of 330 min, than the cell with HRT of 280 min (89% reduction in peak) [54]. Akratos and Tsihrintzis [39] studied the impact of various HRT (6, 8, 14, and 20 days) on the pollutant removal from HSFCW and found that removal efficiency of organic matter, TKN and phosphate enhanced at the 8-day HRT. Furthermore, mercury (Hg) removal in planted CW cell increased from 50 to 95% with increase in HRT from 6 to 24 h indicating that the longer HRT increase the removal efficiency of Hg [55].

4 Operation and Maintenance

4.1 Feeding Mode

The performance of CW is substantially influenced by the feeding mode such as continuous feeding, batch or intermittent feeding [25]. The contaminant removal in CW largely depends on the oxidation and the reduction potential in system which is further influenced by the mode of operation (batch or continuous) [24]. For instance, the system with intermittent feeding mode provides an oxidized environment conducive for ammonia removal whereas continuous feeding mode supports redox zone for sulphate removal. In batch or intermittent-fed mode, the system is alternatively flooded and drained resulting in the air entrainment in the substrate (soil or gravel) pores. The air in the substrate pores enhance the oxidation of carbon (C) and Nitrogen (N) in the CW than in the continuous feeding mode, thus enhancing the overall performance of the system in batch feeding mode [19]. Caselles-Osorio and García [56] studied the impact of intermittent feeding and continuous feeding on the removal efficiency of HFCW and found that the ammonia removal enhanced significantly (average range: 80–99%) in intermittent feeding mode (range: 0.58–0.67 g N/m^2 day) than in the continuous feeding mode (range: 0.52–0.58 g N/m^2 day). Moreover, the sulphate removal efficiency was found to be higher (average: 76–79%) in the continuous feeding mode (51–58%). Zhang et al. [57] studied the impact of intermittent feeding and continuous feeding on removal of eight pharmaceuticals namely, naproxen, ibuprofen, diclofenac, carbamazepine, caffeine, clofibric acid, salicylic acid, and ketoprofen in the mesocosm-scale CW. The study found the high removal efficiency in the intermittent-feeding mode for almost all the pharmaceuticals except naproxen, carbamazepine, and clofibric acid. The intermittent-feeding mode enhance the atmospheric transfer, improves the CW substrate characteristics, and mass transfer mechanism thus enhancing the contaminant removal efficiency of CW [19, 24].

4.2 Clogging Control

FWS are low-cost, easy to maintain, and are less prone to clogging than other CW [18]. Clogging in SFCW is mainly associated with the solid accumulation due to high retention, microbial biofilm growth, and unmanaged vegetation [24]. Poor hydraulic conductivity and reduced porosity leads to clogging of the CW and performance decline, this further reduces the designed life of the system [24]. The degree and distribution of clogging in the system depends on the hydraulic behavior and solid accumulation, respectively. Pedescoll et al. [50] studied the impact of hydraulic loading rate on clogging and found that solid accumulation rate of 0.31 kg/m^2 for TSS and 0.08 kg/m^2 for BOD at the HLR of 50 mm/day and 100 mm/day, respectively. Plant species such as *Typha latifolia* and *Typha augustifolia* produce high biomass whose litter and decayed leaves could clog of the system if biomass harvesting is untimely [26]. The below-ground biomass of the *Typha augustifolia* is two times more than the *Phragmites australis* thus develops clogging in the system. The presence of plants and organic load result in the loss of hydraulic conductivity by 20 and 10% [50]. The CW system with soil and sand substrate (small diameter media) are often prone to clogging than the other substrate resulting in ponding (overflowing in SFCW) and reduction in removal efficiency. It is recommended to use the substrate of 40 and 80 mm at the inlet and outlet zones to minimize clogging. Furthermore, the length of the inlet and outlet zones should be 2 m and 1 m, respectively for even distribution and collection of influent and effluent without clogging [4, 29]. Selection of appropriate filter media or substrate and proper design of the CW is essential to avoid clogging of the system.

4.3 Plant Biomass Harvesting

The CW plants are crucial for the effective functioning of the system. Most of these plants are self-maintaining, they grow, die, and then regrow every year. Harvesting of plants is largely site-specific and depends on the objective of the CW. However, periodic biomass harvesting is required to maintain the desired plant community, enhance hydraulic capacity, prevent mosquito breeding within the CW and remove the undesired species [29, 30]. Unwanted harvesting may disturb the CW performance. A well-designed and operated system may not require the regular biomass harvesting. Constant load of pollutants in the wetland affects the diversity of the plant species and sometimes toxic to plant leading to their die-off and decay in the system further increasing the nutrient load within the treatment system [4]. Contaminant removal within the plants can through absorption, and microbial degradation (biofilm development) and plant uptake. In general, the uptake of the plant is not permanent, once they die, they release back the pollutants thus reduces the efficiency of the system. Biomass harvesting may increase the cost but aid in nutrient removal from the system. Furthermore, high biomass plants such as *Typha latifolia* can also cause clogging of the CW system, therefore routine harvesting is required to maintain the growth [26]. Plant harvesting enhance the growth of the plant and their contaminant removal efficiency. In some cases, harvesting plants such as *Phragmites* at the end of the growing season reduces the N and P [29]. However, for metal removal the suggested harvesting period is beginning of senescence and the end of the growth period, depending on the quantity of metal uptake [10].

5 Conclusion

The chapter illustrate the criteria and various aspects of Constructed Wetland design and its effective functioning such as site selection, sizing, water depth, hydraulic parameters (HRT and HLR), vegetation and substrate selection. Appropriate design, regular operation, and maintenance ensure the enhanced performance and sustainability of the CW for wastewater treatment. Although there are various design criteria and guidelines, the chapter emphasize that the design criteria vary with location and must always be site-specific. The use of locally available materials is highlighted for ease of construction and cost-efficiency. A well-design treatment system requires low maintenance than a poorly designed system. The chapter will be beneficial for the concerned authority, planners, scientists, students, and other stakeholders in designing, operating, and maintaining their CW system.

Conflicts of Interest The authors declare no conflict of interests.

References

1. Pratap B, Kumar S, Nand S, Azad I, Bharagava RN, Romanholo-Ferreira LF, Dutta V (2023) Wastewater generation and treatment by various eco-friendly technologies: possible health hazards and further reuse for environmental safety. Chemosphere 313:137547. https://doi.org/10.1016/j.chemosphere.2022.137547
2. ENRD (2022) Wastewater as a resource. European Investment Bank Projects Directorate. https://www.eib.org/attachments/publications/wastewater_as_a_resource_en.pdf
3. Malyan SK, Yadav S, Sonkar V, Goyal VC, Singh O, Singh R (2021) Mechanistic understanding of the pollutant removal and transformation processes in the constructed wetland system. Water Environ Res 93:1882–1909. https://doi.org/10.1002/wer.1599
4. Davis L (1996) A handbook of constructed wetlands: general considerations. United States Environmental Protection Agency. Region III. https://doi.org/10.1016/j.ecoleng.2005.07.010

5. Vymazal J (2010) Constructed wetlands for wastewater treatment. Water (Switzerland) 2:530–549. https://doi.org/10.3390/w2030530
6. Kadlec RH (2016) Large constructed wetlands for phosphorus control: a review. Water (Switzerland) 8. https://doi.org/10.3390/w8060243
7. Brix H (1994) Functions of macrophytes in constructed wetlands. Water Sci Technol 29:71–78. https://doi.org/10.2166/wst.1994.0160
8. Ghermandi A, Bixio D, Thoeye C (2007) The role of free water surface constructed wetlands as polishing step in municipal wastewater reclamation and reuse. Sci Total Environ 380:247–258. https://doi.org/10.1016/j.scitotenv.2006.12.038
9. Díaz FJ, Ogeen AT, Dahlgren RA (2012) Agricultural pollutant removal by constructed wetlands: implications for water management and design. Agric Water Manag 104:171–183. https://doi.org/10.1016/j.agwat.2011.12.012
10. Guittonny-Philippe A, Masotti V, Höhener P, Boudenne JL, Viglione J, Laffont-Schwob I (2014) Constructed wetlands to reduce metal pollution from industrial catchments in aquatic Mediterranean ecosystems: a review to overcome obstacles and suggest potential solutions. Environ Int 64:1–16. https://doi.org/10.1016/j.envint.2013.11.016
11. Gholipour A, Zahabi H, Stefanakis AI (2020) A novel pilot and full-scale constructed wetland study for glass industry wastewater treatment. Chemosphere 247:125966. https://doi.org/10.1016/j.chemosphere.2020.125966
12. Ventura D, Barbagallo S, Consoli S, Ferrante M, Milani M, Licciardello F, Cirelli GL (2019) On the performance of a pilot hybrid constructed wetland for stormwater recovery in Mediterranean climate. Water Sci Technol 79:1051–1059. https://doi.org/10.2166/wst.2019.103
13. Wei F, Shahid MJ, Alnusairi GSH, Afzal M, Khan A, El-Esawi MA, Abbas Z, Wei K, Zaheer IE, Rizwan M, Ali S (2020) Implementation of floating treatment wetlands for textile wastewater management: a review. Sustainability 12:1–29. https://doi.org/10.3390/su12145801
14. Vymazal J, Zhao Y, Mander Ü (2021) Recent research challenges in constructed wetlands for wastewater treatment: a review. Ecol Eng 169:106318. https://doi.org/10.1016/j.ecoleng.2021.106318
15. Yadav S, Kumar J, Malyan SK, Singh R, Singh O, Goyal VC, Singh J, Negi R (2023) Evaluating pilot-scale floating wetland for municipal wastewater treatment using *Canna indica* and *Phragmites australis* as plant species. Sustainability 15. https://doi.org/10.3390/su151813601
16. Singh S, Sharma PK, Rani A, Alotaibi KM (2024) Assessing lab-scale hybrid wetland performance for pollutant and pathogen removal from high organic loading septage. Groundw Sustain Dev 25:101097
17. Gikas GD, Pérez-Villanueva M, Tsioras M, Alexoudis C, Pérez-Rojas G, Masís-Mora M, Lizano-Fallas V, Rodríguez-Rodríguez CE, Vryzas Z, Tsihrintzis VA (2018) Low-cost approaches for the removal of terbuthylazine from agricultural wastewater: constructed wetlands and biopurification system. Chem Eng J 335:647–656. https://doi.org/10.1016/j.cej.2017.11.031
18. Colares GS, Dell'Osbel N, Barbosa CV, Lutterbeck C, Oliveira GA, Rodrigues LR, Bergmann CP, Lopez DR, Rodriguez AL, Vymazal J, Machado EL (2021) Floating treatment wetlands integrated with microbial fuel cell for the treatment of urban wastewaters and bioenergy generation. Sci Total Environ 766:142474. https://doi.org/10.1016/j.scitotenv.2020.142474
19. Nuamah LA, Li Y, Pu Y, Nwankwegu AS, Haikuo Z, Norgbey E, Banahene P, Bofah-Buoh R (2020) Constructed wetlands, status, progress, and challenges. The need for critical operational reassessment for a cleaner productive ecosystem. J Clean Prod 269:122340
20. Malyan SK, Bhatia A, Kumar A, Gupta DK, Singh R, Kumar SS, Tomer R, Kumar O, Jain N (2016) Methane production, oxidation and mitigation: a mechanistic understanding and comprehensive evaluation of influencing factors. Sci Total Environ 572:874–896
21. Malyan SK, Singh O, Kumar A, Anand G, Singh R, Singh S, Yu Z, Kumar J, Fagodiya RK, Kumar A (2022) Greenhouse gases trade-off from ponds: an overview of emission process and their driving factors. Water 14(6):970
22. Wu H, Zhang J, Ngo HH, Guo W, Liang S (2017) Evaluating the sustainability of free water surface flow constructed wetlands: methane and nitrous oxide emissions. J Clean Prod 147:152–156. https://doi.org/10.1016/j.jclepro.2017.01.091

23. UN-HABITAT (2008) Constructed wetlands manual. Agriculture
24. Wu H, Zhang J, Ngo HH, Guo W, Hu Z, Liang S, Fan J, Liu H (2015) A review on the sustainability of constructed wetlands for wastewater treatment: design and operation. Bioresour Technol 175:594–601. https://doi.org/10.1016/j.biortech.2014.10.068
25. Rahman ME, Halmi MIEB, Samad MYBA, Uddin MK, Mahmud K, Shukor MYA, Abdullah SRS, Shamsuzzaman SM (2020) Design, operation and optimization of constructed wetland for removal of pollutant. Int J Environ Res Public Health 17:1–40. https://doi.org/10.3390/ijerph17228339
26. Shukla A, Parde D, Gupta V, Vijay R, Kumar R (2022) A review on effective design processes of constructed wetlands. Int J Environ Sci Technol 19:12749–12774. https://doi.org/10.1007/s13762-021-03549-y
27. Vymazal J, Brix H, Cooper PF, Haberl R, Perfler R, Laber J (1998) Removal mechanisms and types of constructed wetlands. WRc Publ 50
28. Vymazal J (2018) Constructed wetlands for wastewater treatment. Encycl Ecol 14–21. https://doi.org/10.1016/B978-0-12-409548-9.11238-2
29. US-EPA (2000) EPA manual constructed wetlands treatment of municipal wastewaters. United States Environmental Protection Agency, p 165
30. Crites RW (1994) Design criteria and practice for constructed wetlands. Water Sci Technol 29:1–6. https://doi.org/10.2166/wst.1994.0144
31. Barrett J, Barry G, Carew D, Correia L, Flanagan M, Harvey L, Jordan B (2019) Wetlands design manual—design, construction and establishment of wetlands. In: Manual Melbourne water, Melbourne
32. Nivala J, Boog J, Headley T, Aubron T, Wallace S, Brix H, Mothes S, van Afferden M, Müller RA (2019) Side-by-side comparison of 15 pilot-scale conventional and intensified subsurface flow wetlands for treatment of domestic wastewater. Sci Total Environ 658:1500–1513. https://doi.org/10.1016/j.scitotenv.2018.12.165
33. Kato K, Inoue T, Ietsugu H, Koba T, Sasaki H, Miyaji N, Kitagawa K, Sharma PK, Nagasawa T (2013) Performance of six multi-stage hybrid wetland systems for treating high-content wastewater in the cold climate of Hokkaido, Japan. Ecol Eng 51:256–263. https://doi.org/10.1016/j.ecoleng.2012.12.002
34. Ali Z, Mohammad A, Riaz Y, Quraishi UM, Malik RN (2018) Treatment efficiency of a hybrid constructed wetland system for municipal wastewater and its suitability for crop irrigation. Int J Phytoremediation 20:1152–1161. https://doi.org/10.1080/15226514.2018.1460311
35. Ren B, Wang T, Zhao Y (2021) Two-stage hybrid constructed wetland-microbial fuel cells for swine wastewater treatment and bioenergy generation. Chemosphere 268:128803. https://doi.org/10.1016/j.chemosphere.2020.128803
36. Millot Y, Troesch S, Esser D, Molle P, Morvannou A, Gourdon R, Rousseau DPL (2016) Effects of design and operational parameters on ammonium removal by single-stage French vertical flow filters treating raw domestic wastewater. Ecol Eng 97:516–523. https://doi.org/10.1016/j.ecoleng.2016.10.002
37. CPCB (2019) Manual on constructed wetland as an alternative technology for sewage management in India
38. Tournebize J, Passeport E, Chaumont C, Fesneau C, Guenne A, Vincent B (2013) Pesticide de-contamination of surface waters as a wetland ecosystem service in agricultural landscapes. Ecol Eng 56:51–59. https://doi.org/10.1016/j.ecoleng.2012.06.001
39. Akratos CS, Tsihrintzis VA (2007) Effect of temperature, HRT, vegetation and porous media on removal efficiency of pilot-scale horizontal subsurface flow constructed wetlands. Ecol Eng 29:173–191. https://doi.org/10.1016/j.ecoleng.2006.06.013
40. Singh S, Chakraborty S (2022) Impact of seasonal variation on the treatment response of constructed wetlands receiving acid mine drainage in a subtropical region. J Water Process Eng 49:103182. https://doi.org/10.1016/j.jwpe.2022.103182
41. Singh S, Soti A, Kulshreshtha NM, Brighu U, Gupta AB (2022) Customized design of horizontal flow constructed wetlands employing secondary datasets. Bioresour Technol Rep 18:101037. https://doi.org/10.1016/j.biteb.2022.101037

42. Díaz FJ, O'Geen AT, Dahlgren RA (2012) Agricultural pollutant removal by constructed wetlands: implications for water management and design. Agric Water Manag 104:171–183. https://doi.org/10.1016/j.agwat.2011.12.012
43. Ilyas H, Masih I (2017) Intensification of constructed wetlands for land area reduction: a review. Environ Sci Pollut Res 24:12081–12091. https://doi.org/10.1007/s11356-017-8740-z
44. Singh S, Maithani C, Malyan SK, Soti A, Kulshreshtha NM, Singh R, Brighu U, Gupta AB, Kumar J, Yadav S, Singh O, Goyal VC (2023) Comparative performance and 16S amplicon sequencing analysis of deep and shallow cells of a full scale HFCW having sequentially decreasing depths reveals vast enhancement potential. Bioresour Technol Rep 22:101404
45. Vymazal JHB (2006) Removal mechanisms and types of constructed wetlands
46. Malyan SK, Kumar SS, Fagodiya RK, Ghosh P, Kumar A, Singh R, Singh L (2021) Biochar for environmental sustainability in the energy-water-agroecosystem nexus. Renew Sustain Energy Rev 149:111379
47. García J, Aguirre P, Barragán J, Mujeriego R, Matamoros V, Bayona JM (2005) Effect of key design parameters on the efficiency of horizontal subsurface flow constructed wetlands. Ecol Eng 25:405–418. https://doi.org/10.1016/j.ecoleng.2005.06.010
48. Romain V, Sylvie D, David B (2015) Water residence time and pesticide removal in pilot-scale wetlands. Ecol Eng 85:76–84. https://doi.org/10.1016/j.ecoleng.2015.09.040
49. Kumar D, Malyan SK, Kumar A, Kumar J (2021) Microbiological removal of heavy metals from the environment: an eco-friendly approach. In: Nanobiotechnology for green environment. CRC Press, pp 139–164
50. Pedescoll A, Sidrach-Cardona R, Sánchez JC, Carretero J, Garfi M, Bécares E (2013) Design configurations affecting flow pattern and solids accumulation in horizontal free water and subsurface flow constructed wetlands. Water Res 47:1448–1458. https://doi.org/10.1016/j.watres.2012.12.010
51. Wu J, Li Z, Wu L, Zhong F, Cui N, Dai Y, Cheng S (2017) Triazophos (TAP) removal in horizontal subsurface flow constructed wetlands (HSCWs) and its accumulation in plants and substrates. Sci Rep 7:1–8. https://doi.org/10.1038/s41598-017-05874-0
52. Bastviken SK, Weisner SEB, Thiere G, Svensson JM, Ehde PM, Tonderski KS (2009) Effects of vegetation and hydraulic load on seasonal nitrate removal in treatment wetlands. Ecol Eng 35:946–952. https://doi.org/10.1016/j.ecoleng.2009.01.001
53. Li Y, Wang S, Li Y, Kong F, Xi H, Liu Y (2018) Corn straw as a solid carbon source for the treatment of agricultural drainage water in horizontal subsurface flow constructed wetlands. Water (Switzerland) 10. https://doi.org/10.3390/w10040511
54. Elsaesser D, Blankenberg AGB, Geist A, Mæhlum T, Schulz R (2011) Assessing the influence of vegetation on reduction of pesticide concentration in experimental surface flow constructed wetlands: application of the toxic units approach. Ecol Eng 37:955–962. https://doi.org/10.1016/j.ecoleng.2011.02.003
55. Singh RP, Wu J, Fu D (2019) Purification of water contaminated with Hg using horizontal subsurface constructed wetlands. Environ Sci Pollut Res 26:9697–9706. https://doi.org/10.1007/s11356-019-04260-9
56. Caselles-Osorio A, García J (2007) Impact of different feeding strategies and plant presence on the performance of shallow horizontal subsurface-flow constructed wetlands. Sci Total Environ 378:253–262. https://doi.org/10.1016/j.scitotenv.2007.02.031
57. Zhang DQ, Gersberg RM, Zhu J, Hua T, Jinadasa KBSN, Tan SK (2012) Batch versus continuous feeding strategies for pharmaceutical removal by subsurface flow constructed wetland. Environ Pollut 167:124–131. https://doi.org/10.1016/j.envpol.2012.04.004

Constructed Wetlands: Insights and Future Directions in Sustainable Approach for Wastewater Treatment

Varun Shukla, Isha Hiwrale, Rita S. Dhodapkar, and Sukdeb Pal

Abstract In recent years, the presence of contaminants of emerging concern (CECs) has raised serious concerns as they pose significant threats to water quality, aquatic ecosystems, and human health. To address this issue, various treatment technologies have been employed to remove CECs from wastewater. Among these technologies, constructed wetlands have gained popularity as an effective and sustainable approach for remediation. Constructed wetlands (CWs), an engineered system designed employing combination of wetland soils, plants, and associated microbial communities emerged as vital tools for treating different types of wastewater including domestic, mining, industrial, and agricultural effluents. In CWs, selection of plant, microbial media, Hydraulic Retention Time (HRT), and Hydraulic Loading Rate (HLR) play crucial roles in the removal of CECs through mechanisms such as filtration, sedimentation, adsorption, volatilization, microbial activity, and phyto-accumulation along with phytoremediation. Wetlands are categorized into subsurface flow and free water surface types, with the aim of optimizing the elimination of organic pollutants from wastewater. Constructed wetlands exhibit superior performance in removing organic nutrients, heavy metals, pharmaceuticals, and personal care products. Notably, these systems demonstrate higher efficiency when operating under low hydraulic loading rates (HLR). This chapter provides a comprehensive overview of constructed wetlands and their significance in phytoremediation, factors affecting the system's removal efficiency, and highlights the potential of this eco-friendly approach for environmental remediation. It serves as a valuable resource for researchers, professionals, and decision-makers, offering information on the current

V. Shukla · I. Hiwrale · R. S. Dhodapkar · S. Pal
Academy of Scientific and Innovative Research (AcSIR), Ghaziabad 201002, India
e-mail: rs_dhodapkar@neeri.res.in

V. Shukla · I. Hiwrale · S. Pal (✉)
Wastewater Technology Division, CSIR-National Environmental Engineering Research Institute, Nagpur 440020, India
e-mail: s_pal@neeri.res.in

R. S. Dhodapkar
CSIR-National Environmental Engineering Research Institute, Nagpur 440020, India

M. M. Ghangrekar et al. (eds.), *Biological and Hybrid Wastewater Treatment Technology*, Earth and Environmental Sciences Library,
https://doi.org/10.1007/978-3-031-63046-0_13

state of the art and potential future directions for effective and sustainable pollution control.

Keywords Constructed wetland · Phytoremediation · Contaminants of emerging concerns · Hydrological · Biological · Hydraulic retention time (HRT) · Hydraulic loading rate (HLR)

1 Introduction

1.1 Background

The current global scenario is marked by pressing water environmental challenges, including water pollution, scarcity, and degradation of water resources [1]. The situation is exacerbated by a surge in environmentally detrimental activities and a burgeoning population, especially in developing nations. Urbanization, industrial growth, and population expansion have significantly escalated water consumption rates in cities, exerting unprecedented pressure on water resources. This surge poses challenges like water scarcity and effectively managing water pollution [2]. Furthermore, extensive use of water for both industrial and domestic purposes has resulted in a substantial increase in wastewater production. The decontamination of wastewater to meet local regulatory water quality standards before discharge into water bodies or reuse for non-potable purposes remains a significant challenge. Because of the increase in population, there has been a noticeable increase in the need for wastewater reclamation [3]. The environment and human health are at risk because a significant amount of wastewater in poor countries is dumped straight into water bodies without being treated. It is imperative to address this problem since the discharge of industrial wastewater that has either not been treated at all or only partially is harmful to human health and disturbs ecological equilibrium [4]. The composition of pollutants found in wastewater, both in terms of quantity and quality, is primarily determined by their sources. Consequently, effective treatment methods are imperative, albeit challenging due to the intricate composition of industrial wastewater. Common pollutants identified in wastewater encompass nutrients (phosphorus and nitrogen), non-biodegradable organic materials (including detergents, greases, pesticides, solvents, fats, oils), heavy metals (such as Cu, Pb, Cd, Hg, Cr, and Ni), biodegradable organic materials, and microorganisms (comprising viruses and pathogenic bacteria) [5]. Levels of total nitrogen (TN), total suspended solids (TSS), and total phosphorus (TP) in municipal wastewater range from 30 to 100 mg/L, 250 to 600 mg/L, and 6 to 25 mg/L, respectively. Heavy metals are typically present within the range of 0.003 to 1.0 mg/L, while microorganisms like coliforms are found at concentrations ranging from 10^{11} to 10^{13} colony forming units per milliliter (CFU/100 ml) [6].

Historically, many countries have successfully relied on traditional centralized sewage treatment systems to combat water pollution. However, these conventional wastewater treatment technologies, including processes like the activated sludge method, membrane separation and membrane bioreactors, come with significant costs and are not always practical for broad implementation in rural areas. Additionally, as water and wastewater treatment standards become increasingly stringent, these methods may prove limited and insufficient [7].

In light of these challenges, the importance of identifying cost-effective and efficient alternative treatment technologies, particularly in developing regions, cannot be overstated. CWs have emerged as an appealing option for wastewater treatment due to their lower costs and reduced maintenance and operation requirements [8]. The concept of CWs is believed to have originated in Germany. In CWs, wastewater undergoes treatment as it flows through a connected system of plant roots, porous media, and attached biofilms. This intricate network allows for the removal of pollutants through various chemical, physical, and biological processes [9, 10]. As a result, they are garnering significant attention as a viable and sustainable solution. This book chapter delves into various techniques, encompassing both biological and physico-chemical methods, commonly employed in industrial wastewater treatment. Notably, biological methods stand out for their energy efficiency and relatively low operational costs. The chapter explores these methods in-depth, aiming to shed light on sustainable solutions for wastewater treatment and its vital role in mitigating environmental and public health risks.

1.2 Overview of Various Treatment Technologies

Typically, traditional wastewater treatment involves employing a mix of biological, physical, and/or chemical processes and activities to eliminate solids such as colloids, soluble pollutants, organic substances, (organics, metals, etc.) and nutrients from wastewater [11]. Chemical Precipitation involves the uptake of pollutants and separation of the resulting products. It is characterized by its technological simplicity, making use of straightforward equipment. This integrated physicochemical process proves to be both economically advantageous and efficient, particularly when dealing with high pollutant loads. It excels in the removal of metals and fluoride, although it is not selective to specific metals. The process significantly reduces chemical oxygen demand (COD) but necessitates chemical consumption, including lime, oxidants, and H_2S [12]. Monitoring the effluent's pH is crucial. However, it is ineffective at removing metal ions when they are present at low concentrations. When dealing with complexed metals, an oxidation step becomes necessary. A drawback of chemical precipitation is the substantial production of sludge, leading to challenges in its handling and disposal, including management and cost considerations [13].

Coagulation/Flocculation also involves the pollutants uptake and separation of the resultant products. It is characterized by its process simplicity and integration of physicochemical aspects. The process benefits from a wide array of commercially available chemicals and boasts an inexpensive capital cost [14]. Coagulation/ flocculation is highly efficient in dealing with SS and colloidal particles, showcasing excellent dewatering properties and sludge settling. Moreover, it leads to a notable reduction in COD, total organic carbon, BOD, and organic halogens which can be adsorbed (especially in the paper and pulp industry). The technique also demonstrates efficient removal of insoluble contaminants like pigments. However, coagulation/flocculation necessitates the addition of non-reusable chemicals such as aid chemicals, coagulants, and flocculants. Monitoring the pH of effluent is crucial, and the process results in an increased volume of sludge, posing challenges in management, treatment, and cost. Additionally, it exhibits lower effectiveness in removing arsenic [15].

Biological methods encompass various bioreactors such as Biological Activated Sludge (BAS), enzymatic decomposition, microbiological treatments, and lagoon systems. These methods involve the use of biological cultures, either mixed or pure, for the organic contaminants' biodegradation. They are celebrated for their simplicity, cost-effectiveness, and widespread acceptance by the public. A wide array of species, including consortiums of mixed cultures and pure cultures like white-rot fungi, are utilized. White-rot fungi, in particular, produce a range of extracellular enzymes with high biodegradability capacity. These biological methods effectively eliminate biodegradable organic matter, NH_3, NH_4^+, and iron. They are also proficient at attenuating color, achieving high removal of BOD and SS, especially in the case of BAS [16]. The future of water treatment technologies for removing emerging contaminants greatly relies on microbiological processes. However, creating an optimally favorable environment is essential for the success of biological methods. Maintenance and management of microorganisms and potential physicochemical pre-treatment are often required, especially when dealing with non-degradable or toxic compounds. The processes can be slow due to kinetic challenges, and some molecules, such as certain dyes, exhibit low biodegradability. BAS may face issues like poor decolonization, sludge bulking, foaming, and the generation of uncontrolled degradation products and biological sludge. Due to the complexity of microbiological mechanisms and the possibility that mixed cultures' composition may alter during decomposition, a thorough understanding of the enzymatic processes regulating substance degradation is essential [17].

Membrane Filtration encompasses various techniques like Emulsion Liquid Membranes (ELM), Ultrafiltration (UF), Electro-electrodialysis (EED), Microfiltration (MF), Electrodialysis (ED), Reverse Osmosis, Nanofiltration (NF), Dialysis, and Supported Liquid Membranes. In order to separate materials without causing damage, these techniques employ semipermeable barriers. They obtain from a large selection of commercial membranes that are offered by several manufacturers, with applications encompassing numerous industries and flexible module configurations [18]. Advantages of membrane filtration includes minimal space requirements, simplicity, rapid and efficient processes, even at high concentrations, and the production of high-quality treated effluent without the need for additional chemicals. Additionally, it generates low solid waste, effectively eliminating dyes, salts, mineral derivatives, particles, suspended solids, microorganisms, nonvolatile and volatile organics, dissolved inorganic matter, phenols, cyanide, and zinc. It also provides the possibility of being metal-selective and offers a wide range of practical applications

[19]. However, there are challenges associated with membrane filtration, such as high investment costs, substantial energy requirements, the need for tailored system designs, high maintenance and operation expenses, high concentrations can cause rapid membrane fouling, limited flow rates, low throughput, and limited efficiency at low solute feed concentrations. The choice of membrane depends on the specific application, such as reduction in hardness, total or particulate organic carbon removal, or production of potable water, and addressing the issue of concentrate disposal [20].

Adsorption/Filtration involves the use of solid materials like Commercial Activated Alumina (CAA), Commercial Activated Carbons (CAC), mixed materials, sand, and silica gel. This process is nondestructive and utilizes a solid material, allowing for adaptability to various treatment formats. It is technologically simple, requiring basic equipment, and offers a wide array of commercial products to choose from. Advantages of this method include its high effectiveness with fast kinetics, producing an excellent quality of treated effluent [21]. It can eliminate contaminants globally, especially when utilizing CAC, which is highly efficient and effective for a broad range of pollutants, particularly refractory molecules. The CAC is efficient for the removal of COD and shows exceptional treatment efficiency when coupled with coagulation to reduce COD, color, and SS. Sand is efficient for SS and turbidity removal, while alumina is efficient for the removal of fluoride. However, there are challenges, including relatively high investment costs for CAC and material expenses for both CAC and CAA. The processes are non-selective, and their performance is contingent on the specific material used [22]. Multiple types of adsorbents may be needed, and chemical derivatization may be necessary to enhance the adsorption capacity. Reactors can experience clogging and rapid saturation, making regeneration expensive. Additionally, the method may not be efficient with some metals and certain dyestuffs, necessitating the adsorbent elimination through incineration, replacement, or regeneration which can be economically non-viable for certain industries like paper and pulp and textiles [23].

Chemical Oxidation involves the use of oxidants like ClO_2, O_3, H_2O_2, Cl_2, and $KMnO_4$ and is an integrated physicochemical process. It is known for its simplicity, rapid efficiency, and the capability to generate ozone on-site, avoiding storage-related hazards. This process ensures high-quality treated water by effectively destroying pollutants and significantly reducing color. Ozone, in particular, excels in eliminating odor and color. It is efficient for removing cyanide and sulfide and initiates azo bond cleavage in the case of hypochlorite treatment [24]. Additionally, chemical oxidation enhances the biodegradability of the treated products, boasts high throughput, generates no sludge, offers water recycling potential, and provides disinfection against bacteria and viruses. However, chemical oxidation requires the use of chemicals and involves the transport, production, and management of oxidants (except for ozone). The pre-treatment is often necessary, and process's effectiveness is influenced by the type of oxidant used. Ozone has a short half-life, and some dyes may require high ozone doses due to their resistance to treatment. The process may lead to the formation of unknown intermediates and may not significantly reduce COD values or impact salinity (in the case of ozone). Hypochlorite treatment can release aromatic amines and volatile compounds and generate sludge [25].

Various techniques fall under the umbrella of conventional methods, established recovery processes, and emerging removal methods includes Aerated lagoon, Aerobic oxidation pond, Trickling filter (TF), Activated sludge process (ASP), Up flow anaerobic sludge blanket reactor (UASB), Membrane bioreactor (MBR), Sequential batch reactor (SBR), Anaerobic baffled reactor (ABR), Fluidized bed bioreactor (FBB), Expanded granular sludge bed reactor (EGSB), Rotating biological contactor (RBC Fixed-film bioreactor (FFB), etc. [26]. Each of these treatment methods varies in terms of land area requirements, expenses, payback periods, and efficacy in pollutant removal. Among the aforementioned technologies, constructed wetlands (CWs) have emerged as a cost-effective and sustainable wastewater treatment solution that demands less in terms of operational and maintenance efforts [27]

1.3 Constructed Wetlands Application in Various Wastewater Remediation

CWs have demonstrated their effectiveness in treating a broad spectrum of wastewater, ranging from very low COD levels to 20,000 mg/l, and Nitrogen (N) concentrations reaching up to 4000 mg/l. This technology has gained widespread adoption in various countries, including Australia, Czech Republic, Austria, China, Hungary, Thailand, India, Finland, Argentina, Mexico, Italy, Greece, Sweden, Germany, Denmark, Brazil, the Netherlands, and many parts of the United States [28]. Notably, in recent years, the application of CWs has been expanding to address diverse wastewater sources, including industrial, farm, and agro-industrial effluents [29–32]. This evolution underscores the adaptability and versatility of CW treatment systems in handling a wide range of wastewater compositions and underscores their role in sustainable wastewater management practices worldwide.

Municipal Wastewater

The primary and widely adopted use of CWs technology is in treating domestic or municipal wastewater. Municipal wastewater typically contains various pollutants such as salts, Biochemical Oxygen Demand (BOD), diverse hydrocarbons including suspended solids (SS), pharmaceuticals and personal care products, a variety of heavy metals, and coliforms, making it a common and essential application area for CW technology. Typically, municipal wastewater exhibits concentration ranges for various parameters, including TP at 6–20 mg/l, BOD at 100–300 mg/l, COD at 500 mg/l, TSS at 100–350 mg/l, Total Dissolved Solids (TDS) at 350–1200 mg/l, TN at 20–85 mg/l, and Ammonium Nitrogen (NH_4^+–N) at 25 mg/l [33]. CW have been achieved significant reductions in wastewater parameters, including a BOD reduction of 60–93%, COD reduction of 34–93%, TSS reduction of 74–98%, nitrogen reduction of 14–82%, and phosphorous reduction of 35–82% [28]. CW can provide a cost-effective alternative to conventional treatment methods, yielding similar results in reducing BOD, COD, TSS, nitrogen, phosphorous, and pathogens in wastewater.

Industrial Wastewater

The treatment of industrial effluents from a variety of industries, including distilleries, cork boiling, petrochemicals, coffee, potato, and seafood processing, acid mine drainage, tanneries, dairy parlours, slaughterhouses, steel production, coke ovens, pulp and paper mills, oil fields, floriculture, textile manufacturing, and agro-food production, has been successfully accomplished using CW technology [28]. Industrial effluents typically exhibit distinctive characteristics, such as high organic loads, elevated COD, extreme acidity or alkalinity, coloration, turbidity, nutrient loads, TSS, high salt content, colloidal substances, and specific toxic contaminants [34]. CW technology has been shown to achieve results comparable to biological tertiary and conventional treatment systems, with reductions exceeding 75% in COD and BOD, a 50–60% decrease in organic matter, and significant reductions in turbidity and colour in industrial effluent. Furthermore, CW systems offer effective means for controlling the alkalinity and pH of industrial wastewater [35].

Farm and Agricultural Wastewater

Farm and agricultural wastewater arising from various farming activities necessitates prior treatment before being released into the environment. Common contaminants found in agricultural wastewater include organic matter, nitrogen, phosphorus, total solids, fats, grease, oils, and pathogens [35, 36]. The primary objective of treating this wastewater is to eliminate organic and biodegradable substances, excessive nutrient levels, and certain pesticides and herbicides. CWs are predominantly used in agricultural and farm wastewater treatment to reduce the high levels of organic matter and nutrients associated with it, ensuring compliance with discharge standards. Especially in remote farm areas where there may be no connection or just limited access to traditional wastewater treatment facilities, CWs constitute a practical treatment option [37]. Some reviews focusing on the utilization of CWs for livestock wastewater sources in North America revealed that average decrease in removal efficiencies of 65% for BOD_5, 48% for NH_4^+–N, 53% for TSS, 42% for TP, and 42% for TN, depending on influent concentrations and hydraulic loading rates. Dairy wastewater showed the highest reduction in pollutant concentration following CW treatment [29].

Another study revealed that the highest pesticide removal rates were observed in CWs for organochlorine, strobilurin/strobin, organophosphate, and pyrethroid pesticide classes, while triazinone, aryloxyalkanoic acid, and urea group pesticides had lower removal rates [38]. CWs have also proven effective in treating effluents containing a variety of veterinary drugs. These applications of CWs underscore their significance in mitigating pollution in agricultural wastewater and addressing the unique challenges posed by this type of effluent [39].

1.4 Advantages of Using Constructed Wetlands for Wastewater Treatment

CWs stand as an ecologically sound alternative to conventional wastewater treatment methods like activated sludge systems. While conventional systems have been in use for over a century, relying on centralized facilities and extensive sewer networks, they come with substantial construction, operation, and maintenance costs [40]. These conventional facilities, often hidden away from residential areas, have an industrial look and require significant mechanical equipment, such as pumps and ventilators for aeration. Additionally, they extensively use non-renewable materials like concrete and steel. In contrast, CWs represent decentralized, point-of-use treatment systems that offer multiple advantages over conventional approaches. CWs are easy to construct, easy to operate, and resilient to fluctuations in flow and pollutant concentrations. They are especially well suited for onsite treatment, small villages, and settlements where there is a significant demand for efficient wastewater treatment [41].

CW facilities have lower operational costs as they require minimal energy input, mainly for pumps and are often powered by renewable energy sources like solar and wind. Maintenance needs are also minimal, with typical visits occurring monthly. Moreover, specialized personnel are less necessary for CW facility operation compared to conventional plants, further reducing costs. Global experience has demonstrated that the operational costs of CW facilities may be up to 90% cheaper than those of conventional plants [42]. CWs operate without the use of chemical additives, which are commonly used in conventional treatment plants. Moreover, they do not generate any by-products, as any produced sludge remains within the system. In contrast, conventional plants produce significant amounts of excess sludge daily, which requires management, dewatering, and stabilization, often accounting for up to 50% of total facility costs. While CWs do require occasional sludge removal, this is infrequent compared to the constant generation of excess sludge in conventional plants. CWs excel in minimizing by-product production, offering a significant advantage. The occasional maintenance cost associated with sludge removal is much lower than the ongoing costs of handling excess sludge in conventional systems [43].

The comparison between CWs and conventional treatment systems reveals key differences. Conventional systems feature many complex mechanical parts, while CWs have minimal mechanical components, mainly pumps [41]. CWs typically have lower construction costs than conventional systems, especially if land is readily available. Conventional systems have high operational costs, whereas CWs have very low to almost zero operational costs [39, 41, 44]. Conventional water treatment systems use non-renewable materials like concrete and steel, whereas CWs predominantly rely on renewable energy sources with minimal energy requirements, such as solar and wind, aligning with an environmentally friendly approach. Conventional systems produce high greenhouse gas emissions, while CWs emit very low levels and can operate autonomously with minimal maintenance, requiring occasional checks. Conventional systems, while CWs. Conventional systems produce significant daily volumes of sludge, necessitating daily management and handling, may experience breakdowns due to toxic pollutants whereas CWs generate no by-products and resilience to certain toxic constituents. Conventional facilities are industrial and unattractive, while CWs offer a more aesthetically pleasing, green environment [41].

2 Constructed Wetlands

2.1 Definition and Description of Constructed Wetlands

CWs, also referred to as reed bed treatment systems, provide an energy-efficient alternative to mechanical treatment systems for reclaiming wastewater by efficiently removing pollutants and pathogens, resulting in appreciable operational cost and carbon footprint reductions [45]. The endorsement of CWs by the United Nations Water Development Report is founded on their capacity to replicate natural processes, effectively address contemporary water management concerns, and provide essential advantages crucial to all dimensions of sustainable development [46]. These engineered systems primarily composed of plants, soil, substrates, microorganisms and water utilize natural processes within a controlled environment to enhance the quality of water [47]. These natural processes encompass a spectrum of biological processes, including biodegradation and plant absorption, as well as various physicochemical processes like adsorption, chemical degradation, oxidation, volatilization, and photo degradation. This amalgamation of processes underscores the effectiveness of CWs in elevating water quality, positioning them as a valuable component in sustainable water management strategies [48].

CWs are physical structures resembling impermeable canals filled with substrate and enveloped by wetland plants, known as macrophytes, aimed at filtering wastewater. A vital aspect of CW technology lies in the capability of macrophytes to transfer oxygen from the environment to the substrates, encouraging microbial growth around their roots. This microbial activity facilitates the breakdown of harmful substances within the wastewater [49]. The efficacy of CWs in treating wastewater is evident in the reduction of various contaminants, including TSS, TDS, hardness, BOD, TN, COD, heavy metals, TP, microbial load, as well as other pollutants like pesticides, personal care products, antibiotics, and oil [50]. The principal mechanisms responsible for pollutant removal within a CW encompass biochemical transformations, adsorption, volatilization, precipitation, and the uptake of pollutants by the plants themselves.

They are also now utilized more frequently to efficiently dry sludge from wastewater treatment plants, leading to a reduction in water volume of up to 97.0% [51]. Apart from water and wastewater treatment, CWs serve as flood control systems by effectively receiving and storing excess runoff in urban regions [52]. Additionally, CWs offer numerous ecosystem services and ecological advantages, such as enriched biodiversity, heightened environmental aesthetic appeal, better air quality, carbon capture, and diminished urban heat island effects. CW systems have fewer land requirements and operating and maintenance costs than conventional wastewater treatment methods [53].

2.2 Components of Constructed Wetlands

The key components of CWs which play a vital role in successfully implementing the technology include macrophytes, substrates and biofilm [54]. The role of individual components has been explained in detail as follows.

2.2.1 Impact of Macrophyte in Constructed Wetland

Macrophytes are the larger aquatic plant that represents the vital biotic component of CW. Morphologically, macrophytes are specialized to thrive in waterlogged conditions through the development of an extensive internal lacunal system of air spaces. This system aids in the transport of oxygen to the rhizomes and roots, effectively meeting the respiratory needs of the submerged tissues. Various species of macrophytes are considered based on location, survival rates, uptake capacity, availability, productivity, and tolerance [55]. The vegetation within a CW system can be classified into the following types:

i. **Free-Floating**: These macrophytes have most of their photosynthetic parts above the water surface, while their roots extend below.
ii. **Floating-Leaved**: This category includes macrophytes rooted in the substrate, with their leaves floating on the water's surface.
iii. **Submerged**: These macrophytes grow entirely underwater, with roots either attached to the substrate or absent.
iv. **Emergent**: In this class of macrophytes, the vegetative portions of the plants emerge above the water's surface and are rooted in shallow water.

Macrophytes within CW have distinct features essential for water treatment, making them a vital design aspect (see Figs. 1 and 2). They also play a pivotal role in oxygen provision within these wetlands, particularly in the root zone, facilitated by a phenomenon referred to as radial oxygen loss (ROL). ROL is a vital mechanism as it fosters an aerobic micro-environment, expediting the removal of pollutants. In contrast, pollutant removal is less effective in anaerobic conditions, which are predominant in CW [56]. A recent study conducted a comparison between high-density (32 plants per square meter) and low-density (16 plants per square meter) CW systems. The results revealed that high-density systems achieved double the nitrogen compound removal compared to their low-density counterparts, underscoring the significance of plant presence in such systems [57]. Furthermore, the removal rates of TN and TP exhibited a positive correlation with the ROL exhibited by wetland plants. This relationship was established through a study involving different plant species [58].

Plant roots serve as a habitat for a multitude of microorganisms due to their capacity to offer surfaces for microbial attachment. Moreover, roots release exudates, a carbon-rich secretion that supports the process of denitrification, enhancing the pollutants removal efficiency in anaerobic conditions [59]. In aquatic environments, plant tissue brings about various physical effects, including a decrease in water flow velocity, facilitation of sedimentation, prevention of resuspension, and absorption of nutrients. Specifically, when roots and rhizomes are embedded in the sediment, they stabilize the sediment surface, mitigate erosion, absorb nutrients, prevent medium clogging (especially in subsurface conditions), and enhance hydraulic conductivity. On the other hand, the above-water plant tissue contributes to light attenuation, reducing photosynthesis growth. Additionally, it lowers wind velocity, stores nutrients, and adds to the aesthetic appeal of the system [60] (Fig. 2).

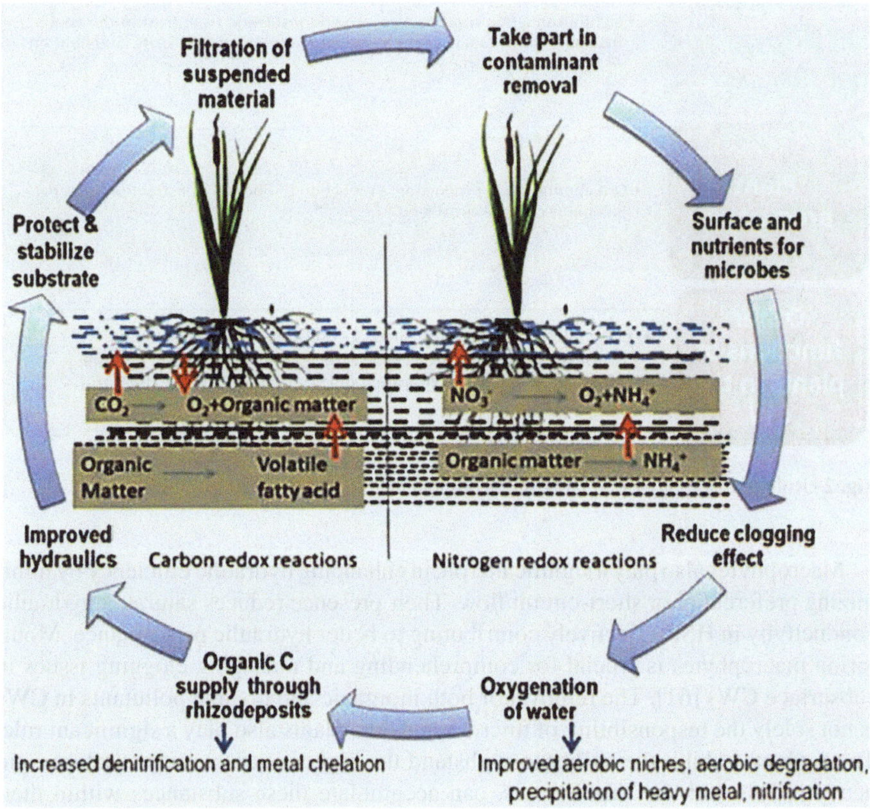

Fig. 1 Mechanism involved in macrophytes in constructed wetland. Adapted from [28]

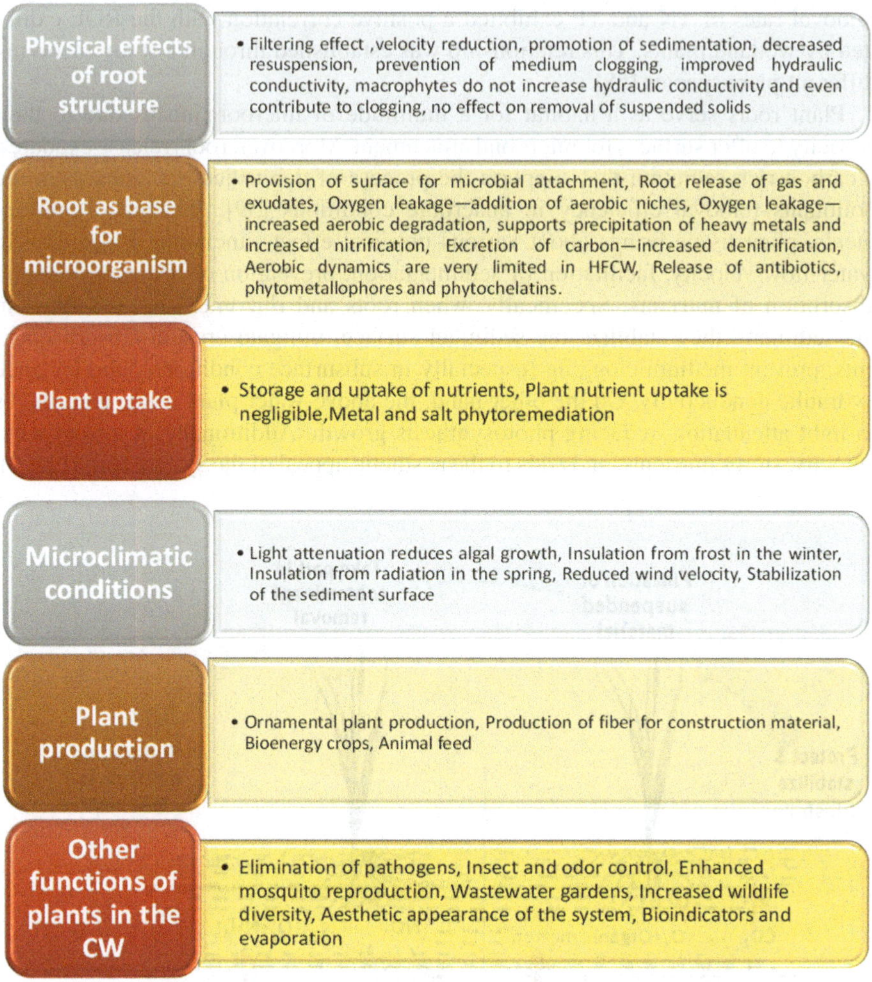

Fig. 2 Role of macrophytes on individual components of CW. Adapted from [60]

Macrophytes also play a significant role in enhancing hydraulic efficiency by minimizing preferential or short-circuit flow. Their presence reduces saturated hydraulic conductivity in HSF, effectively contributing to better hydraulic performance. Monitoring macrophytes is crucial for comprehending and managing clogging issues in subsurface CWs [61]. The removal of both inorganic and organic pollutants in CWs is not solely the responsibility of microorganisms; plants also play a significant role. These plants exhibit the ability to withstand the higher nutrient concentrations and heavy metals and, in certain cases, can accumulate these substances within their tissues. Research indicates that CW plants can remove approximately 15–32 mg of TN and 2–6 mg of TP per gram of dry mass, primarily from aboveground biomass [62].

The most commonly used emergent species include *Scirpus* spp. (Cyperaceae), *Phragmites* spp. (Poaceae), *Eleocharis* spp. (Spikerush) *Typha* spp. (Typhaceae), *Juncus* spp. (Juncaceae), and *Iris* spp. (Iridaceae). The most commonly utilised submerged plants include *Ceratophyllum demersum*, *Potamogeton crispus*, *Hydrilla verticillata*, *Myriophyllum verticillatum*, and *Vallisneria natans*. The floating leafed plants are mostly *Nymphaea tetragona*, *Marsilea quadrifolia*, and *Trapa bispinosa*. Some of the free-floating plants are *Salvinia natans*, *Eichhornia crassipes*, *Hydrocharis dubia*, and *Lemna minor* [63].

2.2.2 Impact of Substrate in Constructed Wetland

The substrate, also known as the media, serves as the fundamental and vital element supporting the entire CW system. Beyond its role as a structural foundation, the substrate significantly facilitates the growth of macrophytes and the development of biofilms. Moreover, it acts as the platform for reactions, allowing the immobilization of contaminants [34]. In a CW, the selection of bed materials offers considerable flexibility (see Fig. 3), allowing the utilization of locally available low-cost materials based on factors such as media size, texture, hydraulic conductivity, and porosity. An ideal substrate should possess specific properties, including the capacity to create a balance between aerobic and anoxic/anaerobic spaces within the matrix. This balance enhances processes like nitrification, denitrification, and removal of organic matter. Moreover, the substrate should act as an internal carbon source, fostering more efficient denitrification metabolism, especially in wastewater with low carbon content [64]. The substrates within CW operate through purification mechanisms including physical actions like ion exchange, filtration and sedimentation, sorption, precipitation, and complexation which are influenced by the composition of colloids. Moreover, these mechanisms involve transformation, microbial degradation, and bio-immobilization within the substrate. Additionally, the substrate supports metabolism and uptake by macrophytes in the wetland system [65].

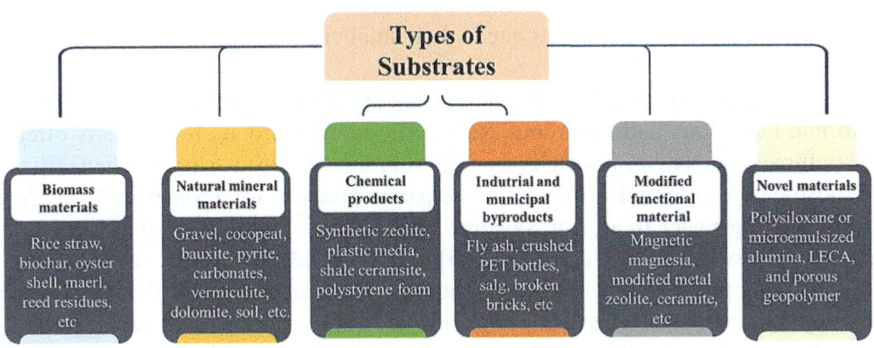

Fig. 3 Types of substrates and their environmental condition. Adapted from [10]

The physical characteristics of the substrate, such as the distribution of particle sizes, the spaces between particles, effective particle sizes, irregularity levels, and permeability coefficients, plays a vital role in influencing the effectiveness of the system [66]. On the other hand, the chemical properties of the substrate that primarily impact nutrient removal efficiency include pH, cation exchange capacity, organic matter content, and electrical conductivity (EC). The pH of the substrate directly affects the speciation, availability, and retention of nutrients and heavy metals. The cation exchange capacity of the substrate determines its ability to bind and retain nutrients [10]. Gravel, calcite, clay, sand, vermiculite, marble, fly ash, dolomite, slag, bentonite, zeolite, limestone, shell, wollastonite, light weight aggregates, and activated carbon are some of the natural, artificial and industrial byproduct material used as a substrate. The study found that natural materials, along with industrial and agricultural waste materials, were deemed more suitable and economically efficient compared to artificial materials [67]. Furthermore, a combination of substrates, such as dolomite and sand, was employed for phosphate removal. Additionally, a blend of substrates including ceramsite, vermiculite, calcium silicate hydrate, and gravel was used for the treatment of surface water with low nutrient concentrations. These mixtures of substrates provide reactive surfaces for microbial adhesion and strong hydraulic conductivity, which successfully prevents short-circuiting problems in CWs [68].

Some of the common adsorbents used as substrate in CW are gravel, soil and light expanded clay aggregate (LECA). The utilization of gravel substrate demonstrated no detrimental impact on the performance of CW. In fact, the gravel substrate proved to be highly effective in sorption-based removal of pharmaceuticals that exhibit resistance to biodegradation, particularly those with relatively high hydrophobic properties [69]. LECA serves as an effective sorbent for acidic as well as neutral organic compounds such as pharmaceuticals and personal care products. In recent times, specific biosorbents like rice husk, pine bark and granulated cork have gained attention as compelling substitutes for conventional substrate materials in CWs. This is attributed to their cost-effectiveness, potential for reuse, and simple disposal through incineration [70]. Materials sourced from the natural environment are abundant, easily accessible, and cost-effective. However, their treatment capacity is limited for high-concentration contaminants due to their smaller specific surface area and lower active constituents [71].

By-products generated from municipal and industrial sources can be categorized into non-hazardous and hazardous waste. The non-hazardous by-products offer a cost-efficient and environmentally friendly solution for waste disposal when utilized as substrates. Various industrial and municipal by-products like broken brick, residual from drinking water treatment, sawdust furnace slag, and discarded polyethylene can be effectively employed in CWs [65]. The by-products possess notable disparities in both physical attributes and chemical composition, contributing to their distinct characteristics when utilized as substrates. One example is using aggregates from basic oxygen furnace steel slag as substrates for removing phosphorus. The main removal method was precipitation of calcium phosphate, and these aggregates showed adsorption capabilities ranging from 0.12 to 8.78 mg/g [72].

2.2.3 Impact of Biofilm Development in Constructed Wetland

In CW, aquatic plants support diverse microorganisms, either as individual entities or as part of biofilms adhering to the substrate or plant surfaces [73]. Typically, a biofilm comprises a variety of microbial community, encompassing bacteria, fungi, algae, protozoa, and metazoans coexisting harmoniously within a complex and well-coordinated community, providing resilience in adverse conditions such as extreme chemical environments, nutrient scarcity, and exposure to antimicrobial agents [74]. The biofilms attached to surfaces aid in biomineralizing nutrients, which are then transported by plants to their above-ground parts via the vascular bundles and epidermis in the root zone [75]. Bacteria within biofilms produce exopolymers that facilitate their adherence to surfaces, cohesion among themselves, and protect the surrounding environment. The structure and spatial arrangement of the biofilm are initially influenced by the ratio and makeup of the bacteria, but the microbial populations within the biofilm are also influenced by the physical traits and material properties of the developing matrix, such as the availability of nutrients, shear stresses, and competing organisms [55]. In the oxygen-rich biofilm of the root zone, the prevalent microorganisms are *nitrosomonas, methanotrophs, and pseudomonas sp*. These organisms play a crucial role in aerobically breaking down the pollutants. On the other hand, in the deeper regions where oxygen levels are lower, methanogenic and sulphur-reducing bacteria dominate, adapting to the oxygen-deficient conditions [76].

The existence of plant growth-promoting rhizobacteria within the biofilm offers numerous valuable benefits to the plant. These include controlling the availability of nutrients, promoting root growth by regulating hormones, and providing protection against both pathogens and abiotic stress. Approximately 40–55% of the biofilm consists of organic matter, with the remaining portion comprising finely trapped inorganic particles. Biofilms possess the capacity to absorb contaminants using diverse mechanisms, including bioprecipitation, biosorption, redox immobilization, and intracellular accumulation [77]. Biofilms exhibit a greater rate of nutrient uptake compared to macrophytes. They assist in capturing suspended solids, a crucial aspect for the effective removal of nitrogen and phosphorus. Additionally, biofilms serve as effective reservoirs for retaining phosphates and play a vital role in facilitating denitrification and nitrification processes [78]. Variability in biofilm diversity, distribution, and composition is an intrinsic feature observed in CWs [79]. The arrangement of microbial communities within the biofilm is influenced by their location inside the system-whether in the plant roots, effluent, or substrate-as well as the specific plant species inhabiting the environment. The plant species influence the composition of bacteria in biofilms and is distinct from the composition found in the substrate [80].

To maintain consistent performance in CW, regulating the biofilm's thickness is essential. Achieving optimal nutrient uptake by biomass in CWs operational conditions involves managing biofilm growth to prevent excess while enhancing biofilm diversity [81]. Several factors influence the biofilm's optimal activity, including pH, temperature, salinity, nutrient/contaminant type and strength, presence of macrophytes, substrate surface characteristics, hydrodynamic conditions or turbulence, and the presence of divalent cations. Biofilm density tends to be highest in areas with significant carbon deposition, such as at the influent end where loading occurs. Additionally, biofilm growth may be constrained in wastewater with high colour content [82]. Microorganisms residing within biofilms can adapt to diverse local environmental conditions using multiple strategies. These include modifying physiological activities or gene-expression patterns, enhancing populations of the most well-adapted species in response to current conditions, and selecting mutant strains that are best suited for survival. Plants play a crucial role in facilitating effective biofilm activities by providing more active sites than aggregates alone. Research shows that TN removal was lower in a system utilizing only suspended biomasses (25%) and aggregate-biofilm compared to a system involving suspended biomasses (38%)and plant-biofilm [83].

In CW with submerged macrophytes, the regulation of epiphytic biofilm development, bacterial community diversities, and distribution on the leaf interface can be achieved by using water flow. This flow directly influences the availability of oxygen around leaves [84]. High nutrient loading can negatively impact the biofilm, potentially affecting CWs performance. Strategies involving changes in nutrient load and varying flow rates have been proposed to maintain a thin and active biofilm. These alterations affect the aerobic state of the system, placing stress on the biofilm. This stress decouples respiration from bacterial growth, minimizing biofilm formation while still effectively removing organic carbon [85]. Utilizing biofilm systems for wastewater treatment offers several advantages, including operational flexibility, reduced space requirements, shorter hydraulic retention time (HRT), resilience to environmental changes, extended high active biomass concentration, biomass residence time, and improved ability to degrade challenging compounds. Therefore, future research efforts aimed at comprehending biofilm community function, composition, interactions, and optimal manipulation of biofilm composition can significantly contribute to the optimization of CWs efficiency [86].

2.3 *Types of Constructed Wetlands*

CWs are typically categorized into two main types based on wetland hydrology (see Fig. 4) Subsurface flow (SSF) CWs and Free water surface (FWS) CWs. Depending on the direction of flow, SSF CWs can be further subcategorized into horizontal flow (HF) and vertical flow (VF) CWs [27]. Another design, known as hybrid CWs, combines various wetland systems for wastewater treatment. Hybrid CWs typically involve two stages of several parallel CWs in series, such as HF–VF CWs, FWS-HF CWs, VF–HF CWs, and HF-FWS CWs. Additionally, multi-stage CWs, which consist of more than three stages of CWs, have also been utilized. In recent years, various improved designs have been proposed to enhance removal processes in CWs. These include artificial aerated CWs, hybrid towery CWs, baffled flow CWs, circular flow corridor CWs, and step feeding CWs, all aiming to boost the wastewater treatment performance [87].

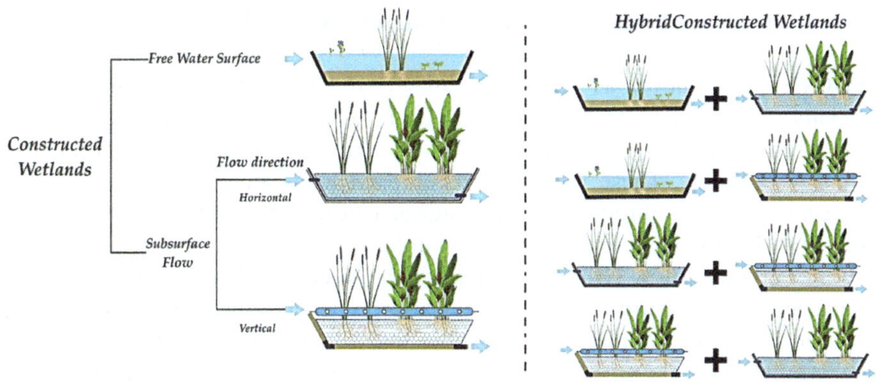

Fig. 4 Categorization of constructed wetland based on hydrology. Adapted from [88]

2.3.1 Free Water Surface Constructed Wetlands (FWS-CWs)

This setup closely emulates a typical natural wetland system, primarily comprising channels or basins where water flows shallowly over a supporting medium with saturated substrates. An impervious barrier, often made of materials like gravel (coarse or fine) or clay, or combinations thereof, lies beneath and provides attachment surfaces for emergent vegetation [89]. Simultaneously, it helps manage the percolation and contaminants seepage. Water flow is regulated by maintaining a characteristic shallow low flow velocity, water depth, and the presence of petioles (stalks) and plant litter from aquatic macrophytes [90]. These aquatic macrophytes serve several roles, including regulating hydraulic conditions, acting as hyper-accumulators for nutrient uptake and stabilizing heavy metals, serving as insulating surfaces that aid biodiversity in absorbing environmental shocks, releasing electron acceptors and root exudates, providing substrates, and offering surfaces for bacterial attachment [91].

Occasionally, the performance of FWS-CWs is enhanced by introducing plug-flow scenarios to minimize the risk of failure. FWS-CWs excel at removing organic matter (OM) and suspended solids, but their nitrogen (N) and phosphorus (P) removal capabilities are typically lower. FWS systems exhibit particular strength in TN removal compared to other types of CWs [92]. However, FWS-CWs are highly susceptible to climatic variations, particularly cold climates, which can severely limit their effectiveness. Furthermore, the decomposition of emergent vegetation, which falls and decays over time, can significantly reduce the removal efficiency in FWS-CWs. While considering structural optimization cost, FWS-CWs are described by a low construction budget, making them a popular choice for rural settlement systems [36].

2.3.2 Subsurface Flow Constructed Wetlands (SSF-CWs)

SSF-CWs, also known as vegetated submerged beds (VSB), typically involve specially constructed trenches with impermeable linings made of materials like rocks, sand, or crushed stone gravel, with a relatively uniform diameter ranges from 10 to 20 cm respectively. These trenches are equipped with saturated substrate and an emergent vegetation through which wastewater flows either horizontally or vertically [93]. In simpler terms, SSF-CWs can be categorized as either vertical flow (VF) or horizontal flow (HF), depending on the direction of wastewater flow. The choice between these types of SSF-CWs primarily depends on the specific contaminant removal goals. For instance, if the objective is to enhance oxidation of nitrogen to nitrate-nitrogen via nitrification for the uptake by the plant, vertical subsurface flow CWs (VSSF-CWs) are preferred [94]. On the other hand, when the goal is to prioritize denitrification over nitrification, converting available nitrogen species to nitrogen gas and releasing them into the atmosphere, horizontal surface flow CWs (HSSF-CWs) are often chosen [95].

The vegetation plays an important role in facilitating the transfer of oxygen to the rhizosphere and plant roots, particularly in the saturated and anaerobic subsurface zone. Unlike free water surface CWs (FWS-CWs), SSF-CWs are well-suited for addressing underground contaminants and exhibit a wide range of contaminant removal capabilities [66]. Consequently, they are highly efficient in removing suspended solids, organic matter, pathogenic microbes, and heavy metals. SSF-CWs are also less sensitive to cold climates. However, the removal of nitrogen (N) in SSF-CWs is constrained by the availability of carbon and oxygen because these systems are typically waterlogged [96]. Therefore, nitrogen removal efficiency depends on the extent of oxygen transfer to the rhizomes and subsurface zones. Furthermore, the efficiency of phosphorus removal in SSF system can be improved by incorporating high-sorption media [97].

2.3.3 Hybrid Constructed Wetlands (HF-CWs)

The hybrid-CW system, a distinct type of wetland, combines features from other wetland types (SSF or FWS and their variants) either in series or parallel at different filter stages. This approach optimizes CW system performance by leveraging type-specific functions under a single remedial design to enhance wastewater treatment [98]. Common technological combinations include HF-VF, FWS-HF, HF-FWS, VF-HF, VF-FWS, and FWS-VF, although more complex or multi-stage combinations have also been reported [99]. Among these, VF-HF is widely utilized for treating municipal sewage and, industrial wastewater, particularly excelling in NH_4^+–N removal. On the other hand, HF-VF has a narrower application and is primarily suitable for purifying municipal sewage [100]. Hybrid systems represent more intricate CWs designs with significant functional enhancements but come with a higher capital investment. For instance, when comparing domestic wastewater treatment using intensified and conventional SSF CWs, it was found that pollutant removal rates increased with design complexity in the order: HF < VF hybrid system [101]. High levels of dissolved oxygen (DO) in VF-CW systems considerably inhibit denitrification, resulting in larger concentrations of nitrate-nitrogen (NO_3^-–N) compared to ammonium-nitrogen (NH_4^+–N) than are required for macrophytes. Contrarily, the restricted oxygen diffusion capability of HSSF-CWs prevents nitrifying bacteria and archaea from oxidizing (NH_4^+–N) [102].

A traditional HSSF system loses significant volumes of flow to the atmosphere through evaporation, technically removing pollutants while concentrating them over time. A hybrid-CW system, such as HSSF-VSSF, is often adopted to address these type-specific limitations effectively. This combined system simultaneously addresses nitrification and denitrification issues, making it the most efficient CW system for TN removal from wastewater [99]. In practice, hybrid-CWs are adapted systems that merge different types of wetland systems based on the nature of the wastewater and treatment requirements. The studies have demonstrated remarkable removal efficiencies of organic content in wastewater treatment, with some achieving over 90% removal [103].

The capabilities of hybrid-CW systems have also been extended to address emerging contaminants such as antibiotics and antibiotic-resistant genes (ARGs). The substantial removal of total ARGs (from 87.8 to 99.1%) and total antibiotics (from 87.4 to 95.3%) in a four hybrid built wetland system in South China have been reported [104]. For peri-urban regions and big towns, hybrid systems offer outstanding performance capabilities, successfully treating a variety of toxins in wastewater, provided there is adequate arable land and sufficient financial resources, which are typical operating restrictions. However, the added expense of preprocessing for avoiding substrate clogging might render the technology unaffordable, deterring its adoption in rural regions with little financial resources [105].

2.4 Mechanisms Involved in Constructed Wetlands

Wastewater treatment in CWs system primarily relies on three internal components: enriched microorganisms, biofilms or vegetated plants, within the and filter media or substrate and plant root zone. These systems employ a combination of physicochemical and biological processes to effectively remove various pollutants [76]. The processes involved in pollutant removal encompass sedimentation, adsorption, filtration, chemical transformation, decomposition, biological assimilation, volatilization, and biodegradation. Sedimentation, sorption, and filtration are typically the primary mechanisms for removing solids such as TSS [106]. Nutrients like phosphorus and nitrogen are removed through both biological and physicochemical processes. In the case of nitrogen removal, multiple processes come into play, including adsorption, ammonia volatilization, mineralization (ammonification), nitrification, desorption, denitrification, anaerobic oxidation of ammonia, nitrogen fixation, as well as microbial and plant uptake [107]. The two main biological processes for nitrogen removal are nitrification and denitrification assisted by certain bacteria. Denitrification is the process of lowering nitrate to nitrogen gas, while nitrification is the process of turning ammonia into nitrate [108].

Phosphorus removal primarily relies on physicochemical processes like adsorption/desorption and chemical precipitation. Plants can also take up phosphorus through their roots and store it within various plant parts [109]. Unlike nitrogen, the occurrence of phosphate accumulating organisms (PAOs) in wastewater allows for enhanced biological phosphorus removal (EBPR). Recent research has demonstrated that aerobic PAOs (APAOs) and denitrifying PAOs (DPAOs) play essential roles in this process [110]. Sedimentation is the primary mechanism for heavy metal removal, whether in natural systems or CW setups. Organic pollutant removal involves a variety of biological processes, including anaerobic processes such as sulfate reduction, fermentation, and methanogenesis and aerobic respiration (biodegradation) [111]. Plants can also aid in the removal of organic pollutants by taking them up and metabolizing them. Aerobic and anaerobic organic degraders, such as Zobellella, Thauera, Aeromonas, Proteobacteria, Pseudomonas, Methanobacteriales, Methanomicrobiales, and Methanosarcinales, play crucial roles in these processes [112]. Microbial contaminants in CWs are eliminated through filtration, sedimentation, natural die-off, and adsorption due to factors like starvation and predation by higher organisms [47]. The possible mechanisms to eliminate both microbial and chemical pollutants from wastewater are outlined (see Fig. 5).

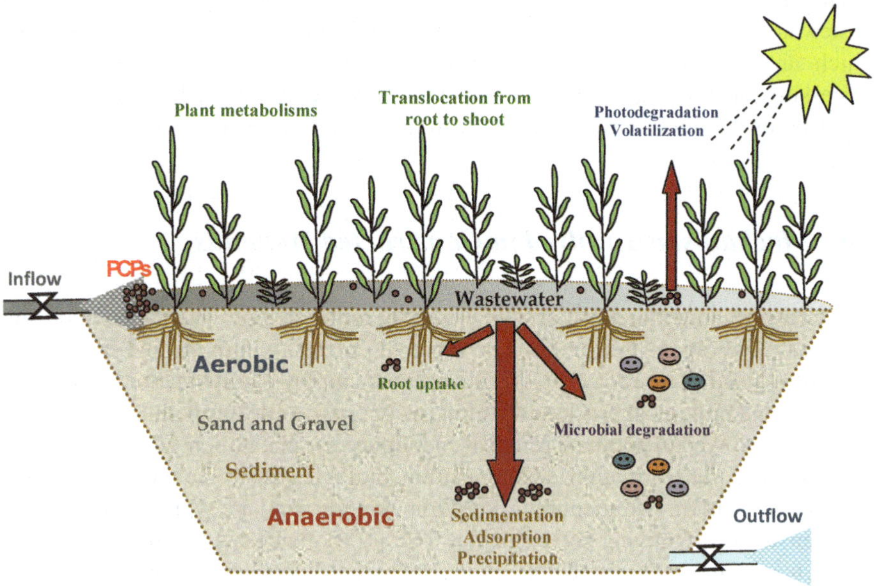

Fig. 5 Various mechanism involved in constructed wetland. Adapted from [113]

3 Efficiency of Constructed Wetlands in Removing Specific Contaminants

3.1 *Removal of Organic Nutrients*

CWs are efficient systems for sewage treatment, employing shallow gravel-filled beds planted with emergent macrophytes. These plants, along with attached aerobic bacteria on porous media and roots, facilitate organic matter removal in SFCWs [114]. Reed roots provide a surface for bacterial growth, supply oxygen, and aid bacteria. Nitrogen removal occurs through bacterial processes, plant uptake, and adsorption, where ionized ammonia reacts with SF wetland media. Additionally, volatilization transforms ammonia into free nitrogen [115]. Phosphorus removal in SF wetlands is the result of bacterial removal, plant uptake, media adsorption, and precipitation reactions with minerals like ferric oxyhydroxide and carbonate. Bacterial processes and plant uptake remove ($P–PO_4^{3-}$), while adsorption and precipitation eliminate all phosphorus forms [116]. HFCWs offer reliable secondary treatment for BOD and TSS but may lack efficiency in nitrogen removal, unless longer retention times and sufficient oxygenation are provided [117]. In contrast, VFCWs are highly effective for BOD, TSS, and nitrification even at higher loading rates [118]. Therefore, VFCWs are becoming more popular [119] and currently they are subject to intensive research in order to optimize their basic design parameters [120]. Wetland plants are vital components of these systems, playing crucial roles in degrading and removing nutrients and pollutants. Through microbial degradation, plant uptake, substrate adsorption, media filtration, and biological predation, a variety of pollutants such as COD, BOD, total phosphorus (TP),TSS, total coliforms, total keljdahl nitrogen (TKN), and metals are removed from wastewater in CWs [121]. Many CWs exhibit high removal efficiencies, surpassing 80% for BOD and TSS. Different studies involving uses of diverse plant species in small scale HFCWs and VFCWs have shown substantial pollutant removal, although challenges remain with TSS, NO_3, and TKN [122].

3.2 *Heavy Metals Removal*

In developing countries, industrialization has led to environmental contamination by heavy metals (HMs), posing challenges due to the lack of advanced and affordable treatment technologies. CWs offer a cost-effective and promising solution for treating industrial effluents. Past research has highlighted CWs' effectiveness in HM removal [123–125]. The previous study focused on a Free Surface Flow Wetland comprising 7 cells, covering an area of 4145.71 m^2, with a total storage capacity of 1305.58 m^3 and a 40-h retention time from inlet to outlet. This CW demonstrated significant HM removal, with Cd > Cr > Fe > Pb > Cu > Ni order of effectiveness [126].

Alternatively, certain research findings propose that phytoremediation could serve as a practical solution for eliminating dioxane from polluted soils. This method should be explored for other water-loving contaminants, and plants with the ability to absorb industrial heavy pollutants in CWs should also be taken into account [127]. CWs systems were effectively employed to address pollutants found in airport runoff, such as heavy metals and glycols, as well as the treatment of stormwater runoff from a residential town-home complex in northern Virginia. These systems were well-designed and successfully reduced the levels of pollutants to minimal levels [128]. CWs utilize solubilization and reprecipitation processes to eliminate metals like Fe, Cu, Zn, Mn, and Al from wastewater [129]. In constructed wetlands, specific plant species like Water Hyacinth demonstrated the capacity to eliminate as much as 95% of bioavailable mercury within the wetland system over a three-day period. These plants have the ability to absorb the ionic form of mercury present in wastewater, and there is a consistent pattern of mercury accumulation in the plant roots [130]. In thatstudy, authors focused on mercury removal, with an initial mercury concentration of 50 ppb in the solution, Water Hyacinths exhibited remarkable efficiency. The average mercury content in Water Hyacinth roots was 3.5 times higher than that in Reeds. Within the first three hours, Water Hyacinth roots accumulated 110.55 mg/g of mercury, significantly surpassing the 28.9 mg/g accumulated in Reeds roots [131]. Recent studies have integrated artificial intelligence methods into CWs to enhance the uptake of highly bioavailable mercury concentrations, specifically in Water Hyacinth. This optimization involved adjusting certain factors: initial total mercury concentration ranging from 1×10^{-4} to 1×10^{-3} mol/l, chloride concentration between 1×10^{-8} and 1×10^{-6} mol/l, and pH values maintained between 5.36 and 6.5. By adhering to these specified conditions, the anticipated bioavailable mercury concentration to be achieved is 6.7×10^{-6} mol/l. These conditions are regarded as optimal removal parameters, ensuring higher bioavailable mercury uptake by plants [132].

3.3 Removal of Contaminants of Emerging Concerns

CWs have demonstrated their viability as a cost-efficient alternative technology to traditional wastewater treatment plants (WWTPs) for small communities. In addition to addressing standard wastewater quality parameters like BOD, TSS, phosphorus and, nitrogen. CWs have exhibited the potential to eliminate emerging contaminants of concern [133]. Contaminant removal in CWs occurs due to intricate microbial and physico-chemical interactions. The efficiency of these processes relies on multiple design and operational factors, including bed depth, substrate composition, hydraulic and organic loading rates, feeding methods, and artificial aeration. CWs are often designed using a black box approach, but omitting certain parameters can decreased treatment effectiveness [134, 135].

In terms the removal of emerging contaminants, initial studies indicated that CWs exhibit comparable or even higher efficiency in removing Pharmaceutical and Personal Care Products (PhACs) when compared to traditional wastewater treatment plants (WWTPs) [70, 136]. Over 60% of Pharmaceutically Active Compounds

(PhACs) have been successfully removed in CWs functioning as primary or secondary treatment for municipal wastewater and hybrid CW systems. Typically, these compounds exhibit removal rates ranging from 80 to 100%. However, certain PhACs resist degradation, with overall removal rates dropping below 60%, falling below 25% in specific cases, particularly in single-stage CW systems or hybrid CW setups [136, 137]. CWs employ intricate and diverse removal mechanisms, including sorption, photodegradation, sedimentation, volatilization aerobic biodegradation, anoxic and anaerobic biodegradation, phytoremediation [70, 137, 138]. Hence, optimal outcomes are obtained through hybrid systems that integrate various aerobic and anaerobic environments and exposure to light [139]. The enhanced removal of certain Pharmaceutical Compounds (PhCs) in Horizontal Constructed Wetlands (HCW) is attributed to the simultaneous presence of anaerobic and aerobic conditions and the extended HRT in HCW. This environment boosts the elimination of substances like diclofenac, acetaminophen, sulfamethoxazole, sulfapyridine, trimethoprim, and atenolol. These compounds are effectively removed through both anaerobic and aerobic conditions, along with sorption or adsorption processes [140]. In CWs, certain Pharmaceutical Compounds (PhCs) are eliminated through the formation of their Transformation Products (TPs). The removal of PhCs like diclofenac, ketoprofen, ibuprofen, tramadol, naproxen, carbamazepine, sulfamethoxazole, and venlafaxine via biotransformation pathways underscores the significance of the TPs' characteristics in PhC removal [141, 142].

4 Effect of Hydrological and Chemical Factors on Constructed Wetlands

Hydrology is a fundamental determinant of wetland functionalities, and managing flow rate is crucial for achieving optimal treatment efficiency [143]. Hydraulic Retention Time (HRT), sometimes referred as hydraulic residence time, represents the average duration a soluble compound remains within a constructed reactor [51]. HRT can be mathematically expressed using following equation:

$$\text{HRT} = \frac{V}{\theta * Q}$$

where: V represents the reactor volume; θ denotes the amount of feed inside the reactor, and X signifies the number of cycles per day.

Volume is typically measured in cubic meters (m^3), influent flow rate is measured in cubic meters per hour (m^3/h). Adjusting the flow rate and ensuring an appropriate HRT are vital strategies for effectively managing the treatment performance [144].

Hydraulic Loading Rate (HLR) in a CWs refer to the volume of wastewater or influent applied to a specific area of the wetland surface over a given period of time. It is a critical parameter in designing and operating CWs for wastewater treatment. Mathematically, HLR is calculated using the following formula:

$$\mathrm{HLR} = \frac{A}{Q}$$

where: Q is the volume of wastewater or influent applied to the constructed wetland per unit of time (e.g., cubic meters per day); A is the area of the wetland over which the influent is applied (e.g., square meters) [145].

Essentially, HLR indicates how much wastewater is being distributed across a given area of the constructed wetland. It helps in determining the appropriate size of the wetland and ensuring that the wastewater is evenly distributed, allowing for effective treatment and absorption by the substrate and plants within the wetland [146]. Optimal HLR is critical to avoid overloading the system, which could lead to inadequate treatment and potential environmental issues. Conversely, too low in HLR may result in underutilization of the wetland's treatment capacity. Balancing HLR with the wetland's design, substrate characteristics, vegetation, and local climate is essential for achieving efficient and effective wastewater treatment in a CWs [147].

4.1 Effect of HLR and HRT in Improving Removal Efficiency

The effective design of HLR and HRT is a critical factor in determining the CWs removal efficiency [148]. A higher HLR encourages a faster flow of wastewater through the media, which, in turn, reduces the ideal contact time for treatment. Conversely, a longer HRT allows for the development of a suitable microbial community within CWs, ensuring sufficient contact time to effectively remove contaminants [149]. When the loading rates are too high, it results in increased loads of pollutants and shorter the retention times. Consequently, this can cause an accumulation of organic matter and a reduction in effective void space, ultimately leading to a decrease in removal efficiencies reported in recent studies [150]. Also, the increase in HLR notably hinders the removal process of organic pollutant such as pharmaceuticals and personal care products (PPCPs). This occurs because microbial consortia tend to prioritize the consumption of sucrose in the added loads as an energy source over targeting the pollutants of interest [151]. The removal efficiency of ibuprofen, a nonsteroidal anti-inflammatory drug (NSAID), decreased with increase in rate of HLR in various types of CW which has been reported in the recent research article [152].

A noticeable trend was observed in hybrid subsurface flow-based constructed wetland (HSF-CW) systems treating a mix of stormwater and wastewater.,. As the HLR gradually increased (31, 62, 104, and 146 mm/day), both TN and TP removal consistently decreased. Specifically, removal efficiency of TN dropped from 84 to 16%, and TP decreased from 99 to 72% in response to the increase in HLR [153]. In a pilot-scale plant of HSF-CW system, efficient removal of TP, COD, and TN was noted, reaching 95% for each at an HLR of 0.025 m/day. However, when the HLR doubled to 0.05 m/day, the removal efficiency experienced a decline to 89%, 91%, and 87% for TP, COD, and TN respectively [154]. Whereas it has also been found that increase in HLR increases the rate of removal of nitrogen and organic matter. As the HLR rises, there is a gradual increase in the average effluent concentrations of biological oxygen demand (BOD_5), COD, and Total Kjeldahl Nitrogen (TKN). Interestingly, this rise in HLR did not significantly affect TSS removal [155, 156].

Increased HRT, will increase the contact duration between wastewater and microorganisms in the rhizosphere which has been found to enhance the interaction between wastewater and constructed wetland components, resulting in improved removal efficiency [157]. It has been found that the efficiency of removing COD, TSS, nutrients, BOD, pharmaceuticals, and estrogen typically improves as the HRT increases. This is because a longer HRT notably amplifies the interaction among plants, water, and soil. Nevertheless, there exists a point beyond which further extension of the HRT has marginal impact on pollutant removal efficiency [141, 158]. Longer HRTs, on the other hand, often necessitate greater land sizes and higher capital expenses [159]. The study demonstrated a notable linear relationship between HRT and the removal efficiency of pharmaceuticals like diclofenac, ibuprofen, caffeine, and naproxen in wastewater. The findings reveal that maintaining HRT of over eight days was effective in achieving substantial removal of organic matter up to 91.9%. HRT emerged as a key factor influencing the efficiency of TN removal, indicating that an 8-day HRT was satisfactory for achieving nitrogen removal up to 80 to 90%. Likewise, in the case of TP and $P–PO_4^{3-}$, an HRT of 8 days proved sufficient for temperatures exceeding 15 °C, while lower temperatures necessitated a more extended HRT of either 14 or 20 days [160].

Additionally, the influence of HRT depends on the dominant plant species and temperature, as these factors can also affect the efficiency of hydraulic of wetlands [161]. One of the study revealed that in a series of connected CWs where *Typha sp.* and *Scirpus sp.* were cultivated, a prolonging HRT of 10 days led to a superior removal efficiency of COD at 80%, compared to a shorter HRT of 5 days resulting in 76% COD removal [162]. However, HRT did not directly affect the removal of COD and TSS in a CW treating effluent from a fish farm, even at a surface load of around 1–5 l min^{-1}. Nevertheless, a negative correlation has been noted between HRT and the removal efficiency of TN and TP. A reduced in HRT may lead to incomplete denitrification of wastewater. It has been documented that achieving effective nitrogen removal necessitates an extended HRT compared to what is needed for organic removal [163]. Moreover, the effectiveness of COD removal can be impacted by high organic loading rates. For instance, in a system treating diluted pig wastewater with Vetiveria zizanoides as plant species, a decrease in COD removal efficiency from 90 to 80% was found with a twofold increase in organic loading rate (18 mm d^{-1}) [28]. Furthermore, studies have indicated that the optimal rate of organic removal occurred at lower HLR and higher HRT within specific ranges (150, 75 mm/d and 1.5, 3.0 d, respectively). Moreover, an extended HRT (more than 3.5 d) was found to enhance the removal of organic matter (more than 70%) when the mean temperature ranged between 5 and 15 °C [164].

4.2 Effect of Physicochemical Parameters on Removal Efficiency

In addition to the internal factors or components of CW i.e., substrate, biofilm and macrophytes, the physicochemical parameters include dissolved oxygen, pH, and temperature also affects CWs removal efficiency. The effect of individual parameters has been explained in detail as follows.

4.2.1 Effect of pH

The pH level of wastewater is a critical factor influencing the efficiency of CWs, particularly in terms of removal of nitrogen and organic matter. The efficiency of removing NH_4–N, BOD, nitrate, COD, and TP is considerably influenced by the pH of the wastewater within the range of 4–8 [165]. Both the chemical composition and biological processes in wetland water are notably influenced by pH. Denitrification occurs at a slower pace at a pH value of 5, and at pH values below 4, denitrification is extremely limited. Denitrifying bacteria thrive within the pH range (6.5–7.5) [166]. It is well-established that denitrifying bacteria thrive within a pH range of 6.5–7.5. The anaerobic degradation process is incomplete when the pH of the wastewater is beyond this range, which results in the accumulation of volatile fatty acids in the system. This accumulation can further lower the pH, ultimately eliminates all the methanogens present in the system [167].

In addition to the role of pH in biological processes, it is a critical factor in various chemical reactions. For instance, at a pH of 9.5, there is an increase in the proportion of un-ionized ammonia. Moreover, nitrification is favoured when the water's pH falls between 7.5 and 8, the optimal pH range for nitrification is from 6.6 to 8, respectively. Often, external alkalinity is introduced to maintain a pH above neutral for efficient removal of NH_4–N [168]. The pH of the substrate also plays a significant role NH_4–N and TP removal. It has been observed that diverse filter media exhibited maximum NH_4–N adsorption at a neutral pH. The order of NH_4–N adsorption capacity was pH 5 < pH 9 < pH 7 [169]. As for TP, the capacity for adsorbing broken stone and coarse sand was highest at a pH of 7 (30.9 mg/kg), decreasing at pH 5 and 9. However, for calcite, the maximum TP adsorption capacity was found at a pH of 9 (1542 mg/kg) and decreased as pH decreased. In contrast, it was observed that pH has minimal influence on BOD removal [170].

4.2.2 Effect of Dissolved Oxygen (DO)

Oxygen, a crucial environmental factor, plays a vital role in regulating nitrification and biodegradation processes [171]. In wastewater treatment, artificial aeration is used to augment DO levels. Aeration in CWs proves beneficial for effectively removing ammonia, BOD, and COD. However, phosphate removal is minimally impacted by aeration, and the presence of high DO hampers nitrate removal [172]. The main process recognized for TN removal in wetlands is nitrification–denitrification. HF-CWs often face insufficient oxygen levels, leading to the inhibition of the organic matter degradation and nitrification process. On the contrary, VF-CWs with intermittent wastewater feeding have unsaturated substrates that enhance air diffusion, thereby increasing the availability of oxygen within the system. This promotes aerobic degradation and nitrification of ammonia [173]. The limited oxygen availability makes the nitrification process the primary bottleneck for the removal of nitrogen SFCWs. It has been determined that nitrification can only occur when the DO content is 1.5 mg/L or above [174]. It has been found that high DO levels facilitate superior organic compounds degradation [175]. For effective removal NH_4–N, COD, and TN the aerobic conditions should be maintained. Providing artificial aeration has been associated with increased removal of TN, organic matter, and NH_4–N [176]. A study conducted in Jinan, China, demonstrated increase in removal efficiencies of NH_4–N, COD, and TN when VF-CWs were intermittently aerated for 4 h/ day at a rate of 1.0 L min^{-1}. However, aeration did not significantly impact TP removal [177]

4.2.3 Effect of Temperature

Temperature is a critical factor affecting the optimal functioning of both microorganisms and macrophytes in CW systems. It leads to varying nutrient and organic removal efficiency throughout the seasons. Wetlands, being influenced by climate, experience recurring patterns of evapotranspiration, photosynthesis, and microbial activity, all of which vary with the changing seasons. This seasonal fluctuation significantly affects removal efficiencies, with optimal nutrient removal observed at around 30 °C [178]. Furthermore, research has revealed that the effectiveness of COD removal is better in the summer and spring than in the winter and fall. Nitrogen removal efficiencies have also been found to be higher (6–11%) in CW systems during the summer season compared to winter. The findings underscore the direct relationship between microbial activity, temperature, and subsequent contaminant removal efficiency [179]. Moreover, research has demonstrated that treatment efficiency in CWs is notably higher in tropical regions than in temperate or cold regions. This difference is attributed to temperature variances, which enhance plant growth, leading to improved phytoremediation by macrophytes and heightened microbial activity. Understanding the influence of both abiotic and biotic components, including temperature, on CW performance is critical for effective wastewater treatment [164].

Metabolic processes tend to slow down or stop entirely at lower temperatures, which can suspend the normal growth of vegetation. Plants play essential roles in CWs, such as facilitating the transfer of oxygen to the root zone and enhancing the aerobic degradation. These plant functions are notably affected by the ambient temperature in the CW environment. Therefore, it is crucial to select plant species based on the local climate as it greatly influences the effective operation of CWs [180]. Notably, planted CWs have demonstrated superior performance compared to unplanted ones, underscoring the vital role of macrophytes in wetland-based treatment. Additionally, it was observed that mixed plant species systems (polycultures) generally achieved better results than systems with a single plant species (monocultures). In contrast, unplanted wetlands displayed a consistent temperature-dependent pattern, showing lower COD removal efficiency at colder temperatures [181]. The reduction in NH_4–N removal was noticeably smaller in polyculture setups compared to monoculture systems, highlighting the significance of employing polycultures in CWs.

Moreover, a decrease in temperature had a significant detrimental impact on TN and nitrate removal. This was attributed to the nutrient uptake activity of macrophytes nearing a halt due to plant aging [182]. Research has shown that the metabolic activity of denitrifying bacteria is significantly affected by temperature, with a noticeable decrease when there is drop in temperature below 10 °C. The process of denitrification almost comes to a complete halt when there is fall in temperature below 6 °C. In addition, it has been shown that for every 10 °C change in temperature from 0 to 30 °C, NH_4–N volatilization rises by 1.3–3.5 times, and denitrification rates virtually double [163].

While temperature does not have a direct impact on the media itself, numerous studies suggest that it does have an indirect influence on treatment efficiency of CWs. The clogging of the media primarily results from the wastewater type and the media properties, but temperature also plays a pivotal role in this process. In colder months, if the rate of plant debris build-up surpasses the rate of decomposition, it leads to a net accumulation of plant debris over time [183]. Additionally, excessive sludge production occurs due to an imbalance between the microorganism's growth and the biomass decay. At lower temperatures, the system receives an insufficient oxygen supply, which reduces organic matter mineralization by microorganisms. Accumulation of particles within the media, often decreases effectiveness of subsequent removal processes due to reduced pore space [44].

5 Conclusion

CWs possess distinctive qualities, serving as an environmentally conscious treatment approach that is constantly advancing to enhance the global ecosystem. CWs provide not only a straightforward and eco-friendly treatment solution but also act as habitats for both native and migratory wildlife, fostering a harmonious equilibrium between land development and the imperative for green cover. With the growing stringency of water quality standards for wastewater treatment and water reuse on a global scale, CWs still face limitations that underscore the need for ongoing research and development efforts.

- To facilitate the wider utilization of CWs system in the treatment of wastewater containing diverse pollutants, it is imperative to develop comprehensive guidelines encompassing CW design, installation, operational procedures, and maintenance.

These detailed guidelines should cover various aspects such as CW configuration (dimensions in terms of length, width, and height), the selection of filter materials (comprising particle size distribution, media components, permeability, depth, etc.), the choice of planting schemes (considering density and vegetation diversity), and the operational conditions and maintenance requirements for CW systems throughout their various operational phases, which include startup, vegetation growth, and long-term establishment.

- Further research is essential to assess the effectiveness of CWs facilities in demanding environmental conditions, such as cold climates, arid regions, and tropical environments. Challenging environmental factors, encompassing both cold and warm weather conditions, have the potential to influence the hydrology and ecology of CWs system. As a result, these conditions may alter the pace of abiotic and biotic processes that play a role in pollutant removal.
- Extended field studies are imperative to assess the enduring sustainability and stability of CWs systems. Additionally, these studies help in comprehending the diverse environmental and operational circumstances that influence the fate, transport, and elimination of pollutants within wetlands. Long-term experimental data is crucial for the formulation of comprehensive guidelines governing CW design, operation, installation, and maintenance. Implementing such guidelines enhances construction quality, mitigates the risk of wetland system breakdown, and prolongs the lifespan of CW systems. Furthermore, these guidelines can guide decision-making for local regulatory bodies, aiding in the establishment of new standards or revisions to existing ones. This, in turn, aims to enhance the quality of CW effluent, making it suitable for discharge into aquatic environments or for the purpose of treated effluent reuse across various applications.
- Comprehensive experiments are essential for the innovation of substrates with enhanced adsorption capabilities, minimal leaching of secondary pollutants, and the ability for straightforward, cost-effective regeneration. In CW systems, sand and gravel are prevalent choices as filter materials. To decrease the reliance on natural resources and reduce overall operational expenses, it is crucial to explore the potential use of waste materials, such as construction waste and incineration ash, as alternative filter media in the future. Furthermore, the phenomenon of clogging in porous media within CW systems remains not completely understood. Filter media clogging can impede the CW treatment efficiency, necessitating frequent maintenance activities like media regeneration or replacement. Consequently, future research efforts in this direction are warranted.
- Considering vegetation's crucial role in CW systems, upcoming research should prioritize an array of phytoremediation experiments utilizing the plants that are readily available in the local regions. This research should aim to identify plant species that exhibit optimal phytoremediation potential for a wide range of pollutants, even under extreme climatic conditions like warmer climates, high salinity, and extended drought periods. Apart from native vegetation, it's important to explore screening exotic plants for their phytoremediation capabilities in the future. The combined planting of both exotic and native plants may potentially

demonstrate superior treatment efficacy compared to individual species. Furthermore, it is essential to develop guidelines to manage weed growth and control invasive plants within wetland systems.

- While extensive modeling efforts have been devoted to single CW systems, there is a notable scarcity of data regarding the accuracy of these models in predicting removal performance of pollutant and hydraulic in hybrid CW systems. Hybrid systems generally exhibit superior performance compared to single CW systems. Given the diverse and complex processes within CW systems, the development of a comprehensive and sophisticated model is essential. Such a model would aid in comprehending the interactions among biotic and abiotic components, intricate processes, and the optimization of CW performance. An AI-based model would be particularly beneficial for system operators to adapt operational conditions in response to fluctuations in climatic conditions, pollutant loads, hydrology, and more. A well-calibrated and validated model could also assist in decision making in selecting an appropriate CW unit centered on the specific requirements of the local region.

Acknowledgements Director, CSIR-NEERI is thankfully acknowledged for giving the opportunity to pursue the work in CSIR-NEERI, Nagpur, India. The article is checked for plagiarism using the iThenticate software and recorded in the Knowledge Resource Centre, CSIR-NEERI, Nagpur for anti-plagiarism (KRC No.: CSIR-NEERI/KRC/2023/NOV/WWTD-EBGD/1).

References

1. Larsen TA, Hoffmann S, Lüthi C, Truffer B, Maurer M (2016) Emerging solutions to the water challenges of an urbanizing world. Science 352(6288):928–933. https://doi.org/10.1126/science.aad8641
2. Chen Z, Wu G, Wu Y, Wu Q, Shi Q, Ngo HH et al (2020) Water eco-nexus cycle system (WaterEcoNet) as a key solution for water shortage and water environment problems in urban areas. Water Cycle 1:71–77. https://doi.org/10.1016/j.watcyc.2020.05.004
3. Bauer S, Wagner M (2022) Possibilities and challenges of wastewater reuse-planning aspects and realized examples. Water 14(10):1619. https://doi.org/10.3390/w14101619
4. Pratap B, Kumar S, Nand S, Azad I, Bharagava RN, Ferreira LFR et al (2023) Wastewater generation and treatment by various eco-friendly technologies: possible health hazards and further reuse for environmental safety. Chemosphere 313:137547. https://doi.org/10.1016/j.chemosphere.2022.137547
5. Khalid S, Shahid M, Natasha BI, Sarwar T, Shah AH et al (2018) A review of environmental contamination and health risk assessment of wastewater use for crop irrigation with a focus on low and high-income countries. Int J Environ Res Public Health 15(5):895. https://doi.org/10.3390/ijerph15050895
6. Jamwal P, Carvalho L, Bhattacharyya S, Muttepawar P (2023) The benefits of restoring urban lakes in the tropics. Environ Monit Assess 195(11):1294. https://doi.org/10.1007/s10661-023-11888-1
7. Angelakis AN, Capodaglio AG, Dialynas EG (2022) Wastewater management: from ancient greece to modern times and future. Water 15(1):43. https://doi.org/10.3390/w15010043

8. Capodaglio AG (2017) Integrated, decentralized wastewater management for resource recovery in rural and peri-urban areas. Resources 6(2):22. https://doi.org/10.3390/resources6020022
9. Langergraber G, Giraldi D, Mena J, Meyer D, Peña M, Toscano A et al (2009) Recent developments in numerical modelling of subsurface flow constructed wetlands. Sci Total Environ 407(13):3931–3943. https://doi.org/10.1016/j.scitotenv.2008.07.057
10. Ji Z, Tang W, Pei Y (2022) Constructed wetland substrates: a review on development, function mechanisms, and application in contaminants removal. Chemosphere 286(P1):131564. https://doi.org/10.1016/j.chemosphere.2021.131564
11. Crini G, Lichtfouse E (2019) Advantages and disadvantages of techniques used for wastewater treatment. Environ Chem Lett 17(1):145–155. https://doi.org/10.1007/s10311-018-0785-9
12. Chen Q, Yao Y, Li X, Lu J, Zhou J, Huang Z (2018) Comparison of heavy metal removals from aqueous solutions by chemical precipitation and characteristics of precipitates. J water Process Eng. 26:289–300. https://doi.org/10.1016/j.jwpe.2018.11.003
13. Vidu R, Matei E, Predescu AM, Alhalaili B, Pantilimon C, Tarcea C et al (2020) Removal of heavy metals from wastewaters: a challenge from current treatment methods to nanotechnology applications. Toxics. 8(4):101. https://doi.org/10.3390/toxics8040101
14. Verma AK, Dash RR, Bhunia P (2012) A review on chemical coagulation/flocculation technologies for removal of colour from textile wastewaters. J Environ Manage 93(1):154–168. https://doi.org/10.1016/j.jenvman.2011.09.012
15. Wei H, Gao B, Ren J, Li A, Yang H (2018) Coagulation/flocculation in dewatering of sludge: a review. Water Res 143:608–631. https://doi.org/10.1016/j.watres.2018.07.029
16. Koul B, Yadav D, Singh S, Kumar M, Song M (2022) Insights into the domestic wastewater treatment (DWWT) regimes: a review. Water (Switzerland) 14(21). https://doi.org/10.3390/w14213542
17. Bala S, Garg D, Thirumalesh BV, Sharma M, Sridhar K, Inbaraj BS et al (2022) Recent strategies for bioremediation of emerging pollutants: a review for a green and sustainable environment. Toxics. 10(8):484. https://doi.org/10.3390/toxics10080484
18. Obotey Ezugbe E, Rathilal S (2020) Membrane technologies in wastewater treatment: a review. Membranes (Basel) 10(5):89. https://doi.org/10.3390/membranes10050089
19. Othman NH, Alias NH, Fuzil NS, Marpani F, Shahruddin MZ, Chew CM, et al (2021) A review on the use of membrane technology systems in developing countries. Membranes (Basel) 12(1):30. https://doi.org/10.3390/membranes12010030
20. Xu H, Xiao K, Wang X, Liang S, Wei C, Wen X et al (2020) Outlining the roles of membrane-foulant and foulant-foulant interactions in organic fouling during microfiltration and ultrafiltration: a mini-review. Front Chem 8:417. https://doi.org/10.3389/fchem.2020.00417
21. Ayeleru OO, Modekwe HU, Onisuru OR, Ohoro CR, Akinnawo CA, Olubambi PA (2023) Adsorbent technologies and applications for carbon capture, and direct air capture in environmental perspective and sustainable climate action. Sustain Chem Clim Action 100029. https://doi.org/10.1016/j.scca.2023.100029
22. Karić N, Maia AS, Teodorović A, Atanasova N, Langergraber G, Crini G et al (2022) Biowaste valorisation: agricultural wastes as biosorbents for removal of (in) organic pollutants in wastewater treatment. Chem Eng J Adv. 9:100239. https://doi.org/10.1016/j.ceja.2021.100239
23. da Silva Santos DH, Xiao Y, Chaukura N, Hill JM, Selvasembian R, Zanta CLPS et al (2022) Regeneration of dye-saturated activated carbon through advanced oxidative processes: a review. Heliyon. https://doi.org/10.1016/j.heliyon.2022.e10205
24. Hey G, Grabic R, Ledin A, la Cour JJ, Andersen HR (2012) Oxidation of pharmaceuticals by chlorine dioxide in biologically treated wastewater. Chem Eng J 185:236–242. https://doi.org/10.1016/j.cej.2012.01.093
25. Javeed T, Nawaz R, Al-Hussain SA, Irfan A, Irshad MA, Ahmad S et al (2023) Application of advanced oxidation processes for the treatment of color and chemical oxygen demand of pulp and paper wastewater. Water 15(7):1347. https://doi.org/10.3390/w15071347

26. Zinicovscaia I (2016) Conventional methods of wastewater treatment. In: Cyanobacteria for bioremediation of wastewaters, pp 17–25. https://doi.org/10.1007/978-3-319-26751-7_3
27. Parde D, Patwa A, Shukla A, Vijay R, Killedar DJ, Kumar R (2021) A review of constructed wetland on type, treatment and technology of wastewater. Environ Technol Innov 21:101261. https://doi.org/10.1016/j.eti.2020.101261
28. Kataki S, Chatterjee S, Vairale MG, Dwivedi SK, Gupta DK (2021) Constructed wetland, an eco-technology for wastewater treatment: a review on types of wastewater treated and components of the technology (macrophyte, biolfilm and substrate). J Environ Manage 283:111986. https://doi.org/10.1016/j.jenvman.2021.111986
29. Gomes AC, Silva L, Albuquerque A, Simões R, Stefanakis AI (2018) Investigation of lab-scale horizontal subsurface flow constructed wetlands treating industrial cork boiling wastewater. Chemosphere 207:430–439. https://doi.org/10.1016/j.chemosphere.2018.05.123
30. Saeed T, Muntaha S, Rashid M, Sun G, Hasnat A (2018) Industrial wastewater treatment in constructed wetlands packed with construction materials and agricultural by-products. J Clean Prod 189:442–453. https://doi.org/10.1016/j.jclepro.2018.04.115
31. Wang F, Liu Y, Ma Y, Wu X, Yang H (2012) Characterization of nitrification and microbial community in a shallow moss constructed wetland at cold temperatures. Ecol Eng 42:124–129. https://doi.org/10.1016/j.ecoleng.2012.01.006
32. Tunçsiper B, Drizo A, Twohig E (2015) Constructed wetlands as a potential management practice for cold climate dairy effluent treatment—VT, USA. CATENA 135:184–192. https://doi.org/10.1016/j.catena.2015.07.028
33. Pescod MB (1992) Wastewater characteristics and effluent quality parameters. wastewater treatment and use in agriculture: fAO irrigation and drainage paper, p 47
34. Wu S, Wallace S, Brix H, Kuschk P, Kirui WK, Masi F et al (2015) Treatment of industrial effluents in constructed wetlands: challenges, operational strategies and overall performance. Environ Pollut 201:107–120. https://doi.org/10.1016/j.envpol.2015.03.006
35. Gomes HI, Mayes WM, Whitby P, Rogerson M (2019) Constructed wetlands for steel slag leachate management: partitioning of arsenic, chromium, and vanadium in waters, sediments, and plants. J Environ Manage 243:30–38. https://doi.org/10.1016/j.jenvman.2019.04.127
36. Nuamah LA, Li Y, Pu Y, Nwankwegu AS, Haikuo Z, Norgbey E et al (2020) Constructed wetlands, status, progress, and challenges. the need for critical operational reassessment for a cleaner productive ecosystem. J Clean Prod 269:122340. https://doi.org/10.1016/j.jclepro.2020.122340
37. Wang M, Zhang DQ, Dong JW, Tan SK (2017) Constructed wetlands for wastewater treatment in cold climate—a review. J Environ Sci 57:293–311. https://doi.org/10.1016/j.jes.2016.12.019
38. Vymazal J, Březinová T (2015) The use of constructed wetlands for removal of pesticides from agricultural runoff and drainage: a review. Environ Int 75:11–20. https://doi.org/10.1016/j.envint.2014.10.026
39. Almeida CMR, Santos F, Ferreira ACF, Gomes CR, Basto MCP, Mucha AP (2017) Constructed wetlands for the removal of metals from livestock wastewater–can the presence of veterinary antibiotics affect removals? Ecotoxicol Environ Saf 137:143–148. https://doi.org/10.1016/j.ecoenv.2016.11.021
40. Hazra M, Durso LM (2022) Performance efficiency of conventional treatment plants and constructed wetlands towards reduction of antibiotic resistance. Antibiotics 11(1):114. https://doi.org/10.3390/antibiotics11010114
41. Stefanakis AI (2020) Constructed wetlands: description and benefits of an eco-tech water treatment system. In: Waste management: concepts, methodologies, tools, and applications. IGI Global, pp 503–525. https://doi.org/10.4018/978-1-7998-1210-4.ch025
42. Xu D, Yin X, Zhou S, Jiang Y, Xi X, Sun H et al (2022) A review on the remediation of microplastics using constructed wetlands: bibliometric, co-occurrence, current trends, and future directions. Chemosphere 303:134990. https://doi.org/10.1016/j.chemosphere.2022.134990

43. García-Ávila F, Avilés-Añazco A, Cabello-Torres R, Guanuchi-Quito A, Cadme-Galabay M, Gutiérrez-Ortega H et al (2023) Application of ornamental plants in constructed wetlands for wastewater treatment: a scientometric analysis. Case Stud Chem Environ Eng 7:100307. https://doi.org/10.1016/j.cscee.2023.100307
44. Aguilar-Paredes A, Valdés G, Araneda N, Valdebenito E, Hansen F, Nuti M (2023) Microbial community in the composting process and its positive impact on the soil biota in sustainable agriculture. Agronomy 13(2):542. https://doi.org/10.3390/agronomy13020542
45. Stefanakis AI (2019) The role of constructed wetlands as green infrastructure for sustainable urban water management. Sustainability 11(24):6981. https://doi.org/10.3390/su11246981
46. World Water Development Report 2018 [Internet]. Available from: https://www.unwater.org/publications/world-water-development-report-2018
47. Wu H, Zhang J, Ngo HH, Guo W, Hu Z, Liang S et al (2015) A review on the sustainability of constructed wetlands for wastewater treatment: design and operation. Bioresour Technol 175:594–601. https://doi.org/10.1016/j.biortech.2014.10.068
48. Hijosa-Valsero M, Reyes-Contreras C, Domínguez C, Bécares E, Bayona JM (2016) Behaviour of pharmaceuticals and personal care products in constructed wetland compartments: influent, effluent, pore water, substrate and plant roots. Chemosphere 145:508–517. https://doi.org/10.1016/j.chemosphere.2015.11.090
49. Vymazal J (2018) Constructed wetlands for wastewater treatment. Encycl Ecol 45:14–21. https://doi.org/10.3390/w2030530
50. Sijimol MR, Joseph S (2021) Constructed wetland systems for greywater treatment and reuse: a review. Int J Energy Water Resour. 5(3):357–369. https://doi.org/10.1007/s42108-021-00129-1
51. Biswal BK, Balasubramanian R (2022) Constructed wetlands for reclamation and reuse of wastewater and urban stormwater: a review. Front Environ Sci 1–21. https://doi.org/10.3389/fenvs.2022.836289
52. Schmitt N, Wanko A, Laurent J, Bois P, Molle P, Mosé R (2015) Constructed wetlands treating stormwater from separate sewer networks in a residential Strasbourg urban catchment area: micropollutant removal and fate. J Environ Chem Eng 3(4):2816–2824. https://doi.org/10.1016/j.jece.2015.10.008
53. Tao W, Sauba K, Fattah KP, Smith JR (2017) Designing constructed wetlands for reclamation of pretreated wastewater and stormwater. Rev Environ Sci Biotechnol 16(1):37–57. https://doi.org/10.1007/s11157-016-9419-5
54. Avellan CT, Ardakanian R, Gremillion P (2017) The role of constructed wetlands for biomass production within the water-soil-waste nexus. Water Sci Technol 75(10):2237–2245. https://doi.org/10.2166/wst.2017.106
55. Valipour A, Ahn Y-H (2016) Constructed wetlands as sustainable ecotechnologies in decentralization practices: a review. Environ Sci Pollut Res 23:180–197. https://doi.org/10.1007/s11356-015-5713-y
56. Wang Q, Hu Y, Xie H, Yang Z (2018) Constructed wetlands: a review on the role of radial oxygen loss in the rhizosphere by macrophytes. Water 10(6):678. https://doi.org/10.3390/w10060678
57. Hernández ME, Galindo-Zetina M, Carlos H-HJ (2018) Greenhouse gas emissions and pollutant removal in treatment wetlands with ornamental plants under subtropical conditions. Ecol Eng 114:88–95. https://doi.org/10.1016/j.ecoleng.2017.06.001
58. Lai W-L, Zhang Y, Chen Z-H (2012) Radial oxygen loss, photosynthesis, and nutrient removal of 35 wetland plants. Ecol Eng 39:24–30. https://doi.org/10.1016/j.ecoleng.2011.11.010
59. Martínez NB, Tejeda A, Del Toro A, Sánchez MP, Zurita F (2018) Nitrogen removal in pilot-scale partially saturated vertical wetlands with and without an internal source of carbon. Sci Total Environ 645:524–532. https://doi.org/10.1016/j.scitotenv.2018.07.147
60. Shelef O, Gross A, Rachmilevitch S (2013) Role of plants in a constructed wetland: current and new perspectives. Water 5(2):405–419. https://doi.org/10.3390/w5020405
61. Baptestini GCF, Matos AT, Martinez MA, Borges AC, Matos MP (2017) Hydraulic conductivity variability in horizontal subsurface flow constructed wetlands. Eng Agrícola 37:333–342

62. Sandoval L, Zamora-Castro SA, Vidal-Álvarez M, Marín-Muñiz JL (2019) Role of wetland plants and use of ornamental flowering plants in constructed wetlands for wastewater treatment: a review. Appl Sci 9(4):1–17. https://doi.org/10.3390/app9040685
63. Vymazal J (2013) Emergent plants used in free water surface constructed wetlands: a review. Ecol Eng 61:582–592. https://doi.org/10.1016/j.ecoleng.2013.06.023
64. Yang C, Zhang X, Tang Y, Jiang Y, Xie S, Zhang Y et al (2022) Selection and optimization of the substrate in constructed wetland: a review. J Water Process Eng. 49:103140. https://doi.org/10.1016/j.jwpe.2022.103140
65. Yang Y, Zhao Y, Liu R, Morgan D (2018) Global development of various emerged substrates utilized in constructed wetlands. Bioresour Technol 261:441–452. https://doi.org/10.1016/j.biortech.2018.03.085
66. Khalifa ME, Abou El-Reash YG, Ahmed MI, Rizk FW (2020) Effect of media variation on the removal efficiency of pollutants from domestic wastewater in constructed wetland systems. Ecol Eng 143:105668. https://doi.org/10.1016/j.ecoleng.2019.105668
67. Wang Q, Cao Z, Hu Y, Kong Q, Xu F, Du Y et al (2019) Season effects on subsurface constructed wetlands performance: role of radial oxygen loss of Phragmites Australis. CLEAN–Soil Air Water 47(8):1800428. https://doi.org/10.1002/clen.201800428
68. Rahman ME, Bin Halmi MIE, Bin Abd Samad MY, Uddin MK, Mahmud K, Abd Shukor MY et al (2020) Design, operation and optimization of constructed wetland for removal of pollutant. Int J Environ Res Public Health 17(22):8339. https://doi.org/10.3390/ijerph17228339
69. Wang Y, Cai Z, Sheng S, Pan F, Chen F, Fu J (2020) Comprehensive evaluation of substrate materials for contaminants removal in constructed wetlands. Sci Total Environ 701:134736. https://doi.org/10.1016/j.scitotenv.2019.134736
70. Li Y, Zhu G, Ng WJ, Tan SK (2014) A review on removing pharmaceutical contaminants from wastewater by constructed wetlands: design, performance and mechanism. Sci Total Environ 468:908–932. https://doi.org/10.1016/j.scitotenv.2013.09.018
71. Gikas GD, Pérez-Villanueva M, Tsioras M, Alexoudis C, Pérez-Rojas G, Masís-Mora M et al (2018) Low-cost approaches for the removal of terbuthylazine from agricultural wastewater: constructed wetlands and biopurification system. Chem Eng J 335:647–656. https://doi.org/10.1016/j.cej.2017.11.031
72. Blanco I, Molle P, de Miera LES, Ansola G (2016) Basic oxygen furnace steel slag aggregates for phosphorus treatment. evaluation of its potential use as a substrate in constructed wetlands. Water Res 89:355–365. https://doi.org/10.1016/j.watres.2015.11.064
73. Chatterjee S, Mitra A, Gupta SK, Gupta DK (2019) A review on reed bed system as a potential decentralized wastewater treatment practice. Adv Plant Transgenics Methods Appl. 239–251. https://doi.org/10.1007/978-981-13-9624-3_11
74. Grebenshchykova Z, Frédette C, Chazarenc F, Comeau Y, Brisson J (2020) Establishment and potential use of woody species in treatment wetlands. Int J Phytoremediation 22(3):295–304. https://doi.org/10.1080/15226514.2019.1658712
75. Gupta SK, Kataki S, Chatterjee S, Prasad RK, Datta S, Vairale MG et al (2020) Cold adaptation in bacteria with special focus on cellulase production and its potential application. J Clean Prod 258:120351. https://doi.org/10.1016/j.jclepro.2020.120351
76. Rajan RJ, Sudarsan JS, Nithiyanantham S (2019) Microbial population dynamics in constructed wetlands: review of recent advancements for wastewater treatment. Environ Eng Res 24(2):181–190. https://doi.org/10.4491/eer.2018.127
77. Mendes R, Garbeva P, Raaijmakers JM (2013) The rhizosphere microbiome: significance of plant beneficial, plant pathogenic, and human pathogenic microorganisms. FEMS Microbiol Rev 37(5):634–663. https://doi.org/10.1111/1574-6976.12028
78. Srivastava P, Dwivedi S, Kumar N, Abbassi R, Garaniya V, Yadav AK (2017) Performance assessment of aeration and radial oxygen loss assisted cathode based integrated constructed wetland-microbial fuel cell systems. Bioresour Technol 244:1178–1182. https://doi.org/10.1016/j.biortech.2017.08.026

79. Geng N, Wu Y, Zhang M, Tsang DCW, Rinklebe J, Xia Y et al (2019) Bioaccumulation of potentially toxic elements by submerged plants and biofilms: a critical review. Environ Int 131:105015. https://doi.org/10.1016/j.envint.2019.105015
80. Han Z, Dong J, Shen Z, Mou R, Zhou Y, Chen X et al (2019) Nitrogen removal of anaerobically digested swine wastewater by pilot-scale tidal flow constructed wetland based on in-situ biological regeneration of zeolite. Chemosphere 217:364–373. https://doi.org/10.1016/j.chemosphere.2018.11.036
81. Badhe N, Saha S, Biswas R, Nandy T (2014) Role of algal biofilm in improving the performance of free surface, up-flow constructed wetland. Bioresour Technol 169:596–604. https://doi.org/10.1016/j.biortech.2014.07.050
82. Sehar S, Naz I (2016) Role of the biofilms in wastewater treatment. In: Microb biofilms—importance and applications, pp 121–144
83. Vishnyakova A, Popova N, Artemiev G, Botchkova E, Litti Y, Safonov A (2022) Effect of mineral carriers on biofilm formation and nitrogen removal activity by an indigenous anammox community from cold groundwater ecosystem alone and bioaugmented with biomass from a "warm" anammox reactor. Biology (Basel) 11(10):1421. https://doi.org/10.3390/biology11101421
84. Wijewardene L, Wu N, Fohrer N, Riis T (2022) Epiphytic biofilms in freshwater and interactions with macrophytes: current understanding and future directions. Aquat Bot 176:103467. https://doi.org/10.1016/j.aquabot.2021.103467
85. Sharma S, Mohler J, Mahajan SD, Schwartz SA, Bruggemann L, Aalinkeel R (2023) Microbial biofilm: a review on formation, infection, antibiotic resistance, control measures, and innovative treatment. Microorganisms. 11(6):1614. https://doi.org/10.3390/microorganisms11061614
86. Saini S, Tewari S, Dwivedi J, Sharma V (2023) Biofilm mediated wastewater treatment: a comprehensive review. Mater Adv. https://doi.org/10.1039/D2MA00945E
87. Masoud AMN, Alfarra A, Sorlini S (2022) Constructed wetlands as a solution for sustainable sanitation: a comprehensive review on integrating climate change resilience and circular economy. Water 14(20):3232. https://doi.org/10.3390/w14203232
88. Marín-Muñiz JL, Sandoval Herazo LC, López-Méndez MC, Sandoval-Herazo M, Meléndez-Armenta RÁ, González-Moreno HR et al (2023) Treatment wetlands in Mexico for control of wastewater contaminants: a review of experiences during the last twenty-two years. Processes 11(2):359. https://doi.org/10.3390/pr11020359
89. Guo C, Cui Y, Dong B, Luo Y, Liu F, Zhao S et al (2017) Test study of the optimal design for hydraulic performance and treatment performance of free water surface flow constructed wetland. Bioresour Technol 238:461–471. https://doi.org/10.1016/j.biortech.2017.03.163
90. Wu H, Zhang J, Ngo HH, Guo W, Liang S (2017) Evaluating the sustainability of free water surface flow constructed wetlands: methane and nitrous oxide emissions. J Clean Prod 147:152–156. https://doi.org/10.1016/j.jclepro.2017.01.091
91. Zheng Y, Wang X, Dzakpasu M, Zhao Y, Ngo HH, Guo W et al (2016) Effects of interspecific competition on the growth of macrophytes and nutrient removal in constructed wetlands: a comparative assessment of free water surface and horizontal subsurface flow systems. Bioresour Technol 207:134–141. https://doi.org/10.1016/j.biortech.2016.02.008
92. Ulén B, Geranmayeh P, Blomberg M, Bieroza M (2019) Seasonal variation in nutrient retention in a free water surface constructed wetland monitored with flow-proportional sampling and optical sensors. Ecol Eng 139:105588. https://doi.org/10.1016/j.ecoleng.2019.105588
93. Zhou X, Gao L, Zhang H, Wu H (2018) Determination of the optimal aeration for nitrogen removal in biochar-amended aerated vertical flow constructed wetlands. Bioresour Technol 261:461–464. https://doi.org/10.1016/j.biortech.2018.04.028
94. Rahi MA, Faisal AAH, Naji LA, Almuktar SA, Abed SN, Scholz M (2020) Biochemical performance modelling of non-vegetated and vegetated vertical subsurface-flow constructed wetlands treating municipal wastewater in hot and dry climate. J Water Process Eng 33:101003. https://doi.org/10.1016/j.jwpe.2019.101003

95. Srivastava P, Abbassi R, Garaniya V, Lewis T, Yadav AK (2020) Performance of pilot-scale horizontal subsurface flow constructed wetland coupled with a microbial fuel cell for treating wastewater. J Water Process Eng 33:100994. https://doi.org/10.1016/j.jwpe.2019.100994
96. Abedi T, Mojiri A (2019) Constructed wetland modified by biochar/zeolite addition for enhanced wastewater treatment. Environ Technol Innov 16:100472. https://doi.org/10.1016/j.eti.2019.100472
97. Cancelli AM, Gobas FAPC, Wang Q, Kelly BC (2019) Development and evaluation of a mechanistic model to assess the fate and removal efficiency of hydrophobic organic contaminants in horizontal subsurface flow treatment wetlands. Water Res 151:183–192. https://doi.org/10.1016/j.watres.2018.12.020
98. Zhuang L-L, Yang T, Zhang J, Li X (2019) The configuration, purification effect and mechanism of intensified constructed wetland for wastewater treatment from the aspect of nitrogen removal: a review. Bioresour Technol 293:122086. https://doi.org/10.1016/j.biortech.2019.122086
99. Kraiem K, Kallali H, Wahab MA, Fra-Vazquez A, Mosquera-Corral A, Jedidi N (2019) Comparative study on pilots between ANAMMOX favored conditions in a partially saturated vertical flow constructed wetland and a hybrid system for rural wastewater treatment. Sci Total Environ 670:644–653. https://doi.org/10.1016/j.scitotenv.2019.03.220
100. Nivala J, Abdallat G, Aubron T, Al-Zreiqat I, Abbassi B, Wu G-M et al (2019) Vertical flow constructed wetlands for decentralized wastewater treatment in Jordan: optimization of total nitrogen removal. Sci Total Environ 671:495–504. https://doi.org/10.1016/j.scitotenv.2019.03.376
101. Huang Z, Zhang X, Cui L, Yu G (2016) Optimization of operating parameters of hybrid vertical down-flow constructed wetland systems for domestic sewerage treatment. J Environ Manage 180:384–389. https://doi.org/10.1016/j.jenvman.2016.05.060
102. Chyan JM, Lin CJ, Lin YC, Chou YA (2016) Improving removal performance of pollutants by artificial aeration and flow rectification in free water surface constructed wetland. Int Biodeterior Biodegradation 113:146–154. https://doi.org/10.1016/j.ibiod.2016.04.034
103. Ali M, Rousseau DPL, Ahmed S (2018) A full-scale comparison of two hybrid constructed wetlands treating domestic wastewater in Pakistan. J Environ Manage 210:349–358. https://doi.org/10.1016/j.jenvman.2018.01.040
104. Chen J, Deng W-J, Liu Y-S, Hu L-X, He L-Y, Zhao J-L et al (2019) Fate and removal of antibiotics and antibiotic resistance genes in hybrid constructed wetlands. Environ Pollut 249:894–903. https://doi.org/10.1016/j.envpol.2019.03.111
105. Rai JPN, Saraswat S (2023) Green technologies for waste management: a wealth from waste approach. CRC Press
106. Achak M, Barka N, Lamy E (2023) Constructed wetlands process for treating sewage to improve the quantitative and qualitative management of groundwater resources
107. Wang P, Zhang H, Zuo J, Zhao D, Zou X, Zhu Z et al (2016) A hardy plant facilitates nitrogen removal via microbial communities in subsurface flow constructed wetlands in winter. Sci Rep 6(1):33600. https://doi.org/10.1038/srep33600
108. Grzyb A, Wolna-Maruwka A, Niewiadomska A (2021) The significance of microbial transformation of nitrogen compounds in the light of integrated crop management. Agronomy 11(7):1415. https://doi.org/10.3390/agronomy11071415
109. Bunce JT, Ndam E, Ofiteru ID, Moore A, Graham DW (2018) A review of phosphorus removal technologies and their applicability to small-scale domestic wastewater treatment systems. Front Environ Sci 6:8. https://doi.org/10.3389/fenvs.2018.00008
110. Wang J, Xia L, Chen J, Wang X, Wu H, Li D et al (2021) Synergistic simultaneous nitrification-endogenous denitrification and EBPR for advanced nitrogen and phosphorus removal in constructed wetlands. Chem Eng J 420:127605. https://doi.org/10.1016/j.cej.2020.127605
111. Lyu T, Zhang L, Xu X, Arias CA, Brix H, Carvalho PN (2018) Removal of the pesticide tebuconazole in constructed wetlands: design comparison, influencing factors and modelling. Environ Pollut 233:71–80. https://doi.org/10.1016/j.envpol.2017.10.040

112. Verduzo Garibay M, Fernández del Castillo A, de Anda J, Senés-Guerrero C, Gradilla-Hernández MS (2021) Structure and activity of microbial communities in response to environmental, operational, and design factors in constructed wetlands. Int J Environ Sci Technol 1–26. https://doi.org/10.1007/s13762-021-03719-y
113. Ilyas H, van Hullebusch ED (2020) The influence of design and operational factors on the removal of personal care products by constructed wetlands. Water 12(5):1367. https://doi.org/10.3390/w12051367
114. Vymazal J (2011) Plants used in constructed wetlands with horizontal subsurface flow: a review. Hydrobiologia 674(1):133–156. https://doi.org/10.1007/s10750-011-0738-9
115. Kumar Maharjan A, Mori K, Toyama T (2020) Nitrogen removal ability and characteristics of the laboratory-scale tidal flow constructed wetlands for treating ammonium-nitrogen contaminated groundwater. Water 12(5):1326. https://doi.org/10.3390/w12051326
116. Lan W, Zhang J, Hu Z, Ji M, Zhang X, Zhang J et al (2018) Phosphorus removal enhancement of magnesium modified constructed wetland microcosm and its mechanism study. Chem Eng J 335:209–214. https://doi.org/10.1016/j.cej.2017.10.150
117. Liu W, Dahab MF, Surampalli RY (2005) Nitrogen transformations modeling in subsurface-flow constructed wetlands. Water Environ Res 77(3):246–258. https://doi.org/10.2175/106143005X41825
118. Prochaska CA, Zouboulis AI, Eskridge KM (2007) Performance of pilot-scale vertical-flow constructed wetlands, as affected by season, substrate, hydraulic load and frequency of application of simulated urban sewage. Ecol Eng 31(1):57–66. https://doi.org/10.1016/j.ecoleng.2007.05.007
119. Molle P, Liénard A, Grasmick A, Iwema A (2006) Effect of reeds and feeding operations on hydraulic behaviour of vertical flow constructed wetlands under hydraulic overloads. Water Res 40(3):606–612. https://doi.org/10.1016/j.watres.2005.11.026
120. Brix H, Arias CA (2005) The use of vertical flow constructed wetlands for on-site treatment of domestic wastewater: new danish guidelines. Ecol Eng 25(5):491–500. https://doi.org/10.1016/j.ecoleng.2005.07.009
121. Saeed T, Sun G (2011) A comparative study on the removal of nutrients and organic matter in wetland reactors employing organic media. Chem Eng J 171(2):439–447. https://doi.org/10.1016/j.cej.2011.03.101
122. Ye F, Li Y (2009) Enhancement of nitrogen removal in towery hybrid constructed wetland to treat domestic wastewater for small rural communities. Ecol Eng 35(7):1043–1050. https://doi.org/10.1016/j.ecoleng.2009.03.009
123. Mays PA, Edwards GS (2001) Comparison of heavy metal accumulation in a natural wetland and constructed wetlands receiving acid mine drainage. Ecol Eng 16(4):487–500. https://doi.org/10.1016/S0925-8574(00)00112-9
124. Chen TY, Kao CM, Yeh TY, Chien HY, Chao AC (2006) Application of a constructed wetland for industrial wastewater treatment: a pilot-scale study. Chemosphere 64(3):497–502. https://doi.org/10.1016/j.chemosphere.2005.11.069
125. Al-Omari A, Fayyad M (2003) Treatment of domestic wastewater by subsurface flow constructed wetlands in Jordan. Desalination 155(1):27–39. https://doi.org/10.1016/S0011-9164(03)00236-4
126. Khan S, Ahmad I, Shah MT, Rehman S, Khaliq A (2009) Use of constructed wetland for the removal of heavy metals from industrial wastewater. J Environ Manage 90(11):3451–3457. https://doi.org/10.1016/j.jenvman.2009.05.026
127. Aitchison EW, Kelley SL, Alvarez PJJ, Schnoor JL (2000) Phytoremediation of 1, 4-dioxane by hybrid poplar trees. Water Environ Res 72(3):313–321. https://doi.org/10.2175/106143000X137536
128. Carleton JN, Grizzard TJ, Godrej AN, Post HE, Lampe L, Kenel PP (2000) Performance of a constructed wetlands in treating urban stormwater runoff. Water Environ Res 72(3):295–304. https://doi.org/10.2175/106143000X137518
129. Silver M (2020) Biology and chemistry of generation, prevention and abatement of acid mine drainage. In: Constructed wetlands for wastewater treatment. CRC Press, pp 753–760

130. Fang SC (1978) Sorption and transformation of mercury vapor by dry soil. Environ Sci Technol 12(3):285–288. https://doi.org/10.1021/es60139a004
131. El-Agroudy AA (1999) Investigation of constructed wetlands capability to remove mercury from contaminated waters. Concordia University
132. Qasaimeh A, Elektorowicz M, Balazinski M (2012) GA-fuzzy decision support system for mercury removal in natural waters. http://www.scirp.org/journal/PaperInformation.aspx?PaperID=18612
133. Hijosa-Valsero M, Matamoros V, Sidrach-Cardona R, Martín-Villacorta J, Bécares E, Bayona JM (2010) Comprehensive assessment of the design configuration of constructed wetlands for the removal of pharmaceuticals and personal care products from urban wastewaters. Water Res 44(12):3669–3678. https://doi.org/10.1016/j.watres.2010.04.022
134. Ávila C, Reyes C, Bayona JM, García J (2013) Emerging organic contaminant removal depending on primary treatment and operational strategy in horizontal subsurface flow constructed wetlands: influence of redox. Water Res 47(1):315–325. https://doi.org/10.1016/j.watres.2012.10.005
135. Ávila C, García J, Garfí M (2016) Influence of hydraulic loading rate, simulated storm events and seasonality on the treatment performance of an experimental three-stage hybrid constructed wetland system. Ecol Eng 87:324–332. https://doi.org/10.1016/j.ecoleng.2015.11.042
136. Verlicchi P, Zambello E (2014) How efficient are constructed wetlands in removing pharmaceuticals from untreated and treated urban wastewaters? A review. Sci Total Environ 470:1281–1306. https://doi.org/10.1016/j.scitotenv.2013.10.085
137. Ávila C, Nivala J, Olsson L, Kassa K, Headley T, Mueller RA et al (2014) Emerging organic contaminants in vertical subsurface flow constructed wetlands: influence of media size, loading frequency and use of active aeration. Sci Total Environ 494:211–217. https://doi.org/10.1016/j.scitotenv.2014.06.128
138. Reyes-Contreras C, Matamoros V, Ruiz I, Soto M, Bayona JM (2011) Evaluation of PPCPs removal in a combined anaerobic digester-constructed wetland pilot plant treating urban wastewater. Chemosphere 84(9):1200–1207. https://doi.org/10.1016/j.chemosphere.2011.06.003
139. Sánchez M, Ruiz I, Soto M (2022) The potential of constructed wetland systems and photodegradation processes for the removal of emerging contaminants—a review. Environments 9(9):116. https://doi.org/10.3390/environments9090116
140. Ilyas H, van Hullebusch ED (2020) Performance comparison of different types of constructed wetlands for the removal of pharmaceuticals and their transformation products: a review. Environ Sci Pollut Res 27(13):14342–14364. https://doi.org/10.1007/s11356-020-08165-w
141. Auvinen H, Gebhardt W, Linnemann V, Du Laing G, Rousseau DPL (2017) Laboratory- and full-scale studies on the removal of pharmaceuticals in an aerated constructed wetland: effects of aeration and hydraulic retention time on the removal efficiency and assessment of the aquatic risk. Water Sci Technol 76(6):1457–1465. https://doi.org/10.2166/wst.2017.328
142. Auvinen H, Havran I, Hubau L, Vanseveren L, Gebhardt W, Linnemann V et al (2017) Removal of pharmaceuticals by a pilot aerated sub-surface flow constructed wetland treating municipal and hospital wastewater. Ecol Eng 100:157–164. https://doi.org/10.1016/j.ecoleng.2016.12.031
143. Çakir R, Gidirislioglu A, Çebi U (2015) A study on the effects of different hydraulic loading rates (HLR) on pollutant removal efficiency of subsurface horizontal-flow constructed wetlands used for treatment of domestic wastewaters. J Environ Manage 164:121–128. https://doi.org/10.1016/j.jenvman.2015.08.037
144. Abdelgadir A, Chen X, Liu J, Xie X, Zhang J, Zhang K et al (2014) Characteristics, process parameters, and inner components of anaerobic bioreactors. Biomed Res Int. https://doi.org/10.1155/2014/841573
145. Waly MM, Ahmed T, Abunada Z, Mickovski SB, Thomson C (2022) Constructed wetland for sustainable and low-cost wastewater treatment: review article. Land 11(9):1–17. https://doi.org/10.3390/land11091388

146. Dotro G, Langergraber G, Molle P, Nivala J, Puigagut J, Stein O et al (2017) Treatment wetlands. IWA Publishing. https://doi.org/10.2166/9781780408774
147. Almuktar SA, Abed SN, Scholz M (2018) Wetlands for wastewater treatment and subsequent recycling of treated effluent: a review. Environ Sci Pollut Res 25:23595–23623. https://doi.org/10.1007/s11356-018-2629-3
148. Gorgoglione A, Torretta V (2018) Sustainable management and successful application of constructed wetlands: a critical review. Sustainability 10(11):3910. https://doi.org/10.3390/su10113910
149. Ávila C, Matamoros V, Reyes-Contreras C, Piña B, Casado M, Mita L et al (2014) Attenuation of emerging organic contaminants in a hybrid constructed wetland system under different hydraulic loading rates and their associated toxicological effects in wastewater. Sci Total Environ 470:1272–1280. https://doi.org/10.1016/j.scitotenv.2013.10.065
150. Varma M, Gupta AK, Ghosal PS, Majumder A (2021) A review on performance of constructed wetlands in tropical and cold climate: insights of mechanism, role of influencing factors, and system modification in low temperature. Sci Total Environ 755:142540. https://doi.org/10.1016/j.scitotenv.2020.142540
151. Ramprasad C, Smith CS, Memon FA, Philip L (2017) Removal of chemical and microbial contaminants from greywater using a novel constructed wetland: GROW. Ecol Eng 106:55–65. https://doi.org/10.1016/j.ecoleng.2017.05.022
152. Al-Baldawi IA, Mohammed AA, Mutar ZH, Abdullah SRS, Jasim SS, Almansoory AF (2021) Application of phytotechnology in alleviating pharmaceuticals and personal care products (PPCPs) in wastewater: source, impacts, treatment, mechanisms, fate, and SWOT analysis. J Clean Prod 319:128584. https://doi.org/10.1016/j.jclepro.2021.128584
153. Choi H, Geronimo FKF, Jeon M, Kim L-H (2021) Investigation of the factors affecting the treatment performance of a stormwater horizontal subsurface flow constructed wetland treating road and parking lot runoff. Water 13(9):1242. https://doi.org/10.3390/w13091242
154. Angassa K, Leta S, Mulat W, Kloos H, Meers E (2019) Effect of hydraulic loading on bioremediation of municipal wastewater using constructed wetland planted with vetiver grass, Addis Ababa, Ethiopia. Nanotechnol Environ Eng 4:1–11. https://doi.org/10.1007/s41204-018-0053-z
155. Stefanakis AI, Tsihrintzis VA (2012) Effects of loading, resting period, temperature, porous media, vegetation and aeration on performance of pilot-scale vertical flow constructed wetlands. Chem Eng J 181:416–430. https://doi.org/10.1016/j.cej.2011.11.108
156. Xu J, Shi Y, Zhang G, Liu J, Zhu Y (2014) Effect of hydraulic loading rate on the efficiency of effluent treatment in a recirculating puffer aquaculture system coupled with constructed wetlands. J Ocean Univ China 13(1):146–152. https://doi.org/10.1007/s11802-014-2000-3
157. Muduli M, Sonpal V, Ray S, Haldar S (2022) In-depth performance study of an innovative decentralized multistage constructed wetland system treating real institutional wastewater. Environ Res 210:112896. https://doi.org/10.1016/j.envres.2022.112896
158. Sultana M, Mourti C, Tatoulis T, Akratos CS, Tekerlekopoulou AG, Vayenas DV (2016) Effect of hydraulic retention time, temperature, and organic load on a horizontal subsurface flow constructed wetland treating cheese whey wastewater. J Chem Technol Biotechnol 91(3):726–732. https://doi.org/10.1002/jctb.4637
159. Merino-Solís ML, Villegas E, De Anda J, López-López A (2015) The effect of the hydraulic retention time on the performance of an ecological wastewater treatment system: an anaerobic filter with a constructed wetland. Water 7(3):1149–1163. https://doi.org/10.3390/w7031149
160. Zhang D, Gersberg RM, Ng WJ, Tan SK (2014) Removal of pharmaceuticals and personal care products in aquatic plant-based systems: a review. Environ Pollut 184:620–639. https://doi.org/10.1016/j.envpol.2013.09.009
161. Vera I, Verdejo N, Chávez W, Jorquera C, Olave J (2016) Influence of hydraulic retention time and plant species on performance of mesocosm subsurface constructed wetlands during municipal wastewater treatment in super-arid areas. J Environ Sci Heal Part A 51(2):105–113. https://doi.org/10.1080/10934529.2015.1087732

162. Álvarez JA, Ávila C, Otter P, Kilian R, Istenič D, Rolletschek M et al (2017) Constructed wetlands and solar-driven disinfection technologies for sustainable wastewater treatment and reclamation in rural India: SWINGS project. Water Sci Technol 76(6):1474–1489. https://doi.org/10.2166/wst.2017.329
163. Yan Y, Xu J (2014) Improving winter performance of constructed wetlands for wastewater treatment in northern China: a review. Wetlands 34:243–253. https://doi.org/10.1007/s13157-013-0444-7
164. Ji M, Hu Z, Hou C, Liu H, Hao H, Guo W et al (2020) New insights for enhancing the performance of constructed wetlands at low temperatures. Bioresour Technol 301(72):122722. https://doi.org/10.1016/j.biortech.2019.122722
165. Al-Baldawi IA, Abdullah SRS, Suja F, Anuar N, Mushrifah I (2013) Effect of aeration on hydrocarbon phytoremediation capability in pilot sub-surface flow constructed wetland operation. Ecol Eng 61:496–500. https://doi.org/10.1016/j.ecoleng.2013.10.017
166. Albina P, Durban N, Bertron A, Albrecht A, Robinet J-C, Erable B (2019) Influence of hydrogen electron donor, alkaline pH, and high nitrate concentrations on microbial denitrification: a review. Int J Mol Sci 20(20):5163. https://doi.org/10.3390/ijms20205163
167. Casagli F, Rossi S, Steyer JP, Bernard O, Ficara E (2021) Balancing microalgae and nitrifiers for wastewater treatment: can inorganic carbon limitation cause an environmental threat? Environ Sci Technol 55(6):3940–3955. https://doi.org/10.1021/acs.est.0c05264
168. Tang Y, Zhou C, Ziv-El M, Rittmann BE (2011) A pH-control model for heterotrophic and hydrogen-based autotrophic denitrification. Water Res 45(1):232–240. https://doi.org/10.1016/j.watres.2010.07.049
169. Chen P, Wu J, He Y, Zhang Y, Yu R, Lu X (2022) Enhanced nutrient removal in A2N effluent by reclaimed biochar adsorption. Int J Environ Res Public Health 19(7):4016. https://doi.org/10.3390/ijerph19074016
170. Paing J, Guilbert A, Gagnon V, Chazarenc F (2015) Effect of climate, wastewater composition, loading rates, system age and design on performances of french vertical flow constructed wetlands: a survey based on 169 full scale systems. Ecol Eng 80:46–52. https://doi.org/10.1016/j.ecoleng.2014.10.029
171. Ayiti OE, Babalola OO (2022) Factors influencing soil nitrification process and the effect on environment and health. Front Sustain Food Syst. 6:821994. https://doi.org/10.3389/fsufs.2022.821994
172. Donoso N, van Oirschot D, Kumar Biswas J, Michels E, Meers E (2019) Impact of aeration on the removal of organic matter and nitrogen compounds in constructed wetlands treating the liquid fraction of piggery manure. Appl Sci 9(20):4310. https://doi.org/10.3390/app9204310
173. Zhou X, Wang X, Zhang H, Wu H (2017) Enhanced nitrogen removal of low C/N domestic wastewater using a biochar-amended aerated vertical flow constructed wetland. Bioresour Technol 241:269–275. https://doi.org/10.1016/j.biortech.2017.05.072
174. Hassan I, Chowdhury SR, Prihartato PK, Razzak SA (2021) Wastewater treatment using constructed wetland: current trends and future potential. Processes 9(11):1917. https://doi.org/10.3390/pr9111917
175. Liu H, Hu Z, Zhang J, Ngo HH, Guo W, Liang S et al (2016) Optimizations on supply and distribution of dissolved oxygen in constructed wetlands: a review. Bioresour Technol 214:797–805. https://doi.org/10.1016/j.biortech.2016.05.003
176. Li F, Lu L, Zheng X, Zhang X (2014) Three-stage horizontal subsurface flow constructed wetlands for organics and nitrogen removal: effect of aeration. Ecol Eng 68:90–96. https://doi.org/10.1016/j.ecoleng.2014.03.025
177. Wu H, Fan J, Zhang J, Ngo HH, Guo W, Hu Z et al (2016) Optimization of organics and nitrogen removal in intermittently aerated vertical flow constructed wetlands: effects of aeration time and aeration rate. Int Biodeterior Biodegradation 113:139–145. https://doi.org/10.1016/j.ibiod.2016.04.031
178. Zhang D-Q, Jinadasa K, Gersberg RM, Liu Y, Tan SK, Ng WJ (2015) Application of constructed wetlands for wastewater treatment in tropical and subtropical regions (2000–2013). J Environ Sci 30:30–46. https://doi.org/10.1016/j.jes.2014.10.013

179. Mietto A, Politeo M, Breschigliaro S, Borin M (2015) Temperature influence on nitrogen removal in a hybrid constructed wetland system in Northern Italy. Ecol Eng 75:291–302. https://doi.org/10.1016/j.ecoleng.2014.11.027
180. Zhou Q, Zhu H, Bañuelos G, Yan B, Liang Y, Yu J et al (2017) Impacts of vegetation and temperature on the treatment of domestic sewage in constructed wetlands incorporated with ferric-carbon micro-electrolysis material. Int J Phytoremediation 19(10):915–924. https://doi.org/10.1080/15226514.2017.1303811
181. Choudhury MI, Segersten J, Hellman M, Mckie BG, Hallin S, Ecke F (2019) Importance of plant species for nitrogen removal using constructed floating wetlands in a cold climate. Ecol Eng 138:126–132. https://doi.org/10.1016/j.ecoleng.2019.07.012
182. Zhou Q, Zhu H, Bañuelos G, Yan B, Liang Y, Yu X et al (2017) Effects of vegetation and temperature on nutrient removal and microbiology in horizontal subsurface flow constructed wetlands for treatment of domestic sewage. Water Air Soil Pollut 228:1–13. https://doi.org/10.1007/s11270-017-3280-1
183. Katibi KK, Yunos KF, Che Man H, Aris AZ, Mohd Nor MZ, Azis RS et al (2021) Contemporary techniques for remediating endocrine-disrupting compounds in various water sources: advances in treatment methods and their limitations. Polymers (Basel) 13(19):3229. https://doi.org/10.3390/polym13193229

Biogas and Nutrient Recovery from Algal Biomass

Ranjeet Kumar Singh, Aditya Chaturvedi, Shailendra Kumar Singh, Sudarshan Singh Rathore, Rinku D. Jaiswal, Sameer Shekhar, Bhanu Pandey, Sidharth Singh, and Raj S. Singh

Abstract The rise of energy demand is likely to be twice by 2050 due to the massive growth of the population forcing several researchers, academicians, and scientists to think of alternative sources of energy by utilizing food waste, cow manure, plant leaves, and algae. This chapter deals with the significance of microalgae as the best source of generating green biogas/biofuels which not only save rapidly declining stocks of fossil fuels but also mitigates environmental problems. Low or no competition with human food or crops, low amounts of lignin, and rapid growth rate are the benefits of biogas production through Algae. Conversely, issues like chemical content changes, high amounts of moisture, and other inhibition practices hinder in generation of cost-effective biogas. However, it is still more economically and environmentally affable than ordinary fossil fuels. Anaerobic digestion of algal biomass suffers from low digestibility due to cell wall resistance and inappropriate carbon-to-nitrogen ratio. A short-span method comprising fungal crude enzyme-based pre-treatment of algal biomass is introduced and the effect of it was qualitatively assessed through visual and microscopic observations and quantitatively through measuring algal biomass solubilization. Up to 50% biomass COD solubilization was observed within 150 min of pretreatment under optimal conditions. A higher production is observed in subsequent anaerobic digestion of pretreated algal biomass from untreated algal biomass. Interestingly, methane yield upsurged when pretreated algal biomass co-digested with cattle dung while sugarcane bagasse harmed algal biomass co-digestion due to its poor digestibility. Overall, the present attempt showed promising results by improving methane yield from algal biomass though pretreatment and co-digestion.

R. K. Singh (✉) · S. K. Singh · S. S. Rathore · S. Shekhar · B. Pandey · S. Singh · R. S. Singh
CSIR-Central Institute of Mining and Fuel Research, Barwa Road, Dhanbad 826015, India
e-mail: ranjeetenvironment@gmail.com

R. K. Singh
Academy of Scientific and Innovative Research, Ghaziabad, Uttar Pradesh 201002, India

A. Chaturvedi · R. S. Singh
Central Pollution Control Board, Delhi, India

R. D. Jaiswal · R. S. Singh
CSIR-Central Institute of Mining and Fuel Research, Bilashpur Unit, Bilashpur, India

M. M. Ghangrekar et al. (eds.), *Biological and Hybrid Wastewater Treatment Technology*, Earth and Environmental Sciences Library,
https://doi.org/10.1007/978-3-031-63046-0_14

Keywords Biogas · Microalgae · Anaerobic digestion · Algae cultivation · Water treatment

1 Introduction

Globally, the increasing population and demand for large amounts of energy, were forced to think of alternative sources to generate energy likewise food waste to biogas [1]. The generated biogas not only fulfills the energy requirement but also saves our limited resources. Because of the above problem number of researchers, academicians, and scientists engaged in searching for alternative sources such as food waste, cow dung, leaf manure sludge, and algae. Microalgae is considered a renewable and sustainable biomass to produce bio-energy and other high-value products [2]. Besides, the cultivation of microalgae does not need any fertile land and it provides opportunities for climate change mitigation by sequestering atmospheric carbon dioxide (CO_2), facilitating nutrient recovery from wastewater, and regulating industrial pollutions/emissions [1]. Algal biomass harvested from different technologies is unique in its physio-chemical properties and its contents of proteins, lipids, and carbohydrates which are easily consumed through anaerobic digestion. Biomass from microalgae is digested easily to consume waste as well as generated energy [3] and it's also profitable for the environment [4, 5].

In this chapter, we elaborate on the importance of cell wall weakening followed by pretreatment as a key process step and strategy to reduce the energy cost of converting algal biomass into bio-energy as well as biogas. From the energy calculations, it was measured that the cell wall weakening significantly improves the net-energy ratio from 0.68 to 1.02 [6]. This approach could be integrated with any pre-treatment options, while it reduces the time of pre-treatment and costs of energy/chemicals required for the hydrolysis of algal biomass [1].

It appears that you have provided a comprehensive overview of the potential benefits and challenges associated with using microalgae as a feedstock for biogas production. Microalgae indeed offer numerous advantages over terrestrial biomass sources, such as higher photosynthetic efficiency, less land requirement, and the ability to grow in non-arable and low-quality water [7].

Anaerobic digestion is a promising process for converting microalgae biomass into biogas, primarily composed of methane and carbon dioxide. This process can be resource-efficient and has been widely used to treat organic waste streams due to its high energy output-to-input ratio and environmental benefits [5]. Since microalgae have negligible lignin content, harsh pretreatment is not necessary, making anaerobic digestion a potentially viable option for wet algae biomass feedstock with minimal dewatering.

However, the utilization of microalgae for biogas production is not without its challenges. Significant nutrient utilization, high energy input for harvesting and dewatering, and complex downstream conversion processes for usable fuels are some of the major obstacles that need to be addressed to ensure commercial feasibility.

To assess the economic viability of the biogas production process from microalgae, Techno-Economic Analysis (TEA) is a crucial tool [8]. TEA helps establish capital and operating cost profiles, determining the potential economic feasibility of the production process and aiding in making informed decisions during technology development and investment [8].

While there have been numerous techno-economic assessments conducted to evaluate the economic feasibility of biodiesel derived from microalgae, there seems to be a lack of similar analyses on anaerobic digestion of microalgae for biogas production, especially at full-scale production levels, considering the characteristics of different algae species [9–11].

2 Anaerobic Digestion Process

Anaerobic digestion is a biological process that involves the breakdown of organic matter [12], such as food waste, agricultural residues, sewage sludge, and other organic materials, in the absence of oxygen. It is a natural process that occurs in environments without oxygen, such as the depths of swamps or landfills. During anaerobic digestion, complex organic compounds are broken down by a consortium of microorganisms, mainly bacteria, into simpler substances, including biogas and digestive. The key components of anaerobic digestion are substrate, anaerobic digester, and microbial consortia (Fig. 1).

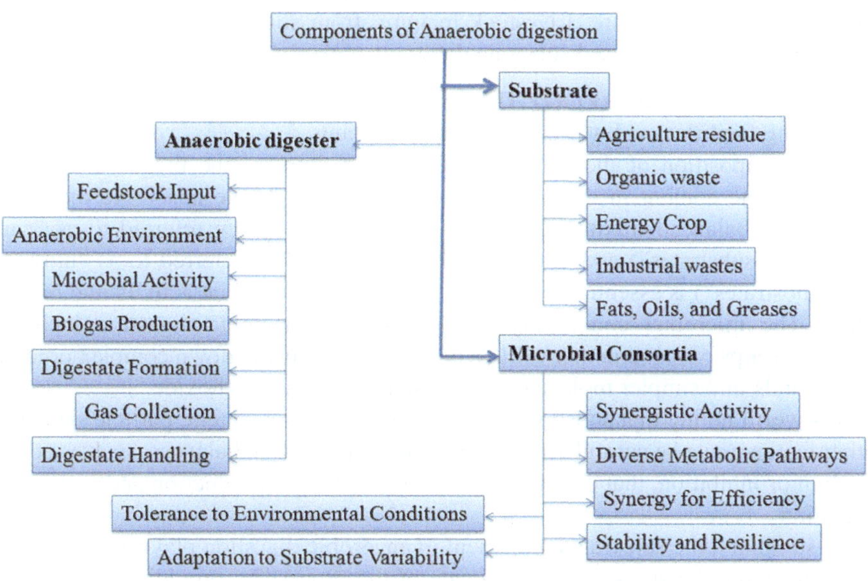

Fig. 1 Important component of anaerobic digestion

3 Substrate

In the context of anaerobic digestion, a "substrate" refers to the organic material that is fed into the anaerobic digester for the production of biogas through microbial decomposition in the absence of oxygen. The substrate component is a crucial aspect of anaerobic digestion, as it determines the efficiency and effectiveness of the process. Different substrates can yield varying amounts of biogas and digestate, which is the solid and liquid material left behind after the digestion process.

4 Common Substrate Components Used in Anaerobic Digestion Include

Agricultural Residues: These include crop residues (such as corn stover, and wheat straw), animal manures, and other agricultural byproducts.

Organic Wastes: Municipal solid waste, food waste, kitchen scraps, yard waste, and sewage sludge are examples of organic waste materials that can be used as substrates.

Energy Crops: Plants specifically grown for biogas production like algae, maize (corn), switch grass, and paddy husk, are also used as substrates.

Industrial Wastes: Certain industrial byproducts, such as brewery waste, food processing residues, and paper mill sludge, can serve as substrates for anaerobic digestion. Animal Manure: Livestock manure, from sources like cows, pigs, and poultry, is a commonly used substrate due to its high organic content.

Fats, Oils, and Greases (FOG): These substrates can come from restaurants, food processing facilities, and wastewater treatment plants.

The composition of the substrate significantly impacts the biogas production rate, methane content in the biogas, and the stability of the anaerobic digestion process. A balanced mixture of carbon-rich materials (often referred to as "carbonaceous substrates") and nitrogen-rich materials (often referred to as "nitrogenous substrates") is essential for maintaining optimal microbial activity and biogas yield. The microbial community responsible for anaerobic digestion includes various groups of bacteria that work synergistically to break down complex organic compounds into simpler molecules like methane and carbon dioxide.

It's worth noting that substrate composition, moisture content, particle size, and other factors must be carefully managed to avoid process imbalances, such as acidification or inhibition, that could lead to decreased biogas production and inefficient digestion.

In summary, the substrate component in anaerobic digestion plays a critical role in determining the success and efficiency of the process by providing the organic material necessary for microbial degradation and biogas production.

5 Anaerobic Digester

An anaerobic digester is a specialized system designed to break down organic materials in the absence of oxygen through a biological process called anaerobic digestion. This process involves the activity of microorganisms that convert organic matter into biogas and digestate.

Here's an overview of how an anaerobic digester works:

1. **Feedstock Input**: Organic materials, known as substrates, are introduced into the anaerobic digester. These substrates can include agricultural residues, organic waste, animal manure, energy crops, and more.
2. **Anaerobic Environment**: The digester is kept in an oxygen-free environment. This is important because anaerobic microorganisms thrive in such conditions, allowing them to carry out the digestion process.
3. **Microbial Activity**: Microorganisms, primarily bacteria, perform a series of biochemical reactions to break down complex organic molecules present in the substrates. This breakdown produces various intermediate compounds, including volatile fatty acids and alcohols.
4. **Biogas Production**: One of the main products of anaerobic digestion is biogas. Biogas is a mixture of methane (CH_4), carbon dioxide (CO_2), and smaller amounts of other gases. Methane is the primary component of biogas and can be used as a renewable energy source.
5. **Digestate Formation**: As the microbial digestion process progresses, the remaining material transforms into digestate. Digestate is a mixture of solid and liquid residues that can be used as a nutrient-rich fertilizer.
6. **Gas Collection**: The biogas produced in the digester is collected and can be used for various purposes. It can be used as a source of renewable energy for electricity and heat generation, or it can be refined for use as vehicle fuel or injected into natural gas pipelines.
7. **Digestate Handling**: The digestate can be separated into solid and liquid fractions, both of which have agricultural value. The solid fraction can be applied as a soil conditioner, while the liquid fraction can be used as a nutrient-rich liquid fertilizer.

Anaerobic digesters come in various designs and configurations, such as batch, continuous, plug-flow, and complete-mix digesters. The choice of digester type depends on factors like the type of feedstock, desired biogas output, available space, and operational preferences.

Anaerobic digestion offers several benefits, including waste reduction, renewable energy production, and nutrient recycling. It also helps mitigate greenhouse gas emissions by capturing methane, a potent greenhouse gas, that would otherwise be released into the atmosphere from decomposing organic matter. Overall, anaerobic digesters play a significant role in sustainable waste management, renewable energy production, and circular economy initiatives.

6 Microbial Consortia

Microbial consortia are a key component of anaerobic digestion processes. A microbial consortium refers to a group of different microorganisms, such as bacteria, archaea, and fungi, that work together synergistically to carry out complex biochemical reactions involved in the breakdown of organic matter and the production of biogas. In anaerobic digestion, these microbial consortia are responsible for the decomposition of organic materials and the generation of methane-rich biogas.

Here's how microbial consortia contribute to anaerobic digestion:

1. **Synergistic Activity**: Anaerobic digestion is a multi-step process that involves a series of biochemical reactions. Different microorganisms within the consortium specialize in various stages of this process, working together to efficiently break down complex organic molecules into simpler compounds.
2. **Diverse Metabolic Pathways**: Anaerobic microorganisms possess a range of metabolic pathways that allow them to convert complex organic matter into biogas components like methane and carbon dioxide. Some microorganisms break down complex carbohydrates, proteins, fats, and other organic molecules into intermediate products like volatile fatty acids and alcohols.
3. **Synergy for Efficiency**: Microbial consortia in anaerobic digesters exhibit a mutualistic relationship. Some microorganisms produce substances that serve as food for others, creating a balanced ecosystem that maximizes the efficiency of organic matter degradation. For example, methanogenic archaea rely on the products produced by other bacteria, such as acetate, to produce methane.
4. **Tolerance to Environmental Conditions**: Different microorganisms within the consortium might have varying tolerance levels to factors like temperature, pH, and substrate composition. This diversity ensures that the anaerobic digestion process remains stable even in fluctuating conditions.
5. **Adaptation to Substrate Variability**: Anaerobic digestion systems often receive a mix of organic substrates with varying compositions. The presence of a diverse microbial consortium enables the system to adapt and efficiently process different types of feedstock.
6. **Stability and Resilience**: The presence of multiple microbial species provides resilience to the system against potential disruptions, such as the introduction of toxic compounds or changes in operating conditions. This helps maintain the stability of the anaerobic digestion process.

Managing and maintaining a balanced and diverse microbial consortium is crucial for optimal anaerobic digestion performance. Various factors, such as feedstock type, temperature, pH, and retention time, can influence the composition and activity of the microbial community. Operators of anaerobic digestion systems often monitor these factors and make adjustments to ensure the health and activity of the microbial consortia.

In summary, microbial consortia are central to anaerobic digestion processes, as they drive the biochemical transformations that result in the production of biogas.

Understanding and managing these microbial communities are essential for efficient and sustainable biogas generation in anaerobic digesters.

The process typically involves four stages: hydrolysis, acidogenesis, acetogenesis, and methanogenesis.

a. Hydrolysis: Complex organic compounds are broken down into simpler molecules like sugars, amino acids, and fatty acids.
b. Acidogenesis: The simpler molecules are further broken down into volatile fatty acids (VFAs) and alcohols.
c. Acetogenesis: VFAs and alcohols are converted into acetic acid, hydrogen, and carbon dioxide.
d. Methanogenesis: Methanogenic archaea convert acetic acid and hydrogen into methane (CH_4) gas, the primary component of biogas.

Biogas: The end product of anaerobic digestion is biogas, which mainly consists of methane (50–75%) and carbon dioxide (25–50%). It may also contain traces of other gases like hydrogen sulfide (H_2S) and water vapor [12].

Digestate: The solid and liquid residue left behind after the digestion process is called digestate. It is rich in nutrients and can be used as a nutrient-rich fertilizer or soil amendment.

Anaerobic digestion has various benefits, such as:

Energy Production: Biogas can be used as a renewable energy source to generate electricity and heat.

Waste Reduction: Anaerobic digestion helps to manage and reduce organic waste, diverting it from landfills and reducing greenhouse gas emissions.

Nutrient Recovery: The digestate can be used as a nutrient-rich fertilizer, closing the loop in nutrient cycling.

Environmental Impact: It helps in reducing the release of methane, a potent greenhouse gas, which is otherwise released when organic waste decomposes anaerobically in landfills. Anaerobic digestion is widely used in biogas plants, wastewater treatment facilities, agricultural operations, and industrial settings to convert organic waste into valuable resources while contributing to sustainable waste management and energy production.

Algae Activation: Algae activation usually refers to the process of stimulating the growth and metabolic activity of algae cells. Algae are photosynthetic microorganisms that can convert sunlight and nutrients into organic matter through photosynthesis. Proper activation involves providing the algae with optimal conditions for growth, including suitable light, temperature, nutrients (such as nitrogen, phosphorus, and micronutrients), and pH levels. Activation can be achieved in various ways, including adjusting environmental conditions, adding growth-promoting substances, and maintaining a controlled culture environment.

Biomass Recovery: Biomass recovery pertains to the collection and harvesting of algae biomass after it has grown and accumulated. The harvested biomass can then be processed and used for various applications. Efficient biomass recovery is crucial to maximizing the yield of valuable compounds that can be extracted from algae, such as lipids for biofuel production, pigments, proteins, and carbohydrates for food or feed, and even potential pharmaceutical compounds.

Methods for algae biomass recovery include mechanical separation (centrifugation, filtration), flocculation (using chemicals to clump algae cells together for easier separation), sedimentation, and flotation. The choice of method depends on factors like the algae species, the desired end products, and the scale of production.

7 The Applications of Algae Activation and Biomass Recovery Are Diverse

Algae activation and biomass recovery are key processes in the cultivation and harvesting of algae for various applications, including biofuel production, wastewater treatment, and nutritional supplements. Here's an overview of these processes:

1. Algae Activation:

   a. Cultivation: Algae are typically grown in controlled environments like photobioreactors, open ponds, or closed-loop systems. The cultivation process involves the following steps:

      Inoculation: Algae strains are introduced into the growth medium (usually water) to start the culture.

      Nutrient Addition: Nutrients like nitrogen, phosphorus, and micronutrients are added to the culture medium to promote algae growth.

      Light and CO_2: Adequate light and carbon dioxide are provided to support photosynthesis. Temperature Control: Maintaining optimal temperatures for the specific algae species being cultivated.

   b. Monitoring and Optimization: Algae growth is closely monitored, and various parameters like pH, temperature, nutrient levels, and CO_2 concentration are optimized to maximize biomass production.

   c. Harvesting: When the algae culture reaches the desired biomass concentration, it's ready for harvesting. Harvesting methods can vary and include centrifugation, filtration, flocculation, and more. The choice of method depends on the type of algae and the intended application.

2. Biomass Recovery:

   a. Dewatering: Algae cells have a high water content, which needs to be reduced to obtain a concentrated biomass. Dewatering methods include centrifugation, filtration, and sedimentation.

   b. Cell Disruption: In many applications, it's essential to break open the algae cells to access their valuable contents. Techniques like mechanical shear, ultrasonication, and chemical treatments can be used for cell disruption.

c. Separation: After cell disruption, the desired components (e.g., lipids, proteins, carbohydrates) can be separated from the cell debris. This can involve additional filtration, centrifugation, or chromatography steps.
d. Product Extraction: The extracted components can be used for various purposes. For example, lipids can be converted into biodiesel, proteins can be used as food supplements, and carbohydrates can be fermented into biofuels or biochemicals.
e. Residue Management: The remaining biomass residue after extraction may be used for various purposes, such as animal feed or fertilizer, to minimize waste.

It's worth noting that the specific methods and technologies used in algae activation and biomass recovery can vary widely depending on the type of algae, the intended end product, and the scale of the operation. Additionally, ongoing research and development are continuously improving these processes to make algae cultivation and biomass recovery more efficient and sustainable.

Biofuel Production: Algae are a promising source of biofuels due to their high lipid content. Activating algae for increased lipid production and efficiently recovering the biomass can enhance the biofuel yield (Fig. 2).

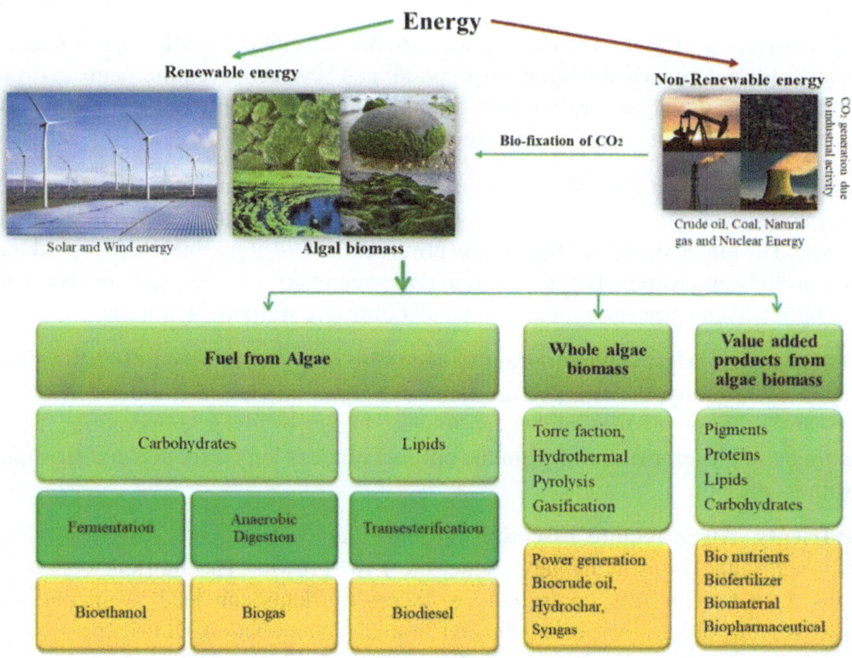

Fig. 2 Applications of algae activation and biomass recovery

Nutritional Supplements: Some algae species are rich in proteins, omega-3 fatty acids, vitamins, and minerals. Activated algae biomass can be harvested and processed to create nutritional supplements for humans and animals.

Wastewater Treatment: Algae can be used to treat wastewater by absorbing nutrients like nitrogen and phosphorus. Activation of specific algae strains can lead to more effective nutrient removal.

Carbon Sequestration: Algae can capture carbon dioxide from the atmosphere through photosynthesis. Optimizing algae activation and biomass recovery can aid in carbon sequestration efforts.

Bioremediation: Certain algae can accumulate heavy metals and other pollutants from water. Effective activation and biomass recovery can facilitate the use of algae for environmental cleanup.

In recent years, research and development efforts have been focused on improving the efficiency of both algae activation and biomass recovery processes to make algae-based technologies more economically viable and environmentally sustainable.

8 Growth Requirements for Algae Cultivation

The cultivation of algae requires careful attention to their growth requirements to ensure optimal growth and biomass production. The specific requirements can vary depending on the algae species and the intended application (e.g., bio-fuels, food, bioremediation), but here are some key factors to consider:

Light Intensity and Quality: Algae are photosynthetic organisms that require light for photosynthesis, which is the process of converting light energy into chemical energy. The intensity and quality of light are crucial. Most algae thrive under moderate to high light intensities, but the ideal intensity can vary between species. Blue and red light wavelengths are typically the most effective for photosynthesis.

Temperature: Algae have different temperature preferences based on their natural habitats. Generally, the temperature range for algae cultivation is around 20–30 °C (68–86 °F). Some extremophiles can tolerate higher or lower temperatures. Maintaining a stable temperature within the optimal range is important to avoid stress and promote growth.

Nutrients: Algae require essential nutrients such as nitrogen, phosphorus, and micronutrients (iron, manganese, zinc, etc.) for growth. The nutrient concentrations must be carefully controlled, as excess nutrients can lead to uncontrolled growth (eutrophication) or imbalanced biomass composition. Nutrient ratios can also influence the production of specific compounds like lipids or proteins.

pH Level: The pH of the culture medium affects nutrient availability and biochemical reactions within algae cells. Most algae species grow well in a slightly alkaline to neutral pH range of 7–9. Regular monitoring and adjustment of pH are important for maintaining optimal growth conditions.

Carbon Source: Algae utilize carbon dioxide (CO_2) for photosynthesis. Providing a consistent source of CO_2 is important for promoting growth. In some cases, CO_2 supplementation may be necessary to enhance growth rates.

Mixing and Aeration: Proper mixing and aeration of the culture medium ensure uniform distribution of nutrients, gases, and light within the culture. This prevents sedimentation of algae cells and allows for efficient gas exchange.

Salinity: Salinity requirements can vary widely among different algae species. Some algae are freshwater species, while others thrive in brackish or even marine environments. It's crucial to match the salinity level to the specific algae species being cultivated.

Cultivation System: Algae can be grown in various systems, such as open ponds, closed photo-bioreactors, or raceway ponds. The choice of system impacts factors like light exposure, temperature control, and nutrient distribution.

Contamination Control: Preventing contamination from other microorganisms (bacteria, fungi) is vital. Proper sterilization of culture equipment and maintenance of sterile conditions can minimize the risk of unwanted species outcompeting the desired algae.

Harvesting and Biomass Recovery: Consideration should be given to the methods and timing of algae biomass harvesting. This can impact the overall efficiency of the cultivation process.

It's important to note that successful algae cultivation often involves a balance between these factors. Detailed understanding of the specific algae species being cultivated and careful monitoring of the culture conditions are essential for achieving optimal growth and biomass production.

9 Conversion Technologies for Algal Biomass in the Extraction of Value-Added Product

Utilization of bacteria and other microorganisms is one of the important steps in the anaerobic digestion of algae biomass with organic feedstock and fermentation of bio-plastics, bio-ethanol, and bio-hydrogen. Reasons behind the difficulty of algae cell walls, during anaerobic digestion hydrolyzing of enzymes is very challenging, and its effects directly show with very low productions. To address this issue various types of breakdown and preprocessing techniques are applied [13]. For the production of biofuels, numerous biochemical processes are applied to convert algal biomass. Microalgae cultivated in municipal wastewater produced about 17 mL/g COD during a process of anaerobic digestion [14] whereas, [15], developed biohydrogen in their

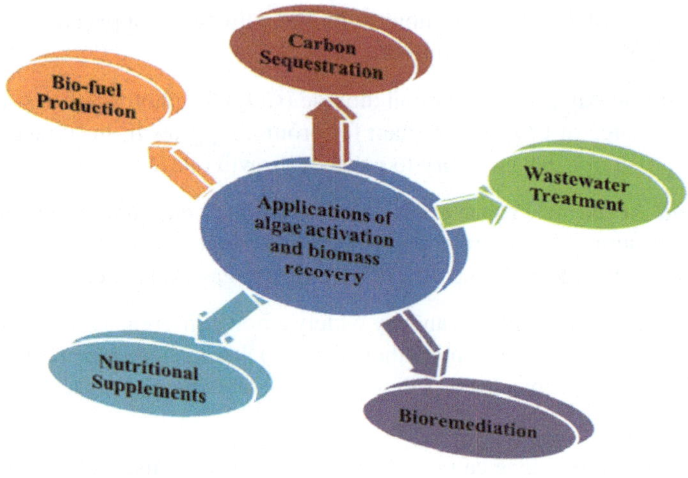

Fig. 3 A role of algal biomass in energy cycle and value-added product extraction

laboratory to utilized various microalgal species biomass and their concentration of about 2.50 g/L of algal biomass. Numerous researchers also engaged to boost productivity and production of biohydrogen at a pilot scale. As, at high-rate algal ponds, one gram of microalgal biomass generated approx 0.3 L of VS of biomethane [16]. Biofuel produced by the microalgal species biomass depends on the concentration of biomass and cultivation conditions at varying degrees. A study done by ref. [17], developed synthetic growth media at the laboratory level, where a 1 L reactor was utilized to grow the microalga Chlorella vulgaris, it generated about 1.35 g/L of biomass and 3.50 g/L of bioethanol. In the same cultivation media 2 L laboratory scale reactor generates 0.26 g/g bioethanol with the help of Scenedesmus dimorphus. Figure 3, shows algal biorefineries integrated with upstream and downstream bioreactions. It may be possible to stimulate waste CO_2 and wastewater streams through three different processing routes by the growth of algae cultivation [18].

Algae activation and biomass recovery are important processes in the field of biotechnology and environmental science, particularly in the context of algae cultivation for various purposes, including biofuel production, wastewater treatment, and nutritional supplements. Let's break down these concepts.

10 Algae Cultivation and Wastewater Treatment

Algae cultivation can play a significant role in wastewater treatment due to its ability to remove nutrients and pollutants from water. This process is known as algae-based wastewater treatment or algal bioremediation. Here's how algae cultivation can be integrated into wastewater treatment:

Nutrient Removal: Algae have a natural capacity to absorb nutrients like nitrogen and phosphorus from the water. These nutrients are typically found in excess in wastewater, especially in agricultural runoff and municipal wastewater. Algae can take up these nutrients for growth, effectively reducing their concentration in the water. This helps mitigate eutrophication, a process that leads to excessive plant and algae growth, oxygen depletion, and ecological imbalances in water bodies.

Algae Ponds or Raceways: Wastewater treatment facilities can use open ponds or raceway systems to cultivate algae. Wastewater is directed into these systems, providing nutrients for algae growth. As the algae grow, they remove nutrients from the water. The treated water, with reduced nutrient levels, can then be discharged or further processed.

Photobioreactors: Closed systems like photobioreactors can also be used for algae cultivation in wastewater treatment. These systems offer better control over environmental factors such as light, temperature, and CO_2 supply. They are particularly useful for wastewater treatment in locations with limited space.

Harvesting and Biomass Recovery: After algae have absorbed nutrients from the wastewater, the algae biomass can be harvested. This harvesting process effectively removes the nutrients from the wastewater stream. Depending on the specific algae species and wastewater composition, the harvested biomass can be used for various purposes, such as biofuel production, animal feed, or fertilizer.

Microbial Community Balance: While algae play a central role in nutrient removal, it's essential to maintain a balanced microbial community within the system. Bacteria also play a role in breaking down organic matter and can aid in the overall wastewater treatment process.

Challenges and Considerations: Implementing algae-based wastewater treatment requires careful planning and management. Some challenges to consider include the need to control algal species to avoid potential invasive species introduction, dealing with fluctuating environmental conditions, and optimizing the balance between algae growth and nutrient removal.

Efficiency and Scaling: The efficiency of algae-based wastewater treatment depends on factors such as wastewater composition, algae species selection, and system design. Research is ongoing to improve the effectiveness of nutrient removal and optimize the overall treatment process.

Algae-based wastewater treatment offers a sustainable and environmentally friendly approach to managing nutrient-rich wastewater streams. It not only helps clean up water bodies but also has the potential to generate valuable biomass for various applications. As technology and research in this field continue to evolve, algae cultivation for wastewater treatment could become an integral part of water management strategies.

11 Limitations in Algae Use for Biogas Production Processes

While algae hold promise as a potential feedstock for biogas production, several limitations and challenges need to be addressed to optimize their use in biogas production processes. Some of the key limitations include:

Low Methane Content: Biogas is primarily composed of methane (CH_4) and carbon dioxide (CO_2). While certain organic materials, like food waste and agricultural residues, have higher methane content, algae typically have a lower methane yield per unit of biomass. This can impact the overall energy efficiency of biogas production.

Cell Wall Complexity: Algae often have complex cell walls made of cellulose and other compounds that can be challenging to break down during the anaerobic digestion process. Effective digestion requires the breakdown of these cell walls to release the organic content and produce methane.

Nutrient Imbalance: While algae are rich in nutrients, their composition can vary widely based on factors such as species, growth conditions, and harvesting methods. An imbalance of nutrients like carbon, nitrogen, and phosphorus can affect the efficiency of anaerobic digestion and the biogas yield.

Lack of Lignocellulosic Enzymes: Algae do not naturally contain the lignocellulosic enzymes (such as cellulases and hemicellulases) needed to efficiently break down their cell walls. This requires pre-treatment processes or the addition of external enzymes to enhance digestion.

Harvesting and Dewatering costs: Efficient harvesting and dewatering of algae biomass can be energy-intensive and costly. The water content of algae can be high, leading to increased transportation and processing costs.

Contamination Risk: Algae cultivation systems can be susceptible to contamination by other microorganisms, which could interfere with the digestion process. Proper management and sterilization practices are needed to minimize this risk.

Scale and Infrastructure: Scaling up algae cultivation for biogas production requires significant infrastructure investment and operational considerations. Large-scale cultivation systems need to be carefully designed and managed to ensure consistent and reliable biomass production.

Species Selection: Not all algae species are suitable for biogas production. Selecting the right species that are well-suited to anaerobic digestion and have favorable nutrient compositions is crucial.

C: N Ratio: The carbon-to-nitrogen (C: N) ratio is an important factor influencing the efficiency of anaerobic digestion. Algae with extremely low or high C: N ratios can lead to process imbalances and reduced biogas yield.

Research and Development: While there has been significant research into optimizing algae for biogas production, the technology is still evolving. Advances in algae genetics, cultivation techniques, and digestion processes are necessary to overcome these limitations.

In light of these limitations, it's important to conduct thorough research and feasibility assessments before implementing algae-based biogas production processes. Addressing these challenges may require innovative approaches, such as genetic engineering for improved digestibility, optimized cultivation practices, and integration with other waste streams to balance nutrient content.

12 Potential Benefits of Using Algae Biomass for Biogas Production Purposes

Using algae biomass for biogas production offers several potential benefits, despite the challenges and limitations. Here are some of the key advantages:

Renewable Energy Generation: Biogas, predominantly composed of methane (a potent greenhouse gas), can be used as a renewable energy source. Algae-based biogas production contributes to reducing reliance on fossil fuels and mitigates greenhouse gas emissions.

Carbon Sequestration: Algae capture carbon dioxide (CO_2) during photosynthesis. By cultivating and using algae for biogas production, CO_2 is captured from the atmosphere and then released as biogas when the algae biomass is digested. This creates a closed carbon loop, making the process more environmentally friendly.

Efficient Nutrient Recycling: Algae can utilize nutrients like nitrogen and phosphorus present in wastewater or agricultural runoff. Integrating algae cultivation into biogas production can help remove excess nutrients from water bodies, preventing eutrophication and improving water quality.

Waste Valorization: Algae can be cultivated using various organic waste streams, such as agricultural residues or wastewater nutrients. This turns waste materials into valuable biomass, which can be used for energy generation.

Diversification of Feedstock: Algae provide an alternative feedstock for biogas production, reducing competition with food crops and avoiding potential food security concerns associated with biofuel production.

Co-products and value-added products: Beyond biogas, algae biomass can be used to produce value-added products like bio-fertilizers, animal feed, and high-value compounds such as lipids, proteins, and pigments for various industries.

Local Production Potential: Algae can be cultivated in various locations, including non-arable land, wastewater treatment facilities, and even industrial sites with excess CO_2 emissions. This allows for local energy generation and waste treatment.

Short Growth cycle: Some algae species have rapid growth rates and can be harvested frequently, allowing for consistent biogas production and quicker turnover compared to traditional crops.

Flexible cultivation methods: Algae can be grown using a variety of cultivation systems, from open ponds to closed photobioreactors. This flexibility allows for adaptation to different environmental and operational conditions.

Research and Innovation: The field of algae-based biogas production is an area of ongoing research and innovation. Advances in genetics, cultivation techniques, and digestion processes could lead to improved efficiency and reduced limitations.

Economic opportunities: Algae-based biogas production can create economic opportunities in terms of job creation, technological innovation, and the development of new markets for algae-based products.

Energy Independence: Using locally grown algae biomass for biogas production can contribute to energy independence by reducing reliance on imported fossil fuels.

While there are challenges to overcome, the potential benefits of using algae biomass for biogas production make it an area of interest for researchers, engineers, and policymakers seeking sustainable energy and waste management solutions.

13 The Use of Macroalgae in Processes of Methane Fermentation

The use of macroalgae in processes of methane fermentation, also known as anaerobic digestion, involves the breakdown of macroalgae biomass by microorganisms in an oxygen-free environment to produce biogas, which is primarily composed of methane (CH_4) and carbon dioxide (CO_2). Here's how macroalgae can be utilized in methane fermentation processes:

Macroalgae Biomass Collection and Preparation: Macroalgae, commonly referred to as seaweeds, are large multicellular algae that grow in marine environments. They can be collected from coastal areas or cultivated in seaweed farms. The collected macroalgae need to be properly washed, cleaned, and potentially dried to reduce moisture content before entering the digestion process.

Anaerobic Digestion: The prepared macroalgae biomass is introduced into an anaerobic digester, which is a controlled environment where microorganisms break down the organic matter in the absence of oxygen. During this process, complex organic compounds in the macroalgae are broken down by a consortium of microorganisms, including bacteria and archaea.

Biogas Production: As the macroalgae biomass is digested, microorganisms produce biogas as a metabolic byproduct. This biogas consists of methane (CH_4) and carbon dioxide (CO_2), with methane being the primary component. The biogas can be collected and stored for energy generation or other applications.

Methane Capture and Utilization: The produced biogas, rich in methane, can be captured and used for various purposes. It can be burned for heat or electricity generation, used as a fuel for vehicles, or upgraded to natural gas quality for injection into the natural gas grid.

Digestate as a Byproduct: After the anaerobic digestion process is complete, a nutrient-rich residual material called digestate remains. This digestate can be used as a fertilizer due to its high content of nutrients that are present in the macroalgae biomass.

14 Benefits of Using Macroalgae for Methane Fermentation

Abundant Feedstock: Macroalgae are abundant in coastal regions and can provide a readily available and renewable feedstock for methane fermentation.

Carbon sequestration: Similar to land-based plants, macroalgae sequester carbon dioxide during their growth. Using them for methane fermentation can help capture and store carbon.

Nutrient Recycling: Macroalgae can absorb nutrients like nitrogen and phosphorus from the water, which can be beneficial for water quality improvement.

Reduced Competition with Land Use: Using macroalgae for biogas production avoids competing with land-based crops for resources like arable land and freshwater.

15 Challenges and Considerations

High Moisture Content: Macroalgae have a high water content, which can impact the energy efficiency of the digestion process due to the need to heat and maintain the reactor at the desired temperature.

Nutrient Balance: The nutrient composition of macroalgae can vary, and achieving an optimal carbon-to-nitrogen (C: N) ratio in the feedstock is crucial for efficient digestion.

Pre-treatment: Macroalgae cell walls can be resistant to breakdown, necessitating pre-treatment methods such as mechanical disruption or chemical pretreatment to enhance digestion.

Microbial Adaption: The microorganisms in the digestion process need to adapt to the specific characteristics of macroalgae biomass, which might require an acclimation period.

Scale and Infrastructure: Scaling up macroalgae-based methane fermentation requires appropriate infrastructure for biomass collection, transportation, and digestion.

Overall, utilizing macroalgae in methane fermentation processes has the potential to contribute to renewable energy generation, waste reduction, and sustainable nutrient management, provided that the challenges associated with macroalgae characteristics and the anaerobic digestion process are effectively addressed.

16 The Use of Microalgae in Processes of Methane Fermentation

Using microalgae in processes of methane fermentation, also known as anaerobic digestion, involves the breakdown of microalgae biomass by microorganisms in an oxygen-free environment to produce biogas, primarily composed of methane (CH_4) and carbon dioxide (CO_2). Here's how microalgae can be utilized in methane fermentation processes:

Microalgae Biomass cultivation: Microalgae are microscopic photosynthetic organisms that grow in aquatic environments. They can be cultivated in controlled conditions such as open ponds, closed photobioreactors, or other cultivation systems. During cultivation, microalgae photosynthetically fix carbon dioxide (CO_2) and accumulate organic matter.

Microalgae Harvesting and Biomass Preparation: Once the microalgae biomass has reached a suitable concentration, it needs to be harvested and prepared for anaerobic digestion. Harvesting methods include mechanical separation, flocculation, centrifugation, and filtration. After harvesting, the biomass is typically dewatered to reduce moisture content.

Anaerobic Digestion: The prepared microalgae biomass is introduced into an anaerobic digester, which is an environment devoid of oxygen. In the digester, microorganisms break down the organic matter present in the microalgae biomass. The complex organic compounds are converted into simpler molecules, including biogas.

Biogas Production: As the microalgae biomass undergoes anaerobic digestion, microorganisms produce biogas as a metabolic byproduct. The biogas generated consists mainly of methane (CH_4) and carbon dioxide (CO_2). This biogas can be collected and stored for energy generation or other applications.

Methane Capture and Utilization: The produced biogas, particularly the methane component, can be captured and utilized for various purposes. It can be burned for heat or electricity generation, used as a fuel for vehicles, or upgraded to natural gas quality for injection into the natural gas grid.

Digestate as a Byproduct: After the anaerobic digestion process, a nutrient-rich residue called digestate remains. This digestate can be used as a soil amendment due to its high nutrient content.

Benefits of Using Microalgae for Methane Fermentation.

Fast growth and high yield: Microalgae can have rapid growth rates and high biomass yields, making them a potentially productive feedstock for biogas production.

Carbon sequestration: Microalgae fix carbon dioxide (CO_2) during photosynthesis. Using them for methane fermentation can help capture and store carbon.

Nutrient Recycling: Microalgae can absorb nutrients such as nitrogen and phosphorus from water. Using them for biogas production can help remove excess nutrients and improve water quality.

Reduced competition with land use: Microalgae cultivation can occur in non-arable land or aquatic environments, reducing competition with land-based crops.

17 Challenges and Considerations

Nutrient Balance: Achieving the optimal carbon-to-nitrogen (C: N) ratio in the microalgae biomass is important for efficient digestion. Nutrient content can vary depending on cultivation conditions.

Cell wall composition: Some microalgae have complex cell walls that might require pre-treatment to improve digestibility.

Scale and Infrastructure: Scaling up microalgae-based methane fermentation requires appropriate infrastructure for biomass cultivation, harvesting, transportation, and digestion.

Energy Balance: The energy input required for microalgae cultivation, harvesting, and digestion should be considered in the overall energy balance.

Microbial Adaptation: The microorganisms in the digestion process need to adapt to the specific characteristics of microalgae biomass.

Overall, utilizing microalgae in methane fermentation processes has the potential to contribute to renewable energy generation, waste reduction, and sustainable nutrient management. Addressing challenges related to biomass composition, nutrient balance, and process optimization will be key to maximizing the benefits of this approach.

18 Conclusion

In conclusion, utilizing microalgae for biogas production through anaerobic digestion holds promise, but it requires careful consideration of various factors, including cost drivers and overall economic feasibility. Conducting thorough techno-economic analyses specific to the microalgae species used and the full-scale production process is necessary to determine its commercial viability and potential contributions to meeting energy demands sustainably. A detailing of the conclusion withdrawn from this chapter is summarized below:

1. Algae-based wastewater treatment offers a sustainable and environmentally friendly approach to managing nutrient-rich wastewater streams. It not only helps clean up water bodies but also has the potential to generate valuable biomass for various applications. Thus algae cultivation for wastewater treatment could become an integral part of water management strategies.
2. On the use of algae for biogas production, CO_2 is captured from the atmosphere and released as biogas when the algae biomass is digested which creates a closed carbon loop, making the process more environmentally friendly.
3. Cultivation of Algae using various organic waste streams, such as agricultural residues or wastewater nutrients turns waste materials into valuable biomass, which can be used for energy generation.
4. Algae provide an alternative feedstock for biogas production, and further algae biomass can be used to produce value-added products like biofertilizers, animal feed, and high-value compounds such as lipids, proteins, and pigments for various industries.
5. Using locally grown algae biomass for biogas production can contribute to energy independence by reducing reliance on imported fossil fuels.
6. Utilizing macro algae in methane fermentation processes has the potential to contribute to renewable energy generation, waste reduction, and sustainable nutrient management, provided that the challenges associated with macroalgae characteristics and the anaerobic digestion process are effectively addressed.
7. Microalgae is the best source of generating green biogas/biofuels cost-effectively through the method of cell wall weakening followed by pre-treatment. Microalgae have negligible lignin content, thus without harsh pre-treatment, making anaerobic digestion for wet algae biomass feedstock with minimal dewatering a potentially viable option.
8. Utilizing microalgae in methane fermentation processes has the potential to contribute to renewable energy generation, waste reduction, and sustainable nutrient management. Addressing challenges related to biomass composition, nutrient balance, and process optimization will be key to maximizing the benefits of this approach.

References

1. Kannah RY, Kavitha S, Karthikeyan OP, Rene ER, Kumar G, Banu JR (2021) A review on anaerobic digestion of energy and cost effective microalgae pretreatment for biogas production. Bioresour Technol https://doi.org/10.1016/j.biortech.2021.125055
2. Khan MI, Shin JH, Kim JD (2018) The promising future of microalgae: current status, challenges, and optimization of a sustainable and renewable industry for biofuels, feed, and other products. Microb Cell Fact. https://doi.org/10.1186/s12934-018-0879-x
3. Barsanti L, Gualtieri P (2018) Is exploitation of microalgae economically and energetically sustainable? Algal Res 31:107–115. https://doi.org/10.1016/j.algal.2018.02.001
4. Kannah RY, Velu C, Banu JR, Heimann K, Karthikeyan OP (2018) Food waste valorization by microalgae BT—waste to wealth. In: Singhania RR, Agarwal RA, Kumar RP, Sukumaran RK (eds) Springer Singapore, Singapore, pp 319–342
5. Kavitha S, Jayashree C, Kumar SA, Yeom IT, Banu JR (2014) The enhancement of anaerobic biodegradability of waste activated sludge by surfactant mediated biological pretreatment. Bioresour Technol 168:159–166
6. Kavitha S, Schikaran M, Kannah RY, Gunasekaran M, Kumar G, Banu JR (2019) Nanoparticle induced biological disintegration: a new phase separated pretreatment strategy on microalgal biomass for profitable biomethane recovery. Biores Technol 289:121624
7. Benedetti M, Vecchi V, Barera S et al (2018) Biomass from microalgae: the potential of domestication towards sustainable biofactories. Microb Cell Fact 17:173. https://doi.org/10.1186/s12934-018-1019-3
8. Wu N, Moreira CM, Zhang Y, Doan N, Yang S, Phlips EJ, Svoronos SA, Pullammanappallil PC (2019) Techno-economic analysis of biogas production from microalgae through anaerobic digestion. Intech Open. https://doi.org/10.5772/intechopen.86090
9. Barlow J, Sims RC, Quinn JC (2016) Techno-economic and life-cycle assessment of an attached growth algal biorefinery. Biores Technol 220:360–368
10. Manganaro JL, Lawal A, Goodall B (2015) Techno-economics of microalgae production and conversion to refineryready oil with co-product credits. Biofuels, Bioprod Biorefin 9(6):760–777
11. Hoffman J, Pate RC, Drennen T, Quinn JC (2017) Techno-economic assessment of open microalgae production systems. Algal Res 23:51–57
12. Singh RS, Singh RK, Tripathi N (2021) Chapter 11. Food waste utilization for the production of Biogas by anaerobic digestion: a case study in coal capital of India. In: Yadav S, Negm A, Yadava RN, (eds) Environmental management in India: waste to wealth. Springer
13. Chen Z, Rao Y, Usman M, Chen H, Białowiec A, Zhang S, Luo G (2021) Anaerobic fermentation of hydrothermal liquefaction wastewater of dewatered sewage sludge for volatile fatty acids production with focuses on the degradation of organic components and microbial community compositions. Sci Total Environ 777:146077
14. Bhatia SK, Jagtap SS, Bedekar AA, Bhatia RK, Patel AK, Pant D, Banu JR, Rao CV, Kim Y-G, Yang YH (2020) Recent developments in pretreatment technologies on lignocellulosic biomass: effect of key parameters, technological improvements, and challenges. Biores Technol 300:122724
15. Kumar G, Nguyen DD, Sivagurunathan P, Kobayashi T, Xu K, Chang SW (2018) Cultivation of microalgal biomass using swine manure for biohydrogen production: impact of dilution ratio and pretreatment. Bioresour Technol 260:16–22
16. Passos F, Ferrer I (2015) Influence of hydrothermal pretreatment on microalgal biomass anaerobic digestion and bioenergy production. Water Res 68:364–373
17. Ho S-H, Huang S-W, Chen C-Y, Hasunuma T, Kondo A, Chang JS (2013) Bioethanol production using carbohydrate-rich microalgae biomass as feedstock. Bioresour Technol 135:191–198
18. Cheng C-L, Che P-Y, Chen B-Y, Lee W-J, Lin C-Y, Chang J-S (2012) Biobutanol production from agricultural waste by an acclimated mixed bacterial microflora. Appl Energy 100:3–9

Integrated and Hybrid Bioelectrical Systems (BES) for Wastewater Treatment

Athar Hussain and Richa Madan

Abstract Industrial and agricultural wastewater is one of the most crucial sources of water pollution. Traditional system of wastewater, for instance, trickling filter, activated sludge and other aerobic technologies, are not cost and resource effective owing to population expansion and ever-increasing demand for clean water. The integration of state-of-the-art technologies such as bioelectrical systems (BES) with anaerobic digestion, dark fermentation and constructed wetlands is increasingly promising for industrial and domestic wastewater treatment. Microbial fuel cells (MFCs) are an upcoming technology where microbial catalytic reactions convert chemical energy stored in organic fraction of waste to electricity. This technique is also gradually applied for removal of a variety of pollutants namely pharmaceuticals, heavy metals, decolorization of dyes and xenobiotics. Hence, these systems provide benefits in two ways of electric power generation and wastewater treatment, thereby making the whole procedure sustainable and economically feasible. This chapter discusses the principles of microbial fuel cells and their subsequent utilization for removal of pollutants from wastewater, alongwith bioelectricity generation performance. Further, the chapter discusses integration of microbial fuel cells with current treatment technologies to scale up MFCs for sustainable wastewater treatment. The concluding remarks comprise of future applications and challenges faced in the real scale systems working for wastewater treatment and bioelectricity generation.

Keywords Bioelectrical system · Wastewater treatment · Microbial fuel cells · Dark fermentation · Bioelectricity

A. Hussain
Civil Engineering Department, Netaji Subhas University of Technology, West Campus, New Delhi, India

R. Madan (✉)
Gurukula Kangri (Deemed to be University), Haridwar, Uttarakhand, India
e-mail: richa.madan.92@gmail.com

M. M. Ghangrekar et al. (eds.), *Biological and Hybrid Wastewater Treatment Technology*, Earth and Environmental Sciences Library,
https://doi.org/10.1007/978-3-031-63046-0_15

1 Introduction

Owing to increased industrialization and urbanization, there happens to be a huge stress on the natural resources along with tremendous growth in the generation of waste materials. For long, humans have been exploiting the reserve of fossil fuels for the generation of energy to the extent that they might not be available to our future generations. Also, the intensifying issues related to global warming and climate change have accelerated the efforts towards relying more on renewable energy sources. Henceforth, it is vital to look for alternative sources of energy that are sustainable, economical and environment friendly.

In the last two decades, bioelectrical systems (BES) have become increasingly popular which can provide sustainable solutions to heavy metals, organic pollutants and nutrients. The main upcoming technologies is microbial fuel cells (MFCs) which exploits bio-catalytic potential of microorganisms which convert chemical energy in organic matter to electric current [9, 43]. MFCs can also help in the remediation of nutrients, removal of heavy metals, pharmaceuticals, decolorization of dyes alongwith generation of biofuels such as biomethane and biohydrogen [41]. Hence, these cells provide the two-fold advantage of treatment of wastewater and generation of electricity making them sustainable and economically feasible [13].

MFCs pose as bioreactors in which microbes act as biocatalysts and convert chemical energy from organic matter to electrical energy. The crucial components of an MFC include electrochemically active microbes, proton exchange membrane (PEM), cathode and anode electrodes and oxidizable waste [40]. The electricity in MFCs is generated due to a potential difference created due to electrons produced in the anode and electrons accepted in cathode [51]. In the anaerobic anodic chamber, electrochemically active microbes that can donate electrons to anode oxidize the organic matter in the waste and energy is produced. The protons that are generated in anode diffuse through PEM into cathodic chamber where oxygen is the oxidizing agent used due to its high reduction potential and abundance. Figure 1 shows the schematic depiction of MFC. Microbes capable of donating (exoelectrogens) and accepting (exoelectrotrophs) electrons are known as electroactive bacteria (EAB) [50]. To be categorized as an MFC, the cell should have a source of fuel such as wastewater which has to be replenished otherwise it is called a biobattery [37].

With an upsurge in research and innovation, the trend of hybrid technologies has been on the rise in recent years. This incorporates the integration of two or more treatment technologies to give shape to a new methodology which encompasses the advantages of the original processes. Such new integrated technologies possess reliable features to overcome the demerits of a single technology along with increased system efficiency [10, 46, 55–59, 74]. Although MFCs are effective in decreasing the amount of organic matter of wastewater, their capital, operation and maintenance expenses must be drastically lowered in order for such a revolutionary technology to be successfully field-tested. This may be accomplished by combining conventional techniques of wastewater treatment with MFCs, that provide treatment of wastewater at a reduced operational cost by lowering the total system cost while providing more

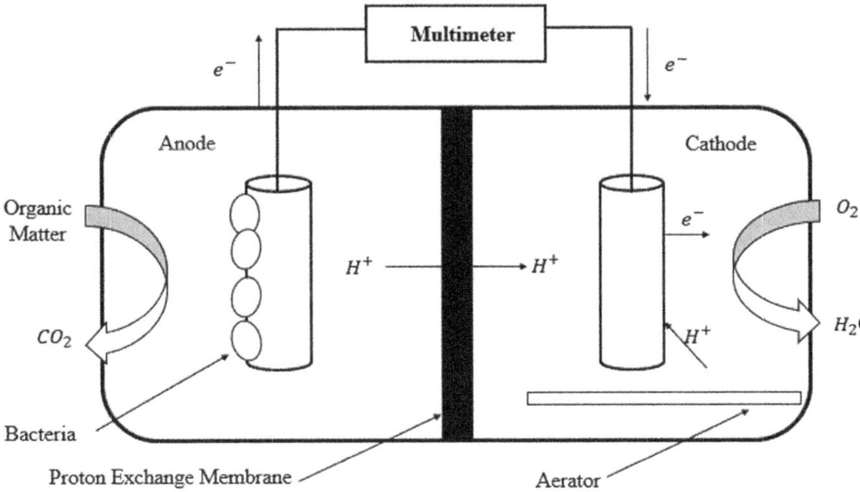

Fig. 1 Schematic representation of MFC

efficient treatment. Hence, MFCs have been combined with aerobic and anerobic systems, constructed wetlands, ASP, dark fermentation etc. to aid in treatment of sewage and wastewater and also help in solving the energy crisis.

2 Anaerobic Digestion—Microbial Fuel Cell

Anaerobic digestion (AD) remains a renowned technology where organic fraction of waste is stabilized with help of anaerobic microorganisms resulting in major end products of carbon dioxide and methane. The main procedural steps in anaerobic degradation are hydrolysis in which fats, proteins and carbohydrates are transformed to sugar, amino acids and fatty acids followed by their simultaneous conversion to volatile fatty acids and hydrogen by acidogenesis and acetogenesis with the end products being CO_2 and CH_4 by methanogenesis [62, 86]. It has the advantages of low energy requirement, waste stabilization, resource recovery and biogas production. However, to enhance the performance of this system some operational features have to be improved upon such as unbalanced C:N ratio, substrate inhibition, high pH variations due to organic loading and low methane production rate [19, 34, 66, 64]. Also, generation of sulfide in the anaerobic digestion adds to total expenditure of additional treatment of sulfide treatment. Hence, by integrating MFC with AD many such issues can be addressed. The major benefits of MFCs also include aspects such as environmental (low carbon footprint, water reclamation), energy (electricity generation, low sludge yield, low energy consumption due to lack of aeration) and economic (low operational costs, self-generation of microorganisms) [4, 22].

Some of the features that are essential to improve the AD process include; the rate of microbial substrate conversion and metabolism, the application of an exchange membrane allows to separate the waste and product as well as the elimination or recovery of repressing or valuable constituents. AD effluents often include a high quantity of residual organics which are mainly biodegradable and constitute a reservoir of unrecoverable chemical energy that might be transformed straight into electrical energy via MFC. Electricity generation might also counterbalance a portion of energy required in process of treatment if adequate energy-harvesting technologies are implemented which makes MFC as a cleansing step to purify AD discharge [6].

Many anaerobic reactors have long been used to integrate with MFCs. Upflow anaerobic sludge blanket (UASB) reactor is one such system which has been most widely applied for the treatment of industrial wastewater and can work at high loading rates such as 10–20 kgCOD/m^3.d [25, 26, 31, 36]. For the treatment of oilfield wastewater, a combined application of UASB and MFC was utilized and efficiency was evaluated across three distinct HRTs at 6, 13 and 20 h [16]. The results revealed that at hydraulic retention time (HRT) of 20 h, removal efficiencies of chemical oxygen demand (COD) and NH_3N were found to be highest at 93.5% and 83.4% respectively. However, MFC system at HRT of 13 h was revealed to achieve a maximum power output of 93 mW/m^2. MFCs were utilized for assessing elimination of total ammonia nitrogen (TAN) from swine effluent in batch mode. Total TAN and COD removal achieved was 5.8 and 80.5% in the anaerobic digestor leading with COD/TAN ratio of 3.0. However, in MFC, TAN removal efficiency was found to be 77.5% with COD/TAN ratio rising to 8.1. This study demonstrated that for wastewater rich in nitrogen, COD/TAN ratio happens to be an essential factor for the successful operation of AD-MFC system [31]. Figure 2 shows the AD-MFC coupled system.

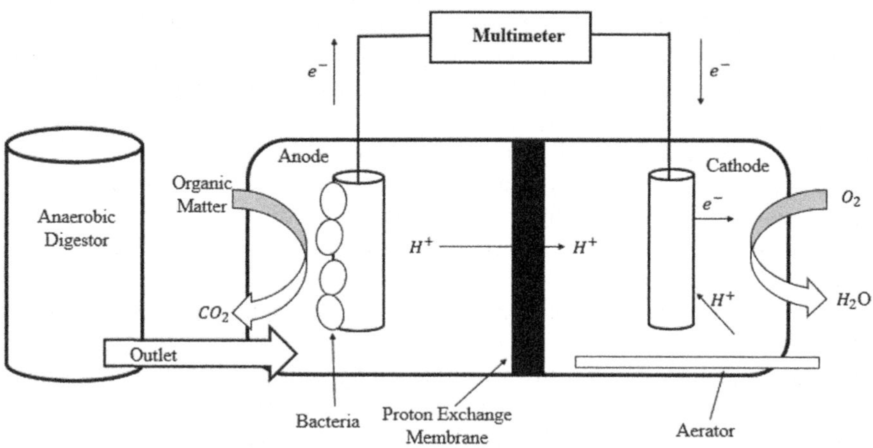

Fig. 2 Schematic representation of AD-MFC coupled system

3 Constructed Wetland—MFC Coupled System

Constructed wetlands (CWs) are designed sewage treatment systems which use natural mechanisms of soil, plants, microorganisms and substrates to perform a harmonious mix of physico-chemical and biological activities [14, 70]. A CW enables the biological elimination of contaminants and other organic waste from sewage and wastewater via the processes of deterioration, plant accumulation, filtration, conversion by microorganisms and adsorption. CW arrangements have advanced swiftly over the last two to three decades and are now widely used in the treatment of a wide range of sewage, including household, industrial wastewater, river water, rainwater and raw drinking water [3, 5, 49, 61]. CWs provide a dependable, acceptable, and environmentally friendly treatment procedure for established and economically disadvantaged locations owing to its low cost, operational feasibility and low maintenance. Though CWs are proven to be efficient for both typical and recalcitrant pollutants, they also pose some limitations, including substrate blockage, low removal efficiency, and inability to remove certain refractory pollutants [39]. Multiple strategies are being contemplated upon to boost CW efficiency, involving modifications in design and operation and wastewater recirculation [35], bioaugmentation, artificial aeration [42] and combinations with some other techniques for improving proton and electron transfer [71, 77, 78].

The bed of a CW is divided into two zones, aerobic being in the upper region while anaerobic in lower, identical to the MFC's anaerobic and aerobic compartments. Because MFCs and CW have physical similarities, they are appropriate for unification and may be used to gather power using MFCs. The coupled CW-MFC approach is known to be efficient not merely for eliminating phosphorus (P), total nitrogen (TN), BOD and COD but also for the removal of hazardous substances comprising of heavy metals, reactive dyes, hydrocarbons, pharmaceuticals, nutrients, pesticides and others, alongside generating electrical energy [52, 68]. As a result, artificial wetlands combined with microbial fuel cells provide a viable option for long-term, cost-effective, and energy-efficient treatment of wastewater with correspondingly generating bio-electricity. Past investigations have shown that the degradation of pollutants and performance of CW-MFC for generating bioelectricity relies on structure, electrode materials, and functioning factors as in temperature and residence time [15, 33, 84, 85]. The mechanisms employed by CWs for the removal of pollutants comprise of ion exchange, sedimentation, filtration, adsorption, plant uptake, co-precipitation and microbial degradation.

3.1 Components Affecting the Performance of CW-MFC System

(a) Role of Microorganisms

The major component, microbial flora, has a substantial impact on the detoxification and energy production efficiency of CW-MFC systems. EAB are abundant in anaerobic sludge and sediment found in wastewater treatment plants. EAB proliferate on electrode surfaces and roots of plants and produces energy by decomposing the organic materials, thereby releasing electrons. Pollutant oxidation is catalyzed by microorganisms on the anode surface, which produces metabolites like as protons, electrons and CO_2 [12, 76, 84]. Some microbial communities are more frequently enhanced around the electrode in CW-MFC systems. The anode has a greater quantity of microbes for electricity generation than the cathode [27, 69]. The type of active microbes at both poles also varies. Nitrite-oxidizing ammonia-oxidizing and anammox microbes are primarily assembled at anode, while phosphorus-accumulating and denitrifying bacteria are primarily concentrated near cathode. Table 1 shows the common electrogenic bacteria.

(b) Role of Plants

CW-MFC systems containing wetland plants tend to have an enhanced pollution removal efficacy than those which do not have plants. The presence of macrophytes frequently increases the size and diversity of bacterial population, alongwith its influence on pollution elimination and generation of bioelectricity [67, 69]. The roots of these macrophytes offer a large area for growth and adhesion of microbes, aiding in decomposition and pollutants absorption from sewage and wastewater [55]. During the process of phytoremediation and bioaccumulation aquatic florae play a crucial indirect or direct part in the absorption, storage and elimination of nutrients in their biomass. They absorb soluble inorganic nutrients in order to increase their biomass. Hence, the main criteria in choosing the macrophyte for elevated levels of pollutant elimination would be the one with an enhanced efficacy in absorbing and eliminating nutrients along with transforming them to organic biomass [17, 28, 60].

(c) Role of electrodes

For any CW-MFC system, the biocompatibility of material used for electrodes is critical for electrons passage from microorganisms to the electric circuit. The electrode's constituent material, influence the treatment efficiency, production of bioelectricity and operational stability of CW-MFC system [58, 79]. In order to optimize the use of the artificial wetland's natural redox gradient, cathode electrode is retained in aerobic zone while anode is concealed in the anaerobic zone. The properties of an ideal electrode comprise of strong conductance, high mechanical strength and low

Table 1 Common exo-electrogenic bacteria (EAB)

Denitrifying bacteria	Ochrobactrum, Pseudomonas
Dissimilatory metal-reducing bacteria	Geobacter, Geopsychrobacter, Shewanella, Geothrix
Sulfate-reducing bacteria	Desulfolobus, Desulfuromonas
Fermentative	E. coli, Clostridium

internal resistance. The other factors include wide surface area, porosity, biocompatibility, non-corrosivity and being chemical stable [29]. Carbon-based electrodes consisting of graphite rods, activated carbon, stainless steel, carbon felt, titanium mesh, and others can be utilized. Anode is immersed in specific chemicals to sustain anaerobic environment, which allows oxidation process to take place. EAB and other bacteria colonize the anode surface and decompose organic matter, creating protons and electrons. However, reduction takes place at cathode which is exposed freely to oxygen. The passage of electrons and protons from anode to cathode generates bioelectricity. For many circumstances, the generation of bioelectricity is directly associated with pollutant reduction expressed in COD and nitrogen inside a system [21, 72]. The electrode location also influences growth as well as reproduction of microorganisms. The amount of EAB also increases as depth of the chamber that houses the electrodes increases [2]. To enable a good cathodic reaction, the anode should be located in an anaerobic atmosphere whereas the cathode electrode must be located in aerated chamber [80]. The oxygen transport to the cathode is affected by the gap between the water surface and cathode. Corbella et al. [7] established the fact that the best location of cathode was found to be 1–2 cm above the surface of water to enhance performance of the system.

(d) Role of Substrate

Contaminants in wastewater from various sources impact the treatment efficacy of CW-MFC system. Domestic effluent contains a high organic matter load, which tends to reduce the effectiveness of wastewater treatment and also power density. Textile industry effluent contains resistant colors that must be decolorized during treatment. Heavy metals are present in effluent water from oil refineries and the smelting sector, and they must be removed and recovered [63]. The most important component of CW-MFCs are the substrates that promote development of macrophytes as well as wetland microorganisms. The physico-chemical mechanisms that are involved in CW-MFCs wastewater treatment include ion exchange, filtration and adsorption [8, 54, 83].

(e) Organic Loading Rate (OLR) and Hydraulic Retention Time (HRT)

In a MFC, the OLR affects quantity of substrate which is available to bacteria for breakdown and metabolic activity, reproduction and electron production. Aguirre-Sierra et al. [1] investigated surface flow CW-MFC at OLR of 2–24 BOD_5/m^2.d at HRT of 0.5–4 days and found that NH_3 and TN removal efficiencies were 97 and 69% at 3.4 days HRT. The interaction between the bacteria and the substrate and is also altered by HRT. Increasing the HRT improves treatment efficiency by increasing the pollutant removal rate and generation of electricity. [73] for CW-MFC operating at OLR of 27.6 gCOD/m^2.d, power output of 10.77 W/m^2 was obtained. [82] investigated operation of CW-MFC at various HRTs of 6, 12, 18, 24 and 48 h. The examination found a relation that as HRT is increased, the internal resistance and time required for MFC-CW system to reach a stable power output also rises.

3.2 CW-MFC Performance

Xu et al. [72, 75] investigated the ability of CW-MFC in treating municipal effluent with generation of bioelectricity. The study found that TN, COD and P removal was 82.4%, 82.3% and 95% respectively. After three days of continuous operation, the combined system gave the maximum power output was 3.71 W/m^2, which was greater than that of individual manmade wetlands. Yang [81] revealed the consequence of numerous parameters on performance optimization of CW-MFCs. It was established that CW-MFCs showed higher TP, TN, COD removal efficiency than typical CWs by 3.68%, 3.7%, 6.06% respectively. When run under ideal conditions of COD (200 mg/L), HRT (24 h) and external resistance of 1000 W, the highest power output was 107.54 mW/m^3. Likewise, Mu et al. [45] investigated many impacts of several parameters on removal efficiency of Cr(VI) with generation of bioelectricity, including effluent Cr(VI) and COD concentrations, HRT and spacing of electrodes in an unplanted CW-MFC. The starting Cr(VI) concentration was 40 mg/mL with 10 cm of electrode spacing and HRT of 3 days, they observed an optimal removal rate of Cr(VI) of 93.4%. However, when Cr(VI) was 60 mg/mL and COD was 500 mg/L, the maximum power output was 458.2 mW/m^3 with COD and Cr(VI) elimination efficacy at 92.5% and 90.7% respectively.

3.3 Challenges

The CW-MFC provides a revolutionary technology for addressing the present challenge of water scarcity and rising power demands. This is provided by not just reclaiming water from wastewater but additionally generating bioelectricity. As a result, CW-MFC is a cheaper, ecologically welcoming and sustainable way of treatment of wastewater than traditional technologies, that are still significantly energy and cost inclusive and also emit secondary pollutants. On top of eliminating phosphorus, COD, nitrogen and BOD from polluted effluents, CW-MFCs are also increasingly responding to lower amount of resistant pollutants like dyes, heavy metals and xenobiotics. One significant shortcoming of this technology is its inability to generate enough electricity for any practical use. As a result, deeper and radical investigations are needed to comprehend the flexible ecology of CW-MFCs systems in order to adjust and fine-tune them to generate more power output while also effectively treating wastewater. Furthermore, research on optimizing many factors such as HRT, power generation, substrate feed, pollutant removal efficacy, electrode materials, macrophyte species and microflora may also be enhanced.

4 Dark Fermentation – Microbial Electrolysis Cell

Dark fermentation (DF) occurs in the acidogenic phase of the anaerobic digestion that involves transformation of organic matter to hydrogen and CO_2. In lack of light and oxygen, facultative anaerobes and obligate anaerobes such as Enterobacter and Clostridium perform dark fermentation. Dark fermentation feeds on lignocellulosic biomass, carbohydrate sources such as industrial effluent, sugar-containing agricultural wastes, and municipal solid waste. A microbial electrolysis cell (MEC) is a novel bio-electrochemical equipment which employs organic matter of wastewater to create H_2. There are two types of (MECs): single-chamber and dual-chamber. The major difference between single and dual chamber cells is that in the single chamber cell both the electrodes are arranged in a single compartment without any ion exchange membrane making them small and inexpensive. Whereas, in the dual chamber MEC, the electrodes are divided by the proton exchange membrane making it more spacious [11]. The operational parameters that affect the hydrogen fermentation procedure are temperature, HRT, pH, inorganic content and volatile fatty acids. Microbes deliver more hydrogen under extreme thermophilic temperatures than mesophilic surroundings. At high thermophilic settings (70 °C), the maximum output of hydrogen was 4 mol/mole glucose [18, 47]. HRT is another critical parameter for dark fermentation biohydrogen generation. Low HRT is usually utilized in a standard continuous stirred-tank reactor (CSTR) for clear up of methanogens. A short HRT of less than 3 days in a CSTR system, according to Kim et al. might enhance hydrogen generation [32].

4.1 DF-MEC Performance

Under thermophilic conditions, generation of biohydrogen using palm oil mill wastewater was explored employing DF and microbial electrolysis. The pH and optimal COD content for DF were 6.5 and 66 g/L respectively, including a biohydrogen production of 73 mL-H_2g/COD. The best voltage for microbial electrolysis was found to be 0.7 V [30]. [24] evaluated DF-MEC sequential system for generation of hydrogen from food waste. It was found out that biogas generation rate was 0.83 L/L.d with hydrogen proportion of 92.58%. Jia et al. [56] investigated hydrogen generation from distillery effluent using DF-MEC. The maximum hydrogen production was found to be 99.7 mL-H_2/g VS_{added} at applied voltage of 1 V.

4.2 Challenges

MEC technique has been demonstrated to be an outstanding method for recovering resources from effluent. Some advantages of MEC include that they work with a variety of substrates, their H_2 yields are substantially higher and they improve the overall efficiency of the system by recovering chemicals. However, some disadvantages of the system include reactor configuration and type of substrate to be used. The system might be expensive as external power is required and overtime the generation of hydrogen declines [53, 65].

5 Microbial Carbon Capture Cell (MCC)

This happens to be a developing technique aiding in C-sequestration employing photosynthetic microorganisms alongwith generating electricity during wastewater treatment. In this cell, CO_2 produced during decomposition of organic material in anodic compartment by electrogenic microbes can essentially be used to generate bioelectricity and sequester CO_2. Numerous functional parameters which support microbial development in anodic compartment, such as analyte pH and electrode materials, alongwith cathodic factors as in photosynthetic microorganism development and concentration of CO_2, influence the power generated in MCC. Several additional environmental parameters, such as carbon supply, light intensity, influence microorganism development, which in turn influences biomass amount and hence CO_2 production [44].

5.1 Performance of MCC

Jadhav et al. [23] made use of *Anabaena ambigua* and *C. pyrenoidosa* inside the cathodic chamber of a MCC, with achieving energy densities of 4.29 W/m^3 and 6.36 W/m^3, respectively. Lu et al. [38] used photosynthetic microbes in cathodic compartment of MCC and could recover 80 - 93% of CO_2 and removed 99% and 56% COD from synthetic and real effluent respectively. Hou et al. [20] achieved an energy output of 6.3 W/m^3 for MCC, as well as COD elimination of 44% by utilizing *Golenkinia sp* in cathodic compartment. Onyeaka and Ekwebelem [48] also reviewed the engineered bacteria for enhanced carbon capture and storage.

5.2 Challenges

The concept of microbial carbon capture is still in the nascent stage owing to technological hurdles, limited knowledge and equipment. The energy efficacy has to be beefed up for wide scale application of such a system. Also, the operational parameters comprising of temperature, substrate type, pH and electrode materials requires more attention for proper application of this system.

6 Future Outlook and Conclusion

MFC is undeniably an intriguing technology for meeting sustainable goals, and it has the potential to drastically alter the wastewater treatment landscape. However, because MFCs do currently limit themselves by low energy output, high construction and operating costs, it appears to be an improved approach to incorporate MFCs with traditional techniques of treating wastewater to overcome the latter's inadequacies, including elimination of aromatic and recalcitrant compounds, endocrine disruptors and pharmaceuticals. In contrast to conventional technologies of wastewater treatment which have proven to attain exceptional treatment performance and functional constancy via extensive analysis and through optimizing many mechanism parameters, the MFC technique is relatively in infant stages. Furthermore, in any combined system, it happens to be of vital importance to assure effective operation of both contributing technologies, therefore appropriate consideration ought to be to paid it.

Significant development and technological advancement have occurred in the arena of MFC for wastewater treatment. MFCs' potential advantages over traditional wastewater treatment systems include its environmentally friendly and carbon–neutral tactic to waste decomposition and concurrent creation of a clean and eco-friendly supply of electricity. Production of energy happens to be a one-step procedure with steady functioning and sustainability due to self-proliferation of anodic microbiota, which provides strong endurance to ecological stress. Nevertheless, efficiency and also cost-related difficulties with PEM, cathode catalyst and upscaling must be overcome before this emerging technology can be used in the real world. However, MFCs ought to be associated with treatment units for their widespread application and more lies in this technology need to be explored.

References

1. Aguirre-Sierra A, Bacchetti-De Gregoris T, Berná A, Salas JJ, Aragón C, Esteve-Núñez A (2016) Microbial electrochemical systems outperform fixed-bed biofilters in cleaning up urban wastewater. Environ Sci Water Res Technol 2(6):984–993. https://doi.org/10.1039/C6EW00172F
2. An J, Kim B, Nam J, Ng HY, Chang IS (2013) Comparison in performance of sediment microbial fuel cells according to depth of embedded anode. Biores Technol 127:138–142. https://doi.org/10.1016/j.biortech.2012.09.095
3. Avellan CT, Ardakanian R, Gremillion P (2017) The role of constructed wetlands for biomass production within the water-soil-waste nexus. Water Sci Technol 75(10):2237–2245. https://doi.org/10.2166/wst.2017.106
4. Cao TND, Chen SS, Ray SS, Le HQ, Chang HM (2019) Application of microbial fuel cell in wastewater treatment and simultaneous bioelectricity generation. In: Bui XT, Chiemchaisri C, Fujioka T, Varjani S (eds) Water and wastewater treatment technologies. Energy, environment, and sustainability. Springer, Singapore, pp 501–526. https://doi.org/10.1007/978-981-13-3259-3_23
5. Chatterjee S, Mitra A, Gupta SK, Gupta DK (2019) A review on reed bed system as a potential decentralized wastewater treatment practice. In: Advances in plant transgenics: methods and

applications. Springer, Singapore, pp 239–251. https://doi.org/10.1007/978-981-13-9624-3_11
6. Cheng KY, Kaksonen AH (2017) Integrating microbial electrochemical technologies with anaerobic digestion for waste treatment: possibilities and perspectives. Curr Develop Biotechnol Bioeng 191–221. https://doi.org/10.1016/B978-0-444-63664-5.00009-5
7. Corbella C, Garfí M, Puigagut J (2016) Long-term assessment of best cathode position to maximise microbial fuel cell performance in horizontal subsurface flow constructed wetlands. Sci Total Environ 563:448–455. https://doi.org/10.1016/j.scitotenv.2016.03.170
8. Das B, Thakur S, Chaithanya MS, Biswas P (2019) Batch investigation of constructed wetland microbial fuel cell with reverse osmosis (RO) concentrate and wastewater mix as substrate. Biomass Bioenerg 122:231–237. https://doi.org/10.1016/j.biombioe.2019.01.017
9. Du Z, Li H, Gu T (2007) A state of the art review on microbial fuel cells: a promising technology for wastewater treatment and bioenergy. Biotechnol Adv 25(5):464–482. https://doi.org/10.1016/j.biotechadv.2007.05.004
10. Dwivedi KA, Huang SJ, Wang CT (2022) Integration of various technology-based approaches for enhancing the performance of microbial fuel cell technology: a review. Chemosphere 287:132248. https://doi.org/10.1016/j.chemosphere.2021.132248
11. El-Sayad RM, Khalil AS (2023) Wastewater treatment and hydrogen production via Microbial electrolysis cells (MECs) and fermentation methods: a comparative review. J Contemp Technol Appl Eng 2(1):29–41. https://doi.org/10.21608/jctae.2023.218803.1014
12. Fang Z, Cao X, Li X, Wang H, Li X (2017) Electrode and azo dye decolorization performance in microbial-fuel-cell-coupled constructed wetlands with different electrode size during long-term wastewater treatment. Biores Technol 238:450–460. https://doi.org/10.1016/j.biortech.2017.04.075
13. Flimban SG, Ismail IM, Kim T, Oh SE (2019) Overview of recent advancements in the microbial fuel cell from fundamentals to applications: Design, major elements, and scalability. Energies 12(17):3390. https://doi.org/10.3390/en12173390
14. Fu G, Wu J, Han J, Zhao L, Chan G, Leong K (2020) Effects of substrate type on denitrification efficiency and microbial community structure in constructed wetlands. Biores Technol 307:123222. https://doi.org/10.1016/j.biortech.2020.123222
15. Ge X, Cao X, Song X, Wang Y, Si Z, Zhao Y et al (2020) Bioenergy generation and simultaneous nitrate and phosphorus removal in a pyrite-based constructed wetland-microbial fuel cell. Bioresour Technol 296:122350. https://doi.org/10.1016/j.biortech.2019.122350
16. Gong D, Qin G (2012) Treatment of oilfield wastewater using a microbial fuel cell integrated with an up-flow anaerobic sludge blanket reactor. Desalin Water Treat 49(1–3):272–280. https://doi.org/10.1080/19443994.2012.719336
17. Greenway M (2007) The role of macrophytes in nutrient removal using constructed wetlands. Environ Bioremediat Technol 331–351. Springer, Berlin. https://doi.org/10.1007/978-3-540-34793-4_15
18. Van Groenestijn JW, Hazewinkel JHO, Nienoord M, Bussmann PJT (2002) Energy aspects of biological hydrogen production in high rate bioreactors operated in the thermophilic temperature range. Int J Hydrogen Energy 27(11–12):1141–1147. https://doi.org/10.1016/S0360-3199(02)00096-4
19. Hobbs SR, Landis AE, Rittmann BE, Young MN, Parameswaran P (2018) Enhancing anaerobic digestion of food waste through biochemical methane potential assays at different substrate: inoculum ratios. Waste Manage 71:612–617. https://doi.org/10.1016/j.wasman.2017.06.029
20. Hou Q, Nie C, Pei H, Hu W, Jiang L, Yang Z (2016) The effect of algae species on the bioelectricity and biodiesel generation through open-air cathode microbial fuel cell with kitchen waste anaerobically digested effluent as substrate. Biores Technol 218:902–908. https://doi.org/10.1016/j.biortech.2016.07.035
21. Huang S, Zhu G, Gu X (2020) The relationship between energy production and simultaneous nitrification and denitrification via bioelectric derivation of microbial fuel cells at different anode numbers. Environ Res 184:109247. https://doi.org/10.1016/j.envres.2020.109247

22. Imoro AZ, Acheampong NA, Oware S, Okrah H, Coulibaly VT, Ali AG et al (2022) The potential benefits of microbial fuel cells in the context of the sustainable development goals. In: Microbial fuel cells for environmental remediation. Springer Nature, Singapore, pp 167–182. https://doi.org/10.1007/978-981-19-2681-5_9
23. Jadhav DA, Jain SC, Ghangrekar MM (2017) Simultaneous wastewater treatment, algal biomass production and electricity generation in clayware microbial carbon capture cells. Appl Biochem Biotechnol 183:1076–1092. https://doi.org/10.1007/s12010-017-2485-5
24. Jia X, Li M, Wang Y, Wu Y, Zhu L, Wang X, Zhao Y (2020) Enhancement of hydrogen production and energy recovery through electro-fermentation from the dark fermentation effluent of food waste. Environ Sci Ecotechnol 1:100006. https://doi.org/10.1016/j.ese.2019.100006
25. Jia H, Liu W, Wang J, Ngo HH, Guo W, Zhang H (2018) Optimization of sensing performance in an integrated dual sensors system combining microbial fuel cells and upflow anaerobic sludge bed reactor. Chemosphere 210:931–940. https://doi.org/10.1016/j.chemosphere.2018.07.119
26. Jia H, Yang G, Wang J, Ngo HH, Guo W, Zhang H, Zhang X (2016) Performance of a microbial fuel cell-based biosensor for online monitoring in an integrated system combining microbial fuel cell and upflow anaerobic sludge bed reactor. Biores Technol 218:286–293. https://doi.org/10.1016/j.biortech.2016.06.064
27. Jiang X, Shen J, Lou S, Mu Y, Wang N, Han W et al (2016) Comprehensive comparison of bacterial communities in a membrane-free bioelectrochemical system for removing different mononitrophenols from wastewater. Bioresour Technol 216:645–652. https://doi.org/10.1016/j.biortech.2016.06.005
28. Jingyu H, Miwornunyuie N, Ewusi-Mensah D, Koomson DA (2020) Assessing the factors influencing the performance of constructed wetland–microbial fuel cell integration. Water Sci Technol 81(4):631–643. https://doi.org/10.2166/wst.2020.135
29. Kalathil S, Patil SA, Pant D (2018) Microbial fuel cells: electrode materials. Encycl Interfacial Chem 309:318. https://doi.org/10.1016/B978-0-12-409547-2.13459-6
30. Khongkliang P, Jehlee A, Kongjan P, Reungsang A, Sompong O (2019) High efficient biohydrogen production from palm oil mill effluent by two-stage dark fermentation and microbial electrolysis under thermophilic condition. Int J Hydrogen Energy 44(60):31841–31852. https://doi.org/10.1016/j.ijhydene.2019.10.022
31. Kim T, An J, Jang JK, Chang IS (2015) Coupling of anaerobic digester and microbial fuel cell for COD removal and ammonia recovery. Biores Technol 195:217–222. https://doi.org/10.1016/j.biortech.2015.06.009
32. Kim IS, Hwang MH, Jang NJ, Hyun SH, Lee ST (2004) Effect of low pH on the activity of hydrogen utilizing methanogen in bio-hydrogen process. Int J Hydrogen Energy 29(11):1133–1140. https://doi.org/10.1016/j.ijhydene.2003.08.017
33. Kim C, Kim JR, Heo J (2019) Enhancement of bioelectricity generation by a microbial fuel cell using Ti nanoparticle-modified carbon electrode. J Chem Technol Biotechnol 94(5):1622–1627. https://doi.org/10.1002/jctb.5931
34. Latif MA, Mehta CM, Batstone DJ (2017) Influence of low pH on continuous anaerobic digestion of waste activated sludge. Water Res 113:42–49. https://doi.org/10.1016/j.watres.2017.02.002
35. Lian-Sheng H, Hong-Liang L, Bei-Dou X, Ying-Bo Z (2006) Effects of effluent recirculation in vertical-flow constructed wetland on treatment efficiency of livestock wastewater. Water Sci Technol 54(11–12):137–146. https://doi.org/10.2166/wst.2006.845
36. Liu W, Yang G, Jia H, Wang J (2020) A novel UASB-MFC dual sensors system for wastewater treatment: on-line sensor recovery and electrode cleaning in the long-term operation. Chemosphere 246:125751. https://doi.org/10.1016/j.chemosphere.2019.125751
37. Logan BE, Hamelers B, Rozendal R, Schröder U, Keller J, Freguia S et al (2006) Microbial fuel cells: methodology and technology. Environ Sci Technol 40(17):5181–5192. https://doi.org/10.1021/es0605016
38. Lu L, Huang Z, Rau GH, Ren ZJ (2015) Microbial electrolytic carbon capture for carbon negative and energy positive wastewater treatment. Environ Sci Technol 49(13):8193–8201. https://doi.org/10.1021/acs.est.5b00875

39. Lutterbeck CA, Machado ÊL, Sanchez-Barrios A, Silveira EO, Layton D, Rieger A, Lobo EA (2020) Toxicity evaluation of hospital laundry wastewaters treated by microbial fuel cells and constructed wetlands. Sci Total Environ 729:138816. https://doi.org/10.1016/j.scitotenv.2020.138816
40. Ma Z, Song H, Wang H, Xu P (2017) Improving the performance of microbial fuel cells by reducing the inherent resistivity of carbon fiber brush anodes. J Power Sources 348:193–200. https://doi.org/10.1016/j.jpowsour.2017.02.083
41. Mathuriya AS, Yakhmi JV (2016) Microbial fuel cells–Applications for generation of electrical power and beyond. Crit Rev Microbiol 42(1):127–143. https://doi.org/10.3109/1040841X.2014.905513
42. Merlin, G., & Cottin, N. (2012). Lab-scale performance evaluation of vertical flow reed beds for the treatment of chlorobenzene contaminated groundwater. J Environ Protect 3:847.https://doi.org/10.4236/jep.2012.328099
43. Mohan SV, Velvizhi G, Modestra JA, Srikanth S (2014) Microbial fuel cell: critical factors regulating bio-catalyzed electrochemical process and recent advancements. Renew Sustain Energy Rev 40:779–797. https://doi.org/10.1016/j.rser.2014.07.109
44. Mohan SV, Rohit MV, Subhash GV, Chandra R, Devi MP, Butti SK, Rajesh K (2019) Algal oils as biodiesel. In: Biofuels from algae. Elsevier, Amsterdam, pp 287–323. https://doi.org/10.1016/B978-0-444-64192-2.00012-3
45. Mu C, Wang L, Wang L (2020) Performance of lab-scale microbial fuel cell coupled with unplanted constructed wetland for hexavalent chromium removal and electricity production. Environ Sci Pollut Res 27:25140–25148. https://doi.org/10.1007/s11356-020-08982-z
46. Nguyen DD, Ngo HH, Yoon YS (2014) A new hybrid treatment system of bioreactors and electrocoagulation for superior removal of organic and nutrient pollutants from municipal wastewater. Biores Technol 153:116–125. https://doi.org/10.1016/j.biortech.2013.11.048
47. Van Niel EWJ, Budde MAW, De Haas GG, Van der Wal FJ, Claassen PAM, Stams AJM (2002) Distinctive properties of high hydrogen producing extreme thermophiles, *Caldicellulosiruptor saccharolyticus* and *Thermotoga elfii*. Int J Hydrogen Energy 27(11–12):1391–1398
48. Onyeaka H, Ekwebelem OC (2023) A review of recent advances in engineering bacteria for enhanced CO_2 capture and utilization. Int J Environ Sci Technol 20(4):4635–4648. https://doi.org/10.1007/s13762-022-04303-8
49. Parde D, Patwa A, Shukla A, Vijay R, Killedar DJ, Kumar R (2021) A review of constructed wetland on type, treatment and technology of wastewater. Environ Technol Innov 21:101261. https://doi.org/10.1016/j.eti.2020.101261
50. Parkash A (2016) Microbial fuel cells: a source of bioenergy. J Microb Biochem Technol 8(3):247–255. https://doi.org/10.4172/1948-5948.1000293
51. Rahimnejad M, Adhami A, Darvari S, Zirepour A, Oh SE (2015) Microbial fuel cell as new technology for bioelectricity generation: a review. Alex Eng J 54(3):745–756. https://doi.org/10.1016/j.aej.2015.03.031
52. Rahman ME, Bin Halmi MIE, Bin Abd Samad MY, Uddin MK, Mahmud K, Abd Shukor MY et al (2020) Design, operation and optimization of constructed wetland for removal of pollutant. Int J Environ Res Public Health 17(22):8339. https://doi.org/10.3390/ijerph17228339
53. Segundo-Aguilar A, González-Gutiérrez LV, Payá VC, Feliu J, Buitrón G, Cercado B (2021) Energy and economic advantages of simultaneous hydrogen and biogas production in microbial electrolysis cells as a function of the applied voltage and biomass content. Sustain Energy Fuels 5(7):2003–2017
54. Sgroi M, Pelissari C, Roccaro P, Sezerino PH, García J, Vagliasindi FG, Ávila C (2018) Removal of organic carbon, nitrogen, emerging contaminants and fluorescing organic matter in different constructed wetland configurations. Chem Eng J 332:619–627. https://doi.org/10.1016/j.cej.2017.09.122
55. Sierra MAA (2017) Integrating microbial electrochemical systems in constructed wetlands: a new paradigm for treating wastewater in small communities. Doctoral dissertation, Universidad de Alcalá, Spain, pp 100–165

56. Sittijunda S, Baka S, Jariyaboon R, Reungsang A, Imai T, Kongjan P (2022) Integration of dark fermentation with microbial electrolysis cells for biohydrogen and methane production from distillery wastewater and glycerol waste co-digestion. Fermentation 8(10):537. https://doi.org/10.3390/fermentation8100537
57. Srivastava P, Abbassi R, Garaniya V, Lewis T, Yadav AK (2020) Performance of pilot-scale horizontal subsurface flow constructed wetland coupled with a microbial fuel cell for treating wastewater. J Water Process Eng 33:100994. https://doi.org/10.1016/j.jwpe.2019.100994
58. Srivastava P, Abbassi R, Yadav AK, Garaniya V, Asadnia M (2020) A review on the contribution of electron flow in electroactive wetlands: electricity generation and enhanced wastewater treatment. Chemosphere 254:126926. https://doi.org/10.1016/j.chemosphere.2020.126926
59. Srivastava P, Abbassi R, Yadav AK, Garaniya V, Khan F (2020) Microbial fuel cell–integrated wastewater treatment systems. In: Integrated microbial fuel cells for wastewater treatment. Butterworth-Heinemann, pp 29–46. https://doi.org/10.1016/B978-0-12-817493-7.00002-3
60. Srivastava P, Yadav AK, Garaniya V, Abbassi R (2019) Constructed wetland coupled microbial fuel cell technology: development and potential applications. In: Microbial electrochemical technology. Elsevier, Amsterdam, pp 1021–1036. https://doi.org/10.1016/B978-0-444-64052-9.00042-X
61. Stefanakis AI (2019) The role of constructed wetlands as green infrastructure for sustainable urban water management. Sustainability 11(24):6981. https://doi.org/10.3390/su11246981
62. Uddin MM, Wright MM (2022) Anaerobic digestion fundamentals, challenges, and technological advances. Phys Sci Rev. https://doi.org/10.1515/psr-2021-0068
63. Villasenor J, Capilla P, Rodrigo MA, Canizares P, Fernández FJ (2013) Operation of a horizontal subsurface flow constructed wetland–microbial fuel cell treating wastewater under different organic loading rates. Water Res 47(17):6731–6738. https://doi.org/10.1016/j.watres.2013.09.005
64. Wang W, Chang JS, Lee DJ (2022) Integrating anaerobic digestion with bioelectrochemical system for performance enhancement: a mini review. Biores Technol 345:126519. https://doi.org/10.1016/j.biortech.2021.126519
65. Wang L, Trujillo S, Liu H (2019) Selective inhibition of methanogenesis by acetylene in single chamber microbial electrolysis cells. Biores Technol 274:557–560. https://doi.org/10.1016/j.biortech.2018.12.039
66. Wang H, Zhang Y, Angelidaki I (2016) Ammonia inhibition on hydrogen enriched anaerobic digestion of manure under mesophilic and thermophilic conditions. Water Res 105:314–319. https://doi.org/10.1016/j.watres.2016.09.006
67. Wang J, Song X, Wang Y, Bai J, Li M, Dong G et al (2017) Bioenergy generation and rhizodegradation as affected by microbial community distribution in a coupled constructed wetland-microbial fuel cell system associated with three macrophytes. Sci Total Environ 607:53–62. https://doi.org/10.1016/j.scitotenv.2017.06.243
68. Wang S, Kong F (2022) Electricity production and the analysis of the anode microbial community in a constructed wetland-microbial fuel cell. In: Phytoremediation technology for the removal of heavy metals and other contaminants from soil and water. Elsevier, Amsterdam, pp 571–600. https://doi.org/10.1016/B978-0-323-85763-5.00009-X
69. Wang Y, Zhao Y, Xu L, Wang W, Doherty L, Tang C et al (2017) Constructed wetland integrated microbial fuel cell system: looking back, moving forward. Water Sci Technol 76(2):471–477. https://doi.org/10.2166/wst.2017.190
70. Wu H, Zhang J, Ngo HH, Guo W, Hu Z, Liang S et al (2015) A review on the sustainability of constructed wetlands for wastewater treatment: design and operation. Bioresour Technol 175:594–601. https://doi.org/10.1016/j.biortech.2014.10.068
71. Wu S, Kuschk P, Brix H, Vymazal J, Dong R (2014) Development of constructed wetlands in performance intensifications for wastewater treatment: a nitrogen and organic matter targeted review. Water Res 57:40–55
72. Xu F, Cao FQ, Kong Q, Zhou LL, Yuan Q, Zhu YJ, Wang Q (2018) Electricity production and evolution of microbial community in the constructed wetland-microbial fuel cell. Chem Eng J 339:479–486. https://doi.org/10.1016/j.cej.2018.02.003

73. Xu D, Xiao ER, Xu P, Zhou Y, Zhou QH, Xu D, Wu ZB (2017) Effects of influent organic loading rates and electrode locations on the electrogenesis capacity of constructed wetland-microbial fuel cell systems. Environ Prog Sustain Energy 36(2):435–441
74. Xu L, Zhao Y, Doherty L, Hu Y, Hao X (2016) The integrated processes for wastewater treatment based on the principle of microbial fuel cells: a review. Crit Rev Environ Sci Technol 46(1):60–91. https://doi.org/10.1080/10643389.2015.1061884
75. Xu L, Zhao Y, Wang X, Yu W (2018) Applying multiple bio-cathodes in constructed wetland-microbial fuel cell for promoting energy production and bioelectrical derived nitrification-denitrification process. Chem Eng J 344:105–113. https://doi.org/10.1016/j.cej.2018.03.065
76. Xu F, Ouyang DL, Rene ER, Ng HY, Guo LL, Zhu YJ et al (2019) Electricity production enhancement in a constructed wetland-microbial fuel cell system for treating saline wastewater. Bioresour Technol 288:121462. https://doi.org/10.1016/j.biortech.2019.121462
77. Yadav AK (2010) Design and development of novel constructed wetland cum microbial fuel cell for electricity production and wastewater treatment. In: Proceedings of 12th international conference on wetland systems for water pollution control (IWA), pp 4–10
78. Yadav AK, Srivastava P, Kumar N, Abbassi R, Mishra BK (2018) Constructed wetland-microbial fuel cell: an emerging integrated technology for potential industrial wastewater treatment and bio-electricity generation. In: Constructed wetlands for industrial wastewater treatment, pp 493–510
79. Yakar A, Türe C, Türker OC, Vymazal J, Saz Ç (2018) Impacts of various filtration media on wastewater treatment and bioelectric production in up-flow constructed wetland combined with microbial fuel cell (UCW-MFC). Ecol Eng 117:120–132. https://doi.org/10.1016/j.ecoleng.2018.03.016
80. Yan Z, Jiang H, Cai H, Zhou Y, Krumholz LR (2015) Complex interactions between the macrophyte *Acorus calamus* and microbial fuel cells during pyrene and benzo [a] pyrene degradation in sediments. Sci Rep 5(1):10709. https://doi.org/10.1038/srep10709
81. Yang GW (2015) Biological power generation constructed wetland system for treatment of domestic sewage. J Zhejiang Univ (Eng Ed) 49(6):1186–1192
82. Yang H, Chen J, Yu L, Li W, Huang X, Qin Q, Zhu S (2022) Performance optimization and microbial community evaluation for domestic wastewater treatment in a constructed wetland-microbial fuel cell. Environ Res 212:113249. https://doi.org/10.1016/j.envres.2022.113249
83. Yang Y, Zhao Y, Liu R, Morgan D (2018) Global development of various emerged substrates utilized in constructed wetlands. Biores Technol 261:441–452. https://doi.org/10.1016/j.biortech.2018.03.085
84. Yu B, Liu C, Wang S, Wang W, Zhao S, Zhu G (2020) Applying constructed wetland-microbial electrochemical system to enhance NH_4^+ removal at low temperature. Sci Total Environ 724:138017. https://doi.org/10.1016/j.scitotenv.2020.138017
85. Yu J, Park Y, Lee T (2015) Electron flux and microbial community in microbial fuel cells (open-circuit and closed-circuit modes) and fermentation. J Ind Microbiol Biotechnol 42(7):979–983. https://doi.org/10.1007/s10295-015-1629-2
86. Zamri MFMA, Hasmady S, Akhiar A, Ideris F, Shamsuddin AH, Mofijur M et al (2021) A comprehensive review on anaerobic digestion of organic fraction of municipal solid waste. Renew Sustain Energy Rev 137:110637. https://doi.org/10.1016/j.rser.2020.110637

Bioelectrochemical Systems (BES) for Wastewater Treatment

Indrasis Das

Abstract Bioelectrochemical Systems (BESs) exhibit substantial potential for addressing wastewater treatment challenges, emphasizing sustainability and resource management in contemporary contexts. Utilizing microbial activities to mediate electrochemical reactions, BESs offer a spectrum of benefits, including efficient wastewater treatment, electricity recovery, bioresource recovery, and carbon dioxide sequestration. The research in this context started with Microbial Fuel Cells (MFCs) and subsequently, next-generation BES technologies, such as Microbial Electrolysis Cells (MEC), Microbial Desalination Cells (MDC), Microbial Carbon Capture Cells (MCC), Microbial Electrosynthesis (MES), Plant and Wetland-MFC, and bioelectro-Fenton (BEF) evolved. The versatility of BES architectures, including single-chambered and multi-chambered designs are explored, offering personalized advantages for wastewater treatment. Thorough investigations into the operational principles of MFCs and advancements in various technologies, like MEC, MDC, MCC, and MES, emphasize their potential applications in wastewater treatment, energy recovery, and environmental sustainability. Despite over 3000 research articles available as per Scopus, BES technologies are facing challenges during field-scale implementation and successful commercialization as viable alternatives to conventional technological options. This chapter elucidates the wastewater treatment processes of different BES technologies and the suitability of specific BES technologies for specific treatment requirements.

Keywords BES technologies · Microbial electrochemical technology · Microbial fuel cells · Sustainability · Wastewater treatment

---

I. Das (✉)
Environmental Engineering Department, CSIR-Central Leather Research Institute, Adyar, Chennai 600020, India
e-mail: indrasis@clri.res.in

Academy of Scientific and Innovative Research (AcSIR), Ghaziabad 201002, India

M. M. Ghangrekar et al. (eds.), *Biological and Hybrid Wastewater Treatment Technology*, Earth and Environmental Sciences Library,
https://doi.org/10.1007/978-3-031-63046-0_16

1 Introduction

Bioelectrochemical systems (BESs) are a group of multidisciplinary technologies leveraging microbial activities to mediate electrochemical reactions [1]. By proficiently operating BES reactors, a spectrum of benefits can be achieved which includes the efficient treatment [2], recovery of electricity [3], recovery of bioresources or biofuels [1], and sequestration of carbon dioxide [4]. Microbial fuel cells (MFCs), were first explored BES technology (Potter, 1911). The end of the twentieth century and the twenty-first century witnessed a gradual evolution in MFC architectures, with diverse modifications, alongside the design of alternative BES technologies [1]. These advancements in BES technologies highlighted it as a promising option to achieve environmental sustainability and mitigate resource management challenges. In BES, microbial actions and electrochemical processes synergize for efficient wastewater treatment and electricity generation, bioresource recovery, and carbon dioxide sequestration.

The diverse architectures of BESs offer specific advantages personalized to the treatment of wastewater with distinct characteristics. The modification and evolution of these BES technologies play a pivotal role in propelling the growth and development of the broader BES research domain. Notably, BES technologies such as Microbial Electrolysis Cells (MEC), Microbial Desalination Cells (MDC), Microbial Carbon Capture Cells (MCC), Microbial Electrosynthesis (MES), Plant and Wetland-MFC, and bioelectro-Fenton (BEF) are currently at the forefront of the investigation. The ongoing research in these technologies is significant for advancing BES-based wastewater treatment methodologies and expanding the scope of applications. The continuous investigation of these BES technologies emphasize a dynamic improvement where innovative solutions are sought to address the problem associated with specific wastewater characteristics and optimize the overall efficiency of the treatment processes.

Currently, over 3000 research articles, according to the Scopus database, highlight the potential of BES technologies. Despite their promising applications, researchers face challenges in enhancing efficiency during scaled-up investigations, hindering the successful commercialization of these technologies as viable alternatives to conventional methods. The ongoing efforts in addressing these efficiency concerns are crucial for unlocking the full potential of BES technologies in different industrial and environmental applications. This chapter aims to elucidate the intricacies of different bioelectrochemical technologies and their potential applications in wastewater treatment. It offers a comprehensive overview of sustainable solutions for mitigating wastewater treatment challenges using BES.

2 Historical Developments of BES Technologies

More than a century ago in 1911, M. C. Potter introduced the concept of MFCs, marking the inception of the technology. Surprisingly, despite this early insight, contemporary research in the field did not gain significant acceleration. After almost 20 years Cohen made an ambitious attempt to scale up MFCs by stacking 35 MFC units in series to generate approximately 35 V potential difference in open circuit conditions [5]. However, merely 2 mA current recovery was observed in short circuit conditions as the outcome which highlights the challenges in early attempts at large-scale implementation [5]. Subsequent breakthroughs came in 1962 when J. B. Davis investigated electricity recovery by a BES using glucose as a substrate, paving the way for advancements in understanding the basics of the electrochemical processes involved [6].

In the subsequent years, detailed research focused on anodic oxidation mechanisms, exploring substrates like glucose [7] and sucrose [8, 9]. However, it was in the last two decades that MFC-based research truly gained prominence. Rigorous investigations were conducted on multi-dimensional aspects, including electrode modifications [10], Proton Exchange Membrane (PEM) modifications [11, 12], and for low-cost PEMs development. Additionally, investigations on the utilization of different terminal electron acceptors, strategies for suppressing methanogens [13, 14], techniques to enhance electrogenic activity [15], and the application of bio-cathodes [16] were also performed. This comprehensive approach also included investigations on the electron transfer mechanisms by electrogenic microorganisms [17]. Simultaneously, researchers initiated detailed investigations on mathematical modelling of anodic substrate utilization in MFCs, formulating activation losses to optimize the overall performance of this evolving technology [18, 19]. This holistic investigation has not only expanded our understanding of microbial interactions within MFCs but has also paved the way for innovative engineering solutions, boosting the technology toward increased efficiency and broader applications.

3 Components and Mechanisms Involved in the BES Reactors

Traditionally, MFCs and other analogous technologies, except MDCs have been structured as two-chambered systems (Fig. 1). However, researchers have explored specific modifications to the reactor architecture, leading to investigations into both single-chambered and multi-chambered designs to achieve customized advantages. In MFCs, the fundamental process involves the releasing of electrons and protons by electroactive microbes through bioelectrochemical oxidation of substrate in the anodic compartment. These released electrons are then transported to the cathodic compartment via an external conductive circuit and protons are diffused towards the cathodic compartment typically through a membrane in a typical double-chambered

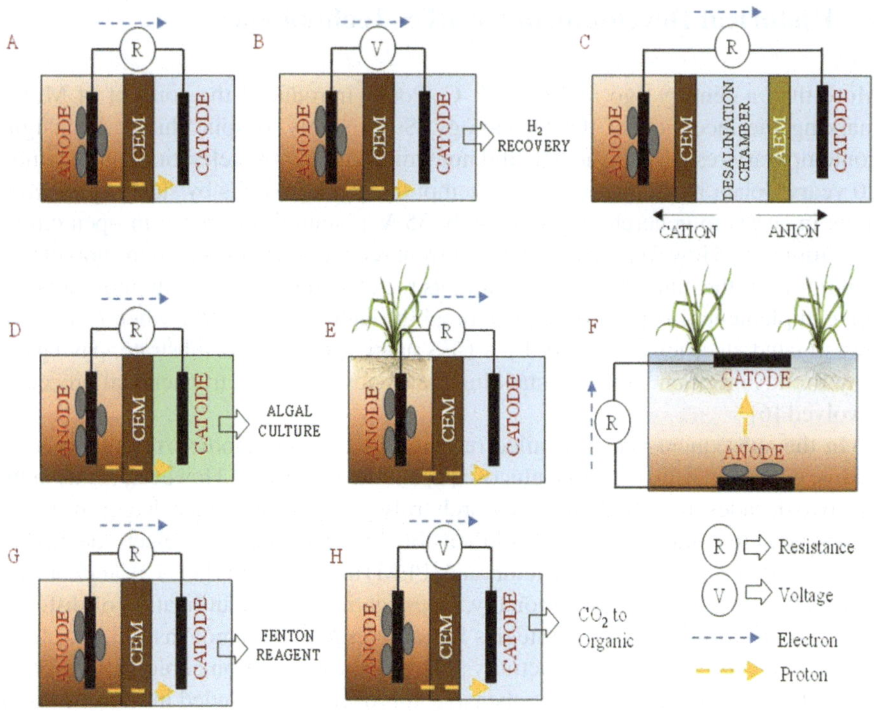

Fig. 1 **a** MFC; **b** MEC; **c** MDC; **d** MCC; **e** Plant-MFC; **f** Wetland-MFC; **g** BEF; **h** MES

MFC setup (Fig. 1a). At the cathode surface, the electrons combine with protons and oxygen to produce water or hydrogen peroxide molecules. The typical configuration of MFC was gradually modified and the operational protocol was updated to design different updated BES based technologies.

The MEC stands out as an environmentally friendly bioelectrochemical technology for hydrogen production, achieved through anodic bio-catalytic oxidation and cathodic reduction processes (Fig. 1b). However, the sustainability of MEC is compromised by its reliance on external electricity imposed between cathode and anode [20]. In contrast, the MDC presents an innovative solution, performing simultaneous desalination and wastewater treatment without requiring external energy input (Fig. 1c). This technology holds promise for providing desalinated water with low-energy input. Additionally, MDC technology has been proposed as a pre-treatment in conventional reverse osmosis plants, aiming to save energy and reduce greenhouse gas emissions associated with typical desalination processes [21]. However, external electricity is occasionally applied to MDC to enhance desalination efficiency when electricity availability is not a limitation [22]. Another notable advancement in BES-based technology is MCC wherein algal biomass facilitates cathodic reduction in the cathodic chamber (Fig. 1d). The key advantage lies in addressing cathodic energy requirements related to aeration and carbon dioxide sequestration

[23]. Microbial electrosynthesis (MES) stands out as a promising technology that facilitates bio-electrochemical CO_2 recycling and transforms carbon dioxide into valuable chemical products such as organic acids, alcohols, and bioplastics, as shown in Fig. 1h [24]. From the perspective of promoting circular economy, the dual benefits of recycling carbon dioxide into organic products and achieving efficient wastewater treatment through MES are helpful [25]. Integrating plants into MFC develops a novel architecture of hybrid Plant-MFC and Wetland-MFC systems which showcase innovative approaches for enhanced wastewater treatment and CO_2 sequestration (Fig. 1e, f). Lastly, the combination of MFC systems with the cathodic electro-Fenton process, introduces Bioelectro-Fenton (BEF) systems (Fig. 1g), demonstrating efficient removal of refractory organic pollutants alongside simultaneous bioelectricity generation [26]. The individual components of the BES reactors are explained in the subsequent sub-sections.

3.1 Anode

Electrons are generated through the anaerobic oxidation of organic compounds within the anodic chamber of MFCs or other similar BES reactors and migrate towards the anode surface. The growth of electrogens on the anode surface is pivotal in facilitating electron shuttling between the anolyte and the anode. Anaerobic oxidation of organic substrate and generation of electrons and protons are the primary mechanisms for the removal of organics from wastewater. This transfer can occur through direct electron transfer involving membrane-bound proteins or lipid structures like pili, c-type cytochrome, and filaments [27, 28]. Alternatively, electrogenic microorganisms may utilize chemical mediators for indirect electron transfer [29–31]. The growth and presence of these electrogenic microbes on the anode surface initiate the primary anodic bioelectrochemical oxidation near the anode surface.

For steady and efficient power recovery, the careful selection of anode material is paramount. An ideal anode material should exhibit biocompatibility, possess a high surface area, be non-corrosive, non-toxic, highly conductive, and maintain physical and chemical stability within the challenging wastewater environment [29]. Additionally, the chosen anode material should be locally available and cost-effective to ensure the feasibility of field-scale applications. This meticulous consideration of anode material characteristics is essential to optimize the performance of MFCs or other BES reactors and enhance their applicability in real-world scenarios.

Various carbon-based materials serve as effective anodes in BES reactors, showcasing their versatility and suitability for different applications. Materials such as carbon felt, graphite felt, graphite rods, and activated carbon have been extensively employed [32–35]. Additionally, conventional electrode materials like platinum rods or platinum-coated carbon-based materials are used as anodes [36, 37]. Stainless steel mesh has also been explored as an alternative anode material in MFC [37]. Carbon nanotubes (CNT) stand out for their efficiency as anode materials, attributed to their high catalytic activity and excellent conductivity [38]. Investigations into graphene-based electrodes as anode materials have demonstrated excellent performance of MFC [38]. Modification of carbon-based anode materials with conductive polymers like polyaniline or various pre-treatments, such as nitric acid or electrochemical oxidation, has been successful in achieving enhanced power recovery [39]. Furthermore, surface modifications involving carbon-based surface modifiers have been shown to increase the specific anodic surface area, providing an enhanced surface area for microbial growth and subsequently improving the performance of BES reactors [40]. Innovative approaches, including the utilization of mining waste materials such as goethite coated on the anode surface, have also demonstrated significant improvements in the performance of anodic electron transfer [41]. These diverse arrays of anode materials and modification techniques underscore the ongoing efforts to optimize the performance of MFC for diverse applications in energy generation and wastewater treatment.

3.2 Cathode

During the anaerobic oxidation of organic matter within the anodic chamber of MFCs, electrons and protons are formed. These generated electrons and protons combine at the cathode surface in the presence of oxygen or other terminal electron acceptors. Among these acceptors, oxygen is the most widely adopted due to its ubiquitous presence in the ambient environment and its excellent oxidation potential [42–44]. Oxygen from air, electrons transported through the external circuit, and protons migrated from the anodic chamber react and either hydrogen peroxide (H_2O_2) or water molecules (H_2O) formed through two or four-electron reduction reactions. The H_2O_2 harvested in the cathodic chamber plays a significant role in carrying out the Fenton or other advanced oxidation process reaction in the cathodic compartment of BEF reactor and may contribute to the degradation of refractory compounds from wastewater. However, during the operation of MFC activation losses during the Oxygen Reduction Reaction (ORR) pose major challenges, particularly in the form of a substantial reduction in cathode potential and a rapid drop in recovered voltage.

To mitigate these challenges, the careful selection of cathode materials becomes pivotal in minimizing ORR overpotential. While many materials used for anodes, as discussed in the preceding section, may also serve as cathode materials. An ideal cathode material should possess characteristics such as high electrical conductivity, robust mechanical strength, and efficient catalysis for the ORR. Moreover, a large active surface area is crucial for facilitating the ORR reaction. Carbon-based materials, including carbon cloth [45], graphite cloth [46], carbon felt [47], graphite felt [48], and carbon fiber [49], stand out as the most commonly employed electrode materials for cathodes. Despite this prevalence, alternative materials like carbon paper [50] and stainless steel [48] have also been explored in certain investigations related to BES. Careful consideration of these material properties ensures the development of efficient and sustainable cathodes for enhanced MFC performance.

3.3 Electrode Catalysts

Researchers employed electrode catalysts, such as cathode and anode catalysts, during the operation of MFCs to mitigate cathodic overpotential and improve anodic electron shuttling. Platinum stands out as the most commonly used electrode catalyst due to its excellent properties; however, the high cost of platinum poses a significant drawback [51]. Extensive investigations have been conducted to achieve a high electrical response using cost-effective electrode materials, and notable progress has been made in recent times [52–54]. Despite significant advancements, further investigations are still necessary to enhance power recovery from a single large MFC for practical field applications. These ongoing efforts aim to address the cost limitations associated with platinum and contribute to the development of more economically viable and efficient electrode catalysts, thereby advancing the feasibility of MFCs for broader implementation.

3.3.1 Anode Catalysts

Enhancing electron shuttling between electrogenic microorganisms and the anode surface plays a crucial role in improving power recovery in MFCs. Significant research efforts have been devoted to increasing electron shuttling, yielding notable progress in this area. A noteworthy example is the utilization of iron-rich waste material from mining mud, specifically goethite, as an excellent anode catalyst. According to an article published by Jadhav et al. [41], a five-fold increase in power was achieved in MFCs after the application of goethite as an anode catalyst [41].

Additionally, the investigation of graphene-based highly conductive anodes has been conducted, leading to an impressive power density (PD) of 1158 mW/m^2 due to superior extracellular electron shuttling [55]. Furthermore, researchers observed an almost two-fold improvement in PD in MFCs after coating the anode surface with conductive glues such as polyaniline and polypyrrole [56]. Optimizing anode catalysts and surface modifications is crucial to maximize electron shuttling efficiency which contributes significantly to the overall performance enhancement of MFCs.

3.3.2 Cathode Catalysts

As previously discussed, the cathodic overpotential loss stands out as the most critical limiting constraint of the cathodic reduction reaction and poses a considerable impact on the performance of BES reactors. The widely used platinum-based cathode catalysts are economically impractical for large-scale, real-life applications. Nearly two hundred research articles have been dedicated to the investigation of non-platinum group, cost-effective cathode catalysts, resulting in notable enhancements in power recovery. Among the different economical cathode materials, researchers predominantly favour transition metal oxides, as highlighted in Table 1. However, alternative approaches include the utilization of bimetallic alloy-based nanoparticles [57, 58] and metal-organic frameworks [59] as cathode catalysts in MFCs. Remarkably, observations from small-volume MFCs (28 mL) revealed a higher PD of approximately ~ 45 W/m^3 [60, 61]. Conversely, as the volume increased to 250 mL, a swift decline in PD (0.13 W/m^3) was noted [62].

Table 1 Application of different cathode catalysts in MFC

Cathode catalyst	Anode volume (mL)	Coulombic efficiency (%)	Volumetric PD (W/m^3)[a]	Reference
Graphene oxide supported magnesium oxide	350	31.6	3.5	[45]
$LiNbO_3$	250	–	0.13	[62]
Mesoporous Co_3O_4 nanoflake	240	–	2.2	[63]
Co/Al_2O_3-rGO nanocomposite	150	–	5.9	[64]
$Co_{0.5}Zn_{0.5}Fe_2O_4$	100	43	21.2	[52]
$LaFeO_3$ perovskite	100	–	8.7	[65]
HfO_2 nanoparticle	28	–	28.8	[66]
Nickel cobaltite	28	–	13.8	[67]
CoO nanosheets/N-doped AC	28	–	41.3	[68]
FeS Anchored (Fe)/Fe_3O_4/N-Doped Graphitic Carbon	28	22.2	23.3	[69]
$Cu_{0.92}Co_{2.08}O_4$	28	–	47.4	[61]
$CoFe_2O_4$ nanoparticles · onto N- doped AC	28	–	44.3	[70]
$CoGa_2O_4$	28	–	47.8	[60]
$NiCo_2O_4$	27	–	9.6	[71]
Nickel-phthalocyanine/MnO_x	19	27.9	8.02	[72]
$CoMn_2O_4$-supported carbon nanotube	18.75	–	17.1	[73]

[a] Volumetric PD is derived from cathode surface area and anodic chamber volume data in articles lacking direct volumetric PD information

3.4 Membrane

In membrane-based MFCs, the membrane serves as a crucial component, acting to separate the cathodic and anodic chambers. The primary purpose of employing a membrane in MFCs is to uphold anaerobic conditions within the anodic chamber while preventing the mixing of anaerobic anolyte and aerobic catholyte. However, the membrane also plays a vital role in facilitating the efficient transport of protons and restricting the diffusion of oxygen from the cathode to the anode. Alternatively, membranes also increase the internal resistance of the BES reactors and therefore in some investigations on BES membrane-less reactor configuration is recomended [74].

The selection of an appropriate membrane, such as PEM, Cation Exchange Membrane (CEM), Anion Exchange Membrane (AEM), or bipolar membrane, is essential and varies based on specific applications in MFCs or other bioelectrochemical systems. Different types of membranes have been adopted during the operation of MFCs, including textile separators [75], J-cloth [76], polymer membranes [77], and ceramic membranes [78]. Textile separators offer advantages in terms of flexibility and ease of use, while J-cloth provides a cost-effective solution. Polymer membranes contribute to the diversity of membrane options with their tunable properties, and ceramic membranes offer durability and resistance to harsh conditions. The choice of membrane depends on the specific requirements of the MFC application, balancing factors like cost, performance, and durability.

3.5 Ion Exchange Membrane

Ion Exchange Membranes (IEM) constitute a specialized group of membranes that find widespread adoption in BES. In the realm of BES, two primary types of IEM are prominently utilized: the CEM [46] and the AEM [79]. The distinctive feature of these membranes lies in the presence of specific functional groups within their polymer structure, endowing them with the capability to selectively transport charged ions, such as cations or anions, facilitated by the Grotthuss mechanism.

The Grotthuss mechanism involves the movement of ions through a series of water molecules, enabling efficient ion transport. Alternatively, a vehicular mechanism may be employed during ion transport in IEM. The CEMs are designed to transport positively charged ions or cations, while AEMs specialize in the transport of negatively charged ions or anions. This selectivity arises from the specific functional groups (SO_3^- for CEM and NH_4^+ for AEM) integrated into the polymer matrix, creating an environment conducive to interactions with particular ions. The application of IEMs in BES is particularly advantageous due to their ability to precisely control the direction of flow, ensuring the necessary separation between anodic and cathodic chambers. The functionality of these membranes is crucial for maintaining desired electrochemical conditions within each chamber.

3.6 Ceramic Separator

Clayware or ceramic membranes have gained substantial popularity as ion-selective membranes, predominantly serving as CEM in BES. The utilization of clayware membranes as an excellent membrane in MFC was first proposed by Behera et al. in 2010 [80]. Subsequent advancements in this field include the modification of clayware membranes through the impregnation of 20% montmorillonite, as demonstrated by Ghadge and Ghangrekar in 2015 [81]. Treatment of clayware membrane using acidic or alkaline solution and impregnating clayware membrane using mining mud goethite was also investigated by the researchers and selective transport of protons was achieved [11, 12]. This modification aimed to enhance the proton-selective nature of the membranes, resulting in outstanding cation transport properties. The impregnation process proved to be an effective strategy for improving the performance of clayware membranes in MFC applications.

Beyond this, other research efforts have explored the application of ceramic separators in MFCs, as evidenced by the work conducted by Yousefi et al. in 2017. These studies collectively contribute to the understanding of the suitability and efficiency of ceramic membranes in BESs, with a particular focus on MFCs [82]. Notably, during field trials, the efficient performance of clayware membranes was observed, as reported by Ieropoulos et al. in 2016 [83]. This underscores the practical viability of clayware membranes and highlights their significance in advancing the field of BES and their role in sustainable energy generation.

3.7 Electron Transfer Mechanism in MFC

The performance of a MFC or other BES systems is intricately dependent on the mechanisms governing anodic electron transport which is a critical aspect influencing the efficiency of BES. Two fundamental electron transfer mechanisms facilitate communication between microbes and electrodes, direct electron transfer and mediated electron transfer. In direct electron transfer, microbial cells establish physical contact with the anode of the MFC, allowing electrons to move directly between the microbial cell membrane and the anode surface. An alternative mechanism involves the use of electronically conducting molecular structures such as pili or nanowires, facilitating the transfer of electrons from microbes to the anode over greater distances [84].

Mediated electron transfer, on the other hand, relies on the involvement of redox mediators, including phenazines, phenothiazines, phenoxazines, quinones, and endogenously secreted compounds such as pyocyanine and 2-amino-3-carboxy-1,4-naphthoquinone [84]. These mediators act as electron shuttles, facilitating the transfer of electrons from microbial cells to the anode. The choice between direct and mediated electron transfer mechanisms remains a subject of debate within the scientific community, with the optimal mechanism varying based on specific MFC configurations and operating conditions.

In practice, both electron transfer mechanisms are often concurrently employed within the anodic chamber of an MFC, especially during real wastewater treatment utilizing mixed culture microbial consortia. This simultaneous utilization may be attributed to the diverse metabolic capabilities of microbial communities and the dynamic conditions inherent in real-world applications. Understanding and optimizing these electron transfer mechanisms are crucial for enhancing the performance of MFCs, ensuring efficient and sustainable electricity generation, and advancing the application of MFCs in wastewater treatment and energy production.

3.8 Inoculum

In the treatment of wastewater using MFCs or other BES technologies, the bioelectrochemical oxidation of dissolved waste organics is a crucial process facilitated by electrogenic microorganisms. Prior to the commissioning of an MFC, effective inoculation is essential to enrich the anodic chamber with electrogenic microbes, such as *Geobacter* and *Shewanella* species [29]. The growth of these electrogens on the anode surface plays a pivotal role in enhancing anodic electron transfer and expediting anodic reduction reactions. Both pure and mixed cultures have been extensively studied in MFCs to understand their performance and efficiency in treating wastewater. However, the potential for contamination poses a challenge, limiting the scalability of MFCs inoculated with pure culture electrogenic microorganisms.

The application of pure culture microbes in the operation of MFCs has been a subject of laboratory-based research, revealing promising performance outcomes. In an article published by Gomes et al. in 2011, the performance of MFC was assessed using a pure culture of *Pseudomonas aeruginosa* as the anodic inoculum, resulting in an impressive current density of 153 mA/m^2 [85]. Further insights into the dynamics of MFCs with pure culture microbes were gained through mathematical modelling [86], focusing on a pure culture of *Shewanella*. This modelling successfully mapped the dynamic utilization of substrates and the corresponding current production. To compare the effectiveness of pure and mixed culture microorganisms, while a mixed electrogenic microbial environment may deliver excellent performance of the MFC, however MFCs based on pure culture microbes offered the advantage of predictability and ease of understanding of the performance of the MFCs [87, 88].

4 Treatment of Organics in MFC or Other BES Reactors

The substrates present in the anodic chamber of a MFC serve as the fuel, where the chemical energy stored in organic substrates is harnessed by electrogenic microorganisms to produce electricity. The characteristics, composition, and loading rate of these substrates are critical factors influencing the performance of an MFC. Extensive research has been conducted using a diverse range of substrates, ranging from simple compounds like sucrose or acetate to more complex wastewater compositions. Investigations, by Yadav et al. [89] and Tiwari and Ghangrekar [90], have explored the treatment of simple substrates in MFCs, providing valuable insights into the fundamental processes involved [89, 90].

Furthermore, MFCs have demonstrated their efficacy in treating complex wastewater, including those containing non-biodegradable compounds such as azo dye, 1,2-Dichloroethane, and phenol [91–93]. The versatility of MFCs is evident in their successful treatment of municipal wastewater, human and animal wastes (e.g., human faeces, urine, cow urine, and swine wastewater), and industrial wastewater from various sectors, including chemical, milk, soya-based food, soft-drink, and laundry industries [83, 94–96]. Remarkably, Pant et al. [97] observed excellent wastewater treatment efficiency coupled with simultaneous electricity recovery in MFCs treating wastewater from the milk industry, attributed to the presence of in-situ mediators [97].

Despite the remarkable capacity of MFCs to treat a wide range of wastewaters with diverse characteristics, challenges in scaling up remain a significant impediment to successful commercialization. Scaling up MFCs for practical applications poses technical and economic challenges, limiting their widespread implementation. Current research efforts are focused on addressing scalability issues, exploring substrates like distillery or brewery wastewater, wastewater from biodiesel production units, coal-tar wastewater, and molasses wastewater, to expand the application potential of MFCs [98]. The ongoing research in this field aims to overcome the scaling-up challenges for diverse substrate loading, composition and unlock the full potential of MFCs for sustainable wastewater treatment and electricity generation.

5 Treatment Mechanisms of Different Pollutants in Different BES Reactors

Different municipal, agricultural, and industrial wastewaters encompass a diverse array of pollutants, broadly classified into four categories: biodegradable compounds, hydrocarbons or non-biodegradable refractory organics, inorganic salts, and heavy metals. Biodegradable organics, primarily carbonaceous compounds, can undergo bioelectrochemical aerobic oxidation facilitated by electroactive microorganisms. This process is prevalent in BES reactors, making BES-based technologies well-suited for the removal of biodegradable organics [99]. However, hydrocarbons and bio-refractory organics that are challenging to degrade under normal conditions present a significant challenge. Acclimatized electroactive microbes have shown promise in the degradation of these challenging compounds [91–93, 98]. Additionally, in BEF technology, the cathodic Fenton reaction, facilitated by the generation of H_2O_2 through a two-electron reduction reaction pathway, can contribute to the degradation of non-biodegradable organics using cathodic advanced oxidation processes [100].

For inorganic salts, MDC provides a potential solution where a central desalination chamber is present alongside the cathodic and anodic chambers. The voltage difference generated between the cathode and anode creates an electrical drag force in the desalination chamber, leading to the separation of cations and anions which causes desalination of water. However, the efficiency of desalination in MDC gradually decreases with a decrease in salinity. Therefore, researchers suggest using MDC in combination with other desalination technologies, such as reverse osmosis, to enhance overall desalination efficiency [101, 102]. The remediation or removal of heavy metals is also feasible in Plant-MFCs, Wetland-MFCs, or MCC systems. In Plant, Wetland-MFCs and MCCs, plants and algae can uptake heavy metals and facilitate their removal through phytoremediation or algal bioremediation [103]. Cathodic reduction of heavy metals is also utilized for the reduction of toxic heavy metals, such as Cr(VI), to less toxic forms, like Cr(III) [104]. In these processes, toxic heavy metal-bearing wastewater is used as catholyte instead of aerated water. The multifaceted applications of bioelectrochemical technologies showcase their potential in managing diverse pollutants present in different composition of wastewater.

6 Future Scope and Challenges

Although considerable scopes are there in wastewater treatment by different BES technologies however several challenges were observed during commercial application of BES technologies. Major challenges during BES operation include voltage reversal during stacking of MFCs, the effect of temperature on the microbial activities in BES reactors, inorganic fouling, biofouling of cathode, architectural challenges during scaling up of BES reactors, and requirement of expertise manpower [105]. In addition, specific BES technologies have specific challenges such as the desalination rate of MDC reactors being less than the conventional RO process. Aeration requirement in water cathode MFC enhanced energy consumption [105]. The instability of H_2O_2 in the BEF cathode reduces the efficiency of wastewater treatment. Furthermore, the elevated cost of the raw materials used in MFCs and other BES technologies presents a significant obstacle to the large-scale production and commercial viability of these technologies [105]. The development of low-cost cathode catalysts, and membranes in the recent decade reduced the challenge however considerable efforts are still required and considerable future scope is available in this regard.

7 Conclusion

In conclusion, the BESs present a transformative nexus of microbial activities and electrochemical reactions, demonstrating remarkable potential across diverse applications. Through the proficient operation of BES reactors, the simultaneous achievement of efficient wastewater purification, electricity recovery, bioresource retrieval, and carbon dioxide sequestration underscores the versatility of these systems. The evolutionary trajectory of MFCs and the emergence of next-generation BES technologies, such as MEC, MDC, MCC, and MES, highlight the dynamic landscape of ongoing research. The modification and evolution of BES architectures, including single-chambered and multi-chambered designs, demonstrate adaptability for addressing specific wastewater characteristics. The diverse applications of these technologies, such as hybrid Plant-MFC and Wetland-MFC systems and Bioelectro-Fenton (BEF) systems, showcase innovative approaches for enhanced wastewater treatment and pollutant removal. Despite promising applications, challenges in efficiency during scaled-up studies pose obstacles to commercialization. This comprehensive overview of bioelectrochemical technologies provides insights into their intricacies and potential applications, offering sustainable solutions for addressing environmental challenges in the quest for a more sustainable future.

Acknowledgements CSIR-CLRI Communication No. 1946.

References

1. Das I, Das S, Chakraborty I, Ghangrekar MM (2019) Bio-refractory pollutant removal using microbial electrochemical technologies: a short review. J Indian Chem Soc 96:493–497
2. Das I, Das S, Das S, Ghangrekar MM (2021) Proficient sanitary wastewater treatment in laboratory and field-scale microbial fuel cell with anti-biofouling $Cu_{0.5}Mn_{0.5}Fe_2O_4$ as cathode catalyst. J Electrochem Soc 168:054519
3. Das I, Noori MT, Bhowmick GD, Ghangrekar MM (2018) Application of low-cost transition metal based $Co_{0.5}Zn_{0.5}Fe_2O_4$ as oxygen reduction reaction catalyst for improving performance of microbial fuel cell. MRS Adv 3:3171–3179
4. Das S, Das I, Ghangrekar MM (2020) Role of applied potential on microbial electrosynthesis of organic compounds through carbon dioxide sequestration. J Environ Chem Eng 8:104028
5. Cohen B (1931) The bacterial culture as an electrical half-cell. J Bacteriol 21:18–19
6. Davis JB, Yarbrough HF (1962) Preliminary experiments on a microbial fuel cell. Science 137:615–616
7. Thurston CF, Bennetto HP, Delaney GM, Mason JR, Roller SD, Stirling JL (1985) Glucose metabolism in a microbial fuel cell. Stoichiometry of product formation in a thionine-mediated *Proteus vulgaris* fuel cell and its relation to coulombic yields. J Gen Microbiol 131:1393–1401
8. Bennett HP, Stirling JL, Tanaka K, Vega CA (1983) Anodic reactions in microbial fuel cells. Biotechnol Bioeng 5:559–568
9. Bennetto HP, Delaney GR, Rason JR, Roller SD, Stirling JL, Thurston CF (1985) The sucrose fuel cell: efficient biomass conversion using a microbial catalyst. Biotechnol Lett 7:699–704
10. Das I, Noori MT, Bhowmick GD, Ghangrekar MM (2018) Synthesis of tungstate oxide/bismuth tungstate composite and application in microbial fuel cell as superior low-cost cathode catalyst than platinum. J Electrochem Soc 165:G146–G153

11. Das I, Das S, Sharma S, Ghangrekar MM (2020) Ameliorated performance of a microbial fuel cell operated with an alkali pre-treated clayware ceramic membrane. Int J Hydrogen Energy 45:16787–16798
12. Das I, Das S, Dixit R, Ghangrekar MM (2020) Goethite supplemented natural clay ceramic as an alternative proton exchange membrane and its application in microbial fuel cell. Ionics (Kiel) 26:3061–3072
13. Das I, Ghangrekar M M, Satyakam R, Srivastava P, Khan S and Pandey HN (2020) On-site sanitary wastewater treatment system using 720-L stacked microbial fuel cell: case study. J Hazardous Toxic Radioact Waste 24:04020025
14. Rajesh PP, Noori MT, Ghangrekar MM (2014) Controlling methanogenesis and improving power production of microbial fuel cell by lauric acid dosing. Water Sci Technol 70:1363–1369
15. Das S, Das S, Ghangrekar MM (2019) Quorum-sensing mediated signals: a promising multi-functional modulators for separately enhancing algal yield and power generation in microbial fuel cell. Bioresour Technol 294:122138
16. Jeremiasse AW, Hamelers HVM, Buisman CJN (2010) Microbial electrolysis cell with a microbial biocathode. Bioelectrochemistry 78:39–43
17. Gorby YAY, Yanina S, McLean JS, Rosso KM, Moyles D, Dohnalkova A, Beveridge TJ, Chang IS, Kim BH, Kim KS, Culley DE, Reed SB, Romine MF, Saffarini DA, Hill EA, Shi L, Elias DA, Kennedy DW, Pinchuk G, Watanabe K, Ishii SS, Logan B, Nealson KH, Fredrickson JK, Yanina S, McLean JS, Rosso KM, Moyles D, Dohnalkova A, Beveridge TJ, Chang IS, Kim BH, Kim KS, Culley DE, Reed SB, Romine MF, Saffarini DA, Hill EA, Shi L, Elias DA, Kennedy DW, Pinchuk G, Watanabe K, Ishii SS, Logan B, Nealson KH, Fredrickson JK (2006) Electrically conductive bacterial nanowires produced by *Shewanella oneidensis* strain MR-1 and other microorganisms. Proc Natl Acad Sci USA 106:11358–11363
18. Katuri KP, Scott K (2011) On the dynamic response of the anode in microbial fuel cells. Enzyme Microb Technol 48:351–358
19. Marcus AK, Torres CI, Rittmann BE (2007) Conduction-based modelling of the biofilm anode of a microbial fuel cell. Biotechnol Bioeng 98:1171–1182
20. Rahimnejad M (2023) Chapter 13—Biohydrogen generation and MECs. Biol Fuel Cells Fundam Appl, 321–349
21. Ramírez-Moreno M, Rodenas P, Aliaguilla M, Bosch-Jimenez P, Borràs E, Zamora P, Monsalvo V, Rogalla F, Ortiz JM, Esteve-Núñez A (2019) Comparative performance of microbial desalination cells using air diffusion and liquid cathode reactions: study of the salt removal and desalination efficiency. Front Energy Res 7:135
22. Aber S, Shi Z, Xing K, Rameezdeen R, Chow CWK, Hagare D, Jindal T (2023) Microbial desalination cell for sustainable water treatment: a critical review. Glob Challenges (Hoboken, NJ) 7:2300138
23. Jadhav DA, Neethu B, Ghangrekar MM (2019) Microbial carbon capture cell: advanced bio-electrochemical system for wastewater treatment, electricity generation and algal biomass production (eds: Gupta SK, Bux F). Springer International Publishing, Cham, pp 317–338
24. Dessì P, Rovira-Alsina L, Sánchez C, Dinesh GK, Tong W, Chatterjee P, Tedesco M, Farràs P, Hamelers HMV, Puig S (2021) Microbial electrosynthesis: towards sustainable biorefineries for production of green chemicals from CO_2 emissions. Biotechnol Adv 46:107675
25. Dange P, Pandit S and Pant D (2022) Chapter 4—Microbial electrosynthesis (MES)-based wastewater treatment and resource recovery. In: Mungray A, Mungray A, Sonawane S (eds) Novel approaches towards wastewater treatment and resource recovery technologies. Elsevier, pp 51–72
26. Soltani F, Navidjouy N, Khorsandi H, Rahimnejad M, Alizadeh S (2021) A novel bio-electro-Fenton system with dual application for the catalytic degradation of tetracycline antibiotic in wastewater and bioelectricity generation. RSC Adv 11:27160–27173
27. Ueki T, Nevin KP, Rotaru AE, Wang LY, Ward JE, Woodard TL, Lovley DR (2018) Geobacter strains expressing poorly conductive pili reveal constraints on direct interspecies electron transfer mechanisms. MBio 9:e01273-e1318

28. Subramanian P, Pirbadian S, El-Naggar MY, Jensen GJ (2018) Ultrastructure of *Shewanella oneidensis* MR-1 nanowires revealed by electron cryotomography. Proc Natl Acad Sci USA 115:E3246–E3255
29. Logan BE (2007) Microbial fuel cells. Wiley, United States of America
30. Logan BE (2009) Exoelectrogenic bacteria that power microbial fuel cells. Nat Rev Microbiol 7:375–381
31. Yong YC, Yu YY, Yang Y, Li CM, Jiang R, Wang X, Wang JY, Song H (2012) Increasing intracellular releasable electrons dramatically enhances bioelectricity output in microbial fuel cells. Electrochem Commun 19:13–16
32. Paul D, Noori MT, Rajesh PP, Ghangrekar MM, Mitra A (2017) Modification of carbon felt anode with graphene oxide-zeolite composite for enhancing the performance of microbial fuel cell. Sustain Energy Technol Assessments 26:77–82
33. Pinto D, Coradin T, Laberty-Robert C (2018) Effect of anode polarization on biofilm formation and electron transfer in *Shewanella oneidensis*/graphite felt microbial fuel cells. Bioelectrochemistry 120:1–9
34. Wang Y, Feng C, Li Y, Gao J, Yu C (2016) Enhancement of emerging contaminants removal using Fenton reaction driven by H_2O_2-producing microbial fuel cells. Chem Eng J 307:679–686
35. Li X, Hu B, Suib S, Lei Y, Li B (2010) Manganese dioxide as a new cathode catalyst in microbial fuel cells. J Power Sources 195:2586–2591
36. Yang E, Chae K-J, Kim IS (2016) Assessment of different ceramic filtration membranes as a separator in microbial fuel cells. Desalin Water Treat 3994:1–9
37. Pocaznoi D, Calmet A, Etcheverry L, Erable B, Bergel A (2012) Stainless steel is a promising electrode material for anodes of microbial fuel cells. Energy Environ Sci 5:9645
38. Yoshida N, Miyata Y, Doi K, Goto Y, Nagao Y, Tero R, Hiraishi A (2016) Graphene oxide-dependent growth and self-aggregation into a hydrogel complex of exoelectrogenic bacteria. Sci Rep 6:21867
39. Qiao Y, Li CM, Bao SJ, Bao QL (2007) Carbon nanotube/polyaniline composite as anode material for microbial fuel cells. J Power Sources 170:79–84
40. Scott K, Yu EH (2015) Microbial electrochemical and fuel cells: fundamentals and applications
41. Jadhav DA, Ghadge AN, Ghangrekar MM (2015) Enhancing the power generation in microbial fuel cells with effective utilization of goethite recovered from mining mud as anodic catalyst. Bioresour Technol 191:110–116
42. Tang H, Cai S, Xie S, Wang Z, Tong Y, Pan M, Lu X (2015) Metal-organic-framework-derived dual metal-and nitrogen-doped carbon as efficient and robust oxygen reduction reaction catalysts for microbial fuel cells. Adv Sci 3:1–8
43. Wang Z, Mahadevan GD, Wu Y, Zhao F (2017) Progress of air-breathing cathode in microbial fuel cells. J Power Sources 356:245–255
44. You S, Gong X, Wang W, Qi D, Wang X, Chen X, Ren N (2016) Enhanced cathodic oxygen reduction and power production of microbial fuel cell based on noble-metal-free electrocatalyst derived from metal-organic frameworks. Adv Energy Mater 6:1–9
45. Li M, Zhou S, Xu M (2017) Graphene oxide supported magnesium oxide as an efficient cathode catalyst for power generation and wastewater treatment in single chamber microbial fuel cells. Chem Eng J 328:106–116
46. Butler CS, Nerenberg R (2010) Performance and microbial ecology of air-cathode microbial fuel cells with layered electrode assemblies. Appl Microbiol Biotechnol 86:1399–1408
47. Zhu X, Ni J (2009) Simultaneous processes of electricity generation and p-nitrophenol degradation in a microbial fuel cell. Electrochem commun 11:274–277
48. Bajracharya S, Ter Heijne A, Dominguez Benetton X, Vanbroekhoven K, Buisman CJN, Strik DPBTB, Pant D (2015) Carbon dioxide reduction by mixed and pure cultures in microbial electrosynthesis using an assembly of graphite felt and stainless steel as a cathode. Bioresour Technol 195:14–24

49. Moqsud MA, Omine K, Yasufuku N, Hyodo M, Nakata Y (2013) Microbial fuel cell (MFC) for bioelectricity generation from organic wastes. Waste Manag 33:2465–2469
50. Sonawane J, Pant D, Ghosh PC, Adeloju SB (2019) Fabrication of a carbon paper/polyaniline-copper hybrid and its utilisation as an air cathode for microbial fuel cells. ACS Appl Energy Mater 2:1891–1902
51. Das I, Noori MT, Shaikh M, Ghangrekar MM, Ananthakrishnan R (2020) Synthesis and application of zirconium metal–organic framework in microbial fuel cells as a cost-effective oxygen reduction catalyst with competitive performance. ACS Appl Energy Mater 3:3512–3520
52. Das I, Noori MT, Bhowmick GD, Ghangrekar MM (2018) Synthesis of bimetallic iron ferrite $Co_{0.5}Zn_{0.5}Fe_2O_4$ as a superior catalyst for oxygen reduction reaction to replace noble metal catalysts in microbial fuel cell. Int J Hydrogen Energy 43:19196–191205
53. Bhowmick GD, Noori MT, Das I, Neethu B, Ghangrekar MM, Mitra A (2018) Bismuth doped TiO_2 as an excellent photocathode catalyst to enhance the performance of microbial fuel cell. Int J Hydrogen Energy 43:7501–7510
54. Das I, Das S, Ghangrekar MM (2020) Application of bimetallic low-cost CuZn as oxygen reduction cathode catalyst in lab-scale and field-scale microbial fuel cell. Chem Phys Lett 751:137536
55. Zhou S, Lin M, Zhuang Z, Liu P, Chen Z (2019) Biosynthetic graphene enhanced extracellular electron transfer for high performance anode in microbial fuel cell. Chemosphere 232:396–402
56. Sonawane JM, Patil SA, Ghosh PC, Adeloju SB (2018) Low-cost stainless-steel wool anodes modified with polyaniline and polypyrrole for high-performance microbial fuel cells. J Power Sources 379:103–114
57. Noori MT, Bhowmick GD, Tiwari BR, Ghangrekar MM, Mukhrejee CK (2018) Application of low-Cost Cu–Sn bimetal alloy as oxygen reduction reaction catalyst for improving performance of the microbial fuel cell. MRS Adv 3:663–668
58. Noori MT, Bhowmick GD, Tiwari BR, Ghangrekar OM, Ghangrekar MM, Mukherjee CK (2018) Carbon supported Cu-Sn bimetallic alloy as an excellent low-cost cathode catalyst for enhancing oxygen reduction reaction in microbial fuel cell. J Electrochem Soc 165:F621-628
59. Li JC, Wu XT, Chen LJ, Li N, Liu ZQ (2018) Bifunctional MOF-derived Co-N-doped carbon electrocatalysts for high-performance zinc-air batteries and MFCs. Energy 156:95–102
60. Liu D, Mo X, Li K, Liu Y, Wang J, Yang T (2017) The performance of spinel bulk-like oxygen-deficient $CoGa_2O_4$ as an air-cathode catalyst in microbial fuel cell. J Power Sources 359:355–362
61. Wang J, Li K, Zhang L, Ge B, Liu Y, Yang T, Liu D (2017) Temperature-depended $Cu_{0.92}Co_{2.08}O_4$ modified activated carbon air cathode improves power output in microbial fuel cell. Int J Hydrogen Energy 42:3316–3324
62. Touach N, Ortiz-Martínez VM, Salar-García MJ, Benzaouak A, Hernández-Fernández F, de Ríos AP, El Mahi M, Lotfi EM (2017) On the use of ferroelectric material $LiNbO_3$ as novel photocatalyst in wastewater-fed microbial fuel cells. Particuology 34:147–155
63. Kumar R, Singh L, Zularisam AW (2017) Mesoporous Co_3O_4 nanoflakes as an efficient and non-precious cathode catalyst for oxygen reduction reaction in air-cathode microbial fuel cells. J Taiwan Inst Chem Eng 78:329–336
64. Papiya F, Nandy A, Mondal S, Kundu PP (2017) Co/Al_2O_3-rGO nanocomposite as cathode electrocatalyst for superior oxygen reduction in microbial fuel cell applications: the effect of nanocomposite composition. Electrochim Acta 254:1–13
65. Nandikes G, Pathak P, Karthikeyan M, Abahussain AAM, Singh L (2024) Mesoporous $LaFeO_3$ perovskite as an efficient and cost-effective oxygen reduction reaction catalyst in an air cathode microbial fuel cell. Int J Hydrogen Energy 52:627–641
66. Ayyaru S, Jayaraman V, Ahn Y-H (2024) Non-precious HfO_2 nanoparticle as an alternative cathode catalyst for microbial fuel cell applications. Int J Hydrogen Energy 57:679–687
67. Xia W-Y, Tan L, Li N, Li J-C and Lai S-H (2017) Nickel cobaltite@nanocarbon hybrid materials as efficient cathode catalyst for oxygen reduction in microbial fuel cells. J Mater Sci 52:7539–7545

68. Huang Q, Zhou P, Yang H, Zhu L, Wu H (2017) CoO nanosheets in situ grown on nitrogen-doped activated carbon as an effective cathodic electrocatalyst for oxygen reduction reaction in microbial fuel cells. Electrochim Acta 232:339–347
69. Xu X, Dai Y, Yu J, Hao L, Duan Y, Sun Y, Zhang Y, Lin Y, Zou J (2017) Metallic state FeS anchored (Fe)/Fe_3O_4/N-doped graphitic carbon with porous spongelike structure as durable catalysts for enhancing bioelectricity generation. ACS Appl Mater Interfaces 9:10777–10787
70. Huang Q, Zhou P, Yang H, Zhu L, Wu H (2017) In situ generation of inverse spinel $CoFe_2O_4$ nanoparticles onto nitrogen-doped activated carbon for an effective cathode electrocatalyst of microbial fuel cells. Chem Eng J 325:466–473
71. Cao C, Wei L, Wang G, Shen J (2017) In-situ growing $NiCo_2O_4$ nanoplatelets on carbon cloth as binder-free catalyst air-cathode for high-performance microbial fuel cells. Electrochim Acta 231:609–616
72. Tiwari BR, Noori MT, Ghangrekar MM (2017) Carbon supported nickel-phthalocyanine/MnO_x as novel cathode catalyst for microbial fuel cell application. Int J Hydrogen Energy 42:23085–23094
73. Zhu N, Lu Y, Liu B, Zhang T, Huang J, Shi C, Wu P, Dang Z, Wang R (2017) $CoMn_2O_4$-supported functionalized carbon nanotube: efficient catalyst for oxygen reduction in microbial fuel cells. J Nanoparticle Res 19:331
74. Logan BE, Call D, Cheng S, Hamelers HVM, Sleutels THJA, Jeremiasse AW, Rozendal RA (2008) Microbial electrolysis cells for high yield hydrogen gas production from organic matter. Environ Sci Technol 42:8630–8640
75. Dong Y, Qu Y, He W, Du Y, Liu J, Han X, Feng Y (2015) A 90-liter stackable baffled microbial fuel cell for brewery wastewater treatment based on energy self-sufficient mode. Bioresour Technol 195:66–72
76. Yazdi H, Alzate-Gaviria L, Ren ZJ (2015) Pluggable microbial fuel cell stacks for septic wastewater treatment and electricity production. Bioresour Technol 180:258–263
77. Tiwari BR, Noori MT, Ghangrekar MM (2016) A novel low cost polyvinyl alcohol-Nafion-borosilicate membrane separator for microbial fuel cell. Mater Chem Phys 182:86–93
78. Ghadge AN, Ghangrekar MM (2015) Performance of low cost scalable air-cathode microbial fuel cell made from clayware separator using multiple electrodes. Bioresour Technol 182:373–377
79. Zhang X, Cheng S, Huang X, Logan BE (2010) Improved performance of single-chamber microbial fuel cells through control of membrane deformation. Biosens Bioelectron 25:1825–1828
80. Behera M, Jana PS, Ghangrekar MM (2010) Performance evaluation of low-cost microbial fuel cell fabricated using earthen pot with biotic and abiotic cathode. Bioresour Technol 101:1183–1189
81. Ghadge AN, Ghangrekar MM (2015) Development of low cost ceramic separator using mineral cation exchanger to enhance performance of microbial fuel cells. Electrochim Acta 166:320–328
82. Yousefi V, Mohebbi-Kalhori D, Samimi A (2017) Ceramic-based microbial fuel cells (MFCs): a review. Int J Hydrogen Energy 42:1672–1690
83. Ieropoulos IA, Stinchcombe A, Gajda I, Forbes S, Merino-Jimenez I, Pasternak G, Sanchez-Herranz D, Greenman J (2016) Pee power urinal-microbial fuel cell technology field trials in the context of sanitation. Environ Sci Water Res Technol 2:336–343
84. Schröder U (2007) Anodic electron transfer mechanisms in microbial fuel cells and their energy efficiency. Phys Chem Chem Phys 9:2619–2629
85. Gomes AS, La Rotta CE, Nitschke M, González ER (2011) Evaluation of current output in *Pseudomonas aeruginosa* microbial fuel cells using glycerol as substrate and Nafion 117 as proton exchange membrane. ECS Trans 41:2011–2017
86. Esfandyari M, Fanaei MA, Gheshlaghi R, Akhavan Mahdavi M (2017) Mathematical modelling of two-chamber batch microbial fuel cell with pure culture of *Shewanella*. Chem Eng Res Des 117:34–42

87. Jothinathan D, Wilson RT (2017) Comparative analysis of power production of pure, coculture, and mixed culture in a microbial fuel cell. Energy Sources Part A Recover Util Environ Eff 39:520–527
88. Cao Y, Mu H, Liu W, Zhang R, Guo J, Xian M, Liu H (2019) Electricigens in the anode of microbial fuel cells: pure cultures versus mixed communities. Microb Cell Fact 18:1–14
89. Yadav AK, Panda P, Bag B (2013) The performance improvement of microbial fuel cells using different waste-sludge as an inoculum. Energy Sources, Part A Recover Util Environ Eff 35:1828–1835
90. Tiwari BR, Ghangrekar MM (2015) Enhancing electrogenesis by pretreatment of mixed anaerobic sludge to be used as inoculum in microbial fuel cells. Energy Fuels 29:3518–3524
91. Sun J, Hu Y, Bi Z, Cao Y (2009) Simultaneous decolorization of azo dye and bioelectricity generation using a microfiltration membrane air-cathode single-chamber microbial fuel cell. Bioresour Technol 100:3185–3192
92. Pham H, Boon N, Marzorati M, Verstraete W (2009) Enhanced removal of 1,2-dichloroethane by anodophilic microbial consortia. Water Res 43:2936–2946
93. Luo H, Liu G, Zhang R, Jin S (2009) Phenol degradation in microbial fuel cells. Chem Eng J 147:259–264
94. Liang P, Duan R, Jiang Y, Zhang X, Qiu Y, Huang X (2018) One-year operation of 1000-L modularized microbial fuel cell for municipal wastewater treatment. Water Res 141:1–8
95. Jadhav DA, Jain SC, Ghangrekar MM (2016) Cow's urine as a yellow gold for bioelectricity generation in low cost clayware microbial fuel cell. Energy 113:76–84
96. Zhuang L, Zheng Y, Zhoua S, Yuana Y, Yuan H, Chenb Y (2012) Scalable microbial fuel cell (MFC) stack for continuous real wastewater treatment. Bioresour Technol 106:82–88
97. Pant D, Van Bogaert G, Alvarez-Gallego Y, Diels L, Vanbroekhoven K (2016) Evaluation of bioelectrogenic potential of four industrial effluents as substrate for low cost microbial fuel cells operation. Environ Eng Manag J 15:1897–1904
98. Pandey P, Shinde VN, Deopurkar RL, Kale SP, Patil SA, Pant D (2016) Recent advances in the use of different substrates in microbial fuel cells toward wastewater treatment and simultaneous energy recovery. Appl Energy 168:706–723
99. Sharma P, Talekar GV, Mutnuri S (2021) Demonstration of energy and nutrient recovery from urine by field-scale microbial fuel cell system. Process Biochem 101:89–98
100. Zou R, Angelidaki I, Jin B, Zhang Y (2020) Feasibility and applicability of the scaling-up of bio-electro-fenton system for textile wastewater treatment. Environ Int 134:105352
101. He Z (2011) Microbial desalination cells. 002:354
102. Shivakumar T, Razaviarani V (2021) An integrated approach to enhance the desalination process: coupling reverse osmosis process with microbial desalination cells in the UAE. Water Sci and Technol: Water Supply 21(3):1127–1143
103. Fang C, Achal V (2019) The potential of microbial fuel cells for remediation of heavy metals from soil and water—review. Microorganisms 7:697
104. Tandukar M, Huber SJ, Onodera T, Pavlostathis SG (2009) Biological chromium(VI) reduction in the cathode of a microbial fuel cell. Environ Sci Technol 43:8159–8165
105. Das I, Ghangrekar MM (2023) Pilot-scale case performance of bioelectrochemical systems. Microb Electrochem Technol Fundam Appl 2:555–582

Integrated and Hybrid Bioelectrochemical Systems (BES) for Wastewater Treatment

Charu Juneja, Purusottam Tripathy, Om Prakash, Abhishek Sharma, and Sukdeb Pal

Abstract Bioelectrochemical systems (BESs) are cutting-edge technologies in the field of wastewater remediation that offer additional advantages of simultaneous recovery of valuable resources. BESs are characterized by advanced oxidation processes such as bioelectro-Fenton (BEF), which integrates interdisciplinary concepts between biology, biophysics, and biochemistry. The BEF system utilizes electrons generated from bioelectrochemical systems for the treatment of wastewater. Bioelectro-Fenton system is an environmentally benign treatment option with notable advantages such as high efficiency without sludge accumulation, low toxicity, mild operational conditions, and low in energy consumption. However, despite having promising potential, BEF systems encounter several challenges that need to be addressed. These challenges include high power density, H_2O_2 concentration, selection of suitable cathode materials, adequate Fe^{2+} concentration and pH regulation. In addition to BEF, microbial biocatalysts are also integrated within BESs for the effective removal of emerging contaminants from the water matrix. The primary objective of this book chapter is to highlight the recent demonstrations pertaining to the removal of refractory organic pollutants from wastewater using BESs. Traditional wastewater treatment technologies often fail to adequately remove these persistent pollutants, leading to their discharge into the environment without proper treatment. In this chapter application of BESs is discussed in terms of their effectiveness in removing and degrading emerging contaminants. Furthermore, the various factors influencing the performance of BESs are summarized. This book chapter aims to assist researchers in developing more efficient BESs to remove emerging contaminants from wastewater.

Keywords Bioelectrochemical systems (BESs) · Bioelectro-Fenton · Microbial Fuel Cells (MFCs) · Emerging contaminants (ECs)

C. Juneja · P. Tripathy · O. Prakash · A. Sharma · S. Pal (✉)
Wastewater Technology Division, CSIR-National Environmental Engineering Research Institute, Nagpur 440020, India
e-mail: s_pal@neeri.res.in

C. Juneja · A. Sharma · S. Pal
Academy of Scientific and Innovative Research (AcSIR), Ghaziabad 201002, India

M. M. Ghangrekar et al. (eds.), *Biological and Hybrid Wastewater Treatment Technology*, Earth and Environmental Sciences Library,
https://doi.org/10.1007/978-3-031-63046-0_17

1 Introduction

Water is pivotal in sustaining life and holds significant importance within global ecosystems [1]. Unfortunately, the issue of water contamination has escalated due to the expanding population and rapid economic development [2]. This unfortunate trend has increased the occurrence and detection of recalcitrant contaminants, commonly referred to as emerging contaminants (ECs), within water matrices. These ECs encompass a range of synthetic and naturally occurring chemicals, some of which are suspected or known to pose potential risks to both aquatic organisms and humans. Nevertheless, regulatory bodies have refrained from setting precise regulatory limits for these pollutants because of the complex and not fully understood characteristics, occurrence, fate, and ecotoxicological attributes within the environmental matrix [3]. These emerging contaminants constitute a diverse group of organic compounds, including substances like pesticides, phenols, artificial sweeteners, herbicides, dyes, pharmaceuticals and, personal care products pharmaceuticals, herbicides, etc. Importantly, even when present in relatively low concentrations, ECs have the capacity to instigate genetic mutations, bioaccumulation, carcinogenicity (both in humans and aquatic life forms), chronic diseases, biotoxicity, and antibiotic resistance [4]. Henceforth, it is imperative that significant attention and efforts be directed towards the development of efficient wastewater treatment methods, as these hold the key to addressing the pressing challenge posed by emerging contaminants.

Various treatment technologies, including membrane filtration, adsorption, ozonation, electrochemical processes, and several others, have been deployed to eliminate organic pollutants from wastewater and address the environmental issues stemming from the discharge of these pollutants. Nevertheless, it is important to note that these technologies are not without their limitations. In the case of adsorption, there exist certain drawbacks, notably the management of secondary sludge and the need to ensure the reusability of adsorbents to prevent secondary pollution thus can significantly escalate operational costs. Conversely, electrochemical processes, including electrochemical oxidation and electrocoagulation, necessitate external energy inputs for their operation. Consequently, these processes can pose challenges when employed in remote areas where a stable energy supply may not be readily available [5]. Furthermore, the fouling of membranes stands out as a primary issue associated with the commercialization of membrane bioreactors and the long-term operation of reverse osmosis systems. Hence, can impede the effectiveness and efficiency of these membrane-based technologies. Efforts to address these shortcomings and optimize the performance of these removal techniques are crucial in the ongoing pursuit of effective wastewater treatment solutions [6, 7].

In this context, Bio-electrochemical systems (BESs) have witnessed substantial progress over the last two decades [8]. Typical BESs are comprised of anodic chambers, cathodic chambers, and electrodes that enable electrochemical reactions involving microbial communities [9]. Two primary applications within the realm of BESs are Microbial Fuel Cells (MFCs) and Microbial Electrolysis Cells (MECs).

When combined with microbial metabolic processes and electrochemical redox reactions, these systems present a promising technological avenue for the mineralization of refractory pollutants [10]. BESs are extensively installed in wastewater treatment systems, addressing a range of substances, including chloramphenicol, dyes, adsorbable organic halogens, and even wastewater originating from the petroleum industry [11]. However, achieving practical applications for both energy generation and efficient sewage treatment solely through BESs remains a challenging task. Advanced Oxidation Processes (AOPs) are known for producing highly reactive chemical oxidants, particularly hydroxyl radicals (OH·). These radicals possess exceptional and non-selective oxidation capabilities, rendering them powerful agents for the mineralization of organic contaminants [12].

Bio-electrochemical systems offer several notable advantages in comparison to conventional activated sludge treatment methods. These advantages encompass a reduction in the production of excess sludge, faster process kinetics, and lower external power consumption [13, 14]. Furthermore, BESs exhibit higher rates of anaerobic degradation when compared to traditional anaerobic treatment techniques. Presently, there exists a substantial body of research dedicated to exploring the diverse applications of BESs, such as electricity generation, the recovery of valuable components, the establishment of environmentally benign wastewater treatment methods, and electricity generation. These applications have extensively been examined, ranging from laboratory-scale experiments to large-scale implementations [15].

The commencement of this book chapter entails a discussion of the evolution of BES systems. Subsequently, the focus shifts to technical aspects of BES for resource recovery, which includes energy recovery, nutrient recovery, water recovery, metal recovery, and chemical recovery. Moreover, the factors affecting the removal efficiency of the emerging pollutants, such as the type of electrode material, influence of initial concentration of the contaminant, Biocatalyst, type of anolyte, repercussion of salinity of the anolyte on the reaction, and response of externally applied voltage. Furthermore, the chapter summarizes the various applications of BES for the elimination of emerging contaminants in MFC, MEC, Bio-electro Fenton system, and hybrid bio-electrochemical systems.

2 Evolution of BES

Bio-electrochemical systems offer a significant advantage when it comes to wastewater purification, resource recovery, and energy generation. These systems harness reducing equivalents (H^+ and e^-) generated during electrochemical processes catalyzed by microorganisms [16, 17]. Considerable research has been conducted on various aspects of BES, including microbiology, electrochemistry, operation, and diverse applications [18]. Among the different configurations within BESs, the MFCs aroses. The notion oe studied in detail. There are also variations and extensions of MFCs, such as the Microbial Recovery Cell (MRC), Microbial Solar Cell (MSC), Microbial Electrosynthesis (MES), Microbial Desalination Cell (MDC), Microbial Electrolysis Cell (MEC) [19]. Figure 1 offers an outline of some BES configurations and their specific purposes. The notion of the MEC and MFC was initially introduced by Potter in 1910 [20, 21]. Initially, the primary objective of MFC technology was to directly transform the chemical into electrical energy through the involvement of microbial metabolism. This process facilitates a spate of redox reactions, particularly within the treatment process. A typical MFC comprises two compartments: an anode compartment and a cathode compartment, positioned oppositely and sundered through a cation-exchange membrane (CEM) (Fig. 2a).

The anode compartment in this construction serves as the site for microbial degradation of organic substances, which results in the liberation of electrons. On the other hand, the cathode compartment supports the entire electron cycle. More specifically, anaerobic microorganisms are actively involved in the biodegradation of organic compounds within the anode chamber. This process leads to the production of electrons and protons in the anolyte. On the other hand, the organic contaminants are concurrently transformed into CO_2. Through an external electric circuit, these liberated electrons initially reach the anode surface with their subsequent migration to the cathode. The cathode compartment in a MFC typically employs a terminal electron acceptor to exploit the migrated electrons for reduction reactions (Eq. 2). This process not only establishes an electric current across the cell but also generates hydroxyl ions. In order to perpetuate the overall charge neutrality of the cell, the cations present in the anolyte solution travel between the two compartments of the system [22]. The MEC is an extension of the MFC, primarily designed for generating hydrogen gas. To surmount the thermodynamic barrier for the reduction reactions to take place, an external power is applied between the anode and cathode chamber [20] (Fig. 2b). It is important to keep in mind that while the reactions inside the anode electrode for

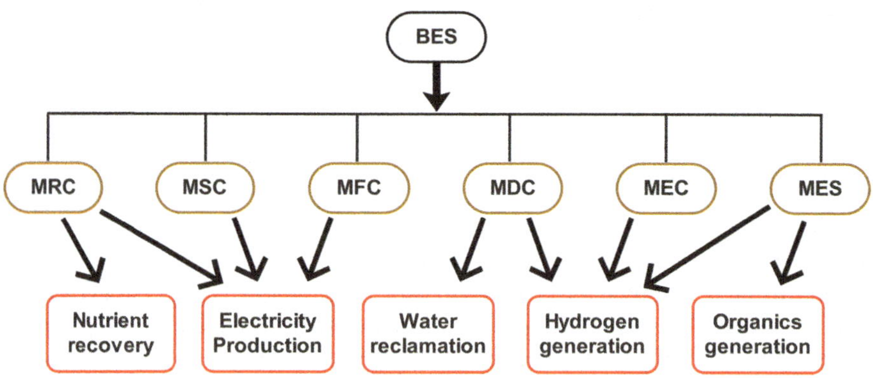

Fig. 1 An outline of the different categories of bioelectrochemical systems (BESs)

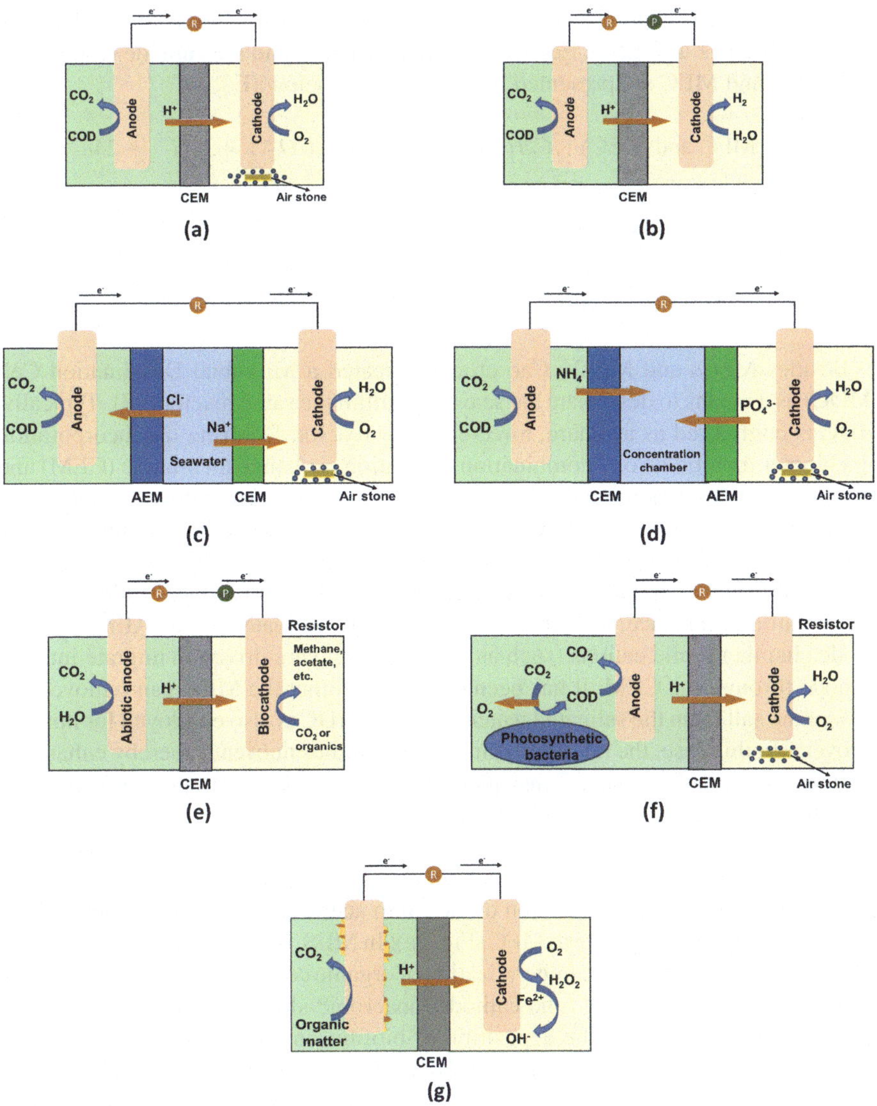

Fig. 2 An illustrative representation of **a** Microbial Fuel Cell (MFC), **b** Microbial Electrolysis Cell (MEC), **c** Microbial Desalination Cell (MDC), **d** Microbial Reverse Electrodialysis Cell (MRC), **e** Microbial Electrosynthesis (MES), and **f** Microbial Sensor Cell (MSC), **g** two-chamber bio-electro-Fenton system. CEM stands for cation-exchange membrane, and AEM stands for anion-exchange membrane

both MFC and MEC are exactly similar, the reactions inside the cathode electrode in both cell types differ (see Eq. 3). The anode reaction and the cathode reaction for both MFC and MEC are presented in the equations below (Eq. 1).

$$\text{Anode (MFC and MEC)}: \ C_6H_{12}O_6 + 6H_2O \rightarrow CO_2 \uparrow + 24H^+ + 24e^- \quad (1)$$

$$\text{Cathode (MFC)}: \ 2H_2O + O_2 + 4e^- \rightarrow 4OH^- \quad (2)$$

$$\text{MEC}: \ 2H_2O + H_2 \uparrow + 2OH^- \quad (3)$$

Besides MFCs and MECs, Cao et al. fabricated a Microbial Desalination Cell (MDC) as a means to desalinate the seawater within the same reactor [23]. Typically, MDC is configured as a square, three-chamber reactor, featuring the incorporation of a bipolar membrane or a combination of a cation exchange membrane (CEM) and an anion exchange membrane (AEM) into MFCs [23]. Furthermore, an additional desalination chamber is positioned between the anode and cathode compartments, separated via AEM and CEM on both sides of the chambers (Fig. 2c). The electric field established over the anodic as well as cathodic compartment induces anion movement (for instance, Cl^- and $SO_4{}^{2-}$) from the saltwater across AEM, into the anode chamber while cations (such as Na^+ and Ca^{2+}) are driven to migrate into the cathode through the CEM. It has been documented that the MDC can remove over 99% of the salt from the salinated water [24]. The MDC is also employed for nutrient recovery. In this case, the middle chamber concentrates nutrients, thereby enhancing nutrient recovery efficiency. Consequently, this particular configuration is referred to as the Microbial Recovery Cell (MRC) (Fig. 2d).

In 2010, Nevin and colleagues introduced the notion of Microbial Electrosynthesis (MES), in which *Sporomusa ovata* played a pivotal role as an external electron acceptor to facilitate the conversion of CO_2 into acetic acid [25]. Importantly, MES operates in a manner opposite to MFC (Fig. 2e). In MES, oxygen is generated through abiotic materials or the biotic mineralization of organic contaminants inside the anode compartment. Meanwhile, in the cathode, anaerobic sludge is introduced into the compartment, resulting in the deposition of biofilm on the surface of the cathode. The biocathode provides electrons to reduce CO_2 or compounds like volatile fatty acids (VFAs), ultimately leading to the generation of organic substances such as acetic acid and methane. Regardless of the mechanisms involved in the transfer of electrons, whether direct or facilitated by substances like H_2 or other redox agents, the cathode remains the exclusive source of reducing power. Critically, MES processes are not thermodynamically spontaneous, which means that an external voltage must be applied to overcome the thermodynamic barriers associated with chemical production. The idea of a Microbial solar cell (MSC) was introduced by Berk and Canfield [26]. The concept of the MSC shares similarities with that of the MFC. In the MSC, photosynthetic microorganisms work in conjunction with electrochemically active bacteria in order to produce chemical compounds like hydrogen and methane or electricity [27]. In a regular MSC configuration, organics are produced under sunlight via photosynthesis. This synthesized organic matter is then moved to the anode compartment, where it undergoes a series of processes to form electricity (Fig. 2f) [28, 29]. The fundamental idea behind the MSC is to change solar energy into chemical energy and, subsequently, into electrical energy.

3 Technical Components of BES for the Recovery of Resources

3.1 Energy Recovery

Bio-electrochemical Systems offer an effective means to harness the energy potential of organic compounds present in wastewater, thereby generating electricity. This electricity generation can help counterbalance the energy utilized during the process. The direct power output realized through BESs may address challenges related to electrical generators and biogas purification compared to energy recovery through anaerobic digestion. The data in Table 1 provides an overview of power outputs reported in various studies for the meaningful contrast substrate's conversion efficiency into electrical energy, expressed either as kilowatt-hours per cubic meter of wastewater treated or kilowatt-hours per kilogram of Chemical Oxygen Demand (COD) removed, referred to as normalized energy recovery (NER). Nevertheless, it is important to note that making precise comparisons can be challenging due to the various factors influencing electricity generation. These factors include operational parameters, BES configurations, and the composition of the microbial inoculum in both the anode and cathode compartments. The data clearly shows that industrial wastewater treated using BES technology has the potential to generate more electricity than sewage. This is primarily attributed to the higher concentration of COD and increased biodegradability of organic matter typically found in industrial wastewater [30].

Namin et al. [36] observed that the concentration of total ammonia nitrogen (TAN) increased from 500 to 4000 mg/L in an air–cathode MFC, leading to a decrease

Table 1 Energy generation from BES through wastewater

BES type	Source of wastewater	%Removal of COD	NER (kWh/m^3)	References
Single chamber	Synthetic wastewater	87.1	0.0146	[31]
Double chamber	Municipal wastewater	85	0.021	[32]
Single chamber	Brewery wastewater	87.6	0.097	[33]
Single chamber	Domestic wastewater	76.8	0.017	[34]
Double chamber	Synthetic wastewater	91.03	2.42×10^{-4}	[35]

in power density from 4.2 to 1.7 W/m^3. This reduction in power density can be ascribed to the inhibition of microbial activity within the anodic compartment at higher ammonium concentrations. Similar results were found in another study that reported a decrease in maximum power density, from 6.1 to 1.4 W/m^3, with an increase in TAN concentration from 3500 to 10,000 mg/L [37]. Contrary to this, Iskander and coworkers ascribed that bacteria in continuously operated MFCs can survive to higher TAN concentrations. Furthermore, it is important to note that the concentration of ammonium can decrease under higher anolyte pH due to hydrolyzation (Eq. 4), leading to a reduction in the inhibitory effect of ammonium. Kuntke et al. studied the reduction pattern and evaluated that the efficiency of dual-chamber MFCs is not affected at neutral pH by varying the concentration of ammonical nitrogen from 70 to 4000 mg/L [38].

$$NH_4^+ \leftrightarrows H^+ + NH_3\,(aq)\,(pK_a = 9.3) \tag{4}$$

The pH of the electrolyte in a BES can be influenced by nitrogenous compounds through a combination of biological and chemical reactions. The process of biological nitrification in a MFC generates protons in the cathodic chamber. These protons can act as a buffer against the hydroxyl ions generated during the reduction of oxygen, thus preventing a significant increase in catholyte pH. For example, when ammonium ions and nitrifying bacteria are introduced inside the cathodic chamber of an MFC, the pH was observed to decrease from 8.8 to 7.1, while the voltage generation increased from 0.30 to 0.57 V [39]. In essence, the NH^{4+}/NH_3 couple functions as a proton shuttle between both compartments, thereby influencing the electrolyte pH. The ammonium ions inside the cathodic chamber could react with the hydroxyl ions produced during the reaction, preventing a significant rise in catholyte pH. Conversely, a neutral pH in the anolyte can be maintained as volatile NH_3 is returned to the anode chamber. These processes collectively contribute to the regulation of the electrolyte pH within the BES. Indeed, in BESs, microorganisms and nitrogenous compounds can fight for electron donors/acceptors [40]. This competition can have significant effects on the BES performance: When higher nitrate concentrations are present, denitrifying bacteria may consume the carbon source (organics) more rapidly, leading to a decrease in the concentration of organics available for bioelectrically active bacteria. As a result, electricity generation can be reduced. During biological nitrification, the nitrifying bacteria and the cathode electrode can compete with one another for oxygen. Elevated ammonium concentrations in the cathodic chamber can reduce the cathode potential, thus, diminishing electricity generation [41]. Some studies have reported that increasing organic loading rates (OLRs) can inhibit electricity generation in BESs [42, 43], while others have found the opposite to be true [44, 45]. These conflicting findings may be attributed to various factors, including BES configurations, the specific species of anaerobic microorganisms employed in wastewater treatment, electrode materials, and other variables that can impact power output. In essence, the performance of BESs is influenced by complex interactions between nitrogenous compounds, microorganisms, and electrodes, and the outcomes can vary depending on the specific conditions and parameters of the system. This underscores the importance of considering multiple factors when designing and optimizing BESs for practical applications.

3.2 *Nutrient Recovery*

Several key steps are involved in the recovery of ammonia recovery process realized in a conventional dual-compartment MFC: (i) Ammonium ions are transferred from the anodic chamber to the cathodic chamber through the Cation Exchange Membrane (CEM). (ii) Ammonium ions accumulate in the cathode chamber. (iii) Ammonium ions are converted to ammonia (NH_3) under high-pH conditions, often facilitated by air or N_2 stripping. (iv) The volatile ammonia is adsorbed by acid solutions, leading to the formation of valuable materials, such as ammonium salts. Dilute sulfuric acid is commonly employed for this purpose, resulting in the production of ammonium sulfate as the final product (Fig. 3) [17, 38]. It is significant to remember that the reduction process results in a high pH inside the cathode electrode [46, 47]. The transport of ammonium involves both current-driven migration and diffusion due to concentration gradients. Importantly, the movement of ammonium ions is not only predicated on concentration gradients. Additionally, the current-driven movement has also been noted in other BES configurations, including MECs and MRCs [47, 48]. Ammonium sulfate, one of the end products of this process, finds wide-ranging applications in industry and agriculture, including its use in food production and fertilizer manufacturing [49]. Additionally, volatile ammonia can react with CO_2 in aqueous solutions to produce NH_4HCO_3, which can potentially be used to draw solutes in Forward Osmosis (FO) processes. This integration of BES with FO enables the simultaneous recovery of ammonium and clean water [50].

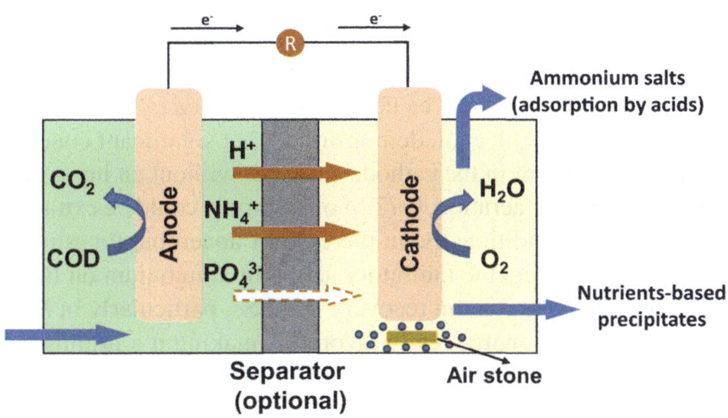

Fig. 3 The recovery of phosphate and ammonium in BES

Ammonium recovery in BESs is influenced by a multitude of factors, and understanding the impact of these factors on BES performance is crucial for enhancing the efficiency of the system. The pH of the cathodic chamber plays a significant role in driving the transformation of ammonium ions (NH^{4+}) to NH_3. Therefore, it is a key factor influencing the recovery of ammonia in BESs. The anolyte's pH could also affect the transfer of ammonium ions (NH^{4+}) transfer due to hydrolysis (Eq. 4). Furthermore, the presence of coexisting cations, such as sodium ions (Na^+) from sodium chloride (NaCl) solution, can influence the pH of the catholyte. This alteration in pH can enhance the transformation of ammonium ions (NH^{4+}) into volatile ammonia (NH_3) inside the cathodic compartment, potentially improving ammonium recovery. The use of buffer solutions, such as phosphate buffer, in the cathode chamber can serve multiple purposes. It can enhance the transfer of ammonium into the cathode compartment by maintaining charge neutrality and also provide additional phosphate ions that can contribute to struvite formation, thereby facilitating ammonium recovery [51]. These factors represent a complex interplay of variables that can impact ammonium recovery in BESs. Understanding and optimizing these factors can lead to improved efficiency and reliability of the ammonium recovery process within BES technology.

In addition, Kim et al. observed that the coexisting cations (such as K^+, Ca^{2+}, and Na^+) in the anolyte had an adverse effect on the diffusion of ammonium across the CEM, even though it is possible to enhance the diffusion at higher OLR. K^+ and H^+ move more quickly through the body than NH^{4+} among the common cations [52]. This demonstrates that ammonium ions migrate to the cathode chamber under low current density because the H^+ and K^+ ions move from the anodic compartment towards the cathodic compartments before they neutralize the large electrons generated by the anode reaction. The salinity of the catholyte is substantially raised by the movement of ions from the mineral salts into the cathodic compartment. This results in the lowering of the catholyte resistance, and ammonium migration and BES's power-generating capability are subsequently enhanced [50].

The movement of ammonium ions is facilitated in a MEC via an external supply of power, which drives the movement of ions within the cell. According to Haddadi et al. the current field may have been accountable for up to 61% of the ammonium transport in an MEC with higher current density. It is worth noting that the removal of ammonium from the catholyte may be negatively impacted by the lack of oxygen in the cathode chamber [53]. Qin et al. demonstrated that significant concentrations of ammonium could be present in the cathode chamber without an input of air. Nevertheless, after adding further aeration, 0.77 M of ammonia could be extracted from the cathodic chamber [50]. Additionally, in the case of anaerobic digestion processes, the ammonia recovery lowers the inhibitory action of ammonium on BES operation [54]. This indicates that ammonium recovery in BESs, particularly in MFCs, can be not only effective but also energetically favorable, making it a promising approach for wastewater treatment and resource recovery.

Phosphate recovery in a BES is typically realized through chemical precipitation, often taking place in the cathodic chamber, where reduction reactions inside the cathode provide a high pH environment [55, 56]. Fischer et al. conducted one of the early studies inside a dual-compartment MFC to extract phosphate from wastewater [57]. In their research, electrons produced by microbial activity in the cathodic chamber reduced $FePO_3$ in digested sludge. As a result, phosphate could be separated from the insoluble $FePO_3$, converting it into a soluble form. Phosphorus recovery from pure ferric phosphate hydrate was 48 and 82% with the incorporation of magnesium and ammonium ions. There has been significant research on the direct recovery of phosphate in BES, often utilizing air–cathode BES configurations, including MFCs and MECs [58]. Phosphate and ammonium-based precipitations in the single-chambered cells typically take place on the cathode surface. As an instance, in a single compartment, MEC, Cusick and Logan accomplished a struvite production rate of 0.3–0.9 g/m^2 h on the cathode surface [59]. In this context, the recovery of phosphate is highly dependent on the increase in pH near the cathode. Nonetheless, the pH elevation inside the catholyte can be disturbed in the single-compartment BES because the anode (which produces protons) and cathode (which produces hydroxyl ions) compartments share the same electrolyte. However, it is still possible to recover 70–82% of the phosphate through struvite precipitation, even when the catholyte pH is only moderately high (about 8) [58]. It is important to consider that cations present in the anodic compartment can potentially move into the cathodic compartment during nutrient recovery inside Bio-electrochemical Systems (BES). As K^+ and Rb^+ have the capacity to replace ammonium ions during the formation of struvite (Eq. 5), this transfer of cations can have a negative impact on struvite formation, which hinders the recovery of ammonium. Moreover, phosphate ions tend to preferentially interact with K^+ over NH_4^+ in the pH range of 8–10, further complicating the struvite formation process [60]. Additionally, chemical precipitation between calcium and phosphate ions (Eq. 6), has adverse effects on struvite formation and nutrient recovery within the BES. These factors emphasize the need for careful control and optimization of BES parameters to achieve effective nutrient recovery while minimizing interference from unwanted reactions.

$$Mg^{2+} + HPO_4^{2-} + K^+ + 6H_2O \rightarrow MgKPO_4 \cdot 6H_2O \downarrow + H^+ \tag{5}$$

$$5Ca^{2+} + 3PO_4^{2-} + OH^- \rightarrow Ca_5(OH)(PO_4)_3 \downarrow \tag{6}$$

3.3 Metal Recovery

Based on earlier research, metal recovery in BES frequently involves a reduction process at the cathode's surface, wherein the metals act as external electron acceptors. In wastewater treatment, there are four fundamental methods for recovering metals through BES [61]. First, taking into account that metals with greater redox potentials than the anode potential, such as Cu (II), Ag (I), and Fe (III), may undergo direct reduction on the surface of the cathode [62] (Fig. 4a) requires significantly less energy due to its favorable thermodynamic mechanism. Secondly, metals with lower redox potentials than the anode, for instance, Cd^{2+}, Pb^{2+}, and, Ni^{2+}, are significantly capable of reduction (Fig. 4b). Consequently, an external power source is necessary to facilitate the flow of electrons from the anode to the cathode. It is worth noting that, in theory, these metals could achieve complete reduction if an ample external power supply is employed. Nonetheless, the energy required from an external source for metal reduction might potentially undermine the primary objective of BES, which is energy efficiency. However, the total energy consumed in the recovery of metal using this method is comparatively lower than a normal electrolysis process, as BES treatment was observed to contribute some of the energy necessary for reducing metal ions [61]. The third approach involves the use of a biocathode to directly reduce specific metals on the cathode's surface, facilitated by specialized microorganisms [63] (Fig. 4c). Metals from the catholyte can be adsorbed into the biofilm of the cathode by maintaining the cathode voltage at a certain level, where bacteria further decrease the metals through microbial respiration. The fourth method combines elements from the second and third approaches. In order to reduce metals with a lower redox potential than the anode, an external source of voltage is specifically provided between the cathode and anode [64] (Fig. 4d). This in turn amplifies the movement of electrons towards the cathodic chamber as well as the cathode's ability to trap metal ions. As a result, if the applied voltage is no longer applied, the biofilm accumulated on the cathode electrode can effectively prevent the metals from detaching. Using a biocathode in BES requires less energy consumption than traditional electrochemical processes for metal removal or reduction since it can quickly recover metals from wastewater under a low applied voltage, utilizing both electrical and biological adsorption. It is important to note that the elevated pH levels near the cathode may also make it easier to recover metals through chemical precipitation in a solid form [65].

Metal recovery contributes to the accumulation of metals and mitigates their harmful effects on microorganisms within BES, thereby enhancing the system's overall performance. Additionally, the preferred setup is the double-compartment configuration to lessen the negative effects of metal on the growth of the microbial community. This configuration allows the anodic chamber to be supplied with wastewater containing organics while the cathode compartment treats the wastewater with metals. In aquatic environments, the metal solubility is largely influenced by the solution pH. The majority of metals can solubilize in an acidic medium. Thus, metals become more accessible when the acidogenic phase contains a lot of volatile fatty acids (VFA). It is essential to recognize that pH isn't the sole factor affecting metal solubility. As an example, when low-molecular-weight molecules are present in the acidogenic phase, the concentration of zinc rises. Consequently, it is crucial to comprehend various metallic forms present in wastewater to optimize metal recovery. Furthermore, the metals recovered from the metal recovery process

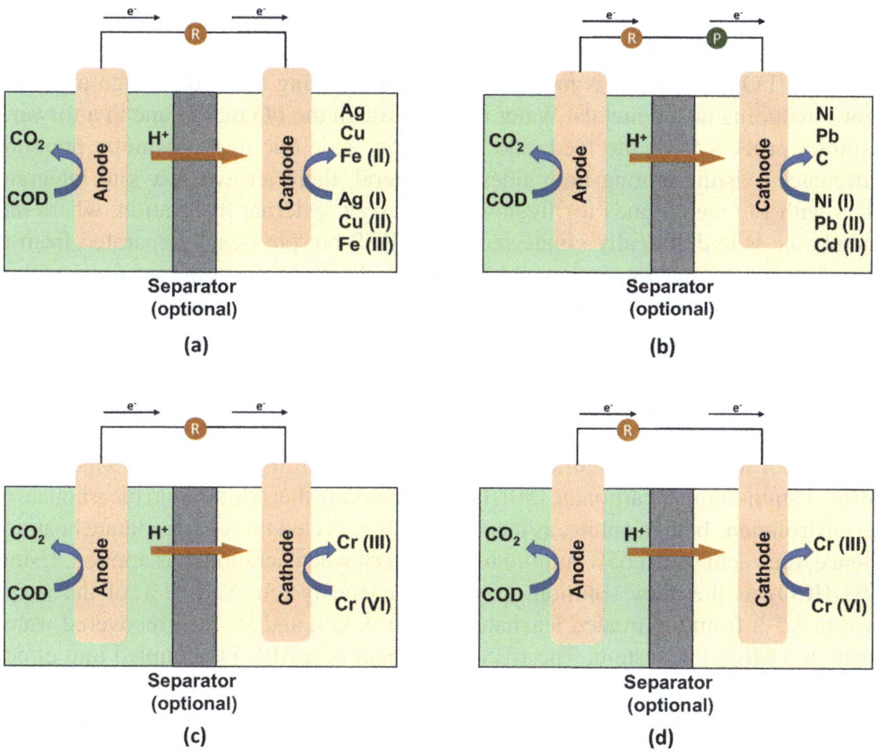

Fig. 4 Four documented methods for metal recovery employing bioelectrochemical systems (BES) include **a** direct metal recovery using abiotic cathodes; **b** metal recovery using abiotic cathodes coupled with supplementary power input; **c** metal transformation using biocathodes; and **d** metal transformation using biocathodes combined with external power input

can be utilized immediately for wastewater remediation. As an example, iron recovered from the metal recovery process, in combination with hydrogen peroxide (H_2O_2) in Fenton-like processes [66].

3.4 *Water Recovery*

BES is often combined with membrane filtration, such as a FO (forward osmosis) membrane, to recover water. Li and coworkers explored this idea by using a dual-compartment MFC to treat wastewater while incorporating an ultrafiltration membrane into the cathodic chamber. In this setup, they were able to significantly reduce COD (by 90%) and turbidity (< 2 NTU), leading to improved water purification. The FO membrane is particularly well-suited for integration with BES due to its advantageous properties, like minimal membrane fouling and low energy requirements [67]. In addition, compared to micro/ultrafiltration techniques, the Forward Osmosis (FO) membrane is more successful at rejecting a broader range of pollutants, including heavy metals. Water moves through the FO membrane in a forward osmosis process from the feed side to the draw side due to the osmotic pressure difference existing among both sides. In general, there are two ways to integrate BES with FO membranes for freshwater recovery: external integration, where the membrane is hydraulically connected to the BES but physically separated from it (Fig. 5a), and internal integration, where the membrane serves as the separator within the BES system itself (Fig. 5b) [68]. Qin and coworkers employed external integration by combining the MEC with the FO process for treating leachate, aiming to recover ammonium and purify water at the same. Their research successfully recovered ammonium from synthetic wastewater using the MEC and converted it into ammonium bicarbonate by reacting it with CO_2. In the FO process, this recovered ammonium was used as the draw solute to pull water from treated wastewater in the form of ammonium bicarbonate [50]. It is worth noting that ammonium bicarbonate is an environment-benign solute, as it can be easily recycled through moderate heating. Hence, they achieved a 65% ammonium recovery when aeration was applied. Using NH_4HCO_3 as the draw solution, they simultaneously retrieved 51% of the water within 3.5 h from the treated leachate. Likewise, Qin and He also recovered water using the MEC-FO system. The BES component of a BES-FO coupled unit effectively removed contaminants and reduced biodegradability, facilitating subsequent FO processes by preventing inorganic scaling and biofouling caused by pollutants from wastewater [69]. Conversely, MDC has shown promise in recovering freshwater from saline water sources. Many researchers have favored using MDCs with low external resistance, typically 1–10 Ω [70]. One plausible rationale behind this preference is that while the generated voltage is constant, ionic transport may be facilitated by a high current density caused by a low external resistance. Additionally, a smaller intermembrane distance proves advantageous in MDC for improving desalination efficiency [71]. Ping and He found that, in a single reactor and under identical conditions, increasing the intermembrane lengths from 3 to 25 mm might accelerate the desalination rate. In their investigation, where the starting concentration of NaCl was 10 g/L, the MDC with a 3-mm intermembrane distance exhibited a desalination rate that was 12 times higher than that observed with a 25-mm intermembrane distance [72].

3.5 Chemical Recovery

MES can transform dissolved CO_2 into valuable organic compounds, including ethanol, methane, acetate, butanol, and formic acid. Researchers have explored laboratory-scale MES in various aspects, such as understanding the biofilm community, microbial pathways, economic feasibility, biocathode design, and materials [73].

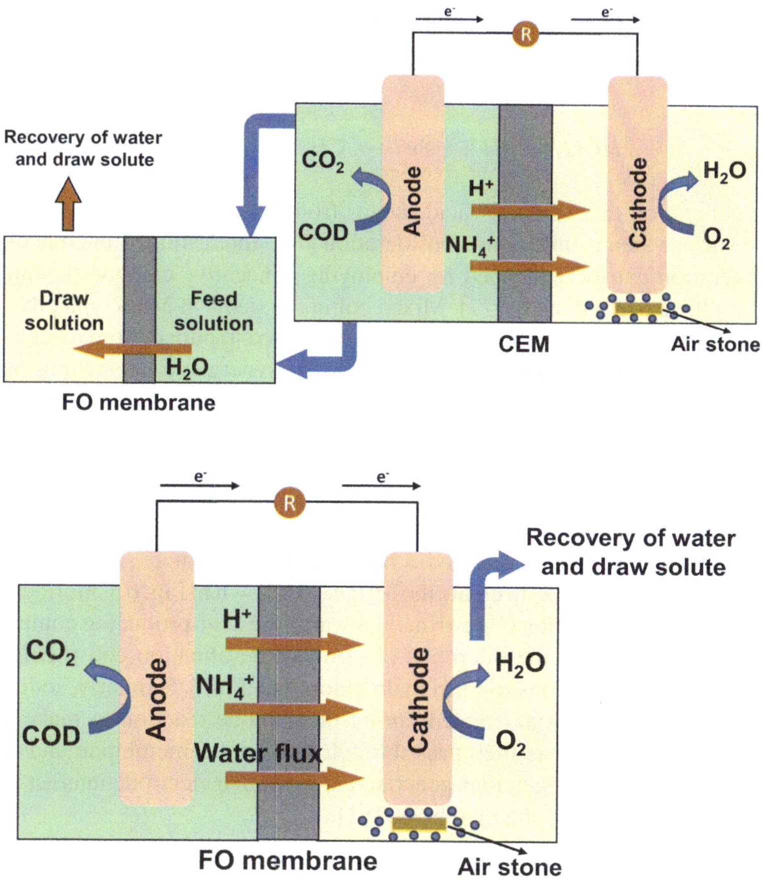

Fig. 5 BES-FO hybrid system: **a** external integration of BES with FO membrane and **b** internal integration of BES with FO membrane

Among the six recognized CO_2 fixation pathways, in MES, methanogens and acetogens, autotrophic microorganisms, have been the primary focus for CO_2 reduction. Methane is a more thermodynamically favorable product of CO_2 reduction compared to other products like acetate and butyrate, primarily due to its higher positive working potential, which is − 0.24 V (vs. SHE) [74] (as shown in Eq. 7).

$$CO_2 + 8H + 8e^- \rightarrow CH_4 + 2H_2O \quad (7)$$

In a study conducted by Cheng's team Nevin, switching from a mixed inoculum to a pure inoculum of *M. palustre* ATCC BAA-1077, the methane production rate reduced from 4.9 to 0.3 L/m^2/day. This could be explained by the synergistic associations exhibited among mixed culture systems and more adaptability towards inward extracellular electron transport [75]. Furthermore, whether inoculated with pure or mixed acetogenic cultures, acetic acid constitutes another significant CO_2 reduction product in such systems. Its working potential is – 0.40 V (vs. SHE), as indicated in Eq. 8 [25].

$$2HCO_3^- + 9H^+ + 8e^- \rightarrow CH_3COO^- + 4H_2O \quad (8)$$

The generation rate of acetic acid ranges from 0.02 to 685 g/m^2/day when the cathode surface area is taken into consideration [76]. Interestingly, the rate of acetic acid generation can be enhanced by employing a negative cathode potential and increasing the current density [77]. Mixed culture systems in MES typically exhibit higher production rates and current densities compared to pure culture reactors at the same cathode potential. This can be ascribed to the synergistic interactions between microbial communities [78]. However, it is important to note that large fluctuations in coulombic efficiency have been observed, which could be attributed to methane consumption again through methanogenesis [79]. On a different note, in BES, the electric field generated can produce hydrogen and hydroxide ions obtained from water. By directing their movement through a bipolar membrane, valuable substances like acids and alkalis can be produced. Chen et al. demonstrated this concept by integrating the MDC with MEC to create the MEDC, as shown in Fig. 6. Chloride ions are moved from the desalination compartment towards the acid-producing compartment by the electrical field in the microbial electrolysis desalination cell (MEDC) and combine with hydrogen ions to form hydrochloric acid [80]. Similarly, sodium ions interact with hydroxide ions created during the cathode reaction to make sodium hydroxide. Additionally, as they pass through the bipolar membrane and into the anode chamber, the hydrogen ions generated at the anode electrode interact with the hydroxide ions formed by the membrane [81].

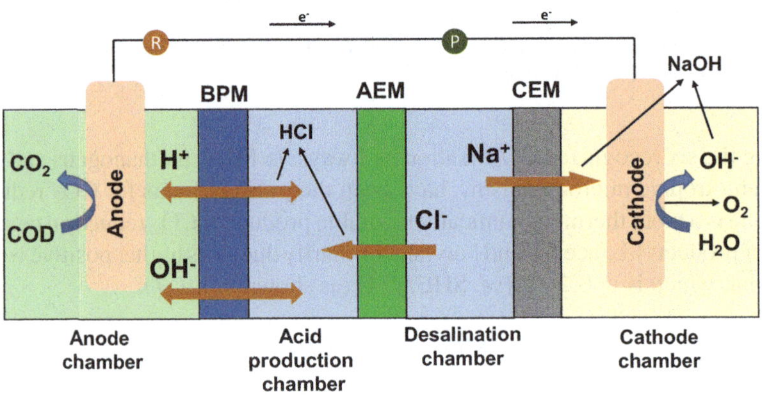

Fig. 6 Schematics of Microbial Electrochemical Desalination Cell (MEDC), CEM (cation-exchange membrane), AEM (anion-exchange membrane), and BPM (bipolar membrane)

4 Factors Affecting the Removal of Emerging Contaminants in BES

The evaluations of the efficiency of BES concerning the removal of emerging contaminants (ECs) from wastewater have uncovered several key factors that influence the performance. Various factors influence the performance of the system. These elements comprise the existence of microorganisms in the anodic chamber, serving as biological catalysts. Additionally, the selection of electrode materials, the nature of the anolyte (e.g., acetate or glucose, which can promote microbial variety), the initial pollutant concentration, anolyte conductivity, and the externally applied voltage all play pivotal roles in this context. These parameters collectively affect the elimination of these intricate organic contaminants. As an example, it's generally observed that as the initial concentration of ECs in the wastewater increases, their removal efficiency in BES tends to decrease. Therefore, based on recent research findings, this section will delve into a comprehensive discussion of how these aforementioned parameters impact the extraction of ECs in BES [82].

4.1 Biocatalyst

The microorganisms that are present in the anodic compartment act as biocatalysts and have a significant influence on the degradation process of emerging contaminants. These microorganisms have the ability to oxidize not only the substrate but also the carbon content within specific pollutants, such as surfactants and pharmaceuticals. This oxidative activity aids in the elimination of these contaminants. Additionally, the microbial consortia generated in this process produce electrons that are transmitted from the anodic compartment to the cathodic compartment via the external electrical circuit. This electron transfer results in the generation of sustainable, green electricity [83]. Inoculating the BES's anodic chamber can be done using both pure and mixed microbial cultures. Pure cultures are typically employed when the goal is the efficient removal of specific ECs like pharmaceuticals and phenols, as these cultures exhibit pollutant-specific metabolic activities. On the other hand, mixed cultures are more suitable for treating composite industrial wastewater carrying various organic pollutants [84]. Microbial diversity, especially the proliferation of microorganisms with electroactive properties on the anode, is a major factor in determining the incubation time of a system. Various strategies can be employed to enhance the diversity of electroactive species and expedite the start-up of bi-anodes. One such strategy uses pure inoculum under anaerobic conditions or recycling the effluent from the anodic compartment of previously functioning BES systems. In this context, Feng et al. demonstrated that introducing a quaternary ammonium compound (QAC), which is a cationic surfactant, into the anolyte can expedite the initiation phase of a BES. When 0.01 M and 0.001 M concentrations of QAC were incorporated into the anolyte, there was a notable 29% reduction in the start-up time (90.5–91.5 h) and 20% (100–102 h), respectively, compared to the control (126–130 h). This acceleration can be attributed to transforming the anode surface from hydrophobic to hydrophilic when a cationic surfactant is added. Consequently, there is an enhancement in covalent and hydrogen bonding interactions between the anode surface and microorganisms. This leads to the anode surface becoming more accessible to exo-electrogens, facilitating their colonization and reducing the start-up time [85].

Over and above that, Chakraborty et al. [86] estimated the impact of the elimination of sodium dodecyl surfactant (SDS) on the biofilm accumulated on the surface of an anode electrode within an MFC. They declared that *Acinetobacter* and *Pseudomonas* were the most abundant genera in MFCs exposed to SDS. These microorganisms were responsible for the degradation of SDS, with more than 70% removal in an anaerobic environment and 10 ppm as the starting SDS level. The extraction process involved the desulphurization of the long alkyl chain of SDS, with the alkyl sulphatase enzyme, recognized in *Pseudomonas* sp., playing a crucial part in breaking down sulfate esters for SDS degradation [87]. Additionally, another exoelectrogen, *Citrobacter* sp., was found to be capable of breaking down the anionic surfactants [88]. Some micro-organisms from the Gram-positive Clostridiaceae family, which are commonly found in sulfate-rich sediments, were identified as exoelectrogens (microorganisms capable of extracellular electron transfer) and were found to assist in the degradation of refractory pollutants such as a pharmaceutical compound, sulfamethoxazole [89]. Likewise, in an MFC specifically engineered for treating wastewater containing an initial phenol concentration of 100 mg/L over a 48-h operational period, the dominant microbes contributing to roughly 97% of phenol degradation were *Pseudomonas* sp., succeeded by *Geobacter* and *Shewanella* [90]. Moreover, various microbes have been identified for their capability to degrade ECs using BES, and some of these microorganisms are listed in Table 2.

4.2 Electrode Material

The choice of electrode material significantly influences the efficiency of BES performance. Electrochemical efficiency, microbial adhesion and electron transfer are all affected by the kind of material used. Typically, uncoated carbon-based electrodes, including carbon fibre, carbon felt, and carbon paper, have frequently been utilized as substrates for the anode in BES [96]. Although, some researchers have gone a step further by investigating the performance benefits of modifying these electrodes through the infusion or decoration of metal or metal oxide to enhance electrochemical behaviour and promote microbial enrichment on the anode [97]. When a BES functions as a MEC, the reduction reaction affects the rate of hydrogen production in the cathode compartment and the energy needed from outside sources. As a result, materials such as iron, platinum, and nickel have been widely employed as cathode catalysts to enhance these processes. Furthermore, the cathode must possess

Table 2 Predominant species of microorganisms for the degradation of emerging contaminants in BES

Contaminant	Predominant species	Observations	References
Nitrobenzene (NB)	*Enterococcus aquimarinus*	• Operation time: less than 24 h • Degradation (%): 88.2% • Initial concentration: 61.5 mg L^{-1}	[91]
Chloramphenicol (CAP)	*Betaproteobacteria*	• Time: less than 12 h • Degradation (%): 84% • Initial concentration: 50 mg L^{-1}	[92]
2,4,6-trichlorophenol	*Acetobacterium*	• Time: 4 days • Degradation (%): 99.5%	[93]
Azo dye	*Pseudomonas aeruginosa*	• Time: 15 days • Decolorization (%): 98%	[94]
Ethynylestradiol (EE2)	*Fusarium proliferatum*	• Time: 15 days • Degradation (%): 97% • Initial concentration: 25 mg L^{-1}	[95]
Sodium dodecyl sulfate (SDS)	*Acinetobacter*	• Time: 12 h • Removal (%): 70%	[86]

high catalytic properties. Both the anode and cathode should ideally possess approximately surface area and porosity for optimal BES performance. However, when the cathode's porosity increases, electrons face more resistance as they move from the anodic compartment toward the cathodic compartment, which tends to decrease the conductivity of the system. This reduction in cathode conductivity can subsequently lower the effectiveness of the overall system [83]. Thus, Xu et al. compared the impact of using metal oxide-modified and metal anodes of a MFC in the treatment of carbamazepine (CBZ) to traditional carbon-based electrodes. They found that after 48 h of operation with an initial CBZ concentration of 5 ppm, it achieved removal rates of approximately 77.3–87.1% with MnO_2, 79.6–87.2% with Fe_3O_4, and 77.7–85.5% with Pd catalysts. These removal rates were higher in comparison with an MFC employing an electrode of carbon black (73.5–83.1%). The increased removal efficiency observed with modified-metal oxide electrodes was attributed to improved electron transfer efficiency resulting from active surface area and enhanced conductivity [97]. Likewise, Shi et al. investigated the impact of removing bisphenol S (BPS) using an anode made up of a composite of nano-zerovalent iron and humic acid (NZVI-HA). They outlined an impressive removal rate of about 92.2% for BPS from a starting concentration of 5 ppm when utilizing the NZVI-HA anode. This rate was approximately three times higher compared to using a carbon cloth anode (31.1%). Conversely, Wu's team demonstrated the degradation of chloramphenicol (CAP) with a starting concentration of 32 ppm and a removal of over 99% by using copper foam as a cathode catalyst. Following the operation of 12 h, the removal efficiency was approximately 6% higher compared to using a carbon rod as the cathode electrode [98]. Additionally, the CAP was mineralized into CO_2 and H_2O on the copper foam electrode, whereas the carbon rod transformed CAP into nitrobenzene and 4-nitrobenzyl alcohol [99]. In summary, modifying electrodes with low-cost metals can significantly enhance the redox reactions that occur during BES. This not only improves the removal of emerging contaminants but also facilitates the complete mineralization of these contaminants into nonharmful by-products.

4.3 Influence of Initial Concentration of the Contaminant

The initial influent concentration of the contaminants indeed influences the efficiency of removing ECs through BES. For instance, Miran et al. investigated the influence of the removal efficiency of the starting concentration of 4-chlorophenols (4-CP). Over a 48-h treatment period, they discovered that the removal efficiency declined from 74.2 to 50.1% when the 4-CP concentration increased from 10 to 100 ppm [100]. This decrease in removal efficiency can be credited to the hindrance in microbial cell growth due to the toxicity of 4-CP, which subsequently reduces microbial activity and slows down the breakdown rate of 4-CP [101]. Likewise, the removal of SDS decreased from 88.6 to 78% over a 24-h hydraulic retention time (HRT) when the loading rate of SDS (sodium dodecyl sulphate) was increased from 10 to 35 ppm per day of a hybrid MFC with a laterite adsorption system [102]. These findings thus show that when EC influent concentration increases, the biodegradability of ECs by microbes is decreased to some extent.

4.4 Type of Anolyte

The choice of a carbon-based substrate can also significantly impact the elimination of emerging contaminants, as microorganisms need both organic matter and energy to fuel all metabolic activities [103]. To investigate this, bench-scale experiments have been conducted to differentiate the effectiveness of BES with and without the incorporation of alternate carbon sources like acetate and glucose. In cases where no alternate carbon source is added, the ECs themselves serve as the sole carbon source for microbial activity. For example, Hou et al. conducted a study in which the wastewater containing Alizarin yellow R was decolorized in a MEC and investigated the impact of co-substrates, namely glucose and sodium acetate on the decolorization. Their findings revealed a decolorization rate of approximately 70 and 75% for AYR when treated with the co-substrates over a period of 4 h. This was observed with an initial dye concentration of 100 ppm [104]. The higher removal rates observed in the presence of sodium acetate and glucose can be credited to the proliferation of non-electroactive bacteria, which thrive by consuming these carbon sources and actively engage in decolorization [105]. Additionally, incorporating an alternative substrate has a significant influence on the power density of an MFC. An examination was conducted by operating one MFC with 10×10^2 ppm of glucose and 5×10^4 ppm of penicillin as co-substrates (referred to as GP-MFC), and another MFC operated with 5×10^4 ppm of penicillin as the sole substrate (referred to as P-MFC). The results showed that the GP-MFC generated approximately 48-fold higher power output (101.2 W m^{-3}) than the P-MFC (2.1 W m^{-3}) [106]. Incorporating penicillin and glucose as co-substrates in the anodic chamber of the MFC enhanced the metabolism of microbial communities, as they had sufficient carbon from the co-substrate. This, in turn, improved the speed of electron transfer, consequently improving the power output [107]. Similarly, when 4-chlorophenol (CP) was employed as the substrate and acetate as a co-substrate, the percentage removal of CP rose from about 53.7 to 93.7% within 12 h of operation, and the power density was additionally increased fourfold compared to an MFC functioning in the absence of acetate [108]. Thus, incorporating co-substrates with organic pollutants as a substrate can enhance power output while simultaneously enabling more effective removal of these compounds. This highlights the interdependence of microbial metabolic activity and the composition of the substrate in a BES.

4.5 *Repercussion of Salinity of the Anolyte*

The anolyte's salinity represents a crucial parameter that significantly impacts the performance of BESs. A rise in substrate salt concentration can enhance the efficiency of BESs, up to the point where exoelectrogens can tolerate it. Beyond this tolerance limit, further increases in salinity have a negative effect, leading to inhibiting bacterial activity and escalated activation losses [109]. In addition to the efficient breakdown of ECs, higher salinity levels also improve the power generation capacity of BESs. This improvement can be ascribed to an enhanced rate of mass transfer of the substrate to the bacteria and variations in microbiological dynamics. Consequently, microbes in the anodic chamber of the BES can more rapidly oxidize organic matter and release more electrons, resulting in higher power output [110]. For instance, in a study by Mousavi and colleagues, the influence of salinity on phenol elimination and power production in MFCs was evaluated. It was found that the anolyte's power density elevated from approximately 31 to 40 mW m^{-3} after 0.5% (W/V) of NaCl was added, in addition to that, the removal of phenol also improved by 4% from 90% (when salt was not added) and an initial level of 300 mg L^{-1} with a processing time of 72 h [111]. In a study, the degradation pattern of chloramphenicol was investigated through the influence of salinity ranging from 0 to 6% NaCl. When the salinity rose from 0 to 0.5%, the elimination efficiency of chloramphenicol enhanced from 88.3 to 92.5%. With an initial chloramphenicol concentration of 50 mg L^{-1}, the percentage removal decreased to 49.5%, further varying the salinity from 0.5 to 6%. This indicates that the efficient elimination of chloramphenicol occurred at a lower concentration of salinity [112]. These investigations provide valuable insights into the efficient biodegradation of ECs in wastewater with elevated salinity using BES treatment.

4.6 *Response of Externally Applied Voltage*

The externally applied potential is a crucial component that remarkably affects the disintegration of Emerging Contaminants in BESs. The degradation of 4-chlorophenol was substantially enhanced by controlling the applied voltage in MECs. Increasing the voltage from 0.2 to 0.7 V led to a rise in 4-chlorophenol removal from 61.8 to 92.5%, with a treatment of 24 h and an initial level of 1 mg L^{-1} [113]. This improvement can be ascribed to the lower cathode potential resulting from increased current density due to the higher applied potential. Consequently, a greater number of electrons became accessible to the microorganisms for the degradation of pollutants, thereby improving elimination efficiency [114]. The breakdown of chloramphenicol using BES rose from 70 to 96% as the cathode potential became more negative, decreasing from − 0.5 to − 1.25 V versus SHE following the operation of 72 h with 50 mg L^{-1} of starting concentration [114]. Yang and coworkers [115] probed the impact of applied voltage on the breakdown of sulfadiazine and found that a more negative cathode potential and increased applied voltage of 0.9 V versus the standard hydrogen electrode (SHE) resulted in a 1.15 times higher degradation of sulfadiazine (91.4%) in comparison with 0 V versus SHE (79.3%) following an operation period of 36 h [115]. In a nutshell, these findings highlight that the elimination of ECs in BESs can be accelerated by a negative cathode potential and a higher negatively biased applied potential.

5 Application of BES for the Elimination of Emerging Contaminants

5.1 *MFC*

A MFC is a versatile technology that transforms the chemical energy present in organic substances into electrical energy while simultaneously treating wastewater through catalytic reactions involving microorganisms. The typical structure of an MFC consists of two chambers, an anodic chamber and a cathodic chamber, partitioned by a proton exchange membrane (PEM). In an MFC, the organic substance within the anodic compartment undergoes oxidation, facilitated through electroactive bacteria. This oxidation process generates electrons and protons [116]. The produced electrons are subsequently transferred to the anode electrode either directly from bacterial cells residing on the surface of the electrode, mediators, or through conductive nanowires. Simultaneously, the protons generated during the oxidation process drift through the PEM, moving from the anodic chamber to the cathodic chamber [117] (Fig. 2a). As a result of the proton migration, an external circuit connecting the cathode and anode experiences an electron flow. The electrons flow through the external circuit to reach the cathode, creating an electrical current in the process. Researchers have explored various configurations of MFCs, including double-chamber and single-compartment MFCs, utilizing either air cathodes or aqueous cathodes, for the elimination of organic pollutants from wastewater. While MFCs are effective in eliminating various contaminants from wastewater, they may face challenges when dealing with recalcitrant compounds like ECs, which can inhibit the activity of exoelectrogenic bacteria due to their toxic nature. Nevertheless, MFCs have shown promise as an environmentally benign methodology for the elimination of ECs, including phenols, dyes, surfactants, and pharmaceuticals from wastewater. In a study conducted by Thung et al. they achieved approximately 80.6% removal of acid orange 7 at a concentration of 50 mg L^{-1}, along with power production of 8.57 $\pm$ 0.2 mW m^{-2}, using a single-chamber up-flow MFC. However, the power generation decreased to 7.07 $\pm$ 1.4 mW m^{-2}, when the concentration of acid orange 7 was increased to 75 mg L^{-1}. Therefore, the anode and the azo dye compounds contended for electrons derived from the oxidized substrate [118]. Chemicals like phenol are frequently utilized in a variety of sectors, including petrochemicals, food, medicine, and chemicals. As a result, it is usually found in industrial wastes. Phenols pose serious risks to aquatic and human life as they enter the food chain and may accumulate in the environment. Consequently, it is essential to treat wastewater that contains phenol. According to Li et al. elevated phenol levels can change the dynamics of the microbial community in the anodic compartment of MFCs, which lowers the generation of biodegradable intermediates and, in turn, negatively affects the performance of MFCs [119]. To address this issue and enhance the operation of MFCs for the degradation of phenol, Shen and colleagues investigated the impact of lignite-activated coke (LAC) loaded in MFCs treating high-concentration phenolic wastewater. The peak power density in the MFCs with LAC packing remained between 252.3 and 278.7 mW m^2, whereas it dropped from 246.5 to 167.5 mW m^2 in the control group without LAC packing as the phenol content increased from 500 to 1000 mg/L [120]. High phenol concentrations have unfavorable effects on cell proteins, which might cause oxidative damage to the electrogenic bacteria. As a result, the MFC produces less power because fewer electrons are transported from the anode to the cathode [121]. Additionally, phenol elimination in MFCs with LAC packing rose from 89.4 to 94.5%, whilst it reduced in the control group from 80.7 to 50.8%. *Pseudomonas*, *Desulfuromonas*, and *Hydrogenophaga*, among other specific species, were found in the LAC-MFCs. These microbes are able to degrade the aromatic rings in the organic contaminants, which facilitates subsequent biodegradation by Shewanella and other microbes. This synergistic activity promoted enhanced phenol removal in the LAC-MFCs compared to the control group [120]. Over 2.5 million tonnes of pesticides are used annually worldwide to control weeds, insects, and other pests in agricultural fields, making them a class of Emerging Contaminants [122]. Due to the ubiquitous presence of these pesticide residues in water and wastewater, there are serious environmental issues. In order to prevent pesticide contamination of water sources, it is essential to remove pesticides from polluted water. MFCs may prove to be vital in the biodegradation of these contaminants in this situation. Zhang and coworkers demonstrated the performance of MFC for the removal of fipronil from the water matrix. With a starting fipronil concentration of 74 mg L^{-1}, they reported an outstanding 94.22% degradation of fipronil over a 72-h operational period at a pH of 7.02, which is close to neutral. Additionally, it was shown that a number of species, including *Pseudomonas* (3.39%), *Chryseobacterium* (4.26%), *Azoarcus* (4.98%), *Sphaerochaeta* (8.95%), *Azospirillum* (51.35%), and a few others, dominated the anodic biofilm of the MFC. Additionally, the MFC treatment effectively eliminated the extremely acute toxicity of fipronil towards zebrafish, resulting in a zero-mortality rate of zebrafish after 96 h of interaction [123]. Pharmaceuticals, such as beta-blockers, antidiabetics, antibiotics, and anti-inflammatory medications, are regularly detected in wastewater, and traditional treatment techniques often prove to be insufficient for successfully eliminating these complex organic contaminants [124]. Particular attention must be paid to antibiotics since they can change the microbial strains participating in biodegradation, which can result in the creation of antibiotic-resistant genes (ARG) and antibiotic-resistant bacteria (ARB) [125]. The conduct of these Emerging Contaminants in single- and dual-compartment MFCs has been thoroughly researched. The results show that within an MFC, degradation is primarily accomplished through hydrophobicity and biodegradability pathways. Additionally, chemicals with a positive charge typically show less degradation than those with a negative charge. This phenomenon is explained by the electrostatic force of attraction, which attracts positively charged substances to the negatively charged biofilm grown in the anodic chamber of the MFC, ultimately promoting biodegradation [126]. As a result, sorption onto the anodic biofilm and subsequent biodegradation are required for the removal of positively charged contaminants. Topu and Taskan investigated the effects of tetracycline (TC) and oxytetracycline (OTC) utilizing Sodium acetate as a carbon source (1.64 g L^{-1}) on the microbial community of MFC. They found that the inhibition proportion in the MFC that was supplied with 0.25 mg L^{-1} of TC was 13%. However, this ratio remarkably rose to 50% with an increase in TC level to 5 mg L^{-1}. Additionally, the inhibition increased to 78% from 50% when the TC dose was raised to 50 mg L^{-1} from 5 mg L^{-1}. On the other hand, the inhibition similarly increased from 16 to 88% when the OTC level was raised from 0.25 to 50 mg L^{-1}. Interestingly, OTC exhibited a higher inhibitory effect compared to TC. The inhibitory concentration (IC) for OTC was 4.35 mg L^{-1}, which was 1.55 times higher than that of TC. These findings highlight the importance of considering the type and concentration of antibiotics when assessing their impact on microbial communities and MFC performance. OTC, in particular, exhibited a stronger inhibitory effect, which could have implications for the use of MFCs in environments contaminated with such antibiotics. The emergence of antibiotic resistance genes (ARGs) in the course of antibiotic treatment is of great concern. Ondon et al. found that norfloxacin (initial concentration of 4 mg/L) was removed from an MFC by 65.5%. The study found that ARG generation in the MFC was approximately 105 copies of DNA per microliter. This level of ARGs is about 10–1000 times lower than that typically observed in conventional wastewater treatment systems for antibiotic removal, which range from 10^6 to 10^8 DNA copies per microlitre. Nonetheless, it is important to note that the duration of this evaluation was not specified in their study [127]. Accordingly, future research should investigate the long-term exposure to antibiotics such as NFX on ARG formation in MFCs. In another study, the pharmaceutical sulfamethoxazole (SMX), was entirely broken down in an MFC, leading to an increase in power production. In another study, Wu et al. explored the effectiveness of an MFC in degrading SMX and generating power. They reported nearly 98% degradation of SMX in the span of 48 h of operation. Additionally, after the introduction of SMX with the initial concentration of 20 mg L^{-1} to the anodic compartment, the power output rose from 33.84 to 39.96 W m^{-3}. This chamber was initially fed with 1 g L^{-1} of sodium acetate as the exclusive carbon source, and this increase in power production was observed when compared to the control group with no incorporation of SMX. This increased electricity generation may be ascribed to the reduction in methanogenesis due to the presence of antibiotics in the anolyte [128]. Hormones and estrogens are part of another group of ECs and are of significant environmental concern due to their potent endocrine-disrupting effects, which can harm both aquatic and human life. Estrogen bioremediation has shown promise in BES. For example, Kumar et al. conducted a study using MFCs to evaluate the removal of α-ethynylestradiol (EE2), a type of estrogen compound. They achieved a removal rate of approximately 44.6% for EE2 from contaminated water by utilizing 3 g L^{-1} of glucose as a substrate over a 48-h period. This process began with an initial EE2 concentration of 1 mg L^{-1}. Additionally, the MFC generated approximately 40.02 mW m^{-2} of electrical power during the same treatment process [129]. While there is finite available literature on the breakdown of surfactants, the majority of researchers have included surfactants inside the anolyte to enhance the power generation in MFC. For instance, Wen and coworkers. Found that increasing the concentration of 0–80 mg L^{-1} significantly improved peak power density, improving it from 0.6 to 5.2 W m^2. This enhancement was ascribed to the improvement in the permeability capacity of the cell membrane to lower electron transfer resistance, which led to a higher electron transfer rate and, as a result, increased power generation [130]. More recently, in the context of SDS degradation, Chakraborty et al. demonstrated the removal of approximately 70% of SDS in an MFC with a starting level of 10 mg L^{-1} and a 12-h retention period. A peak power density of 12.7 W m^{-2} was additionally attained with SDS concentration of 10 mg L^{-1}, which was around 2.65 times lower than a control utilizing L-cysteine as a carbon source. The decrease in power production in the presence of SDS was attributed to its increased toxic effect on exoelectrogenic microbes [86]. These examples illustrate that MFCs can effectively degrade ECs while simultaneously generating energy. Additional research findings related to the use of MFCs for EC removal from wastewater are summarized in Table 3.

Table 3 Application of MFC for emerging contaminants removal from wastewater

MFC configuration	Contaminant	I.C. (mg/L)	Maximum power density ($W\ m^{-2}$)	HRT (h)	(%) Removal efficiency	References
Dual chamber	Atrazine	30	0.25	14	~ 97.9	[131]
	Allura red dye	100	46.42	8	~ 96	[132]
	Triclosan	10	120	40	~ 95	[133]
	2,4-dichlorophenol	10	472.29	48	~ 90	[134]
	Oxyfluorfen	50	–	24	~ 77	[135]
	2-chlorophenol	50	1496	120	~ 75	[136]
	Sulfamethoxazole	200	–	24	~ 70	[137]
Air cathode single chamber	Penicillin	50	1227	24	~ 98	[106]
	Congo red dye	200	23.50	36	~ 88	[138]
	Ceftriaxone sodium	50	1426.76	24	~ 91	[107]

HRT hydraulic retention time, *I.C.* initial concentration (mg/L)

5.2 *MEC*

A modified version of the MFC, defined as a MEC, is used to transform organic materials found in wastewater into useful products like hydrogen gas (H_2) and methane (CH_4), as well as other byproducts including hydrogen peroxide (H_2O_2) and formic acid. Due to the presence of exoelectrogenic bacteria in a MEC, organic waste oxidizes in the anodic chamber. As a result, protons (h^+) and carbon dioxide (CO_2) are released into the solution, and electrons (e^-) are transported from the solution to the anode. These released electrons flow through an external circuit in the direction of the cathode. Protons move concurrently over a PEM from the anodic compartment to the cathodic compartment. When an external voltage is given to the cathodic chamber, typically between 0.2 and 0.8 V in comparison to the SHE [139], these protons are reduced to generate hydrogen gas (H_2) (Fig. 2i). The cathodic compartment of a MEC can produce methane (CH_4) in addition to hydrogen (H_2) through electro-methanogenesis from carbon dioxide (CO_2) [140]. This indicates that EC's cathode can generate both H_2 and CH_4. The removal of numerous contaminants from wastewater, such as dyes, pharmaceuticals, chlorinated chemicals, endocrine-disrupting compounds (EDCs), and heavy metals, has also been found to be extensively useful in MEC. Compared to other wastewater treatment technologies, MEC has many benefits, such as low operating costs, the potential to recover value-added products like H_2 and CH_4, and minimal sludge generation. Because of these benefits, MEC is desirable for effective and sustainable wastewater treatment [141].

A single-chamber MEC using a nickel foam (NF) cathode made of MoS_2-GO was used in a study. The primary objective was to produce hydrogen gas (H_2) while concurrently decolorizing AYR dye. According to the findings, at a starting AYR level of 30 mg L^{-1}, a remarkable 93% decolorization of the dye could be accomplished in just 10 h. Sodium acetate (NaAc) was introduced as an additional substrate with a concentration of 1 g L^{-1} to ameliorate the degradation process. The study also investigated H_2 generation rates with a variety of co-substrates, such as glucose, NaAc, and a mixture of these two. The results showed that for glucose, NaAc, and the combination of glucose and NaAc, H_2 generation rates of 0.119, 0.169, and 0.183 m^3/day m^3 were reached, respectively. Notably, the rate of H_2 generation increased when glucose was added as a co-substrate along with NaAc. This enhancement was attributable to the growth of bacteria that are electrochemically active in NaAc, while glucose improved COD removal, helped AYR decolorize, and increased H_2 synthesis through electrochemical processes [104].

In an additional study, erythromycin (ERY) antimicrobial degradation was demonstrated in a single-compartment MEC equipped with 10 Ω of external resistance with 20 mg L^{-1} as the starting concentration. It was observed that after 48 h of operation, 99% ERY degradation was successfully attained. The reduction peak in the cyclic voltammogram declined as the concentration of erythromycin increased. It was noticed that the activity of the biocatalyst reduced as the ERY concentration was raised from 0 to 30 mg L^{-1}. This is due to the fact that the ERY concentration influences the biocatalyst's electrochemical activity [142]. In the anodic chamber of a MEC, Miran et al. demonstrated a removal of 4-chlorophenol (4-CP) of 42.9% from a starting level of 100 mg L^{-1}. In contrast to Ag/AgCl, they simultaneously produced 13.3 g L^{-1} m^2 of hydrogen peroxide (H_2O_2) inside the cathode compartment with 6 h of treatment time. Sulfate-reducing bacteria (SRB) proved to be the primary driver of the biotransformation of 4-CP and the generation of H_2O_2. The breakdown of 4-CP resulted in the generation of phenol, which in turn broke down into acetate and, ultimately, CO_2 and H_2O [100]. Additionally, a number of other studies have investigated the employment of MECs for the elimination of various forms of ECs, and the findings are summarised in Table 4.

The studies mentioned above have shown the significant potential of MECs for the concurrent elimination of ECs and the retrieval of value products from wastewater. To support the non-spontaneous reactions inherent in microbial electrolysis processes, MECs must, however, be powered by an external source, which must be acknowledged. For the thermodynamic activation barrier to be overcome, this external potential is required. In an MEC, the targeted pollutant's distinctive redox potential is often used to break down the pollutant because it plays a crucial part in the degradation process. The advantage in this instance arises from the reality that the biocatalytic process in the anodic compartment of the MEC uses a lot less energy than traditional electrolysis techniques. Notwithstanding these benefits, a number of constraints, including significantly lower yields of value-added goods, might render MECs less scalable and reproducible. It's crucial to carefully optimize a number of operational factors, such as pH, temperature, initial EC concentration, electrode materials, and applied potential, in order to overcome these difficulties. The efficient

Table 4 Application of MEC for emerging contaminants removal from wastewater

MEC configuration	Pollutant	I.C.	V	HRT (h)	(%) Removal efficiency	References
Single chamber	Diclofenac	1	– 0.4	24	~ 92	[143]
Single chamber	2,4-dichloronitrobenzene	100	1.7	24	~ 92	[144]
Single chamber	Quinoline	250	1	240	~ 91	[145]
Single chamber	2-mercaptobenzothizaole	167.25	0.9	48	~ 73	[146]
Single chamber	2,4,6-trichlorophenol	5	0.8	24	~ 46	[138]
Double chamber	Congo red	200	0.3	24	87.9	[147]
Double chamber	Nitrobenzene	50	0.8	36	98	[148]

HRT hydraulic retention time, *I.C.* initial concentration (mg/L), *V* applied voltage (V) versus SHE

elimination of ECs from wastewater using MEC technology on a wider scale depends on the success of this optimization procedure [82].

5.3 *Bio-Electro Fenton System*

Three different types of reactions-bioelectrochemical, electrochemical, and chemical—are involved in the bio-electro-Fenton process for producing electricity. Equation 9 illustrates a typical electrode reaction using acetate as the substrate [149]. In this process, electrons (e^-) move to the cathode via an external circuit while H^+ ions move across the membrane towards the cathode to generate electrical energy. Table 5 lists some commonly encountered electrochemically active bacteria (EAB).

$$CH_3COO^- + 2H_2O \rightarrow 2CO_2 + 7H^+ + 8e^- \quad (9)$$

EAB serve as the biocatalysts for the bioelectrochemical reaction, which takes place in the anode chamber [150]. Without the need for external redox mediators, these EABs are capable of oxidizing the electron donor (such as organic molecules) and transferring the resultant electrons to the anodic compartment of the MFC [151]. Extracellular electron transfer (EET) processes in EAB can take both direct and indirect routes [152]. Direct EET is mostly mediated by c-type cytochromes, bacterial nanowires, or pili, which are hair-like projections on bacterial cells. In contrast, redox-active substances serve as mediators in indirect electron transfer. Only a few species

Table 5 Electrochemically active microorganisms used in anode chamber

Microbe	Type	References
Arcobacter	Bacterium	[154]
Shewanella oneidensis	Bacterium	[155]
Saccharomyces cerevisiae	Fungus	[156]
Geobacter sulfurreducens	Bacterium	[157]
Firmicutes	Bacterium	[158]
Comamonas	Bacterium	[154]
Desulfobulbaceae	Bacterium	[159]
Geobacter metallireducens	Bacterium	[160]
Hansenula anomala	Fungus	[156]

of EAB have been discovered so far, and each has its own unique mechanism. For instance, the c-type cytochrome OmcZ is used by the *Geobacter* species to enhance electron conduction between cells and transmit electrons to the electrode [153]. *Shewanella* species, in contrast, can engage in indirect electron transfer by employing intermediaries like riboflavin or flavin adenine mononucleotide, as well as direct electron transfer by making direct contact with the electrode through conductive filaments [16].

Electro-Fenton oxidation takes place inside the cathode compartment of a microbial fuel cell (MFC). Equation 10 illustrates how dissolved oxygen (O_2) is reduced by two to produce hydrogen peroxide (H_2O_2) in a continuous process. Typically, in the electro-Fenton process, cathode materials with a wide surface area are employed. Due to their stability, high conductivity, non-toxicity, and low catalytic activity towards the breakdown of H_2O_2, carbon-based cathodes have been found to be superior for this application in prior investigations. The in-situ synthesis of hydroxyl radicals ($OH\cdot$) results from the catalysis of H_2O_2 by ferrous ions (Fe^{2+}) (Fig. 2g) [161]. These hydroxyl radicals are powerful oxidizers that can effectively degrade organic molecules, leading to their complete mineralization as shown in the equation below.

$$O_2 + 2H^+ + 2e^- \rightarrow H_2O_2 \tag{10}$$

$$Fe^{2+} + H_2O_2 \rightarrow Fe^{3+} + OH^{\cdot} + OH^-, \quad \mathrm{k} = 63\,\mathrm{Lmol^{-1}\,s^{-1}} \tag{11}$$

Following the first attempt in 2009 to couple bioelectrochemical systems (BESs) with an electro-Fenton system, bio-electro-Fenton systems are widely used to degrade a variety of persistent pollutants [162]. The major objective of this strategy is to maintain process sustainability by making effective self-sustaining processes that utilize resources more effectively than traditional ones and consume less energy [163]. Inside the cathode compartments of bio-electro-Fenton systems, organic compounds are frequently utilized, whereas biodegradable substrates, such as acetate and glucose, have recently become popular options [164]. Additionally, it was discovered that a number of variables, such as the pH of the solution, the concentration of Orange G, the concentration of catholyte (both high and low concentrations), the flow rate of saltwater, and the flow rate of air, all had a major impact on the degradation of the dye. According to Feng et al. [85] the bio-electro-Fenton process could completely decolorize and mineralize Orange II at pH 7.0 in about 14 h while employing carbon nanotubes/γ-FeOOH as cathode electrodes. These findings demonstrate that bio-electro-Fenton systems are self-sustaining and do not require the addition of Fenton reagents or iron sources on a regular basis. It is useful to use this self-sustaining mechanism for the decolorization and mineralization of various dye wastewater, extremely hazardous polymer solutions and waste solutions [165]. A significant environmental problem is the prevalence of PPCPs (Pharmaceuticals and Personal Care Products) in wastewater. In the pharmaceutical category, PPCPs cover a wide range of substances, including antiepileptic, hormones, cytostatic medicines blood lipid-regulating, anti-inflammatory, and beta-blocking. Additionally, preservatives, insect repellents, antimicrobial agents, sunscreen UV filters, and synthetic musks are commonly found in personal care products, contributing to another class of PPCPs [166]. According to earlier studies, standard wastewater treatment methods frequently fail to entirely eliminate these pollutants, which causes them to build up in water sources [167]. Therefore, it is quite difficult to develop PPCP removal methods that are both efficient and eco-friendly. In a study, the effectiveness of MFC-Fenton systems was evaluated by Wang et al. for the degradation of four pollutants (sulfamethazine, bisphenol A, triclocarban, and estrone) [168]. According to their findings, the fabricated setup successfully degraded all the pollutants that were taken into account. Adsorption onto graphite materials and hydroxyl radical oxidation were the main methods for the degradation of these pollutants. A similar strategy was used by Zhang et al. in another study to degrade paracetamol utilizing MFC-electro-Fenton technology [169]. Notably, this approach did not require an external power source and eliminated the requirement for constant Fenton reagent addition. The researchers maintained the external resistance at 20, the initial pH at 2.0, and the total iron concentration at 5 mg/L. A maximum paracetamol degradation efficiency of 70% was attained in 9 h. However, in this case, the availability of bioelectricity for paracetamol breakdown was the limiting component. Nadais et al. employed MEC-electro-Fenton systems in their laboratory-scale study to address the difficulties and constraints related to the removal of PPCPs. Four non-steroidal anti-inflammatory medicines (NSAIDs) were the focus of their study in order to determine the ideal procedure parameters. The study revealed that a pH of 2.0, a Fe^{2+} concentration of 7.5 mmol/L, an airflow rate of 8 mL/min, and an applied voltage of 0.3 V were the most effective parameters. The study produced encouraging findings, with effective degradation rates of 75–81% for naproxen, 59–61% for ketoprofen, 80–86% for ibuprofen, and 87–97% for diclofenac, attained within 5 h. This work offers insightful information and presents a possible route for creating an effective wastewater treatment device that specifically targets micropollutants like PPCPs [161].

5.4 Hybrid Bioelectrochemical Systems

The utilization of BES offers an innovative solution for treating wastewater containing persistent and resistant ECs while simultaneously generating electrical power. However, independently operating BES faces difficulties when dealing with complicated wastewater compositions at a field-scale level. Whether it be MECs or MFCs, each type of BES has its own restrictions. With regard to MFCs, typical challenges include membrane fouling, poor power output, and the expensive price of catalysts. Nevertheless, MECs are constrained by the requirement for external power during wastewater treatment. To tackle these challenges, researchers have explored the integration of BES with other complementary techniques. These include adsorption, wetland systems, membrane bioreactors, hybrid systems where MECs are operated using the power generated by MFCs and electro-Fenton processes. This integration approach aims to enhance the efficiency of ECs removal, reduce costs, and improve overall treatment performance [170]. In an investigation, the removal of the SMX was examined utilizing a constructed wetland (CW) combined with a Microbial Fuel Cell (CW-MFC) system. The results demonstrated that starting with an initial SMX concentration of 100 μg/L, about 99.7% of SMX was removed within a Hydraulic Retention Time (HRT) of 24 h [171]. Liang et al. employed an independent CW system and a longer HRT of 10 days, however, they were only able to remove 76.3% of the SMX [172]. Additionally, 85% SMX degradation was attained by an MFC system within 60 h of operation [173]. Clearly, the hybrid CW-MFC system demonstrated faster pollutant degradation than standalone CW or MFC systems. Likewise, in a study, nitrobenzene was degraded using MFC in the pretreatment phase, subsequent to a CW (MFC-CW) along with the simultaneous power production. In the MFC-CW system, a maximum power density of 1.53 mW m^2 was achieved, which was nearly 2.6 times more than that attained in the standalone MFC (0.59 mW m^2) [174]. The rise in power output demonstrated in the CW-MFC system can be ascribed to the positive influence of plants within the constructed wetland. The oxygen produced by the photosynthetic activity of plants enhances the oxygen reduction reaction (ORR) at the cathode of the MFC, increasing the concentration of active microorganisms. This improved ORR contributes to the higher power output of the hybrid system [175]. Additionally, the removal of nitrobenzene was slightly enhanced when a CW and MFC were used as pretreatment or post-treatment. In comparison to the MFC alone (81.54%), the CW-MFC hybrid had an average removal rate that was around 2% higher (83.48%) [174]. Sathe and coworkers investigated the possibilities of a photocathode MFC in conjunction with a laterite-based adsorption method for the elimination of SDS surfactant from wastewater. They achieved an impressive 96.3% removal of SDS, under the hydraulic retention time of 24 h and an initial concentration of 10 mg/L. This removal rate was approximately 8% higher than the anodic degradation achieved in the MFC alone (88.6%). This hybrid system produced a maximum power density of 0.73 W m^3, demonstrating the viability of MFC-based hybrid systems for the effective removal of refractory contaminants from

Table 6 Removal of emerging contaminants through hybrid BES

Type of BES system	Contaminant	I.C.	HRT (h)	Removal efficiency (%)	Maximum power density (mW m^{-1})	References
MFC combined with constructed wetland	Nitrobenzene	16	24	92.89	1.53	[174]
Bio-electro Fenton-assisted MFC	Syringic acid	20	22	100	427.81	[176]
Bio-electro Fenton-assisted MFC	Vanillic acid	20	22	100	512.11	[177]
MFC combined with aerobic bioreactor	Acid blue 29	100	120	94 + 1.9	77.3 ± 3.4	[178]
Photo-assisted microbial fuel cell	Ethinylestradiol	1	336	89.82	0.633 ± 0.04	[179]

wastewater [102]. Table 6 summarises several further research on hybrid systems that incorporate BES for the elimination of ECs from contaminated water, highlighting the promise of these methods for water treatment.

I.C. initial concentration (mg/L)

6 Future Outlook

BES have shown promise in effectively detoxifying ECs in wastewater while simultaneously producing valuable products like bioelectricity in MFCs and biohydrogen in MEC systems. A major advantage over conventional treatment systems and advanced treatment technologies like Advanced Oxidation Processes (AOPs) is the recovery of valuable materials. The fundamental issue with BES is its commercial viability and scalability. A number of issues contribute to this challenge, including the high cost of PEM, stability and fouling problems, the high cost of electrode materials, the relatively low energy production, mass transfer constraints, and the presence of methanogens that consume hydrogen ions (H^+). The development of reliable, long-lasting, and economically feasible membranes, electrodes, and electrocatalysts is essential to overcoming these constraints and enabling the scaling-up of BES technology. The exploration of other reactor designs, such as the up-flow mode single-chamber BES, and the use of numerous electrodes may be helpful in overcoming the mass transfer constraints of BES technology. These developments are essential milestones in realizing the full potential of BES for resource recovery and wastewater treatment.

The evaluation of the degradation pattern of emerging contaminants within BES should instead take precedence in future demonstrations, with a stronger emphasis placed on establishing the possible toxicity of the byproducts and end products created during EC degradation. Recently, research has focused on improving and developing novel methods and sensors for EC detection in complicated wastes. As a result, numerous ECs have been identified in wastewater in varying quantities. Consequently, in the coming years, it is essential to conduct more small-scale experiments to clarify how effectively new emerging contaminants are removed in BES. Additionally, more investigation is required to fully understand the microbial dynamics essential for the disintegration of these complex contaminants.

In order to determine the optimum conditions for the effective elimination of contaminants from wastewater, the parameters impacting the performance of BES should be optimized. Additionally, since the contaminants in the class of pharmaceuticals are found in human urine and find their way into sewage, it is important to simultaneously remove EC and recover nutrients like phosphorous and nitrogen from the source itself. To further reduce membrane fouling in BES, a hybrid unit, containing a pretreatment unit followed by BES, may be effective for the treatment of wastewater containing emerging pollutants. As the complete mineralization of these contaminants would not necessarily be possible with a standalone treatment system.

7 Conclusion

For the treatment of ECs in wastewater, it has been found that the novel BES technology offers a practical and energy-efficient option. In addition, BES not only eliminates refractory pollutants but also offers a range of important resource recovery options with the added benefit of less sludge generation. The performance of BES for the removal of refractory pollutants depends on a number of operating factors, including the type of anolyte used, the type of microorganisms used, initial concentration of the pollutant. Subsequent studies should therefore concentrate on the optimization of such factors to ensure the sustainability of bioelectrochemical systems for the field-scale application for the elimination of emerging contaminants.

Acknowledgements Director, CSIR-NEERI is thankfully acknowledged for giving the opportunity to pursue the work in CSIR-NEERI, Nagpur, India. Charu and AS acknowledge the University Grant Commission, New Delhi, India for providing Junior and Senior Research Fellowship, respectively. The article is checked for plagiarism using the iThenticate software and recorded in the Knowledge Resource Center, CSIR-NEERI, Nagpur for anti-plagiarism (KRC No.: CSIR-NEERI/KRC/2023/NOV/WWTD/3).

References

1. Hoque MM, Islam A, Ghosh S (2022) Environmental flow in the context of dams and development with special reference to the Damodar Valley Project, India: a review. Sustain Water Resour Manag 8. https://doi.org/10.1007/s40899-022-00646-9
2. Dai D, Sun M, Xu X, Lei K (2019) Assessment of the water resource carrying capacity based on the ecological footprint: a case study in Zhangjiakou City, North China. Environ Sci Pollut Res 26:11000–11011. https://doi.org/10.1007/s11356-019-04414-9
3. Scaria J, Gopinath A, Nidheesh PV (2021) A versatile strategy to eliminate emerging contaminants from the aqueous environment: heterogeneous Fenton process. J Clean Prod 278:124014. https://doi.org/10.1016/j.jclepro.2020.124014
4. Tran NH, Reinhard M, Gin KYH (2018) Occurrence and fate of emerging contaminants in municipal wastewater treatment plants from different geographical regions—a review. Water Res 133:182–207. https://doi.org/10.1016/j.watres.2017.12.029
5. Yadav AK, Singh L, Mohanty A et al (2012) Removal of various pollutants from wastewater by electrocoagulation using iron and aluminum electrode. Desalination Water Treat 46:352–358. https://doi.org/10.1080/19443994.2012.677560
6. Berkessa YW, Yan B, Li T et al (2020) Treatment of anthraquinone dye textile wastewater using anaerobic dynamic membrane bioreactor: performance and microbial dynamics. Chemosphere 238. https://doi.org/10.1016/j.chemosphere.2019.124539
7. Bhowmick GD, Das S, Ghangrekar MM et al. (2019) Improved wastewater treatment by combined system of microbial fuel cell with activated carbon/TiO_2 cathode catalyst and membrane bioreactor. J Inst Eng (India) Ser A 100:675–682. https://doi.org/10.1007/s40030-019-00406-7
8. Xu L, Yu W, Graham N et al (2019) Application of integrated bioelectrochemical-wetland systems for future sustainable wastewater treatment. Environ Sci Technol 53:1741–1743. https://doi.org/10.1021/acs.est.8b07159
9. Mathuriya AS, Jadhav DA, Ghangrekar MM (2018) Architectural adaptations of microbial fuel cells. Appl Microbiol Biotechnol 102:9419–9432. https://doi.org/10.1007/s00253-018-9339-0
10. Yan W, Xiao Y, Yan W et al (2019) The effect of bioelectrochemical systems on antibiotics removal and antibiotic resistance genes: a review. Chem Eng J 358:1421–1437
11. Wang AJ, Wang HC, Cheng HY et al (2020) Electrochemistry-stimulated environmental bioremediation: development of applicable modular electrode and system scale-up. Environ Sci Ecotechnol 3:100050. https://doi.org/10.1016/j.ese.2020.100050
12. Lyu L, Yu G, Zhang L et al (2018) 4-Phenoxyphenol-functionalized reduced graphene oxide nanosheets: a metal-free Fenton-like catalyst for pollutant destruction. Environ Sci Technol 52:747–756. https://doi.org/10.1021/acs.est.7b04865
13. Brown RK, Harnisch F, Dockhorn T, Schröder U (2015) Examining sludge production in bioelectrochemical systems treating domestic wastewater. Bioresour Technol 198:913–917. https://doi.org/10.1016/j.biortech.2015.09.081
14. Li WW, Yu HQ, He Z (2014) Towards sustainable wastewater treatment by using microbial fuel cells-centered technologies. Energy Environ Sci 7:911–924. https://doi.org/10.1039/C3EE43106A
15. Zhang F, Ge Z, Grimaud J et al (2013) Long-term performance of liter-scale microbial fuel cells treating primary effluent installed in a municipal wastewater treatment facility. Environ Sci Technol 47:4941–4948. https://doi.org/10.1021/es400631r
16. Wang H, Ren ZJ (2013) A comprehensive review of microbial electrochemical systems as a platform technology. Biotechnol Adv 31:1796–1807. https://doi.org/10.1016/j.biotechadv.2013.10.001
17. Kelly PT, He Z (2014) Nutrients removal and recovery in bioelectrochemical systems: a review. Bioresour Technol 153:351–360. https://doi.org/10.1016/j.biortech.2013.12.046

18. Gadkari S, Gu S, Sadhukhan J (2018) Towards automated design of bioelectrochemical systems: a comprehensive review of mathematical models. Chem Eng J 343:303–316. https://doi.org/10.1016/j.cej.2018.03.005
19. Li S, Chen G (2018) Factors affecting the effectiveness of bioelectrochemical system applications: data synthesis and meta-analysis. Batteries 4:34. https://doi.org/10.3390/batteries4030034
20. Logan BE, Call D, Cheng S et al (2008) Microbial electrolysis cells for high yield hydrogen gas production from organic matter. Environ Sci Technol 42:8630–8640. https://doi.org/10.1021/es801553z
21. Ieropoulos IA, Greenman J, Melhuish C, Hart J (2005) Comparative study of three types of microbial fuel cell. Enzyme Microb Technol 37:238–245. https://doi.org/10.1016/j.enzmictec.2005.03.006
22. Logan BE, Hamelers B, Rozendal R et al (2006) Microbial fuel cells: methodology and technology. Environ Sci Technol 40:5181–5192. https://doi.org/10.1021/es0605016
23. Cao X, Huang X, Liang P et al (2009) A new method for water desalination using microbial desalination cells. Environ Sci Technol 43:7148–7152. https://doi.org/10.1021/es901950j
24. Forrestal C, Xu P, Jenkins PE, Ren Z (2012) Microbial desalination cell with capacitive adsorption for ion migration control. Bioresour Technol 120:332–336. https://doi.org/10.1016/j.biortech.2012.06.044
25. Nevin KP, Woodard TL, Franks AE et al (2010) Microbial electrosynthesis: feeding microbes electricity to convert carbon dioxide and water to multi-carbon extracellular organic compounds. MBio 1:10–128. https://doi.org/10.1128/mBio.00103-10
26. Berk RS, Canfield JH (1964) Bioelectrochemical energy conversion. Appl Microbiol 12:10–12. https://doi.org/10.1128/am.12.1.10-12.1964
27. Strik DPBTB, Timmers RA, Helder M et al (2011) Microbial solar cells: applying photosynthetic and electrochemically active organisms. Trends Biotechnol 29:41–49. https://doi.org/10.1016/j.tibtech.2010.10.001
28. Strik DPBTB, Hamelers HVM, Snel JFH, Buisman CJN (2008) Green electricity production with living plants and bacteria in a fuel cell. Int J Energy Res 32:870–876. https://doi.org/10.1002/er.1397
29. Strik DPBTB, Hamelers HVM, Buisman CJN (2010) Solar energy powered microbial fuel cell with a reversible bioelectrode. Environ Sci Technol 44:532–537. https://doi.org/10.1021/es902435v
30. Ge Z, Li J, Xiao L et al (2014) Recovery of electrical energy in microbial fuel cells. Environ Sci Technol Lett 1:137–214. https://doi.org/10.1021/ez4000324
31. Li J, Ge Z, He Z (2014) A fluidized bed membrane bioelectrochemical reactor for energy-efficient wastewater treatment. Bioresour Technol 167:310–315. https://doi.org/10.1016/j.biortech.2014.06.034
32. Xiao L, Ge Z, Kelly P et al (2014) Evaluation of normalized energy recovery (NER) in microbial fuel cells affected by reactor dimensions and substrates. Bioresour Technol 157:77–83. https://doi.org/10.1016/j.biortech.2014.01.086
33. Dong Y, Qu Y, He W et al (2015) A 90-liter stackable baffled microbial fuel cell for brewery wastewater treatment based on energy self-sufficient mode. Bioresour Technol 195:66–72. https://doi.org/10.1016/j.biortech.2015.06.026
34. Ge Z, He Z (2016) Long-term performance of a 200 liter modularized microbial fuel cell system treating municipal wastewater: treatment, energy, and cost. Environ Sci (Camb) 2:274–281. https://doi.org/10.1039/c6ew00020g
35. Puig S, Coma M, Desloover J et al (2012) Autotrophic denitrification in microbial fuel cells treating low ionic strength waters. Environ Sci Technol 46:2309–2315. https://doi.org/10.1021/es2030609
36. Nam JY, Kim HW, Shin HS (2010) Ammonia inhibition of electricity generation in single-chambered microbial fuel cells. J Power Sources 195:6428–6433. https://doi.org/10.1016/j.jpowsour.2010.03.091

37. Kim HW, Nam JY, Shin HS (2011) Ammonia inhibition and microbial adaptation in continuous single-chamber microbial fuel cells. J Power Sources 196:6210–6213. https://doi.org/10.1016/j.jpowsour.2011.03.061
38. Kuntke P, Geleji M, Bruning H et al (2011) Effects of ammonium concentration and charge exchange on ammonium recovery from high strength wastewater using a microbial fuel cell. Bioresour Technol 102:4376–4382. https://doi.org/10.1016/j.biortech.2010.12.085
39. You SJ, Ren NQ, Zhao QL et al (2009) Improving phosphate buffer-free cathode performance of microbial fuel cell based on biological nitrification. Biosens Bioelectron 24:3698–3701. https://doi.org/10.1016/j.bios.2009.05.015
40. Sukkasem C, Xu S, Park S et al (2008) Effect of nitrate on the performance of single chamber air cathode microbial fuel cells. Water Res 42:4743–4750. https://doi.org/10.1016/j.watres.2008.08.029
41. Ryu JH, Lee HL, Lee YP et al (2013) Simultaneous carbon and nitrogen removal from piggery wastewater using loop configuration microbial fuel cell. Process Biochem 48:1080–1085. https://doi.org/10.1016/j.procbio.2013.05.016
42. Villaseñor J, Capilla P, Rodrigo MA et al (2013) Operation of a horizontal subsurface flow constructed wetland—microbial fuel cell treating wastewater under different organic loading rates. Water Res 47:6731–6738. https://doi.org/10.1016/j.watres.2013.09.005
43. Choi J, Ahn Y (2015) Enhanced bioelectricity harvesting in microbial fuel cells treating food waste leachate produced from biohydrogen fermentation. Bioresour Technol 183:53–60. https://doi.org/10.1016/j.biortech.2015.01.109
44. Molognoni D, Puig S, Balaguer MD et al (2016) Multiparametric control for enhanced biofilm selection in microbial fuel cells. J Chem Technol Biotechnol 91:1720–1727. https://doi.org/10.1002/jctb.4760
45. Kim KY, Yang W, Ye Y et al (2016) Performance of anaerobic fluidized membrane bioreactors using effluents of microbial fuel cells treating domestic wastewater. Bioresour Technol 208:58–63. https://doi.org/10.1016/j.biortech.2016.02.067
46. Ye Y, Ngo HH, Guo W et al (2019) Feasibility study on a double chamber microbial fuel cell for nutrient recovery from municipal wastewater. Chem Eng J 358:236–242. https://doi.org/10.1016/j.cej.2018.09.215
47. Shin JW, Seo SJ, Maitlo HA, Park JY (2015) The enhancement of ammonium removal from ethanolamine wastewater using air-cathode microbial fuel cells coupled to ferric reduction. Bioresour Technol 190:466–473. https://doi.org/10.1016/j.biortech.2015.03.048
48. Qin M, Abu-Reesh IM, He Z (2016) Effects of current generation and electrolyte pH on reverse salt flux across thin film composite membrane in osmotic microbial fuel cells. Water Res 105:583–590. https://doi.org/10.1016/j.watres.2016.09.028
49. Matassa S, Batstone DJ, Hülsen T et al (2015) Can direct conversion of used nitrogen to new feed and protein help feed the world? Environ Sci Technol 49:5247–5254. https://doi.org/10.1021/es505432w
50. Qin M, Molitor H, Brazil B et al (2016) Recovery of nitrogen and water from landfill leachate by a microbial electrolysis cell-forward osmosis system. Bioresour Technol 200:485–492. https://doi.org/10.1016/j.biortech.2015.10.066
51. Sotres A, Cerrillo M, Viñas M, Bonmatí A (2015) Nitrogen recovery from pig slurry in a two-chambered bioelectrochemical system. Bioresour Technol 194:373–382. https://doi.org/10.1016/j.biortech.2015.07.036
52. Kim T, An J, Jang JK, Chang IS (2015) Coupling of anaerobic digester and microbial fuel cell for COD removal and ammonia recovery. Bioresour Technol 195:217–222. https://doi.org/10.1016/j.biortech.2015.06.009
53. Haddadi S, Elbeshbishy E, Lee HS (2013) Implication of diffusion and significance of anodic pH in nitrogen-recovering microbial electrochemical cells. Bioresour Technol 142:562–569. https://doi.org/10.1016/j.biortech.2013.05.075
54. Desloover J, Abate Woldeyohannis A, Verstraete W et al (2012) Electrochemical resource recovery from digestate to prevent ammonia toxicity during anaerobic digestion. Environ Sci Technol 46:12209–12216. https://doi.org/10.1021/es3028154

55. Ye Y, Ngo HH, Guo W et al (2019) Microbial fuel cell for nutrient recovery and electricity generation from municipal wastewater under different ammonium concentrations. Bioresour Technol 292:121992. https://doi.org/10.1016/j.biortech.2019.121992
56. Ye Y, Ngo HH, Guo W et al (2019) Effect of organic loading rate on the recovery of nutrients and energy in a dual-chamber microbial fuel cell. Bioresour Technol 281:367–373. https://doi.org/10.1016/j.biortech.2019.02.108
57. Fischer F, Bastian C, Happe M et al (2011) Microbial fuel cell enables phosphate recovery from digested sewage sludge as struvite. Bioresour Technol 102:5824–5830. https://doi.org/10.1016/j.biortech.2011.02.089
58. Hirooka K, Ichihashi O (2013) Phosphorus recovery from artificial wastewater by microbial fuel cell and its effect on power generation. Bioresour Technol 137:368–375. https://doi.org/10.1016/j.biortech.2013.03.067
59. Cusick RD, Logan BE (2012) Phosphate recovery as struvite within a single chamber microbial electrolysis cell. Bioresour Technol 107:110–115. https://doi.org/10.1016/j.biortech.2011.12.038
60. Huang H, Zhang DD, Li J et al (2017) Phosphate recovery from swine wastewater using plant ash in chemical crystallization. J Clean Prod 168:338–345. https://doi.org/10.1016/j.jclepro.2017.09.042
61. Nancharaiah YV, Venkata Mohan S, Lens PNL (2015) Metals removal and recovery in bioelectrochemical systems: a review. Bioresour Technol 195:102–114. https://doi.org/10.1016/j.biortech.2015.06.058
62. Modin O, Wang X, Wu X et al (2012) Bioelectrochemical recovery of Cu, Pb, Cd, and Zn from dilute solutions. J Hazard Mater 235–236:291–297. https://doi.org/10.1016/j.jhazmat.2012.07.058
63. Aryal N, Ammam F, Patil SA, Pant D (2017) An overview of cathode materials for microbial electrosynthesis of chemicals from carbon dioxide. Green Chem 19:5748–5760. https://doi.org/10.1039/c7gc01801k
64. Huang L, Chai X, Chen G, Logan BE (2011) Effect of set potential on hexavalent chromium reduction and electricity generation from biocathode microbial fuel cells. Environ Sci Technol 45:5025–5031. https://doi.org/10.1021/es103875d
65. Wang Z, He Z (2020) Frontier review on metal removal in bioelectrochemical systems: mechanisms, performance, and perspectives. J Hazard Mater Lett 1:100002. https://doi.org/10.1016/j.hazl.2020.100002
66. Jensen DL, Ledin A, Christensen TH (2004) The speciation of heavy metals in landfill-leachate polluted groundwater. Water Res 33:2642–2650. https://doi.org/10.1177/0734242x04042146
67. Li J, Ge Z, He Z (2014) Advancing membrane bioelectrochemical reactor (MBER) with hollow-fiber membranes installed in the cathode compartment. J Chem Technol Biotechnol 89:1330–1336. https://doi.org/10.1002/jctb.4206
68. Yuan H, He Z (2015) Integrating membrane filtration into bioelectrochemical systems as next-generation energy-efficient wastewater treatment technologies for water reclamation: a review. Bioresour Technol 195:202–209. https://doi.org/10.1016/j.biortech.2015.05.058
69. Qin M, He Z (2014) Self-supplied ammonium bicarbonate draw solute for achieving wastewater treatment and recovery in a microbial electrolysis cell-forward osmosis-coupled system. Environ Sci Technol Lett 1:437–441. https://doi.org/10.1021/ez500280c
70. Zuo K, Cai J, Liang S et al (2014) A ten-liter stacked microbial desalination cell packed with mixed ion-exchange resins for secondary effluent desalination. Environ Sci Technol 48:9917–9924. https://doi.org/10.1021/es502075r
71. Davis RJ, Kim Y, Logan BE (2013) Increasing desalination by mitigating anolyte pH imbalance using catholyte effluent addition in a multi-anode bench scale microbial desalination cell. ACS Sustain Chem Eng 1:1200–1206. https://doi.org/10.1021/sc400148j
72. Ping Q, He Z (2014) Effects of inter-membrane distance and hydraulic retention time on the desalination performance of microbial desalination cells. Desalination Water Treat 52:1324–1331. https://doi.org/10.1080/19443994.2013.789406

73. ElMekawy A, Hegab HM, Mohanakrishna G et al (2016) Technological advances in CO_2 conversion electro-biorefinery: a step toward commercialization. Bioresour Technol 215:357–370. https://doi.org/10.1016/j.biortech.2016.03.023
74. Lü F, Guo KJ, Duan HW et al (2018) Exploit carbon materials to accelerate initiation and enhance process stability of CO anaerobic open-culture fermentation. ACS Sustain Chem Eng 6:2787–2796. https://doi.org/10.1021/acssuschemeng.7b04589
75. Cheng S, Xing D, Call DF, Logan BE (2009) Direct biological conversion of electrical current into methane by electro-methanogenesis. Environ Sci Technol 43:3953–3958. https://doi.org/10.1021/es803531g
76. LaBelle EV, May HD (2017) Energy efficiency and productivity enhancement of microbial electrosynthesis of acetate. Front Microbiol 8:262397. https://doi.org/10.3389/fmicb.2017.00756
77. Faraghiparapari N, Zengler K (2017) Production of organics from CO_2 by microbial electrosynthesis (MES) at high temperature. J Chem Technol Biotechnol 92:375–381. https://doi.org/10.1002/jctb.5015
78. Deutzmann JS, Spormann AM (2017) Enhanced microbial electrosynthesis by using defined co-cultures. ISME J 11:704–714. https://doi.org/10.1038/ismej.2016.149
79. Saheb-Alam S, Singh A, Hermansson M et al (2018) Effect of start-up strategies and electrode materials on carbon dioxide reduction on biocathodes. Appl Environ Microbiol 84:e02242-e2317. https://doi.org/10.1128/AEM.02242-17
80. Chen S, Liu G, Zhang R et al (2012) Development of the microbial electrolysis desalination and chemical-production cell for desalination as well as acid and alkali productions. Environ Sci Technol 46:2467–2472. https://doi.org/10.1021/es203332g
81. Liu G, Luo H, Tang Y et al (2014) Tetramethylammonium hydroxide production using the microbial electrolysis desalination and chemical-production cell. Chem Eng J 258:157–162. https://doi.org/10.1016/j.cej.2014.07.025
82. Ahmad A, Priyadarshani M, Das S, Ghangrekar MM (2022) Role of bioelectrochemical systems for the remediation of emerging contaminants from wastewater: a review. J Basic Microbiol 62:201–222. https://doi.org/10.1002/jobm.202100368
83. Chatterjee P, Dessì P, Kokko M et al (2019) Selective enrichment of biocatalysts for bioelectrochemical systems: a critical review. Renew Sustain Energy Rev 109:10–23. https://doi.org/10.1016/j.rser.2019.04.012
84. Gautam S, Pandey LK, Vinayak V, Arya A (2017) Morphological and physiological alterations in the diatom *Gomphonema pseudoaugur* due to heavy metal stress. Ecol Indic 72:67–76. https://doi.org/10.1016/j.ecolind.2016.08.002
85. Feng C, Liu Y, Li Q et al (2016) Quaternary ammonium compound in anolyte without functionalization accelerates the startup of bioelectrochemical systems using real wastewater. Electrochim Acta 188:801–808. https://doi.org/10.1016/j.electacta.2015.12.069
86. Chakraborty I, Bhowmick GD, Nath D et al (2021) Removal of sodium dodecyl sulphate from wastewater and its effect on anodic biofilm and performance of microbial fuel cell. Int Biodeterior Biodegradation 156. https://doi.org/10.1016/j.ibiod.2020.105108
87. Chaturvedi V, Kumar A (2011) Diversity of culturable sodium dodecyl sulfate (SDS) degrading bacteria isolated from detergent contaminated ponds situated in Varanasi city, India. Int Biodeterior Biodegradation 65:961–971. https://doi.org/10.1016/j.ibiod.2011.07.005
88. Yan Q, Min J, Yu Y et al (2017) Microbial community response during the treatment of pharmaceutically active compounds (PhACs) in constructed wetland mesocosms. Chemosphere 186:823–831. https://doi.org/10.1016/j.chemosphere.2017.08.064
89. Buitrón G, Moreno-Andrade I (2014) Performance of a single-chamber microbial fuel cell degrading phenol: effect of phenol concentration and external resistance. Appl Biochem Biotechnol 174:2471–2481. https://doi.org/10.1007/s12010-014-1195-5
90. Lu Z, Chang D, Ma J et al (2015) Behavior of metal ions in bioelectrochemical systems: a review. J Power Sources 275:243–260. https://doi.org/10.1016/j.jpowsour.2014.10.168
91. Wang AJ, Cheng HY, Liang B et al (2011) Efficient reduction of nitrobenzene to aniline with a biocatalyzed cathode. Environ Sci Technol 45:10186–10193. https://doi.org/10.1021/es202356w

92. Zhang Q, Zhang Y, Li D (2017) Cometabolic degradation of chloramphenicol via a meta-cleavage pathway in a microbial fuel cell and its microbial community. Bioresour Technol 229:104–110. https://doi.org/10.1016/j.biortech.2017.01.026
93. Lin XQ, Li ZL, Liang B et al (2019) Accelerated microbial reductive dechlorination of 2,4,6-trichlorophenol by weak electrical stimulation. Water Res 162:236–245. https://doi.org/10.1016/j.watres.2019.06.068
94. Ilamathi R, Merline Sheela A, Nagendra Gandhi N (2019) Comparative evaluation of *Pseudomonas* species in single chamber microbial fuel cell with manganese coated cathode for reactive azo dye removal. Int Biodeterior Biodegradation 144. https://doi.org/10.1016/j.ibiod.2019.104744
95. Shi JH, Suzuki Y, Lee B-D et al (2002) Isolation and characterization of the ethynylestradiol-biodegrading microorganism *Fusarium proliferatum* strain HNS-1. Water Sci Technol 45:175–179. https://doi.org/10.2166/wst.2002.0424
96. Paul D, Noori MT, Rajesh PP et al (2018) Modification of carbon felt anode with graphene oxide-zeolite composite for enhancing the performance of microbial fuel cell. Sustain Energy Technol Assess 26:77–82. https://doi.org/10.1016/j.seta.2017.10.001
97. Xu H, Quan X, Xiao Z, Chen L (2018) Effect of anodes decoration with metal and metal oxides nanoparticles on pharmaceutically active compounds removal and power generation in microbial fuel cells. Chem Eng J 335:539–547. https://doi.org/10.1016/j.cej.2017.10.159
98. Shi C, Xu Y, Liu M et al (2021) Enhanced bisphenol S anaerobic degradation using an NZVI–HA-modified anode in bioelectrochemical systems. J Hazard Mater 403:124053. https://doi.org/10.1016/j.jhazmat.2020.124053
99. Wu D, Sun F, Zhou Y (2017) Degradation of chloramphenicol with novel metal foam electrodes in bioelectrochemical systems. Electrochim Acta 240:136–145. https://doi.org/10.1016/j.electacta.2017.04.059
100. Miran W, Nawaz M, Jang J, Lee DS (2017) Chlorinated phenol treatment and in situ hydrogen peroxide production in a sulfate-reducing bacteria enriched bioelectrochemical system. Water Res 117:198–206. https://doi.org/10.1016/j.watres.2017.04.008
101. Wang L, Liu Y, Ma J, Zhao F (2016) Rapid degradation of sulphamethoxazole and the further transformation of 3-amino-5-methylisoxazole in a microbial fuel cell. Water Res 88:322–328. https://doi.org/10.1016/j.watres.2015.10.030
102. Sathe SM, Bhowmick GD, Dubey BK, Ghangrekar MM (2020) Surfactant removal from wastewater using photo-cathode microbial fuel cell and laterite-based hybrid treatment system. Bioprocess Biosyst Eng 43:2075–2084. https://doi.org/10.1007/s00449-020-02396-4
103. Chen F, Li ZL, Yang J et al (2018) Effects of different carbon substrates on performance, microbiome community structure and function for bioelectrochemical-stimulated dechlorination of tetrachloroethylene. Chem Eng J 352:730–736. https://doi.org/10.1016/j.cej.2018.07.082
104. Hou Y, Tu L, Qin S et al (2021) Dye wastewater treatment and hydrogen production in microbial electrolysis cells using MoS_2-graphene oxide cathode: effects of dye concentration, co-substrate and buffer solution. Process Biochem 102:51–58. https://doi.org/10.1016/j.procbio.2020.12.008
105. Luo H, Liu G, Zhang R, Jin S (2009) Phenol degradation in microbial fuel cells. Chem Eng J 147:259–264. https://doi.org/10.1016/j.cej.2008.07.011
106. Wen Q, Kong F, Zheng H et al (2011) Electricity generation from synthetic penicillin wastewater in an air-cathode single chamber microbial fuel cell. Chem Eng J 168:572–576. https://doi.org/10.1016/j.cej.2011.01.025
107. Wen Q, Kong F, Zheng H et al (2011) Simultaneous processes of electricity generation and ceftriaxone sodium degradation in an air-cathode single chamber microbial fuel cell. J Power Sources 196:2567–2572. https://doi.org/10.1016/j.jpowsour.2010.10.085
108. Yu Y, Ndayisenga F, Yu Z et al (2019) Co-substrate strategy for improved power production and chlorophenol degradation in a microbial fuel cell. Int J Hydrogen Energy 44:20312–20322. https://doi.org/10.1016/j.ijhydene.2019.05.221

109. Guo F, Luo H, Shi Z et al (2021) Substrate salinity: a critical factor regulating the performance of microbial fuel cells, a review. Sci Total Environ 763:143021. https://doi.org/10.1016/j.scitotenv.2020.143021
110. Guo F, Fu G, Zhang Z, Zhang C (2013) Mustard tuber wastewater treatment and simultaneous electricity generation using microbial fuel cells. Bioresour Technol 136:425–430. https://doi.org/10.1016/j.biortech.2013.02.116
111. Mousavi SMS, Ayati B, Ganjidoust H (2016) Phenol removal and bio-electricity generation using a single-chamber microbial fuel cell in saline and increased-temperature condition. Energy Sources Part A Recov Util Environ Effects 38:3300–3307. https://doi.org/10.1080/15567036.2016.1156196
112. Guo N, Wang Y, Tong T, Wang S (2018) The fate of antibiotic resistance genes and their potential hosts during bio-electrochemical treatment of high-salinity pharmaceutical wastewater. Water Res 133:79–86. https://doi.org/10.1016/j.watres.2018.01.020
113. Wen Q, Yang T, Wang S et al (2013) Dechlorination of 4-chlorophenol to phenol in bioelectrochemical systems. J Hazard Mater 244–245:743–749. https://doi.org/10.1016/j.jhazmat.2012.11.002
114. Xie Y, Chen L, Liu R, Tian J (2018) Reduction of AOX in pharmaceutical wastewater in the cathode chamber of bio-electrochemical reactor. Bioresour Technol 265:437–442. https://doi.org/10.1016/j.biortech.2018.06.034
115. Yang XL, Zhang S, Li H et al (2018) Effects of voltage on sulfadiazine degradation and the response of sul genes and microbial communities in biofilm-electrode reactors. Ecotoxicol Environ Saf 151:272–278. https://doi.org/10.1016/j.ecoenv.2018.01.016
116. Jadhav DA, Ghangrekar MM, Duteanu N (2017) Recent progress towards scaling up of MFCs. In: Microbial fuel cell: a bioelectrochemical system that converts waste to watts, vol 151, pp 443–457. https://doi.org/10.1007/978-3-319-66793-5_23
117. Priyadarshini M, Ahmad A, Das S, Ghangrekar MM (2022) Metal organic frameworks as emergent oxygen-reducing cathode catalysts for microbial fuel cells: a review. Int J Environ Sci Technol 19:11539–11560. https://doi.org/10.1007/s13762-021-03499-5
118. Oon YS, Ong SA, Ho LN et al (2018) Disclosing the synergistic mechanisms of azo dye degradation and bioelectricity generation in a microbial fuel cell. Chem Eng J 344:236–245. https://doi.org/10.1016/j.cej.2018.03.060
119. Li B, Liu XN, Tang C et al (2019) Degradation of phenolic compounds with simultaneous bioelectricity generation in microbial fuel cells: influence of the dynamic shift in anode microbial community. Bioresour Technol 291:121862. https://doi.org/10.1016/j.biortech.2019.121862
120. Shen J, Li J, Li F et al (2021) Effect of lignite activated coke packing on power generation and phenol degradation in microbial fuel cell treating high strength phenolic wastewater. Chem Eng J 417:12809. https://doi.org/10.1016/j.cej.2020.128091
121. Zheng M, Han Y, Xu C et al (2018) Discrimination of typical cyclic compounds and selection of toxicity evaluation bioassays for coal gasification wastewater (CGW) based on toxicity mechanism of actions (MOAs). Sci Total Environ 644:324–334. https://doi.org/10.1016/j.scitotenv.2018.06.295
122. Fenner K, Canonica S, Wackett LP, Elsner M (2013) Evaluating pesticide degradation in the environment: blind spots and emerging opportunities. Science 341:752–758. https://doi.org/10.1126/science.1236281
123. Zhang Q, Zhang L, Li Z et al (2019) Enhancement of fipronil degradation with eliminating its toxicity in a microbial fuel cell and the catabolic versatility of anodic biofilm. Bioresour Technol 290:121723. https://doi.org/10.1016/j.biortech.2019.121723
124. Ahmad A, Das S, Ghangrekar MM (2019) Removal of xenobiotics from wastewater by electrocoagulation: a mini-review. J Indian Chem Soc 97:493–500
125. Chakraborty I, Sathe SM, Khuman CN, Ghangrekar MM (2020) Bioelectrochemically powered remediation of xenobiotic compounds and heavy metal toxicity using microbial fuel cell and microbial electrolysis cell. Mater Sci Energy Technol 3:104–115. https://doi.org/10.1016/j.mset.2019.09.011

126. Wang H, Heil D, Ren ZJ, Xu P (2015) Removal and fate of trace organic compounds in microbial fuel cells. Chemosphere 125:94–101. https://doi.org/10.1016/j.chemosphere.2014.11.048
127. Ondon BS, Li S, Zhou Q, Li F (2020) Simultaneous removal and high tolerance of norfloxacin with electricity generation in microbial fuel cell and its antibiotic resistance genes quantification. Bioresour Technol 304:122984. https://doi.org/10.1016/j.biortech.2020.122984
128. Ray G, Noori MT, Ghangrekar MM (2017) Novel application of peptaibiotics derived from *Trichoderma* sp. for methanogenic suppression and enhanced power generation in microbial fuel cells. RSC Adv 7:10707–10717. https://doi.org/10.1039/c6ra27763b
129. Kiran Kumar A, Venkateswar Reddy M, Chandrasekhar K et al (2012) Endocrine disruptive estrogens role in electron transfer: bio-electrochemical remediation with microbial mediated electrogenesis. Bioresour Technol 104:547–556. https://doi.org/10.1016/j.biortech.2011.10.037
130. Wen Q, Kong F, Ma F et al (2011) Improved performance of air-cathode microbial fuel cell through additional Tween 80. J Power Sources 196:899–904. https://doi.org/10.1016/j.jpowsour.2010.09.009
131. Domínguez-Garay A, Esteve-Núñez A (2018) Designing strategies for operating microbial electrochemical systems to clean up polluted soils under non-flooded conditions. Bioelectrochemistry 124:142–148. https://doi.org/10.1016/j.bioelechem.2018.03.006
132. Tacas ACJ, Tsai PW, Tayo LL et al (2021) Degradation and biotoxicity of azo dyes using indigenous bacteria-acclimated microbial fuel cells (MFCs). Process Biochem 102:59–71. https://doi.org/10.1016/j.procbio.2020.12.003
133. Xu W, Jin B, Zhou S et al (2020) Triclosan removal in microbial fuel cell: the contribution of adsorption and bioelectricity generation. Energies (Basel) 13(3):761. https://doi.org/10.3390/en13030761
134. Lu N, Li L, Wang C et al (2021) Simultaneous enhancement of power generation and chlorophenol degradation in nonmodified microbial fuel cells using an electroactive biofilm carbon felt anode. Sci Total Environ 783:147045. https://doi.org/10.1016/j.scitotenv.2021.147045
135. Zhang Q, Zhang L, Wang H et al (2018) Simultaneous efficient removal of oxyfluorfen with electricity generation in a microbial fuel cell and its microbial community analysis. Bioresour Technol 250:658–665. https://doi.org/10.1016/j.biortech.2017.11.091
136. Zahir RM, Hassan H, Nagoor Gunny AA (2021) Application of OPEFB fiber-based electrode in microbial fuel cell system for electricity generation and chlorophenol degradation. IOP Conf Ser Earth Environ Sci 765(1):012096. https://doi.org/10.1088/1755-1315/765/1/012096
137. Wang L, Wu Y, Zheng Y et al (2015) Efficient degradation of sulfamethoxazole and the response of microbial communities in microbial fuel cells. RSC Adv 5:56430–56437. https://doi.org/10.1039/c5ra08438e
138. Dai Q, Zhang S, Liu H et al (2020) Sulfide-mediated azo dye degradation and microbial community analysis in a single-chamber air cathode microbial fuel cell. Bioelectrochemistry 131:107349. https://doi.org/10.1016/j.bioelechem.2019.107349
139. Xue W, Zhou Q, Li F (2019) Bacterial community changes and antibiotic resistance gene quantification in microbial electrolysis cells during long-term sulfamethoxazole treatment. Bioresour Technol 294:122170. https://doi.org/10.1016/j.biortech.2019.122170
140. Guo Z, Thangavel S, Wang L et al (2017) Efficient methane production from beer wastewater in a membrane-less microbial electrolysis cell with a stacked cathode: the effect of the cathode/anode ratio on bioenergy recovery. Energy Fuels 31:615–620. https://doi.org/10.1021/acs.energyfuels.6b02375
141. Priyadarshini M, Ahmad A, Das S, Ghangrekar MM (2021) Application of microbial electrochemical technologies for the treatment of petrochemical wastewater with concomitant valuable recovery: a review. Environ Sci Pollut Res 29(41):61783–61802. https://doi.org/10.1007/s11356-021-14944-w

142. Hua T, Li S, Li F et al (2019) Degradation performance and microbial community analysis of microbial electrolysis cells for erythromycin wastewater treatment. Biochem Eng J 146:1–9. https://doi.org/10.1016/j.bej.2019.02.008
143. De GB, Soetaert M, Hennebel T et al (2012) Catalytic dechlorination of diclofenac by biogenic palladium in a microbial electrolysis cell. Microb Biotechnol 5:396–402. https://doi.org/10.1111/j.1751-7915.2011.00325.x
144. Liu Y, Wang C, Zhang K et al (2020) Rapid degradation of 2,4-dichloronitrobenzene in single-chamber microbial electrolysis cell with pre-acclimated bioanode: a comprehensive assessment. Sci Total Environ 724:138053. https://doi.org/10.1016/j.scitotenv.2020.138053
145. Gao Y, Kong X, Zhou A et al (2020) Enhanced degradation of quinoline by coupling microbial electrolysis cell with anaerobic digestion simultaneous. Bioresour Technol 306:123077. https://doi.org/10.1016/j.biortech.2020.123077
146. San-Martín MI, Escapa A, Alonso RM et al (2020) Degradation of 2-mercaptobenzothizaole in microbial electrolysis cells: intermediates, toxicity, and microbial communities. Sci Total Environ 733:139155. https://doi.org/10.1016/j.scitotenv.2020.139155
147. Huang W, Chen J, Hu Y, Zhang L (2018) Enhancement of Congo red decolorization by membrane-free structure and bio-cathode in a microbial electrolysis cell. Electrochim Acta 260:196–203. https://doi.org/10.1016/j.electacta.2017.12.055
148. Luo H, Hu J, Qu L et al (2019) Efficient reduction of nitrobenzene by sulfate-reducer enriched biocathode in microbial electrolysis cell. Sci Total Environ 674:336–343. https://doi.org/10.1016/j.scitotenv.2019.04.206
149. Du Z, Li H, Gu T (2007) A state of the art review on microbial fuel cells: a promising technology for wastewater treatment and bioenergy. Biotechnol Adv 25:464–482. https://doi.org/10.1016/j.biotechadv.2007.05.004
150. Schröder U, Harnisch F, Angenent LT (2015) Microbial electrochemistry and technology: terminology and classification. Energy Environ Sci 8:513–519
151. Liu X, Shi L, Gu JD (2018) Microbial electrocatalysis: redox mediators responsible for extracellular electron transfer. Biotechnol Adv 36:1815–1827. https://doi.org/10.1039/C4EE03359K
152. Hsu LH, Deng P, Zhang Y et al (2018) Nanostructured interfaces for probing and facilitating extracellular electron transfer. J Mater Chem B 6:7144–7158. https://doi.org/10.1039/C8TB01598H
153. Cao Y, Mu H, Liu W et al (2019) Electricigens in the anode of microbial fuel cells: pure cultures versus mixed communities. Microb Cell Fact 18:1–14. https://doi.org/10.1186/s12934-019-1087-z
154. Debnath D, Gupta AK, Ghosal PS (2019) Recent advances in the development of tailored functional materials for the treatment of pesticides in aqueous media: a review. J Ind Eng Chem 70:51–69. https://doi.org/10.1016/j.jiec.2018.10.014
155. Zhu FJ, Ma WL, Xu TF et al (2018) Removal characteristic of surfactants in typical industrial and domestic wastewater treatment plants in Northeast China. Ecotoxicol Environ Saf 153:84–90. https://doi.org/10.1016/j.ecoenv.2018.02.001
156. Gita S, Shukla SP, Deshmukhe G et al (2021) Toxicity evaluation of six textile dyes on growth, metabolism and elemental composition (c, h, n, s) of microalgae *Spirulina platensis*: the environmental consequences. Bull Environ Contam Toxicol 106:302–309. https://doi.org/10.1007/s00128-020-03074-7
157. Bester K, Theobald N, Schröder HF (2001) Nonylphenols, nonylphenol-ethoxylates, linear alkylbenzenesulfonates (LAS) and bis (4-chlorophenyl)-sulfone in the German Bight of the North Sea. Chemosphere 45(6–7):817–826. https://doi.org/10.1016/S0045-6535(01)00023-6
158. García MT, Campos E, Ribosa I (2007) Biodegradability and ecotoxicity of amine oxide-based surfactants. Chemosphere 69:1574–1578. https://doi.org/10.1016/j.chemosphere.2007.05.089
159. Jiang LL, Li K, Yan DL et al (2020) Toxicity assessment of 4 azo dyes in zebrafish embryos. Int J Toxicol 39:115–123. https://doi.org/10.1177/1091581819898396

160. Lu Z, Smyth SA, De Silva AO (2019) Distribution and fate of synthetic phenolic antioxidants in various wastewater treatment processes in Canada. Chemosphere 219:826–835. https://doi.org/10.1016/j.chemosphere.2018.12.068
161. Nadais H, Li X, Alves N et al (2018) Bio-electro-Fenton process for the degradation of non-steroidal anti-inflammatory drugs in wastewater. Chem Eng J 338:401–410. https://doi.org/10.1016/j.cej.2018.01.014
162. Zhu X, Ni J (2009) Simultaneous processes of electricity generation and p-nitrophenol degradation in a microbial fuel cell. Electrochem Commun 11:274–277. https://doi.org/10.1016/j.elecom.2008.11.023
163. Kahoush M, Behary N, Cayla A, Nierstrasz V (2018) Bio-Fenton and Bio-electro-Fenton as sustainable methods for degrading organic pollutants in wastewater. Process Biochem 64:237–247. https://doi.org/10.1016/j.procbio.2017.10.003
164. Wang W, Lu Y, Luo H et al (2018) A microbial electro-Fenton cell for removing carbamazepine in wastewater with electricity output. Water Res 139:58–65. https://doi.org/10.1016/j.watres.2018.03.066
165. Wang YT, Wang RS (2017) A bio-electro-Fenton system employing the composite FePc/CNT/SS316 cathode. Materials 10(2):169. https://doi.org/10.3390/ma10020169
166. Fu J, Lee WN, Coleman C et al (2019) Removal of pharmaceuticals and personal care products by two-stage biofiltration for drinking water treatment. Sci Total Environ 664:240–248. https://doi.org/10.1016/j.scitotenv.2019.02.026
167. Xu R, Zhang P, Wang Q et al (2019) Influences of multi-influent matrices on the retention of PPCPs by nanofiltration membranes. Sep Purif Technol 212:299–306. https://doi.org/10.1016/j.seppur.2018.11.040
168. Wang Y, Feng C, Li Y et al (2017) Enhancement of emerging contaminants removal using Fenton reaction driven by H_2O_2-producing microbial fuel cells. Chem Eng J 307:679–686. https://doi.org/10.1016/j.cej.2016.08.094
169. Zhang L, Yin X, Li SFY (2015) Bio-electrochemical degradation of paracetamol in a microbial fuel cell-Fenton system. Chem Eng J 276:185–192. https://doi.org/10.1016/j.cej.2015.04.065
170. Koffi NJ, Okabe S (2021) Bioelectrochemical anoxic ammonium nitrogen removal by an MFC driven single chamber microbial electrolysis cell. Chemosphere 274:129715. https://doi.org/10.1016/j.chemosphere.2021.129715
171. Wen H, Zhu H, Yan B et al (2020) Treatment of typical antibiotics in constructed wetlands integrated with microbial fuel cells: roles of plant and circuit operation mode. Chemosphere 250:126252. https://doi.org/10.1016/j.chemosphere.2020.126252
172. Liang Y, Zhu H, Bañuelos G et al (2018) Removal of sulfamethoxazole from salt-laden wastewater in constructed wetlands affected by plant species, salinity levels and co-existing contaminants. Chem Eng J 341:462–470. https://doi.org/10.1016/j.cej.2018.02.059
173. Xue W, Li F, Zhou Q (2019) Degradation mechanisms of sulfamethoxazole and its induction of bacterial community changes and antibiotic resistance genes in a microbial fuel cell. Bioresour Technol 289:121632. https://doi.org/10.1016/j.biortech.2019.121632
174. Xie T, Jing Z, Hu J et al (2018) Degradation of nitrobenzene-containing wastewater by a microbial-fuel-cell-coupled constructed wetland. Ecol Eng 112:65–71. https://doi.org/10.1016/j.ecoleng.2017.12.018
175. Yang Q, Wu Z, Liu L et al (2016) Treatment of oil wastewater and electricity generation by integrating constructed wetland with microbial fuel cell. Materials 9:885. https://doi.org/10.3390/ma9110885
176. Li B, Yan ZY, Liu XN et al (2019) Enhanced bio-electro-Fenton degradation of phenolic compounds based on a novel Fe–Mn/graphite felt composite cathode. Chemosphere 234:260–268. https://doi.org/10.1016/j.chemosphere.2019.06.054
177. Li B, Sun JD, Tang C et al (2021) A novel core-shell Fe@Co nanoparticles uniformly modified graphite felt cathode (Fe@Co/GF) for efficient bio-electro-Fenton degradation of phenolic compounds. Sci Total Environ 760:143415. https://doi.org/10.1016/j.scitotenv.2020.143415
178. Khan MD, Li D, Tabraiz S et al (2021) Integrated air cathode microbial fuel cell-aerobic bioreactor set-up for enhanced bioelectrodegradation of azo dye Acid Blue 29. Sci Total Environ 756:143752. https://doi.org/10.1016/j.scitotenv.2020.143752

179. Sogani M, Pankan AO, Dongre A et al (2021) Augmenting the biodegradation of recalcitrant ethinylestradiol using *Rhodopseudomonas palustris* in a hybrid photo-assisted microbial fuel cell with enhanced bio-hydrogen production. J Hazard Mater 408:124421. https://doi.org/10.1016/j.jhazmat.2020.124421

Application of Mathematical Modelling Techniques in Optimal Design of Wastewater Treatment Plants

Mahendra Kumar Pal and Kirpa Ram

Abstract Wastewater treatment plants (WWTPs) have complex processes and inter-related operations, subsequently poses multi-facet challenges in their planning and designing. Moreover, factors such as effluent concentration, flow rate, dosage of coagulants and treatment techniques influence the efficiency of the wastewater treatment process. Mathematical modeling has evidently proven to be beneficial in designing any WWTP by predicting the trend or behavior of its treatment process in terms of its input variables. Furthermore, state-of-the-art computing power, advanced mathematical models, and user-friendly software have made it effortless to model the complexities and dependencies of the treatment process. This book chapter presents a comprehensive overview of the application of mathematical modeling in the optimal design of economical and efficient wastewater treatment plant. The statistical modeling approach has been demonstrated with a generation of regression prediction models for BOD and TSS removal in tertiary wastewater treatment using coagulant Zetag-4120. Further, chapter presents the formulation of optimization problem statement of WTTP design with emphasis upon selection of objective Function, setting up constraint criterion and choosing suitable trade-offs. Data-driven optimization techniques, including artificial intelligence and machine learning, are briefly discussed. The chapter also touches upon the challenges and path-forward in the application of mathematical modeling in the field of wastewater treatment.

Keywords Mathematical modeling · Polynomial regression model · Optimization method · Conflicting objective · Sensitivity analysis · Wastewater treatment plant design

---

M. K. Pal (✉)
Department of Civil Engineering, Indian Institute of Technology (BHU) Varanasi, Varanasi, India
e-mail: mahendra.civ@iitbhu.ac.in

K. Ram
Institute of Environment and Sustainable Development, Banaras Hindu University, Varanasi, India
e-mail: ram.iesd@bhu.ac.in

M. M. Ghangrekar et al. (eds.), *Biological and Hybrid Wastewater Treatment Technology*, Earth and Environmental Sciences Library,
https://doi.org/10.1007/978-3-031-63046-0_18

1 Introduction

Rapid urbanization and industrialization have become avenues for wastewater (both sewage and industrial waste) generation, which is growing at an exponential rate and polluting not only surface water but also groundwater. Henceforth, urbanization and industrialization have necessitated the treatment of effluent before release [1, 2], which requires planning and designing of wastewater treatment plants (WTTPs). The challenge, however, in WWTP design is a repertoire of suitable technology, which is an amalgamation of multiple unit operations and processes capable of treating effluent such that it is in compliance with the stringent limits set by a municipality or governing body and minimizing the operating cost [3].

The problem of designing WTTP is challenging primarily due to the varying physical and chemical characteristics of effluent, which vary in the amount at different times of the day and at different seasons of the year [4]. Wastewater generated by each user, be it individual, society or industry, brings uncertainty to the effluent as each user has its own operation strategy. More importantly, the quantity and quality of effluent varies temporarily due to different operation cycles.

Furthermore, in SDGs 6.3, the United Nations has mandated improving water quality by reducing pollutants and treating domestic and industrial wastewater in compliance with national and location standards by 2023. To ensure this, the Center Pollution Control Board (CPCB) of the Ministry of Forest, Environment and Climate Change of Indian National Government has determined the emission criterion under the Environmental Protection Act, 1986 [5] by setting up the effluent wastewater quality for different industries located at different municipalities [6]. Henceforth, temporal variation of quantity and quality of effluent has been a bottleneck in the development of advanced wastewater treatment technologies [7–9] in combination with stricter effluent limits. Thus, the primary objective of policymakers is to attain the highest efficiency of WWTP with the lowest operation cost, i.e., optimal treatment effectiveness.

The literature review presents different approaches for handling the optimal operation of WTTPs by considering the various operation strategies and stricter effluent limits. These studies have been done on the network of wastewater systems focusing on social and economic aspects [10–12], planning of regional systems using hydraulic, economic, and technical aspects [13–15]. Furthermore, literature reports the enhancement of the efficiency of WTTP by altering the operational strategy using a different indicator, such as environmental aspects [16, 17], adjusting parameter determining operation cost [18, 19] and optimization modelling with respect to seasonal variation [9].

The aim of this book chapter to present an overview of application of the mathematical modelling in designing any wastewater treatment plants. The chapter briefly presents various stages of wastewater treatment process and need of mathematical modelling in WWTP design in section two and three respectively, Section four describe the various application of mathematical modelling with brief introduction of different approaches of mathematical modelling. A detailed discussion on application of statistical modelling and optimization modelling is presented. In statistical modelling application, generation of predictive models has been demonstrated through an example of regression prediction model for suspended particle removal efficiency. The formulation of an optimization problem with a description of constraint conditions is also detailed out. Data-driven optimization techniques, including artificial intelligence and machine learning, are briefly discussed. The chapter will also touch upon the challenges and path forward in the application of modelling in the field of wastewater treatment.

2 Wastewater Treatment Process: Brief Overview

Primarily, there are two basic stages in wastewater treatment: primary and secondary treatment. In primary treatment, materials which are either floating or settled due to gravity are screened out through physical process of screening, comminution, grit removal and sedimentation.

In the next stage, i.e. secondary treatment, wastewater is purified by eliminating soluble organic matter and suspended solids through a biological process in which microorganisms consume the organic nutrients as food and excrete carbon dioxide, water, and energy through their growth and reproduction. Trickling filter-activated sludge and oxidation ponds are conventional approaches for secondary treatment. Variations in the activated sludge process are introduced by altering the aeration through extension, contact stabilization, and high-purity oxygen aeration. There are numerous technologies for the secondary treatment of wastewater, to name a few: membrane technology, up-flow anaerobic sludge blanket digestion, bioreactor, electrocoagulation, Sequencing Batch Reactor (SBR), and so on. A schematic diagram of the wastewater treatment plant is depicted in Fig. 1.

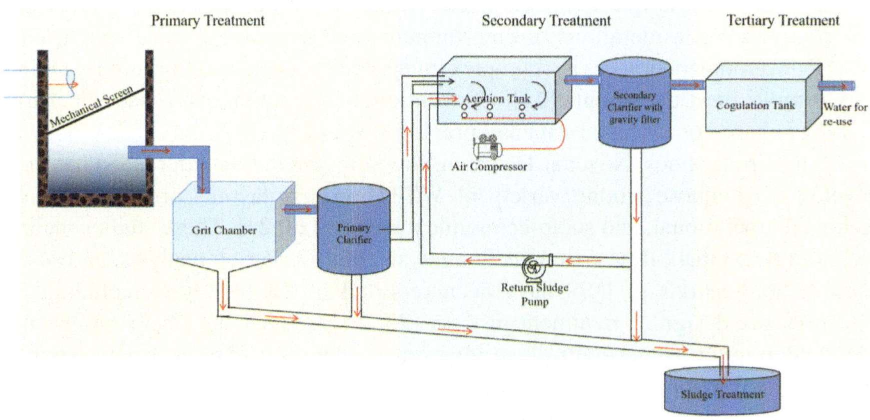

Fig. 1 Schematic diagram of WWTP

There is always a probability that secondary treated effluents have a high quantity of suspended organic particles, nitrogen, phosphorus, and facile due to the high vulnerability of pollutants in effluents. In this circumstance, it is essential to tertiary treat the wastewater so that it can be re-used or released to surface water. Conventionally, tertiary treatment of wastewater is achieved by coagulation and/or flocculation using chemical agents such as alum, PAC, etc.

Sludge is a by-product of wastewater treatment plants, which could be solid, semisolid, or slurry in nature. Sludge is generated because of chemical precipitation, sedimentation, or other process of WTTPs. Disposal and treatment of generated sludge are vital in WTTP operation, which has two basic goals- (i) to reduce the volume and (ii) to stabilize the organic constituents. In stabilizing the organic constituents, it is aimed to eliminate the offensive odour and health hazard of sludge. Sewage sludge is often treated by a combination of thickening, digestion, and dewatering process.

3 Need of Mathematical Modelling

WTTP development program has three phases, namely- Explore, Feasibility, and Plan, in between idea inception and implementation [7, 8, 20]. In this approach, a variety of concepts, ideas, and their combinations are generated and evaluated either in a detailed or simplified way. In the feasibility study of WTTP design, answers to the following questions are sought.

1. Which treatment technology (process) and unit operations to be chosen for the given physical and chemical characteristics of effluents
2. How to verify the higher removal efficiency and cost-effectiveness in compliance with local and national standards for effluents.

Answers to these questions are multi-objective optimal problems with multi-criterion-based considerations of environments and socio-economical constraints. WTTP development process is a complex integrated decision-making activity mainly due to multi-facet constraints [21]. Conventional practice on early-stage decision-making is based on archived data and previous experience [22].

Along similar lines, Novedar Decision Support System (Novedar DSS) has been developed to choose among variety of WTTP process layouts prepared through technical, operational, and socio-economical studies [23, 24]. These studies include technology compatibility, cost–benefit analysis, multi-criterion analysis, and so on. Various applications of DSS have been reported in the literature, including the selection and design of treatment process [25, 26], sequencing of operation units [27], and monitoring and controlling of treatment plans [16]. These decision support systems enable policymakers and planners to choose consistent solutions for the given problem and examine the "what-if" scenario. However, these systems are empirical models coupled with database and analysis tools [28, 29].

An alternative approach to developing a DSS is mathematical modelling, which is a popular thrust area in the field of chemical process synthesis [30]. However, it has been adopted in limited applications in the field of wastewater treatment [31–35]. These studies have presented valuable insights and potential of mathematical modelling in the wastewater treatment process. The multi-disciplinary complexity of wastewater engineering necessitates both competencies and multi-objective optimization problems for formulating a realistic wastewater treatment plant design [36–40].

Thus, the mathematical modelling provides a solution for identifying a suitable, efficient, cost-effective, and sustainable treatment process with combination of operation units for the physical–chemical composition of effluents [41, 42]. In addition to this, constraint conditions ensure that effluent quality follows standards put in place by national and local bodies. Moreover, state-of-the-art computing power, along with advanced mathematical models and user-friendly software, has made it effortless to model the complexities and dependencies of the treatment process.

4 Mathematical Modelling

Mathematical modeling is an approach to comprehend the behavior and interaction of physical and chemical phenomena of a given system. Modeling involves representing phenomena using mathematical equations, which elucidate the relation between various variables of the system. These equations are then incorporated to simulate the behavior of the system under the given conditions. There are two approaches for mathematical modeling namely—physical modeling and statistical modeling. In physical modeling, the law of physics and their corresponding mathematical equation are used in modeling, which analyzes the system response under different conditions. On other hand, statistical models use probability theory and data to predict future events and trends. Physical models are generally more complex than statistical models and require verification and validation. While, the accuracy of the physical model is rooted into accuracy of physical law, the accuracy of statistical modelling is depended upon the size and diversity of historical data. In this chapter, our discussion is limited to statistical modeling.

In statistical modeling, either we maximize the likelihood or minimize the norms of residuals to generate predictive models. The predictive model analyses historical and current data and generates a model for future trends. The accuracy of these predictions is highly dependent on the size and diversity of historical data. In absence of historical data, optimization modeling serves as a suitable alternative to optimize the WTTP design. In this section, both statistical modeling and optimization modeling are discussed.

4.1 Statistical Modelling

Statistical modeling is an approach in which predictive model of an objective is generated by utilizing real world observation data with various statistical models and explicit assumption. It is the process of utilizing available data to extract the features of a given physical system and forecast the trend or future events. In simple words, science of statistics facilitates learning from data to predict the future value accurately. It involves the application of statistical models to the datasets, which helps in identifying the relationship between decision variables and making prediction. There are two types of learning in statistical modeling- supervised learning and unsupervised learning. In supervised learning is based upon regression model with classifiers. On other hand, in unsupervised learning, independence of decision variables is verified, and features are extracted.

In statistical modeling, regression model is linear or non-linear combinations of its parameters, which identify the outliers and pattern of a given data and establishes a relation between its decision variables. Regression models also predict the value target function though interpolation and extrapolation. There are multiple types of regression model such as (1) Linear Regression, (2) Logistic Regression, (3) Polynomial Regression, (4) Ridge Regression, (5) Non-linear Regression, etc.

A polynomial regression model for BOD and TSS removal as a function of parameters- Zetag dose (A), settling time (B) and pH (C) is given in Eqs. (1) and (2). A second order regression model is generated including a constant term as a product of parameters, three first-order effects (terms in A, B & C), three interaction effect (AB, BC & CA) and three second-order effects (A^2, B^2 & C^2) adopting a general form of second order equation. These predictive regression models have been generated through statistical investigation of generated experimental data as are shown in Figs. 2 and 3.

$$\begin{aligned} &\textit{BOD removal } (\%) \\ &\quad = 52.89 + 4.94A + 4.34B + 6.17C \\ &\quad + 0.91AB - 1.40AC - 1.62BC - 9.06A^2 - 3.69B^2 - 6.92C^2 \end{aligned} \tag{1}$$

$$\begin{aligned} &\textit{TSS removal}(\%) \\ &\quad = 92.51 + 14.84A + 8.73B - 1.20C - 10.02AB + 1.16AC + 4.28BC \\ &\quad - 12.85A^2 - 3.5B^2 - 3.71C^2 \end{aligned} \tag{2}$$

With the help of Eqs. 1 and 2, one can easily predict the removal efficiency of organic material in the system for a given value of pH and Zetag dosage. Or Zetage dosage can be determined required for a given removal percentage of organic particles in the wastewater. Thus, these prediction models become instrumental in designing an efficient and economical WWTP.

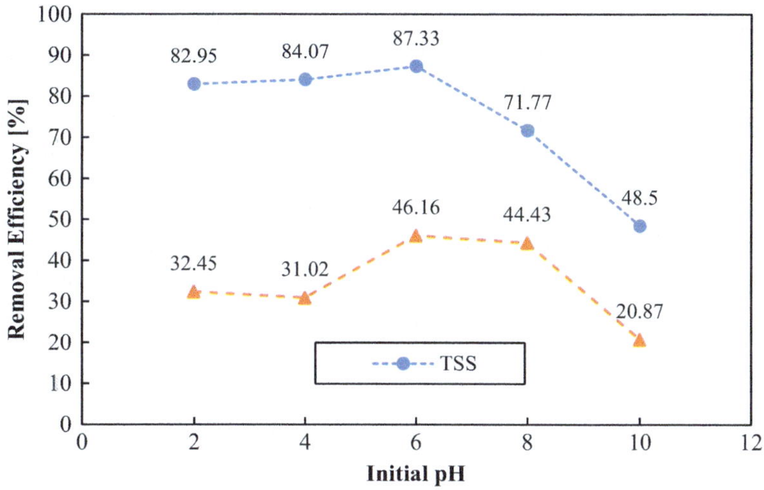

Fig. 2 Effect of initial pH on BOD and TSS removal for Zetag 4120 coagulant

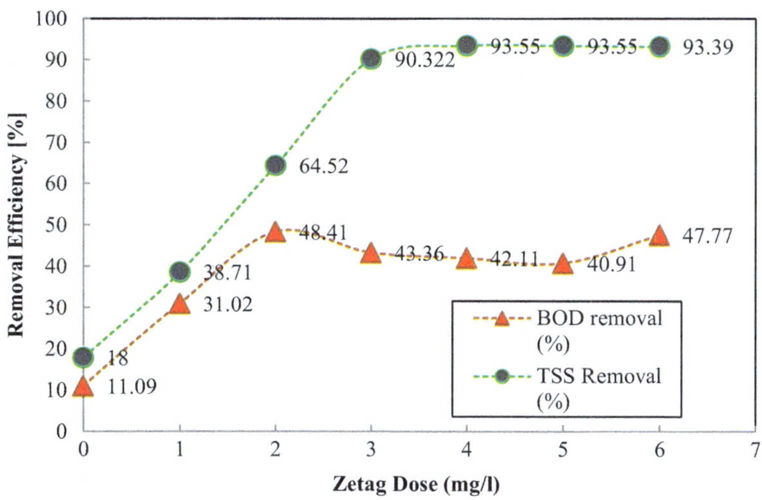

Fig. 3 Effect of Zetag-4120 dose on removal efficiencies of BOD and TSS

Table 1 lists the various examples of statistical modeling where life cycle cost [17], energy cost [43] are estimated in terms of various decision variables. Authors have reported high value of (approximately 98%) reliability of their result. However, it is important to note that the accuracy of statistical modeling is completely dependent upon the size and accuracy of available data. Furthermore, it is important to include independent random variables in your dataset and to eliminate the model dependence in statistical models. Conventionally, ANOVA test is performed to ensure that all the observations are independent of each other's, and variance of each parameter is same [9]. In the cases where data-size is moderate, it is recommended to determine sphericity [44] through Mauchly's test.

In recent years, statistical analysis or modeling have been empowered with advances in search algorithms, machine learning and artificial intelligence. These algorithms models analyze the data, predict the future trend, and determine the efficiency along-with cost–benefit analysis of WWTP. Commonly used computer algorithms are classified into three categories: Artificial Neural Network (ANN), Machine Learning and Search Algorithm. ANN is a mathematical model which imitates the process behavior of treatment plant and can handle a large amount of

Table 1 Example of Statistical modeling in WWTP Design

Author	Goal to be predicted	Modeling Approach	Output	Remark, if Any
Garrido-Baserba et al. [17]	Lifecycle Cost Assessment to quantify the environmental impacts associated with process throughout the entire lifetime using eutrophication potential [45, 46] and global warming [47]	Statistical modeling	Authors compared simulated results with DSS prediction system and found matching in range 55–82%. Further, GWP reliability for the three technology typologies (CAS, OD, and EA) was examined with an average matching close to 98% in the well-operated facilities	For DSS, Innovative Novedar Decision Support System -tool was used in selecting optimal WWTP design from alternative process layouts, which includes analysis of technical, economic, social issue and operation [23, 24]
Castellet-Viciano et al. [43]	Estimation of energy cost of exponential form $EC = AV^{b}e^{(\sum \alpha_i x_i)}$ [48, 49]	Cost estimation modeling using least square regression analysis		In statistical modeling normality, independence and homoscedasticity of modes are ensured
Gulhan Bourget [50]	Adopted two-way (treatment-season) or three-way interaction among observed parameters of various WWTP operating with different technologies	Linear mixed effects model to analyze WWTP data for the first time in this domain		Missing data i.e. Not Applicable (NA) and non-detect (ND) were handled by omitting NA parameters and assigning values to ND through detection limits

complex data [51, 52]. There are various sorts of neural network such as Convolution Neural Network (CNN) [53], Recurrent Neural Network (RNN) [54], Deep Neural Network (DNN) etc. In machine learning approach, algorithms and statistical models are developed in such way that computer itself automatically learns from data without any explicit models [55]. Commonly used machine learning models are decision tree (DT) [56], support vector machine (SVM) [57], Random Forest [58, 59] and so on.

4.2 *Optimization Model*

Optimization problem aims to find best possible solution out of all feasible solutions that satisfies the given constraints. Any optimization problem has five components namely objective Function, mathematical equation of process, initial conditions, constraints conditions and limits on decision variables. These five components construct the statement of optimization problem.

4.2.1 Formulating Optimization Problem Statement

A general problem statement of any optimization problem is stated as follows:

$$\min\left[f_1(X), f_2(X), \ldots\ldots., f_k(X)\right]^T$$

or

$$\max\left[f_1(X), f_2(X), \ldots\ldots., f_k(X)\right]^T$$

Subjected to

$$\text{Process Model} : f(x, \dot{x}, t) = 0$$

$$\text{Initial Condition} : x(t_0) = x_0$$

$$\text{Constraint Condition} : \mathrm{g}(x, \dot{x}, t) \leq 0$$

$$\text{Decision variable bounds} : \dot{x}_L \leq \dot{x} \leq \dot{x}_U$$

(i) **Objective Function (Determining the Goal)**

Objective Function describes the interdependence of decision variables and represented as a mathematical expression. Objective functions are often mathematical expressions of efficiency or risk, which need to be maximized or minimized. Though most of the problems have only one objective, there are possibilities that problems could have multiple objectives. Objective problems with multiple objectives are known as Multi-Objective Optimization (MOO) problems.

Efficiency of WWTP, Operation Cost, Life Cycle Assessment [59, 60] are few examples of objective functions, which have been incorporated in optimal design of wastewater WWTP. Hreiz et al. [61] have used both operation cost and nitrogen discharge as objectives to formulate multi-objective problem. In the cost function, authors have used aeration cost, pumping cost and other allied cost incurred in operation and maintenance of WWTP. In other examples of multi–objective Problems, Flores-Alsina et al. [16] have studied the impact of input uncertainties through Monte-Carlo simulation and considered economical, technical, and legal objectives by comparing six different control strategies by using multiple criterion decision analysis. Guerrero et al. [3] have analyzed mutli-objective problem of objective functions in form of operation cost, effluent quality, and microbiological risks by utilizing random sampling and selecting non-dominated solution in selecting optimal WWTP control set point.

(ii) **Initial Conditions and Physical Phenomena**

A system is an amalgamation of various processes based on specific laws of physics. Representing these processes and their dependencies upon decision variables is known as the Process Model in any optimization problem. Values of decision variables continue to change during any process and have a specific value at the start of these processes, known as initial conditions. The accuracy of any optimization problem is rooted in justifiable initial conditions.

In wastewater treatment optimization problem, physical and chemical characteristics of effluent serve as initial conditions. Process model describes the measure of suspended solids in terms of its contributor, both organic and inorganic. Conservation of total mass (or continuity equation) is another example of process model.

(iii) **Constrain Conditions (Effluent Criterion)**

The restrictions that must be satisfied to find an acceptable value of a given objective functions are collectively called constraint conditions. These conditions are classified into two categories—(i) behavior or functional constraints and (ii) geometrical or side constraints. Behavior or functional constraints are indicative of the performance of the system. On the other hand, geometrical or side constraints describe the physical limitations of decision variables. In an optimization problem, a feasible region is bound by constraint conditions.

In wastewater treatment plant design, stringent effluent criteria are kept as constant conditions. In formulating multi-objective problem, Hreiz et al. [61] have used BOD and TSS of effluent as constrain condition and Hakanen et al. [37] have defined inequality constraint by setting the range of ammonium nitrate, mixed liquor suspended solids (MLSS) in aeration and total nitrogen removal rate. In similar studies, Garrido-Baserba et al. [17] have included environmental criterion by setting the ranges of selected substances which are used for characterization in eutrophication potential [45, 46] and global warming [47].

(iv) **Conflicting Objectives and Trade-Off**

Generally, single objective functions have one objective Function defining the total cost, which may include the operation cost, aeration cost, maintenance cost and other allied cost (i.e. identifying the cheapest design). Despite the simplicity of single objective problems, it has drawback that it may result into depletion of information and inclusion of unnecessary constraints. Therefore, WWTP design and operations are optimized using multiple objectives [62–64] describing each relevant objective, which are optimized separately. In few cases, these multiple objectives could be conflicting such as Hakanen et al. [38] determined the optimal WWTP design by producing representative set of best possible solutions and Beraud et al. [65] found optimal solution for two objective functions, namely maximizing effluent quality, and minimizing energy consumption using genetic algorithm. Such objective functions are of conflicting nature and therefore, both objectives cannot be satisfying simultaneously. Hakanen et al. [38] have presented an interactive multi-objective optimization approach. Other commonly used approach is weighted approach, where all the objectives are combined by positive weighting coefficients [36, 65]. Thus, tradeoffs are accomplished in weighted manner. In another example of tradeoff between conflicting objectives using epsilon-constraint methods [66] known as Pareto-efficient tradeoff, Vadenbo et al. [39, 40] have performed tradeoff between minimizing the impact on climate change and minimization of mineral and fossil energy resource depletion. Qiao et al. [67] have presented a tradeoff between energy required for aeration and effluent quality.

(v) **Sensitivity Analysis**

Sensitivity analysis is output of mathematical model which studies the uncertainty in its input decision variables. Sensitivity analysis is performed to investigate how important a given decision variable is or to identify key parameters. In nutshell, robustness of optimal solution is analyzed through sensitivity analysis. Vadenbo et al. [39, 40] have investigated sensitivity analysis by adopting multiple scenarios namely no energy recovery, substituted district heating supply, increased sludge treatment and so on.

4.2.2 Solution Techniques

Among various classical solution techniques of constraint optimization techniques are penalty approach, feasible direction method and gradient direction methods, which have mathematically complex algorithms. In modern methods of optimization, genetic algorithms [68], simulated annealing [69], fuzzy optimization [18], etc. have gained their due popularity. Furthermore, advancement of computation infrastructures, application of machine learning and artificial intelligence has been trending in last few years (Table 2) [67].

In single objective optimization approaches, operation cost [59, 60] or efficiency of WWTP has been the commonly used objective Function, which furnishes economical solution of WWTP design. However, it is not guaranteed that obtaining an economical solution is also the most efficient solution. Henceforth, researchers have adopted multi-objective approaches and incorporated additional objectives from effluent criterion such as nitrogen discharge [61] or minimum sludge production [67], or minimum human toxicity [39, 40], etc. By doing so, not only economical but also the efficient optimal solutions are obtained.

Table 2 Example of Optimization modeling in WWTP design

Author	Objective Function	Solution Strategy	Output	Remark, if any
Qiao et al. [67]	Multiobjective function comprising of minimum cost, minimum sludge production and sufficient low concentration of biodegradable matter in effluent	Adaptive fuzzy neural network control using adaptive multi objective differential evolution algorithm (based upon stochastic search algorithm)	Control performance of WWTP by adaptive adjustment to select suitable scaling factor and crossover rate	Benchmarking is ensured using International Association on Water Quality Activated Sludge Model-1 [70]
Vadenbo et al. [39, 40]	Minimum climate change, minimum human toxicity, minimum ecotoxicity, Min CExD fossil Energy, Minimum CExD mineral and Minimum ReCiPe H/A	Representing a wider range of objectives by aggregating into weighted objective i,.e single objective using modeling system known as GAMS and solved with CPLEX Optimizer [85]	Concluded that temporal resolution is essential in determining the optimal solution due to seasonal variation in demand and supply. Further, determined the displaced heat supplier, which is a complex problem due to recovery of energy in foreground [72, 73]	GAMS modeling system [71] uses standard monte Carlo simulation to investigate the uncertainty associated with design variables
Bozkurt et al. [60]	Total annual cost (TAC) which is summation of operation and capitol cost	Combinatorial Superstructure approach [20]	Case study using Benchmark wastewater treatment plan [74] Select the optimal combination among the treatment alternatives such that TAC is minimized, and effluent criterion are satisfied	Superstructure is composed of all treatment technologies that are kept under relevant task connected to others in every possible way

(continued)

Table 2 (continued)

Author	Objective Function	Solution Strategy	Output	Remark, if any
Hreiz et al. [61]	Operating cost and nitrogen discharge	Elitist multiobjective genetic algorithm which are population based metahuristic genetic algorithm	Demonstrates the independence of arbitrary guessed initial conditions of effluents and dependency of reduction in oxygen transfer rate upon suspended solid concentration and ensured the good floc settlebility	Authors illustrated a tradeoff between treatment efficiency and operating cost
Francisco et al. [75]	Cost Function including operation cost and other process defined by Alex et al. [59]. Other cost includes safety, Environmental regulation, control limitation and product specification	Model predictive control formulation with closed loop stability Re-optimization has been obtained using maximum gain rule [76], the null-space method [77] and the exact local method [78]	Obtained the optimal controlled variables to give a combination of measurement, which is further utilized to avoid infeasibilities when large load disturbances appears. It further suggests to monitor NCO for dynamic optimization of system	

In doing so, the mathematical formulation of WWTP design results in a complex mathematical expression, which are often mathematically challenging to solve. In addition to that, stringent effluent criterion imposes multiple inequality constraints resulting in over-constrained optimization problem. Under these circumstances, it is prudent to make use of advances in computing tools and infrastructures. In the view of this, various computer algorithms have been developed. Such as Monte-Carlo method [79], genetic algorithm [68], simulated annealing [69], etc. It has further eased with development of artificial intelligence and machine learning [52]. Qiao et al. [67] have applied adaptive fuzzy neural network approach in optimization of benchmark problem. It is just the beginning of the research, and we believe that it will soon overtake the WWTP design work. Selected works of optimization modeling of WWTP design are enlisted in Table 2.

5 Recommendations

The process of the wastewater treatment plant is not only complex due to the varying physical and chemical composition of wastewater [80, 81] and the technologies involved in treating the wastewater [82, 83]. Therefore, there is an urgent need to improve the efficiency and reduce the total cost as well as energy consumption in WWTP [52, 84]. Although various mathematical models and computer algorithms have been developed for design of WWTP, their limited success in optimally designing due to highly non-linear and complex networks involved in WWTP design which cannot be ignored. The challenges and future prospective as well as recommendation of mathematical modeling are summarized as follows.

1. In statistical modeling, regression models have a dependency on the accuracy of observed or recorded data and could have model bias. Moreover, the moderate size of data can potentially predict the wrong trend of future events. Therefore, ANOVA and sphericity tests are recommended, which have been providing satisfactory results but endure multiple stability issues. Big-Data and the Internet of Things (IoT) are recent advances in data science that have introduced a paradigm shift in statistical modeling. These models are applied to different WWTP-related fields with data of different formats and different collection methods [50, 60] leading to poor data management and predictability. Moreover, these computer algorithms are like black boxes, which are developed based on observed data and totally ignore chemical and physical processes occurring in the treatment process. Therefore, we recommend using an integrated black-box model which should include physical as well as statistical modeling to improve the accuracy of the prediction of the efficiency of the WWTP.
2. In optimization models, identifying the single objective Function that can describe all the major aspects of WWTP operation is challenging, and numerous and stringent effluent limits imposes unsolvable optimization problems. Thus, it is wise to have a multi-objective Function to satisfy various goals in one setting, but these goals could be contradicting, and trade-offs will become necessary [40, 61, 67]. Consequently, these optimization models may result in complex and non-linear optimization problems that are challenging to solve with classical methods. In this regard, application of computer algorithms such as machine learning or search algorithm have emerged as promising tools for the optimal designing of WWTP and in solving optimization problems [52, 67]. These methods have shown positive results and therefore, these tools for optimal design of WWTP.
3. Verification and validation (V&V) of any mathematical model is essential. In the absence of any V&V guidelines, mathematical models are generally of no use, as the reliability of these models can not be established with acceptable accuracy. The verification and validations are often done by modeling the benchmark setups and comparing the modeling results with the experimental data. However, due to a lack of benchmark setup, design standard and fair comparison, it is challenging to develop any mathematical model, be it an optimization model or a statistical model. Therefore, bench-marking and standardization should be established for optimal design of WWTP.

6 Conclusions

The wastewater treatment plant involves a complex networks of treatment processes, interrelated operations as well as technologies and thus, poses multi-facet challenges in their planning and designing for the optimal operation of the wastewater treatment process. In addition, complex nature of physical and chemical properties of waste water influences the efficiency of the wastewater treatment process which often need different coagulant and treatment techniques. Therefore, at every step, all of these processes need to be optimized for economy as well as high efficiency. Mathematical modeling is becoming popular in the WWTP design process and has shown several positive results. In this chapter, we have summarized various mathematical modeling approaches of WWTP design, ranging from optimization models to statistical modeling, and the application of computer algorithms. In addition, we have discussed the treatment process of wastewater and presents mathematical formulations of the optimization problem. Moreover, a detailed discussion of the objective Function, constraint conditions, conflicting objectives, and sensitivity analysis has been presented. In the statistical modeling section, regression modeling has been demonstrated with a prediction model of removal efficiency in terms of decision variables such as coagulant dosage and pH of effluent. We have also highlighted challenges and future prospective as well as recommendation of mathematical modeling for the optimal designing of WWTP and in solving optimization problems.

References

1. El-Aswar EI, Ramadan H, Elkik H, Taha AG (2021) A comprehensive review on preparation, functionalization and recent applications of nanofiber membranes in wastewater treatment. J Environ Manage 301:113908. https://doi.org/10.1016/j.jenvman.2021.113908
2. Calderón OAR, Abdeldayem OM, Pugazhendhi A, Rene ER (2020) Current updates and perspectives of biosorption technology: an alternative for the removal of heavy metals from wastewater. Curr Pollution Rep 6:1–20
3. Guerrero J, Guisasola A, Comas J, Rodriguez-Roda I, Baeza J (2012) Multi-criteria selection of optimum wwtp control setpoints based on microbiology-related failures, effluent quality and operating costs. Chem Eng J 188:23–29
4. Seyssiecq I, Ferrasse JH, Roche N (2003) State-of-the-art: rheological characterization of wastewater treatment sludge. Biochem Eng J 16:41–56
5. Environmental Protection Act, 1986
6. Central Pollution Control Board (CPCB), India (2017) Ministry of Environment, Forest and Climate Change notification No. 843
7. Quaglia A, Sarup B, Sin G, Gani R (2013) A systematic framework for enterprise-wide optimization: synthesis and design of processing networks under uncertainty. Comput Chem Eng 59(5):47–62
8. Quaglia A (2013) An integrated business and engineering framework for synthesis and design of processing networks. Ph.D. thesis, Technical University of Denmark
9. Chebor J, Kiprop EK, Mwamburi LA (2017) Effect of seasonal variation on performance of conventional wastewater treatment system. J. App. Environ Microbiol 5(1):1–7

10. Massoud MA, Tarhini A, Joumana A, Nasr (2009) Decentralized approaches to wastewater treatment and management: applicability in developing countries. J Environ Manage 90(1):652–659
11. Libralato G, Volpi Ghirardini A, Avezzù F (2012) To centralise or to decentralise: an overview of the most recent trends in wastewater treatment management. J Environ Manag 94(2012):61–68
12. Eggimann S, Truffer B, Maurer M (2016) Economies of density for on-site wastewater treatment. Water Res 101:476–489
13. Brand N, Ostfeld A (2011) Optimal design of regional wastewater pipelines and treatment plant systems. Water Environ Res 83(1):53–64
14. Jing R, Fusi S, Chan A, Capozzi S, Kjellerup BV (2019) Distribution of polychlorinated biphenyls in effluent from a large municipal wastewater treatment plant: potential for bioremediation? J Environ Sci 78:42–52
15. Zeferino JA, Cunha MC, Antunes AP (2017) Adapted optimization model for planning regional wastewater systems: case study. Water Sci Technol 76(5):1196–1205
16. Flores-Alsina X, Rodriguez-Roda I, Sin G, Gernaey K (2008) Multi-criteria evaluation of wastewater treatment plant control strategies under uncertainty. Water Res 42:4485–4497
17. Garrido-Baserba M, Hospido A, Reif R, Molinos-Senante M, Comas J, Poch M (2014) Including the environmental criteria when selecting a wastewater treatment plant. Environ Model Softw 56:74–82
18. Ebner M, Stalph P, Michel M, Benz R (2010) Evolutionary parameter optimization of a fuzzy controller which is used to control a sewage treatment plant. Water Sci Technol 61(1):53–66
19. Hernández-del-Olmo F, Llanes FH, Gaudioso E (2012) An emergency approach for the control of wastewater treatment plants by means of reinforcement learning techniques. Expert Syst Appl 39(3):2355–2360
20. Quaglia A, Sarup B, Sin G, Gani R (2012) Integrated business and engineering framework for synthesis and design of enterprise-wide processing networks. Comput Chem Eng 38:213–223
21. Hamouda MA, Anderson WB, Huck PM (2009) Decision support systems in water and wastewater treatment process selection and design: a review. Water Sci Technol 60(7):1757–1770
22. Tchobanoglous G, Burton FL, Stensel HD (2003) Wastewater engineering: treatment and reuse. McGraw-Hill Publishing, New York
23. Garrido-Baserba M, Flores-Alsina X, Rodriguez-Roda I, Poch M (2010) Development of a DSS for the generation of WWTP configuration alternatives. Oral contribution to iEMSs 2010. In: International Congress on Environmental Modelling and Software
24. Garrido-Baserba M, Reif R, Rodríguez-Roda I, Poch M (2011) A knowledge management methodology for the integrated assessment of WWTP configurations during conceptual design. Water Sci Technol 66(1):165–172
25. Benedetti L, Dirckx G, Bixio D, Thoeye C, Vanrolleghem PA (2008) Environmental and economic performance assessment of the integrated urban wastewater system. J Environ Manage 88(4):1262–1272
26. Comas J, Alemany J, Poch M, Torrens A, Salgot M, Bou J (2004) Development of a knowledge-based decision support system for identifying adequate wastewater treatment for small communities. Water Sci Technol 48(11–12):393–400
27. Joksimovic D, Kubik J, Hlavinek P, Savic D, Walters G (2006) Development of an integrated simulation model for treatment and distribution of reclaimed water. Desalination 188(1–3):9–20
28. Huang JJ, Hu HY, Lu SQ, Li Y, Tang F, Lu Y, Wei B (2012) Monitoring and evaluation of antibiotic-resistant bacteria at a municipal wastewater treatment plant in China. Environ Int 42:31–36
29. Shim S, Reza A, Kim S, Ahmed N, Won S, Ra C (2020) Simultaneous removal of pollutants and recovery of nutrients from high-strength swine wastewater using a novel integrated treatment process. Animals 10(5):835
30. Grossmann I (2005) Enterprise-wide optimization: a new frontier in process systems engineering. AIChE J 51(7):1846–1857
31. Rigopoulos S, Linke P (2002) Systematic development of optimal activated sludge process designs. Comput Chem Eng 26:585–597

32. Vidal N, Bañares-Alcántara R, Rodríguez-Roda I, Poch M (2002) Design of wastewater treatment plants using a conceptual design methodology. Ind Eng Chem Res 41(20):4993–5005
33. Vidal N, Poch M, Martí E, Rodríguez-Roda I (2002) Evaluation of the environmental implications to include structural changes in a wastewater treatment plant. J Chem Technol Biotechnol Int Res Process Environ Clean Technolo 77(11):1206–1211
34. Alasino N, Mussati MC, Scenna N (2007) Wastewater treatment plant synthesis and design. Ind Eng Chem Res 45:7497–7512
35. Alasino N, Mussati MC, Scenna NJ, Aguirre P (2010) Wastewater treatment plant synthesis and design: combined biological nitrogen and phosphorus removal. Ind Eng Chem Res 49:8601–8612
36. Nakayama H (1995) Aspiration level approach to interactive multi-objective programming and its applications. In: Pardalos P, Siskos Y, Zopoun (eds)
37. Hakanen J, Sahlstedt K, Miettinen K (2011) Wastewater treatment: new insight provided by interactive multi-objective optimization. Decis Support Syst 51:328–337
38. Hakanen J, Sahlstedt K, Miettinen K (2013) Wastewater treatment plant design and operation under multiple conflicting objective functions. Environ Model Softw 46:240–249
39. Vadenbo C, Hellweg S, Guillén-Gosálbez G (2014) multi-objective optimization of waste and resource management in industrial networks—Part I: Model description, Resour Conserv Recycle 2014 (in press)
40. Vadenbo C, Guillén-Gosálbez G, Sanera D, Hellweg S (2014) Multi-objective optimization of waste and resource management in industrial networks—Part II: model application to the treatment of sewage sludge. Resour Conserv Recycl 89:41–51
41. Rivas A, Irizar I, Ayesa E (2008) Model-based optimization of wastewater treatment plants design. Environ Model Softw 23:435–450
42. Souza SM, Araújo OQF, Coelho MAZ (2008) Model-based optimization of a sequencing batch reactor for biological nitrogen removal. Bioresour Technol 99:3213–3223
43. Castellet-Vicianoa L, Torregrossab D, Sanchoa FH (2018) The relevance of the design characteristics to the optimal operation of wastewater treatment plants: energy cost assessment. J Environ Manage 222:275–283
44. Mauchly JW (1940) Significance test for sphericity of a normal n-variate distribution. An Math Stat 11:204–209
45. Corominas L, Foley J, Guest JS, Hospido A, Larsen HF, Morera S, Shaw A (2013) Life cycle assessment applied to wastewater treatment: state of the art. Water Res 47(15):5480–5492
46. Lassaux S, Renzoni R, Germain A (2007) Life cycle assessment of water from the pumping station to the wastewater treatment plant. Int J Life Cycle Assess 12:118–126
47. Larsen HF, Hauschild MZ, Wenzel H, Almemark M2007) Homogeneous LCA methodology agreed by NEPTUNE and INNOWATECH. Deliverable 4.1. EC Project "Neptune", contract No.:036845. https://www.eu-neptune.org
48. Friedler E, Pisanty E (2006) Effects of design flow and treatment level on construction and operation costs of municipal wastewater treatment plants and their implications on policy making. Water Res 40:3751–3758
49. Hernandez-Sancho F, Molinos-Senante M, Sala-Garrido R (2011) Cost modelling for wastewater treatment processes. Desalination 268:1–5
50. Bourget G (2023) Statistical analysis of wastewater treatment plant data. SN Appl. Sci. 5:130
51. Ding S, Li H, Su C, Yu J, Jin F (2013) Evolutionary artificial neural networks: a review. Artif Intell Rev 39:251–260
52. Malviya A, Jaspal D (2021) Artificial intelligence as an upcoming technology in wastewater treatment: a comprehensive review. Environ. Technol. Rev. 10:177–187
53. Alzubaidi L, Zhang J, Humaidi AJ, Al-Dujaili A, Duan Y, Al-Shamma O, Santamaría J, Fadhel MA, Al-Amidie M, Farhan L (2021) Review of deep learning: concepts, CNN architectures, challenges, applications, future directions. J Big Data 8:53
54. Sherstinsky A (2020) Fundamentals of recurrent neural network (RNN) and long short-term memory (LSTM) network. Physica D 404:132306
55. Samuel AL (1959) Machine learning. Technol Rev 62(1):42–45

56. Mining WID (2006) Data mining: concepts and techniques. Morgan Kaufinann 10:559–569
57. Yekkehkhany B, Safari A Homayouni, S, Hasanlou, M (2014) A comparison study of different kernel functions for SVM-based classification of multi-temporal polarimetry SAR data. Int Arch Photogramm Remote Sens Spat Inf Sci 40:281
58. Genuer R, Poggi JM (2010) Tuleau-Malot, C. variable selection using random forests. Pattern Recogn Lett 31:2225–2236
59. Alex JL, Benedetti L, Copp J, Gernaey KV, Jeppsson U, Nopens I, Pons M N, Rosen C, Steyer J P, Vanrolleghem P, Winkler S (2008) Benchmark Simulation Model no.1(BSM1). Technical Report no. LUTEDX/(TEIE- 7229)/1-62/(2008)
60. Bozkurt H, Quaglia A, Gernaey KV, Sin G (2015) A mathematical programming framework for early-stage design of wastewater treatment plants. Environ Model Softw 64:164–176
61. Hreiz R, Rocheb N, Benyahiac B, Latifi MA (2015) multi-objective optimal control of small-size wastewater treatment plants. Chem Eng Res Design 102:345–353
62. Miettinen K (1999) Nonlinear multiobjective optimization. Kluwer Academic Publishers, Boston
63. Miettinen K, Ruiz F, Wierzbicki A (2008) Introduction to multiobjective optimization: Interactive approaches. In: Branke J, Deb K, Miettinen K, Slowinski R (eds) Multiobjective optimization: interactive and evolutionary approaches. Springer, Berlin, pp 27–57
64. Rangaiah G (ed) (2009) Multi-objective optimization: techniques and applications in chemical engineering. World Scientific Publishing
65. Béraud B, Steyer JP, Lemoine C, Latrille E, Manic G, Printemps-Vacquier C (2007) Towards a global multi objective optimization of wastewater treatment plant based on modeling and genetic algorithms. Water Sci Technol 56(9):109–116
66. Haimes Y, Lasdon L, Wismer D (1971) On a bicriterion formulation of the problems of integrated system identification and system optimization. IEEE Trans Syst Man Cybern 1:296–297
67. Qiao JF, Hou Y, Zhang L, Han HG (2018) Adaptive fuzzy neural network control of wastewater treatment process with multiobjective operation. Neurocomputing 275:383–393
68. Goldberg DE (1989) Genetic algorithm. In: Search, optimization and machine learning. Addison-Wesley, New York
69. Granville V, Krivanek M, Rasson JP (1994) Simulated annealing: a proof of convergence. IEEE Trans Pattern Anal Mach Intell 16(6):652–656
70. Henze M, Grady CPL, Gujer W, Marais GR, Matsuo T (1986) Activated sludge model no. 1, IAWQ, London, 1986, IAWQ Scientific and Technical Report No. 1
71. GAMS Development Corporation (2011) GAMS development corporation, Washington DC, USA
72. Fruergaard T, Christensen TH, Astrup T (2010) Energy recovery from waste incineration: assessing the importance of district heating networks. Waste Manage (New York, NY) 30:1264–1272
73. Lundin M, Olofsson M, Pettersson GJ, Zetterlund H (2004) Environmental and economic assessment of sewage sludge handling options. Resour Conserv Recycl 41:255–278
74. Gernaey K V, Jeppsson U, Vanrolleghem PA, Copp JB (eds) (2014) Benchmarking of control strategies for wastewater treatment plants. IWA Scientific and Technical Report No. 23. IWA Publishing, London
75. Franciscoa M, Skogestadb S, Vega P (2015) Model predictive control for the self-optimized operation in wastewater treatment plants: analysis of dynamic issues. Comput Chem Eng 82:259–272
76. Halvorsen IJ, Skogestad S, Morud JC, Alstad V (2003) Optimal selection of controlled variables. Ind Eng Chem Res 42:3273–3284
77. Alstad V, Skogestad S (2007) Null space method for selecting optimal measurement combinations as controlled variables. Ind Eng Chem Res 46:846–855
78. Alstad V, Skogestad S, Hori ES (2009) Optimal measurement combinations as controlled variables. J Process Control 19:138–148

79. Pincus M (1970) A Monte-Carlo method for the approximate solution of certain types of constrained optimization problems. J Oper Res Soc Am 18(6):967–1235
80. Chahal C, Van Den Akker B, Young F, Franco C, Blackbeard J, Monis P (2016) Pathogen and particle associations in wastewater: significance and implications for treatment and disinfection processes. Adv Appl Microbiol 97:63–119
81. Jasim NA (2020) The design for wastewater treatment plant (WWTP) with GPS X modelling. Cogent Eng 7:1723782
82. Zaibel I, Arnon S, Zilberg D (2022) Treated municipal wastewater as a water source for sustainable aquaculture: a review. Rev Aquacult 14:362–377
83. Ren J, Shen W, Man Y, Dong L (2021) Applications of artificial intelligence in process systems engineering. Elsevier, Amsterdam
84. Ma Z, Cheah WY, Ng IS, Chang JS, Zhao M, Show PL (2022) Microalgae-based biotechnological sequestration of carbon dioxide for net zero emissions. Trends Biotechnol 40:1439–1453
85. CIPLEX, IBM ILOG CPLEX Optimization Studio, https://www.ibm.com/products/ilog-cplex-optimization-studio. As of April 2024

Laboratory to Field Scale Applications of Advanced Biological and Hybrid Wastewater Treatment Technologies

Akash Tripathi, Rishabh Raj, Shraddha Yadav, and M. M. Ghangrekar

Abstract Successful pilot-scale operation and commercialisation are critical benchmarks in the development of advanced biological and hybrid technologies. The most challenging issue at the pilot-level operation is performance drop, which is usually inversely proportional to the reactor size. In view of the aforementioned constraints, this chapter aims to elucidate the pivotal factors that determine the commercialisation and economics of hybrid biological systems. In addition, the current technology readiness level (TRL) of different upcoming hybrid technologies has been identified, and suggestions for enhancing the same have been included in future perspectives. Overall, the book chapter presents a comprehensive assessment of the lab-to-field transition of the hybrid biological system and the way forward of these emerging technologies.

Keywords Advanced biological methods · Commercialization · Pilot-scale · Technological readiness level · Wastewater treatment

1 Introduction

Wastewater treatment is an essential component of contemporary infrastructure, serving a crucial function in the preservation of the environment. Traditional approaches, utilising physical, chemical, and biological methods, have historically reduced the harmful effects of wastewater release on both aquatic ecosystems and human well-being. Yet, the constantly changing industrial and human activity has brought forth a wide range of new pollutants, which make it difficult to treat them

A. Tripathi · M. M. Ghangrekar (✉)
Department of Civil Engineering, Indian Institute of Technology Kharagpur, Kharagpur 721302, India
e-mail: ghangrekar@civil.iitkgp.ac.in

R. Raj · S. Yadav · M. M. Ghangrekar
School of Environmental Science and Engineering, Indian Institute of Technology Kharagpur, Kharagpur 721302, India

M. M. Ghangrekar et al. (eds.), *Biological and Hybrid Wastewater Treatment Technology*, Earth and Environmental Sciences Library,
https://doi.org/10.1007/978-3-031-63046-0_19

using conventional approaches [31]. Conventional methods have limits when it comes to dealing with recently discovered long-lasting pollutants, heavy metals, and complex chemicals that are resistant to decompose using traditional biological processes [32].

The need to tackle these issues has resulted in the emergence of hybrid biological technologies, which provide benefits compared to conventional methods in the treatment of wastewater. This chapter examines the scalability of sophisticated biological and hybrid processes, including Membrane Bioreactor (MBR), Moving Bed Biofilm Reactor (MBBR), Upflow Anaerobic Sludge Blanket (UASB) reactor, and Bioelectrochemical processes. Although these technologies demonstrate potential in controlled laboratory environments, the shift to field-scale applications is a critical stage in the study and development of wastewater treatment.

The evaluation of a technology's maturity and preparedness for practical implementation is done by assessing its transition from laboratory to field-scale applications using Technology Readiness Levels (TRL). The scalability of these advanced technologies depends on multiple aspects, including engineering considerations, economic feasibility, environmental consequences, and regulatory adherence [10]. An inclusive comprehension of the interactions between biological and hybrid components is essential for the seamless implementation of large-scale operations. This necessitates the optimisation of reactor configurations, treatment sequences, and the coordination of monitoring and control systems.

This chapter establishes the foundation for a comprehensive examination of the practical execution of advanced biological and hybrid wastewater treatment systems. It also provides a comprehensive examination of the hybrid paradigm, encompassing its mechanics and the obstacles encountered during large-scale operations. This analysis contributes new insights to the field of sustainable wastewater treatment technologies. The following sections will explore the intricate characteristics of hybrid technologies and their interaction with biological treatment, providing crucial insights into the progress of wastewater treatment approaches worldwide.

2 Driving Factors for the Commercialisation

Commercialisation is the ultimate step for converting scientific progress and research discoveries into practical applications that can be easily implemented by government and private institutions. Further, commercialisation also promotes the spread of innovative technologies, making sure they are available to different wastewater-generating sources and encouraging their widespread use. Additionally, commercialisation ensures the financial viability of solutions for attracting investments and promoting long-term sustainability. The section establishes the context for a thorough examination of the factors (Fig. 1) that drive the commercialisation of wastewater treatment technologies, with a focus on the diverse advantages that go beyond scientific advancements.

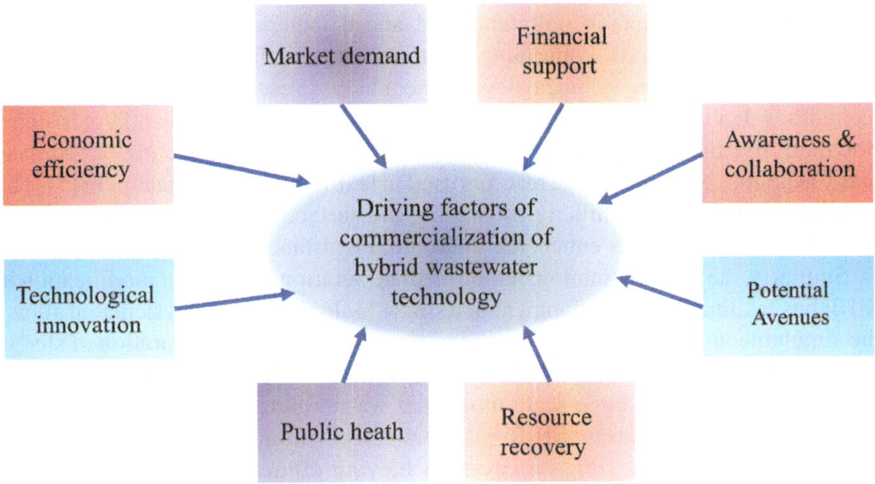

Fig. 1 Factors driving the commercialisation of the hybrid wastewater treatment technologies

2.1 *Public Health Concern*

Pharmaceutical residues and microorganisms are frequently found in municipal wastewater throughout the continents, posing potential risks due to their toxic nature. Introducing these pollutants into the environment occurs promptly through sources such as industrial discharges from manufacturing units, wastewater treatment plant (WWTP) effluents, and hospital effluents. The highest recorded concentrations of toxic compounds, including atenolol (294.7 μg/L), norfloxacin (25.3 μg/L), trimethoprim (162 μg/L), and carbamazepine (21.6 μg/L) were found in the wastewater sample of Asian nations such as India and South Korea [27].

Further, Mohapatra et al. reported that the total concentration of pharmaceutical compounds in the influent of WWTPs was 667 μg/L in the winter season, surpassing the total influent concentrations of 536 μg/L in the summer season [22]. Releasing untreated or poorly treated wastewater presents a direct risk to both public health and the environment. Hence, the commercialisation of advanced biological and hybrid wastewater treatment technology is essential for addressing these hazards, protecting the health of communities, and safeguarding aquatic life.

2.2 *Technological Innovation*

Advancement of wastewater treatment technology is crucial for effectively tackling the emerging contaminants and other hazardous materials in wastewater. Advanced treatment techniques, such as membrane filtration, UV disinfection, and electrochemical approaches, enhance operational efficiency, decrease energy usage, and improve the elimination of pollutants. Advanced materials, including innovative membranes and catalysts, contribute to improving the efficiency and durability of treatment systems; thus, technical development helps in escalating the TRL level of a technology. For example, a unique antifouling coating was created by polymerising a polymerisable bicontinuous microemulsion (PBM) to tackle the fouling of commercially available pilot membrane reactors used in textile wastewater treatment [5]. The coating enhanced hydrophilicity by making the surface smoother compared to the uncoated membrane, thus enhancing the fouling resistance and service life [5].

Similarly, the implementation of intermediate aeration and reactor modification in MBR has facilitated the development of a single biological reactor system that allows the simultaneous removal of carbon and nitrogen, as well as the separation of sludge and liquid. This offers a practical solution to address land scarcity and improve the treatment efficiency in large-scale municipal wastewater treatment plants [17]. In this so-called pilot air-lifting MBR, different zones within the reactor were created for anoxic, oxic, and settling processes, achieving effluent with total phosphorus (0.5 mg/L), COD (117 mg/L) and total nitrogen (10 mg/L) [17]. Innovations like these improve the financial viability of the product and help to make the technology align with industry requirements.

2.3 Economic Efficiency

Economics plays a crucial role in the process of making wastewater treatment technology available for commercial use, particularly for institutions with limited resources, such as hospitals, industries and municipal corporations. The whole cost of owning, which includes both initial investment and operational expenses, impacts the decision-making process. Therefore, technologies that provide a harmonious combination of efficacy and cost-effectiveness have more acceptability. Therefore, the impetus should be on implementing design advancements if the pilot-prototype that decreases energy consumption, minimises chemical use, and optimises maintenance procedures enhances the overall cost-efficiency of the system.

For instance, treatment of wastewater and nitrogen recovery from wastewater through the membrane has been found economically viable compared to traditional technology. The pilot plant treating the 40,000 m^3/day of wastewater and 600 m^3/day of ammonia-rich reject was used for irrigation [26]. Further, Li et al. demonstrated that MBR-nanofiltration hybrid technology for textile wastewater treatment has a 13.7% lesser footprint than the conventional process. Besides, a net revenue of 1.24 million USD per year with a payback period of 3.11 years was estimated for this hybrid system, which showcased the viability of the commercial application [16]. Further technological improvements are necessary to reduce expenses and position the technology to provide value to researchers and industries in selecting target markets strategically.

2.4 Awareness and Collaboration

The increasing recognition of environmental concerns and the need for improved water quality directly affects the process of making wastewater treatment technologies available for reuse. Implementing sustainable and responsible strategies for managing wastewater is a legal obligation and a means to conform to societal demands. In this regard, the collaboration between research institutions, government authorities, and private companies expedites transforming research discoveries into practical applications. Partnerships enable the sharing of knowledge, transfer of technology, and creation of solutions that may be expanded to a larger scale. Collaborative endeavours also aid in tackling implementation obstacles and guarantee that the technologies are curated in line with the requirements and limitations of industries.

Validating the effectiveness of wastewater treatment technology through pilot projects is an essential prerequisite for achieving market approval. Real-world applications provide stakeholders with the opportunity to observe the performance of a technology in practical settings, thereby instilling trust in its efficacy. Effective technology functions as influential endorsements, motivating industries and investors to embrace and allocate resources towards implementing commercialised solutions. In addition, they offer essential data for optimising and perfecting technology before their broader use.

2.5 Financial Support and Demand in the Market

The presence of financial assistance, subsidies, and incentives substantially impacts the landscape of commercialisation. An institution that chooses to employ sophisticated wastewater treatment technologies may be eligible for grants or favourable financing alternatives from governments, NGOs, and other funding bodies. These monetary incentives reduce the initial financial burden and encourage adopting eco-friendly practices. Commercial success frequently relies on aligning with these incentive programmes and showcasing a distinct return on investment over a period of time. The market landscape is also shaped by the overall need for effective and environmentally-friendly wastewater treatment solutions, driven by a growing number of facilities acknowledging the need for responsible wastewater management.

2.6 Potential Avenues for Reclaiming Resources

Within the context of treating wastewater, the investigation of resource reclamation is closely aligned with the fundamental principles of a circular economy, emphasising the need for sustainability and efficient use of resources. There are several ways to recover valuable materials from wastewater, which can significantly contribute to both environmental conservation and economic development. An important aspect of resource recovery involves extracting energy from organic matter found in wastewater using anaerobic digestion, dark fermentation, and bioelectrochemical technologies to produce biogas, electricity and chemicals [8].

Another significant approach involves the recovery of nutrients, specifically nitrogen and phosphorus, present in wastewater. According to the review conducted by Kumar and co-researcher, nutrient removal techniques help to extract and convert nutrients into high-quality fertilisers that may be used in agriculture [14]. For example, nutrient recovery technologies, such as struvite precipitation, can recover phosphorus and other essential nutrients for agricultural purposes [18]. This helps to solve environmental problems and, at the same time, contributes to sustainable green farming practices. Furthermore, the concept of water reuse is becoming a crucial technique, where treated wastewater is purified for non-drinking purposes inside infrastructure projects. Installing such wastewater treatment systems within the industrial or domestic premises guarantees the efficient use of water resources, alleviating the burden on freshwater supplies and promoting sustainable water management practices.

3 Estimating the Technology Readiness Level (TRL) of Hybrid Technologies

The Technology Readiness Level (TRL) is a metrics employed to evaluate the level of development and preparedness of a particular technology. Pioneered by the National Aeronautics and Space Administration (NASA), this ordinal scale has gained widespread adoption across several industries, including wastewater treatment. The TRL scale spans from TRL 1, representing the conceptual stage, to TRL 9, indicating a thoroughly mature and deployed technology (Fig. 2). The main objective of assigning TRL to an evolving technology is to provide insights into the level of development, potential risks, and the likelihood of successful implementation in real-world applications that decision-makers, investors, and researchers can use.

By convention, the conceptualisation of ideas based on a new scientific principle and its experimental validation at the lab-scale is categorised under TRL 1–3. The prototypes and pilot-scale demonstration of technology are assigned TRL levels of 4–6. Whereas, when a technology is successfully operated at an actual site, and it is ready to be commercialised, the TRL level is considered to be 8–9 at this stage. For instance, the idea of producing electricity from organic matter using microbes was first conceptualised by MC Potter back in the early twentieth century, though it was not until 2000 that the idea was first manifested in the form of microbial fuel cells (MFCs). Afterwards, MFC was used to treat real wastewater and was extensively investigated at a lab-scale. Thus, around this time, the TRL of this novel technology

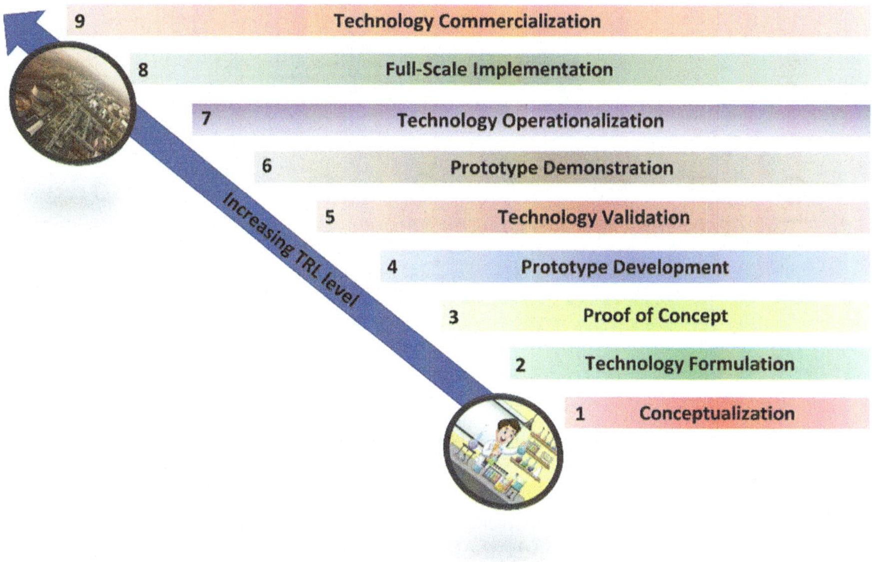

Fig. 2 Schematic representation of Technological Readiness Level

would be around 1–3. After about a decade, Jiang et al. [11] treated domestic wastewater using 20 L single chamber MFC, arguably the first pilot investigation on MFC. Later, a number of pilot-scale prototypes of MFC were tested, with volumes reaching as high as 1000–1500 L [9]. Hence, at this technological development stage, the TRL of MFC would be around 4–7. At present, with many explorations being conducted on the actual site, escalating the TRL of MFC to 7–8, which is often regarded as the pre-commercialisation stage [40]. In the future, if this technology integrates successfully with industrial and commercial platforms, the TRL will be deemed as level 9.

Similarly, among anaerobic bioreactors, anaerobic digestion is the most established and widespread technology, with a TRL of 9 [34]. Meanwhile, the anaerobic membrane bioreactors are at level 4 and are expected to achieve TRL 7 by 2027 [34]. Other emerging technologies, such as microbial electrolysis cells, microbial desalination cells and microbial electrosynthesis systems, are inching closer to a TRL of 4–5 as they progress towards technical validation and demonstration in relevant operating environments [37]. However, there needs to be more guidelines or universally accepted and replicable methods for identifying the specific TRL achieved by a particular technology. Hence, representing the maturity of technology on the TRL scale becomes somewhat subjective (Fig. 3). To highlight, some researchers believe that the MFC technology is advanced enough to be regarded at TRL 8, as few companies have already developed wastewater treatment technologies that fundamentally operate on the MFC principle [40]. On the other hand, one school of thought is that the power production of MFCs in the lab-scale, as well as pilot-scale MFCs, is still meagre, making the TRL of this technology around 3 to 4 [34].

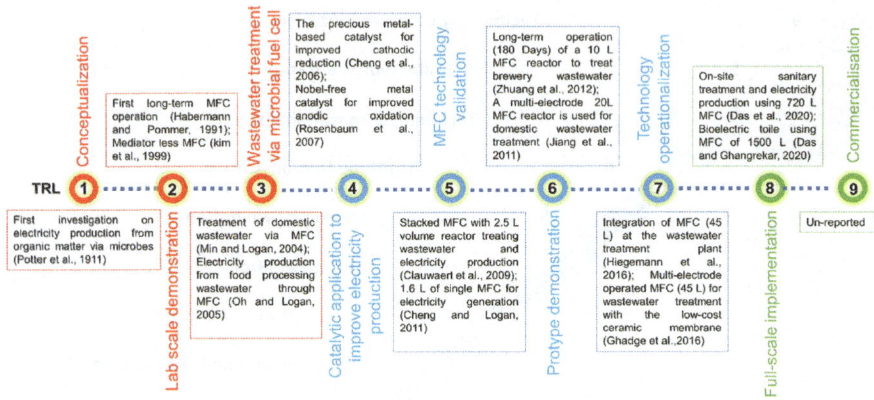

Fig. 3 Evaluation of microbial fuel cell through the lens of technology readiness level

Nevertheless, many techno-economical aspects are considered when establishing TRL indices, and it is still a handy tool to gauge the maturity of a process or technology. In order to appropriately assign a TRL, it is crucial to thoroughly comprehend the technology, including its evolution history and its present development. Involving specialists from other fields can augment the reliability of the assessment process. Using the TRL framework, stakeholders can methodically track the progress of different advanced hybrid biological systems, pinpoint the technical flaws and areas that require rectification and catalogue them into different categories for making a well-informed decision. Some of these hybrid systems and their TRL have been enlisted in Table 1.

Table 1 Emerging hybrid biological systems and their current TRL

S. No.	Hybrid-technology	Technology-readiness level	References
1	Electrochemical membrane bioreactor	6–7	[2, 19]
2	Constructed wetland–microbial fuel cell	6–7	[4, 13]
3	Microbial electrochemical technology-based wetland	8–9	[29, 30]
4	Constructed wetland–photocatalytic oxidation	4–6	[20, 21]
5	Microbial fuel cell-Fenton system	3-4	[33, 42, 43]
6	Algae-based hybrid systems	5–7	[38]

4 Case Study of Commercialised Hybrid Reactors

This section elaborates on the case studies conducted on hybrid reactors that are used in commercial wastewater treatment and examines the operational intricacies and effectiveness that were encountered during the operational period. The main aim is to highlight specifically those biological hybrid systems that are in advanced development that can be a part of mainstream wastewater management sector in the near future.

4.1 Pilot Scale MBBR and MBR

Integrating MBBR with MBR techniques can have symbiotic effect that improves the treatment efficiency by leaps and bounds, as the MBBR facilitates the growth of biofilm, thereby enhancing the process of organic matter oxidation and biological nutrient removal. Meanwhile, the MBR effectively segregates solid and liquid components while preserving biomass [7]. This results in the development of a long-lasting treatment system capable of effectively managing larger volumes of wastewater [36]. Hybrid MBBR and MBR systems minimise sludge generation and increase biomass retention in comparison to conventional treatment methods. This results in a more dependable and enduring wastewater treatment procedure capable of managing alterations in effluent composition and flow. The MBBR and MBR technologies synergistically enhance the efficiency, compactness, and versatility of wastewater treatment systems.

In a pilot-scale exploration on MBBR-MBR coupled system, Diaz et al. operated the hybrid reactor of 445 L of MBBR with polymetric carrier (358 L) and hollow fibre MBR (87 L) for the treatment of domestic wastewater with hydraulic retention time (HRT) of 24 h and membrane flux of 18.96 L/h m^2 [15]. The reactor used the air compressor to facilitate the movement of the carriers within the reactor and ensured the uniformity of the mixed liquor. Further, a tangential air stream was introduced to avoid the accumulation of organic or inorganic particulates on the surfaces of the membranes. During the operation, 95% of COD removal was achieved, and the pilot-scale model achieved a similar COD removal compared to the lab-scale due to higher biofilm density [15].

Apart from the domestic wastewater, the saline wastewater was treated using an innovative hybrid system consisting of an equalising reactor (EQ), rotating hanging MBBR reactor, and a submerged flat sheet-based MBR [23]. The EQ was partitioned into three consecutive compartments, each having a corresponding operating volume of 1.226, 1.197, and 1.9885 m^3; equipped with an agitator and impeller (120 RPM), in order to guarantee thorough mixing of the solution. The excess sludge from MBR was recirculated to EQ to prevent shock loading, perform denitrification, decrease oxygen levels in the internal recycling process, ensure phosphorus elimination, and sustain the biomass concentration in the MBBR system. Further, MBR utilised two flat-sheet modules of the commercial submerged microfilter membranes with an average pore size of 0.25 μm and a total active surface area of 120 m^2. To ensure sufficient levels of dissolved oxygen (DO) for the biomass, coarse bubble aeration was provided using an air diffuser system located beneath each membrane module. The operation achieved 90% of the COD removal with 30% of nutrient removal. Further, treatment costs associated with energy consumption and chemical requirement were found to be reduced (0.055 \$/m^3) due to lower fouling and effective recirculation [23].

4.2 UASB Reactor-MBR

The UASB reactor is a widely used anaerobic wastewater treatment technology that has been installed worldwide since its inception in the 1970s. Primarily, the UASB reactor is used in industrial wastewater treatment because of its effectiveness in handling high organic loading rates (OLRs), considerable COD reduction, and operational flexibility. The UASB reactor is distinguished by its up-flow configuration, which promotes the breakdown of organic substances and the development of anaerobic granular sludge (AGS) [6]. The AGS, which usually measures between 1 and 5 mm in size, demonstrates enhanced sedimentation characteristics in comparison to traditional flocculated sludge due to the increased shear. Moreover, the benefits of this reactor arrangement mainly arise from the improved microbial growth assisted by the clustered consortia microorganisms that facilitate high OLR, short retention time, low nutrient and energy requirement, and less generation of biosolid in comparison to the dispersed systems counterparts.

Despite above-mentioned benefits, the UASB reactor still faces challenges, such as the generation of impure biogas, inability to generate superior effluent with the presence of partially degraded persistent organic compounds and nutrients in the wastewater [35]. In addition, the UASB reactor requires a long Sludge Retention Time (SRT) for stable operation and demands long start-up period [35]. In this veneration, the coupling of UASB reactor and MBR technologies can complement each other and be a viable solution to the existing lacunas in wastewater treatment.

Integration of these two technologies results in the emergence of multiple synergistic effects. The UASB reactor serves as a pre-treatment stage, effectively eliminating a substantial amount of organic pollutants, hence decreasing the organic burden that enters the MBR. This enhances the longevity and efficiency of the MBR unit. Further, the UASB-MBR combination may demonstrate enhanced efficacy in removing nutrients compared to separate systems, hence mitigating concerns over the release of nitrogen and phosphorus in the receiving waters. When establishing the UASB reactor-MBR hybrid system, it is essential to consider the operating parameters, including HRT, OLR, and SRT to extract optimum efficiency. In one such investigation, Rattier and co-researcher operated the pilot UASB reactor and MBR for the municipal wastewater. Both UASB reactor and MBR pilots were fed with identical wastewater and exhibited similar average COD removal of around 50% (HRT = 15 h) and 85% (HRT = 9 h), respectively [35]. Whereas, upon hybrid use, the reactor demonstrates COD removal efficiency higher than 90% with lower HRT (6 h) compared to the individual operation of UASB reactor and MBR for wastewater treatment [35].

An extensive pilot-scale treatment sequence was utilised to evaluate the effectiveness of treating actual winery distillery effluent using a multi-stage approach. The treatment process consisted of three stages: UASB reactor as the first stage, followed by an anoxic-aerobic ultra-filtration membrane bioreactor (UF-MBR), and finally, post-treatment via chemical precipitation with lime or adsorption on granular activated carbons (GAC) [28]. The effluent from the winery distillery was very concentrated and had an acidic pH of 3.8 with 44,600 mg/L of COD, 254 mg/L of total nitrogen, 604 mg/L of total phosphorous, and 660 mg/L of phenols. The UASB reactor, operated with OLR ranging from 3.0 to 11.5 kg-COD/m^3.d, attained a treatment efficiency of up to 97%, along with a methane output rate of 340 L of CH_4 /kg-COD [28]. The MBR system, with an OLR from 0.070 to 0.185 kg-COD/kg-VSS.d, successfully removed 48% of influent COD, 67% of total nitrogen, and 65% of phenols. Nevertheless, the integrated UASB and UF-MBR treatment units proved to be ineffective in removing phosphates and reducing colour. Further, the adsorption unit was used as an advanced treatment to comply with legal regulations for wastewater discharge.

A similar configuration of UASB reactor followed by the anoxic-aerobic MBR-containing ceramic membrane-based reactor was operated for 395 days to treat industrial wastewater with varying COD throughout the operation [25]. The plant was consistent in the COD removal (91%) with the HRT of 6.3 h and 8 h for the UASB reactor and MBR, respectively. Further, the application of ceramic membranes in the MBR reduced irreversible fouling, leading to lower energy loss in the cleaning process. Apart from the COD, the hybrid application of UASB reactor and MBR has the capacity to remove emerging contaminants like sulfamethoxazole and trimethoprim [25]. Primarily, emerging contaminants were removed by absorption and biodegradation mechanism in the hybrid UASB reactor and MBR [1]. Overall, the combined utilisation of UASB reactor and MBR technologies in wastewater treatment exhibits a synergistic impact, improving overall efficiency, enhanced quality of the discharged liquid, and the possibility of energy retrieval. Nevertheless, it is crucial to customise the design and operation of these hybrid systems keeping in view the effluent characteristics to attain the most favourable outcomes.

4.3 Constructed Wetland-Assisted Microbial Fuel Cell

A viable solution for treating wastewater while being ecologically conscious and harnessing energy is the implementation of a pilot-scale constructed wetland-assisted microbial fuel cell (CW-MFC). In order to successfully operate a CW-MFC, crucial factors such as vegetation type, electrode placement, and characteristics of media must be carefully selected. For example, adopting suitable vegetation is crucial due to the varying abilities of different plant species to absorb nutrients and interact with microorganisms [41]. The investigation conducted by Wang et al. demonstrated that the presence of emergent macrophytes, such as *Typha* spp., enhanced the pollutant removal and power generation capabilities of CW-MFC systems [39].

The performance of these bio-electrochemical systems is influenced by the HRT for both the anode compartment and wetland portion. In this regard, Niu et al. [24] conducted a thorough investigation, which showed that the optimal HRT creates a harmonious equilibrium between the rates of microbial decomposition and the effectiveness of electron transfer, resulting in improved treatment in CW-MFC hybrids. Further, electrode materials and their position have a crucial role in determining power generation in CW-MFC. For instance, Estrada et al. [3] conducted an investigation which demonstrated that the use of graphene-modified electrodes and other conductive materials improved the transfer rate of electrons, resulting in increased power production.

Furthermore, in another pilot-scale CW-MFC investigation, Kiran and the co-researcher used a novel reactor configuration consisting of nine separate continuous flow units and dual aerobic chambers organised in a rectangle tank [12]. The reactor had a total working volume of 135.5 L and was planted with *Typha angustifolia* [12]. The CW-MFC demonstrated a remarkable maximum COD removal efficiency of 97.56% in the fed-batch mode. Further, by switching to the continuous up-flow mode, the pilot-scale system successfully processed around 4100 L of wastewater, showcasing an impressive maximum COD removal efficiency of 82.8% with significant peak power and current densities of 41.44 mW/m^2 and 283.3 mA/m^2, respectively. The extensive results highlight the effectiveness of the advanced pilot-scale CW-MFC system compared to simple CW.

5 Social Implications of Commercialisation

Scaling up of the advanced biological and hybrid technologies and their applications can greatly decrease the harmful environmental impact of traditional wastewater treatment. The hybrid units safeguard the environment by eliminating persistent organic compounds from the effluent, reducing energy usage, and enhancing resource efficiency compared to conventional processes. Implementing these hybrid wastewater treatment systems at the field level is also crucial for well-being of downstream communities that depend on these waterbodies for their freshwater supply. Therefore, expanding wastewater treatment technology frequently requires working together with local communities, promoting active involvement, and a sense of responsibility.

Disseminating public awareness campaigns on the advantages of utilising hybrid wastewater treatment technology is vital for getting increasing acceptance and support from the community. Such campaigns serve as crucial tools in educating the public about the positive consequences linked with the adoption of hybrid technologies in wastewater treatment. The emphasis should be on clarifying the environmental benefits, including the decrease of pollution, preservation of water bodies, and conserving aquatic habitats. Moreover, a vital part of these initiatives is stressing the direct correlation between wastewater treatment quality and public health. By demonstrating how hybrid treatment procedures contribute to the removal or neutralisation of hazardous compounds, the message can effectively convey that the treated wastewater is prudent for environmental discharge and reuse, which in turn is linked with increasing public health results. In this regard, incorporating real-world examples and case studies further improves the impact of this message.

Economic factors are another key component that should be integrated into public awareness initiatives. Emphasis should be made on the potential for long-term cost savings and job creation connected with the installation of hybrid water treatment technologies. Quantifying these economic benefits and displaying successful projects with good economic ramifications can appeal to a broader public. Economies of hybrid technologies are typically needed to make it profitable by process optimisation, resource utilisation, and cost reduction per treated wastewater unit. Optimisation ensures that the technology meets economic restrictions, regulatory criteria, and industry benchmarks, promoting its wider use and integration into wastewater treatment processes.

6 Future Recommendations

Future investigations should prioritise optimising hybrid technologies for diverse wastewater compositions and environmental contexts, exploring novel materials, reactor configurations, and treatment sequences to enhance efficiency and effectiveness. Policymakers must integrate advanced biological and hybrid wastewater treatment into regulatory frameworks by incentivising industrial adoption, setting stringent effluent standards, and investing in research and development. Further, collaborative engagement among stakeholders is essential, as well as fostering knowledge exchange and supporting initiatives promoting innovative treatment solutions. Adequate funding is also necessary to support technology development and deployment, with public awareness campaigns informing communities about sustainable wastewater management. In addition, global collaboration is critical for sharing best practices and accelerating progress towards sustainable development goals. Implementing these recommendations will contribute to advancing wastewater treatment, improving environmental and public health outcomes, and promoting global sustainability.

7 Conclusion

The progress of wastewater management relies on the implementation of sophisticated biological and hybrid wastewater treatment methods, spanning from laboratory experiments to pilot applications and ultimately to field usage. Hybrid technology has adaptability and scalability, although their full-scale implementation faces practical obstacles and techno-economic challenges. The challenges and results of the existing pilot exemplify the tenacity and the pressing necessity to tackle the global complexities of wastewater treatment. By embracing the potential and gaining knowledge from these encounters, a durable and adaptable world can be developed utilising advanced biological and hybrid technologies.

References

1. Alvarino T, Suárez S, Garrido M, Lema JM, Omil F (2016) A UASB reactor coupled to a hybrid aerobic MBR as innovative plant configuration to enhance the removal of organic micropollutants. Chemosphere 144:452–458. https://doi.org/10.1016/j.chemosphere.2015.09.016
2. Chen M, Ren L, Qi K, Li Q, Lai M, Li Y, Li X, Wang Z (2020) Enhanced removal of pharmaceuticals and personal care products from real municipal wastewater using an electrochemical membrane bioreactor. Bioresour Technol 311:123579. https://doi.org/10.1016/j.biortech.2020.123579
3. Estrada-Arriaga EB, Guadarrama-Pérez O, Silva-Martínez S, Cuevas-Arteaga C, Guadarrama-Pérez VH (2021) Oxygen reduction reaction (ORR) electrocatalysts in constructed wetland-microbial fuel cells: effect of different carbon-based catalyst biocathode during bioelectricity production. Electrochimica Acta 370:137745. https://doi.org/10.1016/j.electacta.2021.137745
4. Fernandez-Gatell M, Corbella C, Sanchez-Vila X, Puigagut J (2022) Microbial activity enhancement in constructed wetlands operated as bioelectrochemical systems. Chemosphere 287:132383. https://doi.org/10.1016/j.chemosphere.2021.132383
5. Galiano F, Friha I, Deowan SA, Hoinkis J, Xiaoyun Y, Johnson D, Mancuso R, Hilal N, Gabriele B, Sayadi S, Figoli A (2018) Novel low-fouling membranes from lab to pilot application in textile wastewater treatment. J Colloid Interface Sci 515:208–220. https://doi.org/10.1016/j.jcis.2018.01.009
6. Ghangrekar MM, Asolekar SR, Joshi SG (2005) Characteristics of sludge developed under different loading conditions during UASB reactor start-up and granulation. Water Res 396:1123–1133. https://doi.org/10.1016/j.watres.2004.12.018
7. Iliopoulou A, Arvaniti OS, Deligiannis M, Gatidou G, Vyrides I, Fountoulakis MS, Stasinakis AS (2023) Combined use of strictly anaerobic MBBR and aerobic MBR for municipal wastewater treatment and removal of pharmaceuticals. J Environ Manage 343:118211. https://doi.org/10.1016/j.jenvman.2023.118211
8. Jadhav DA, Ghosh Ray S, Ghangrekar MM (2017) Third generation in bio-electrochemical system research—a systematic review on mechanisms for recovery of valuable by-products from wastewater. Renew Sustain Energy Rev 76:1022–1031. https://doi.org/10.1016/j.rser.2017.03.096
9. Jadhav DA, Mungray AK, Arkatkar A, Kumar SS (2021) Recent advancement in scaling-up applications of microbial fuel cells: from reality to practicability. Sustain Energy Technol Assess 45:101226. https://doi.org/10.1016/j.seta.2021.101226
10. Jadhav DA, Park SG, Pandit S, Yang E, Ali Abdelkareem M, Jang J-K, Chae K-J (2022) Scalability of microbial electrochemical technologies: applications and challenges. Bioresour Technol 345:126498. https://doi.org/10.1016/j.biortech.2021.126498
11. Jiang D, Curtis M, Troop E, Scheible K, McGrath J, Hu B, Suib S, Raymond D, Li B (2011) A pilot-scale study on utilizing multi-anode/cathode microbial fuel cells (MAC MFCs) to enhance the power production in wastewater treatment. Int J Hydrogen Energy 361:876–884. https://doi.org/10.1016/j.ijhydene.2010.08.074
12. Kiran Kumar V, Manmohan K, Manangath SP, Gajalakshmi S (2023) Innovative pilot-scale constructed wetland-microbial fuel cell system for enhanced wastewater treatment and bioelectricity production. Chem Eng J 460:141686. https://doi.org/10.1016/j.cej.2023.141686

13. Kumar M, Singh R (2020) Sewage water treatment with energy recovery using constructed wetlands integrated with a bioelectrochemical system. Environ Sci Water Res Technol 63:795–808. https://doi.org/10.1039/C9EW00867E
14. Kumar S, Priyadarshini M, Ahmad A, Ghangrekar MM (2023) Advanced biological and non-biological technologies for carbon sequestration, wastewater treatment, and concurrent valuable recovery: a review. J CO_2 Utiliz 68:102372. https://doi.org/10.1016/j.jcou.2022.102372
15. Leyva-Díaz JC, Martín-Pascual J, González-López J, Hontoria E, Poyatos JM (2013) Effects of scale-up on a hybrid moving bed biofilm reactor—membrane bioreactor for treating urban wastewater. Chem Eng Sci 104:808–816. https://doi.org/10.1016/j.ces.2013.10.004
16. Li K, Liu Q, Fang F, Wu X, Xin J, Sun S, Wei Y, Ruan R, Chen P, Wang Y, Addy M (2020) Influence of nanofiltration concentrate recirculation on performance and economic feasibility of a pilot-scale membrane bioreactor-nanofiltration hybrid process for textile wastewater treatment with high water recovery. J Cleaner Prod 261:121067. https://doi.org/10.1016/j.jclepro.2020.121067
17. Li Y, Liu S, Lu L, Wang J, Huang G, Chen F, Zuo J-E (2023) Non-uniform dissolved oxygen distribution and high sludge concentration enhance simultaneous nitrification and denitrification in a novel air-lifting reactor for municipal wastewater treatment: a pilot-scale study. Bioresour Technol 384:129306. https://doi.org/10.1016/j.biortech.2023.129306
18. Lorick D, Macura B, Ahlström M, Grimvall A, Harder R (2020) Effectiveness of struvite precipitation and ammonia stripping for recovery of phosphorus and nitrogen from anaerobic digestate: a systematic review. Environ Evidence 91:27. https://doi.org/10.1186/s13750-020-00211-x
19. Ma J, Wang Z, He D, Li Y, Wu Z (2015) Long-term investigation of a novel electrochemical membrane bioreactor for low-strength municipal wastewater treatment. Water Res 78:98–110. https://doi.org/10.1016/j.watres.2015.03.033
20. Majumder A, Bhatnagar A, Kumar Gupta A (2023) Simultaneous removal of sulfamethoxazole, 17β-estradiol, and carbamazepine from hospital wastewater using a combination of a continuous constructed wetland-based system followed by photocatalytic reactor. Chem Eng J 466:143255. https://doi.org/10.1016/j.cej.2023.143255
21. Majumder A, Otter P, Röher D, Bhatnagar A, Khalil N, Gupta AK, Bresciani R, Arias CA (2024) Combination of advanced biological systems and photocatalysis for the treatment of real hospital wastewater spiked with carbamazepine: a pilot-scale study. J Environ Manage 351:119672. https://doi.org/10.1016/j.jenvman.2023.119672
22. Mohapatra S, Huang C-H, Mukherji S, Padhye LP (2016) Occurrence and fate of pharmaceuticals in WWTPs in India and comparison with a similar study in the United States. Chemosphere 159:526–535. https://doi.org/10.1016/j.chemosphere.2016.06.047
23. Nguyen DD, Ngo HH, Kim SD, Yoon YS (2014) A specific pilot-scale membrane hybrid treatment system for municipal wastewater treatment. Bioresour Technol 169:52–61. https://doi.org/10.1016/j.biortech.2014.06.087
24. Niu Y, Qu M, Du J, Wang X, Yuan S, Zhang L, Zhao J, Jin B, Wu H, Wu S, Cao X, Pang L (2023) Effects of multiple key factors on the performance of petroleum coke-based constructed wetland-microbial fuel cell. Chemosphere 315:137780. https://doi.org/10.1016/j.chemosphere.2023.137780
25. Niwa T, Hatamoto M, Yamashita T, Noguchi H, Takase O, Kekre KA, Ang WS, Tao G, Seah H, Yamaguchi T (2016) Demonstration of a full-scale plant using an UASB followed by a ceramic MBR for the reclamation of industrial wastewater. Bioresour Technol 218:1–8. https://doi.org/10.1016/j.biortech.2016.06.036
26. Noriega-Hevia G, Serralta J, Seco A, Ferrer J (2021) Economic analysis of the scale-up and implantation of a hollow fibre membrane contactor plant for nitrogen recovery in a full-scale wastewater treatment plant. Sep Purif Technol 275:119128. https://doi.org/10.1016/j.seppur.2021.119128
27. Parida VK, Saidulu D, Majumder A, Srivastava A, Gupta B, Gupta AK (2021) Emerging contaminants in wastewater: a critical review on occurrence, existing legislations, risk assessment, and sustainable treatment alternatives. J Environ Chem Eng 95:105966. https://doi.org/10.1016/j.jece.2021.105966

28. Petta L, De Gisi S, Casella P, Farina R, Notarnicola M (2017) Evaluation of the treatability of a winery distillery (vinasse) wastewater by UASB, anoxic-aerobic UF-MBR and chemical precipitation/adsorption. J Environ Manage 201:177–189. https://doi.org/10.1016/j.jenvman.2017.06.042
29. Prado A, Ramírez-Vargas CA, Arias CA, Esteve-Núñez A (2020) Novel bioelectrochemical strategies for domesticating the electron flow in constructed wetlands. Sci Total Environ 735:139522. https://doi.org/10.1016/j.scitotenv.2020.139522
30. Prado de Nicolás A, Berenguer R, Esteve-Núñez A (2022) Evaluating bioelectrochemically-assisted constructed wetland (METland®) for treating wastewater: analysis of materials, performance and electroactive communities. Chem Eng J 440:135748. https://doi.org/10.1016/j.cej.2022.135748
31. Raj R, Tripathi A, Das S, Ghangrekar MM (2021) Removal of caffeine from wastewater using electrochemical advanced oxidation process: a mini review. Case Stud Chem Environ Eng 4:100129. https://doi.org/10.1016/j.cscee.2021.100129
32. Raj R, Tripathi A, Das S, Ghangrekar MM (2023) Waste coconut shell-derived carbon monolith as an efficient binder-free cathode for electrochemical advanced oxidation treatment of endocrine-disrupting compounds. J Environ Manage 348:119328. https://doi.org/10.1016/j.jenvman.2023.119328
33. Raj R, Sathe SM, Das S, Ghangrekar MM (2023) Nickel–iron-driven heterogenous bio-electro-fenton process for the degradation of methylparaben. Chemosphere 341:139989.https://doi.org/10.1016/j.chemosphere.2023.139989
34. Rani A, Snyder SW, Kim H, Lei Z, Pan S-Y (2022) Pathways to a net-zero-carbon water sector through energy-extracting wastewater technologies. NPJ Clean Water 51:49. https://doi.org/10.1038/s41545-022-00197-8
35. Rattier M, Jimenez JA, Miller MW, Dhanasekar A, Willis J, Keller J, Batstone D (2022) Long-term comparison of pilot UASB and AnMBR systems treating domestic sewage at ambient temperatures. J Environ Chem Eng 106:108489. https://doi.org/10.1016/j.jece.2022.108489
36. Rodriguez-Sanchez A, Leyva-Diaz JC, Poyatos JM, Gonzalez-Lopez J (2019) Influent salinity conditions affect the bacterial communities of biofouling in hybrid MBBR-MBR systems. J Water Process Eng 30:100650. https://doi.org/10.1016/j.jwpe.2018.07.001
37. Roy M, Aryal N, Zhang Y, Patil SA, Pant D (2022) Technological progress and readiness level of microbial electrosynthesis and electrofermentation for carbon dioxide and organic wastes valorization. Curr Opin Green Sustain Chem 35:100605. https://doi.org/10.1016/j.cogsc.2022.100605
38. Saúco C, Cano R, Marín D, Lara E, Rogalla F, Arbib Z (2021) Hybrid wastewater treatment system based in a combination of high rate algae pond and vertical constructed wetland system at large scale. J Water Process Eng 43:102311. https://doi.org/10.1016/j.jwpe.2021.102311
39. Wang J, Song X, Wang Y, Bai J, Li M, Dong G, Lin F, Lv Y, Yan D (2017) Bioenergy generation and rhizodegradation as affected by microbial community distribution in a coupled constructed wetland-microbial fuel cell system associated with three macrophytes. Sci Total Environ 607–608:53–62. https://doi.org/10.1016/j.scitotenv.2017.06.243
40. Yadav RK, Das S, Patil SA (2023) Are integrated bioelectrochemical technologies feasible for wastewater management? Trends Biotechnol 414:484–496. https://doi.org/10.1016/j.tibtech.2022.09.001
41. Zhang L, Liu Y, Lv S, Wang R, Wang Y, Lin K, Hu X, Liu Y, Dong Z, Liu L (2023) An overview on constructed wetland-microbial fuel cell: Greenhouse gases emissions and extracellular electron transfer. J Environ Chem Eng 112:109551. https://doi.org/10.1016/j.jece.2023.109551
42. Zou R, Angelidaki I, Jin B, Zhang Y (2020) Feasibility and applicability of the scaling-up of bio-electro-Fenton system for textile wastewater treatment. Environ Int 134:105352. https://doi.org/10.1016/j.envint.2019.105352
43. Zou R, Angelidaki I, Yang X, Tang K, Andersen HR, Zhang Y (2020) Degradation of pharmaceuticals from wastewater in a 20-L continuous flow bio-electro-Fenton (BEF) system. Sci Total Environ 727:138684. https://doi.org/10.1016/j.scitotenv.2020.138684

Challenges and Opportunities in Enabling Circular Economy for Sustainable Wastewater Treatment

Hema Jha and Brajesh Kumar Dubey

Abstract Globally, the goal of wastewater treatment technologies has shifted from avoiding the spread of water-borne diseases to extracting energy and other resources. The next pursuit in this realm is to solve the challenge of integrating wastewater treatment with circular economy and sustainability at the field scale. Along this line, a plethora of research investigations are available which highlight the reclamation of water, energy, nutrients and other value-added products. This chapter emphasizes the opportunities and challenges in attaining a circular economy in sustainable wastewater treatment. Furthermore, the pivotal role of the wastewater sector in sustainable development is emphasized, exploring its interconnectedness with circularity in other sectors. The chapter introduces the ideas of energy self-sufficiency and resource recovery in wastewater treatment systems, along with the interlinked challenges. The discussion delves into representing the importance of tools such as life cycle assessment (LCA) in assessing treatment technologies and alternate scenarios and making informed decisions. It is concluded that the circularity challenge is subject to local implementation and monitoring administered by global planning and directions in the wastewater sector.

Keywords Wastewater treatment · Circular economy · LCA · Sustainability · Resource recovery · SDGs

H. Jha · B. K. Dubey (✉)
P K Sinha Centre for Bioenergy and Renewables, Indian Institute of Technology Kharagpur, Kharagpur, West Bengal 721302, India
e-mail: bkdubey@civil.iitkgp.ac.in

H. Jha
e-mail: hemajha@kgpian.iitkgp.ac.in

B. K. Dubey
Department of Civil Engineering, Indian Institute of Technology Kharagpur, Kharagpur, West Bengal 721302, India

M. M. Ghangrekar et al. (eds.), *Biological and Hybrid Wastewater Treatment Technology*, Earth and Environmental Sciences Library,
https://doi.org/10.1007/978-3-031-63046-0_20

Abbreviations

AGS	Activated granular systems
AOPs	Advanced oxidation processes
BES	Bioelectrochemical systems
CBA	Cost–benefit analysis
CPE	Circular phosphorus economy
IHP	International hydrological programme
IWRM	Integrated water resource management
LCA	Life cycle assessment
LCC	Life cycle costing
MBR	Membrane bioreactor
MCDM	Multi criteria decision modelling
MEC	Microbial electrolytic cell
MFC	Microbial fuel cell
NZE	Net zero energy
PHA	Polyhydroxyalkanoate
RECEIC	Resource efficiency and circular economy industry coalition
SDGs	Sustainable development goals
SLCA	Social life cycle assessment
SWOT	Strength, weakness, opportunities, threats
WWT	Wastewater treatment
WWTP	Wastewater treatment plant

1 Introduction

The Circular Economy is the new paradigm guiding the progression of human civilization. The economy is dictated by the patterns of production, distribution, trade and consumption of goods and services, and until recently, a linear economic model of extraction of resources and disposal after use has prevailed. However, the circular economy aims to replace the linear model and extend the flow of material and energy within the supply chain indefinitely [1]. The principles of circular economy entail reduction in extraction of natural resources, reuse of the components and recycling of the material [2]. These principles allows to work within the confines of planetary boundaries set for safe operational space for humanity in biophysical earth system [3]. Thus, circular economy has emerged as the strategy of economic development in synergy with environmental considerations.

In the realm of circular economy, wastewater treatment (WWT) can be identified as a simultaneous production process rather than solely a contamination control measure [4]. In this context, wastewater is recognized as the raw material for generating water, energy, nutrients, and additional value-added biomaterials [5]. This approach leads to

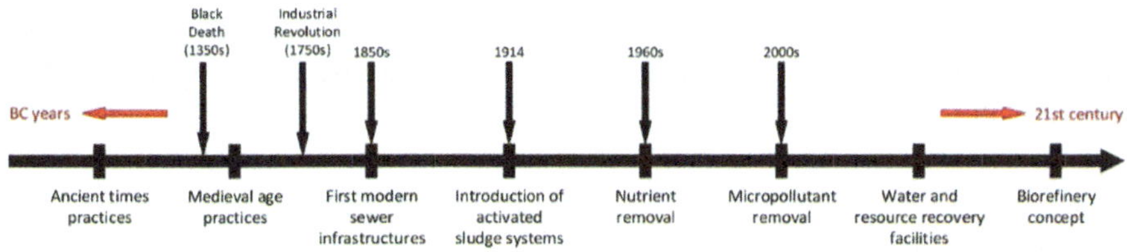

Fig. 1 Important milestones in sanitation and wastewater treatment [8]

a reduced burden on water bodies and partial substitution of other minerals and fossil-based sources by a renewable raw material [6]. There are few instances in history where these approaches were observed, for example, during the Minoan Civilization around the second century BCE the irrigation channels were connected to sewer drainage systems for utilizing the biowaste as fertilizer [7]. However, the focused and intense recovery of resources from wastewater treatment plants (WWTPs) is a modern concept that is still emerging to be implemented effectively on ground level. Figure 1 illustrates the timelines of appearance of circular economy concept and important milestones in WWT.

At present, only 63% of wastewater is collected and 52% undergoes treatment [8]. Moreover, half of the global rivers and lakes are polluted and many of the major rivers do not reach sea due to upstream overexploitation [9]. Consequently, the decisions in wastewater sector is influenced by diverse standpoints including inadequate treatment infrastructure, issues of poor sanitation leading to water borne diseases, scarcity of freshwater resources due to population growth and climate change, as well as the recovery of resources lost in wastewater streams [10, 11]. These factors demand a sustainable approach for water resource management, and indeed a circular approach is essential to facilitate clean water—a primary agenda for governments worldwide [12, 13]. Along this line, numerous research investigations have been dedicated to integrating various aspects of WWT into the circular economy. This involves reclaiming water and other resources, employing solid waste-derived materials in treatment technologies and harnessing renewable energy to achieve self-sufficiency of WWTPs. The implementation of close loop economy in the wastewater sector is expected to foster balance between supply and demand of resources and will ensure environmental and socioeconomic sustainability [14]. Accordingly, the endeavor to establish this circular system is in progress and further demands extensive and collective efforts from water professionals in academia, industry, and government.

This book chapter presents a discussion on the opportunities and challenges in enabling a circular economy in sustainable WWT. The chapter highlights the interconnectedness of circular economy principles and sustainable development goals (SDGs) in the wastewater sector. Further, the opportunities of recovering material and energy from WWT are outlined, and the challenges in the transition towards circular economy is highlighted. The discussion delves into understanding the tools such as life cycle assessment (LCA) and others, which can potentially enhance decision making processes in transitioning towards more circular and sustainable WWT at field scale.

2 Significance of Wastewater Treatment in Sustainable Development and Circular Economy

Circular economy and sustainability are the key drivers of today's global transition in every sector. Sustainability refers to an unceasing presence of resources for the livelihood of humanity and a flourishing environment [15]. It has three pillars of economic, social and environmental sustainability. To attain sustainability in short term, there seems to be a trade-off between economic and environmental schemes, however, in long term there will be a harmony in economic development and environmental cycles, which is expected to address the social imbalances. United Nations SDGs target these three pillars of sustainability through 17 diverged streams with their respective agendas to be attained by 2030.

The advent of SDG in water sector (SDG 6: Clean water and Sanitation) surpasses its predecessor, millennium development goal (MDG) by not only addressing sanitation but also incorporating water resource management aspects [12]. Detailed assessments demonstrates the contribution of SDG 6 to other ten SDGs directly and remaining six SDGs indirectly (as depicted in Fig. 2), emphasizing the significant role of wastewater sectors in accelerating sustainability [13]. In pursuit of sustainability goals, the United Nations' International Hydrological Programme (IHP) introduces 17 international initiatives, including satellite-based monitoring of water levels and quality, as well as the creation of a global hydrological map, among others [9]. Besides, Integrated Water Resource Management (IWRM), a crucial component of SDG 6, emphasizes coordinated management of water in conjunction with land and other associated resources. Notably, IWRM fosters the notion of 'one water', assigning equal importance to all water resources including wastewater streams. An evaluation of the extent of IWRM across 185 countries indicates that 98 nations have reached medium to very high levels of IWRM implementation, whereas 87 countries fall within the medium to lower levels (as illustrated in Fig. 2) [16].

Advancing towards SDG 6 simultaneously contributes to the global aspiration of a circular economy. Specifically, WWT lies at the core of achieving sustainability and circularity in the water sector. Furthermore, the developments in water sector are inherently interconnected with food, and energy sectors giving rise to the concept of water-food-energy nexus [8, 10, 14]. This unification further includes dependencies of carbon sector on the water sector, along with connections to transportation, land usage and sanitation systems [17, 18]. Figure 3 illustrates this interconnectedness of the water sector to other sectors, and depicts the governing role of WWT in enabling a closed loop economy.

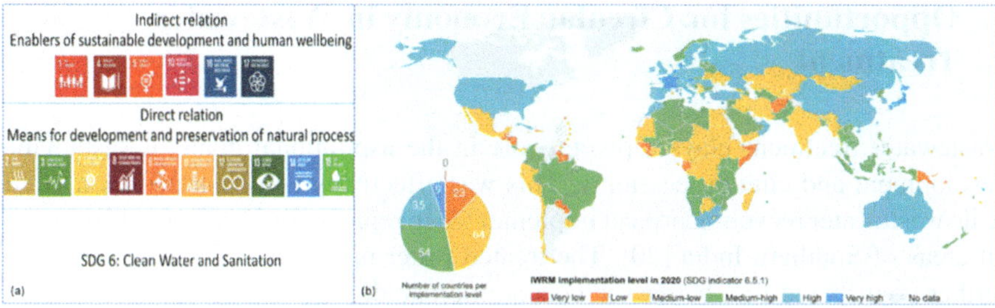

Fig. 2 SDG 6: Clean water and sanitation **a** Direct and indirect relation with other SDGs **b** Implementation of IWRM across globe [13, 16]

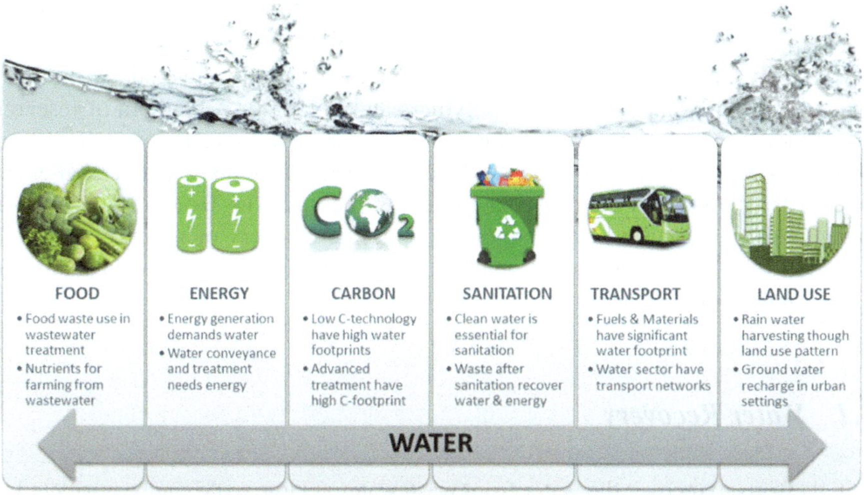

Fig. 3 Interconnectedness of water sector to other sectors indicating the importance of water sector in circular economy

Apparently, the circular economy and SDGs are intrinsically connected, with the wastewater sector playing a pivotal role. The G20 summit of major economies has initiated the Resource Efficiency and Circular Economy Industry Coalition (RECEIC) to expedite the shift towards a worldwide circular economy across various sectors [19]. Moving forward, the wastewater sector is anticipated to be an indispensable part of this coalition. This deliberation sets the foreground for understanding the significance of wastewater sector in sustainable development and circular economy.

3 Opportunities for Circular Economy in Wastewater Treatment

Wastewater treatment plays a pivotal role in the aspirational goals of sustainable development and circular economy. This was effectively demonstrated by a large-scale wastewater recycling project implemented to replenish groundwater in drought-hit areas of Southern India [20]. The treated water met regulatory compliance and further contributed to a better socioeconomic paradigm. Notable improvements in agricultural cropping patterns were observed for water-intensive crops, alongside livestock, milk and fish production increased significantly. Subsequently, there was an increase in land prices, farmers' income and on-site work opportunities for women. A list of many such circular-economy focused WWT projects is provided by Avellán et al. [21]. Focusing on water recovery alone made a substantial contribution to all three pillars of sustainability. However, wastewater has the potential for additional resource recovery, including energy, nutrients, and various value-added products such as cellulose, alginate-like exopolymers and lipids [5]. A plethora of research and reviews are available in literature focusing on advancements of technologies pertaining to resource recovery. A glance of advantages and current challenges offered by the state-of-the-art technologies is presented in Table 1. This section categorizes the discussion into comprehending the recovery of water, energy, and materials from wastewater. Additionally, it explores waste utilization as an additional aspect in the WWT sector.

3.1 Water Recovery

Initial efforts to recover the water from wastewater at large scale can be attributed to the high-water stress situation of Singapore through their attempts at water reclamation and reuse in 1966. Through the advances in membrane technology in 2003, water reclamation was made possible economically and technically feasible in Singapore. Currently, 40% of Singapore's water supply comes from reclaimed water, a proportion expected to grow [33]. At present, membrane technologies, including microfiltration, ultrafiltration, nanofiltration, and forward- and reverse-osmosis, are suggested to be integrated to overcome individual limitations, conserve energy, and enhance overall efficiency in WWT [34]. The circular economy benefits from membrane-based technologies can be maximized through membrane reuse and recycling, though such proficiency in handling exhausted membranes is not available. Various other technologies such as advanced oxidation processes (AOPs) like Fenton oxidation and ozonation, have also been explored for efficient and advanced WWT [35]. Moreover, bioelectrochemical systems (BES), utilizing microbial electrolytic cell (MEC) and microbial fuel cell (MFC), offer promising and eco-friendly alternatives for antibiotic degradation through redox reactions. Bacterial/algal–bacterial activated granular systems (AGS) represent a shift in WWT towards material recycling, water reuse,

Table 1 Opportunities and challenges of processes in resource recovery in wastewater sector

Process	Raw material	Main product	Opportunities	Challenges	References
Conventional activated Sludge process	Wastewater	Treated water	Effective removal of organics; Can treat varieties of wastewater	Excess sludge generation; Sensitive to the influent flow rate	[22]
Sequencing batch reactors	Wastewater	Treated water; Bio-gas (if anaerobic)	Effective for decentralized wastewater treatment; Ability to handle extremely high organic and hydraulic shock loads	Excess sludge generation; Complex Operation and Control	[23]
Membrane bioreactors	Wastewater	High-quality treated water	Reduced sludge generation; Effective for decentralized wastewater treatment	Membrane fouling; High capital cost of installation	[24]
Anaerobic digestion	Wastewater, Microalgae, Sewage sludge	Treated water, Biogas, Digestate	Renewable energy production; Reduced sludge generation	Process control; Sensitivity to toxic substances; Nutrient imbalance	[25]
Constructed wetlands	Wastewater	Treated water; Biomass	Low energy requirements; Cost-effective decentralized wastewater treatment	Efficiency varies with the climate conditions and seasonal changes	[26]
Algal-mediated wastewater treatment	Wastewater	Treated water; Microalgae	Carbon sequestration; Renewable energy production	Harvesting and dewatering of algal biomass; Sensitive to certain contaminants	[27]

(continued)

Table 1 (continued)

Process	Raw material	Main product	Opportunities	Challenges	References
Bioelectrochemical systems	Wastewater	Treated water; Bioenergy; Biochemicals	Carbon sequestration; Renewable energy production; Environment friendly	Scaling-up; Low power density	[28]
Advanced oxidation process	Wastewater	Treated water	Effective removal of recalcitrant pollutants; Minimization of harmful byproducts; Versatility i.e. non-selective oxidation	Safety considerations; High energy consumption; Scaling-up	[29]
Hybrid wastewater treatment	Wastewater	Treated water	Comprehensive treatment; Resource recovery; Improved treatment efficiency	Complexity; Maintenance; Site-specific consideration	[30]
Hydrothermal carbonization	Algae; Sewage sludge	Biochar	Carbon sequestration; Biofuel; Waste valorisation	Economic viability; Scaling-up; Environmental impact assessment	[31]
Transesterification	Algal lipids	Biodiesel	Renewable energy	Economic viability	[32]

energy/nutrients recovery, and biomaterials production. Bacterial AGS, with over three decades of development, has gained global industrial application, particularly through the Nereda® system, demonstrating excellent treatment performance and potential cost reduction in effluent discharge [36].

The use of treated water for agricultural irrigation is a common practice for sustainable development. For instance, Australia witnessed an average of 27.2% recycled water use in urban irrigation in 2009–10 [37]. Despite its benefits, concerns about environmental and health risks arise due to the presence of emerging contaminants in treated water. In the Tula Valley, one of the world's largest agricultural areas reusing wastewater, a comprehensive study found ten emerging contaminants posing environmental risks, including substances never previously quantified in the region [38].

Such instances emphasize the challenge of balancing the nutrient levels in wastewater reuse for agriculture and removing emerging contaminants in WWTPs. The global demand for recycled water in irrigation is expected to rise, particularly in arid regions of developing countries, emphasizing the need for improved technical solutions, infrastructure, and policies to ensure the safe distribution and management of wastewater for irrigation.

Reducing the scale of implementation, the zero-mile system are claimed to be designed within the regenerative circular economy, effectively treating greywater at the household level by employing an engineered microbial consortium in a biofilter [39]. The treated wastewater from the consortium was suitable for plant fertilization for onsite vegetable cropping. Furthermore, three other greywater treatment alternatives: photocatalysis, photovoltaic solar-driven photocatalysis, and membrane bioreactor (MBR) were evaluated for greywater treatment and sustainability performance [40]. The analysis indicates that photovoltaic photocatalysis driven by solar energy is the most environmentally sustainable scenario. These investigations reflect the enormous possibility for innovation and treatment in circular world, however, such demonstrations and perspectives pose challenges for decision-makers to determine the extent of centralization of infrastructure for WWT.

3.2 Energy Recovery

Wastewater treatment has evolved to incorporate advanced technologies that not only address water quality but also focus on energy recovery. The close nexus between water and energy sectors is governed by the energy-intensive nature of water extraction, treatment, and transmission, where the water sector now accounts for 4% of global electricity demand [8]. Traditional processes in WWTPs, utilizing membrane filtration and activated carbon filtration, have been scrutinized for their energy intensity, with specific consumption ranging between 0.5 and 2.0 kWh/m^3 [41]. A sustainable approach to WWT lies in the integration of energy recovery technologies within the treatment schemes.

Integrating anaerobic treatment or BES offers a proven and sustainable approach for efficient WWT, electricity generation, and potential energy self-sufficiency, advancing the path towards a circular economy [41–43]. Particularly, anaerobic-MBR, a novel approach blending anaerobic processing with microfiltration or ultrafiltration membrane separation, emerges as an energy-efficient solution for municipal WWT [44]. Biogas, primarily derived from sludge through anaerobic digestion, serves as the key energy source for WWTPs [45]. Moreover, MFC technology represents a decentralized and sustainable energy solution that efficiently converts organic matter present in wastewater into hydrogen gas or electricity [46]. Situated at the intersection of water, energy, and waste, MFCs harbor transformative potential, turning WWTPs into vital tools for sustainability. At present, despite the challenges in upscaling, enhanced power output for MFC technology promises its implementation as green technology [47, 48]. Microalgae-based technologies, another frontier in WWT, have garnered significant interest due to their sustainable and cost-effective nature. Beyond WWT, the harvested microalgae biomass holds promise for bioenergy production and biomolecules synthesis, contributing to both environmental and economic sustainability [49]. For higher lipid extraction from microalgae leading to enhanced transesterification and biodiesel production, innovations in MFC are devised [50].

Worldwide, several WWTPs have achieved or are close to achieving energy self-sufficiency, primarily through biogas production from anaerobic digestion. Notably, two plants in Austria, Strass TP and Wolfgangsee-Ischl, exhibit a positive energy balance, generating 6.3% and 7% more energy than they consume. The Sheboygan WWTP in the USA is nearly 100% energy self-sufficient through co-digestion and cogeneration [51]. Likewise, analysis of conventional WWTPs in China through a net zero energy (NZE) model reveals that some plants can export excess energy, while others can achieve 100% energy self-sufficiency [52]. The NZE model provides directions for a significant cost offset and underscores the potential for reducing electricity and sludge disposal expenses.

3.3 *Material Recovery*

3.3.1 Nutrient Recovery

In May 2014, the European Commission classified phosphate rock as a critical raw material [53]. The shortage of phosphorus reserve and fossil-based energy usage in extraction is expected to increase the pressure on the fertilizer market in terms of its availability and pricing [8]. Literature advocates for a circular phosphorus economy (CPE) to address these challenges, emphasizing the need for phosphorus recovery [54]. Hydrothermal carbonization of sewage sludge is one of the promising technologies to process sewage sludge for phosphorous recycling [55]. In the coming years, advancements in biological nutrient removal and synergies in nutrient and energy recovery within anaerobic treatment systems are anticipated to enhance nitrogen and phosphorus recovery from WWTPs. Nutrient recovery technologies are growing rapidly to support the Circular Economy's goals, but they are not fully mature. Ongoing research aims to optimize efficiency and enhance the commercial applicability of nutrient recovery technologies [4].

3.3.2 Value-Added Products Recovery

Sludge and microalgae are the reserves of resources in WWTPs offering to develop biorefineries integrated with WWT. Figure 4 represents the potential products from sewage sludge and microalgae along with microalgae-based biorefinery demonstration. Beyond conventional resources such as biogas and nutrient recovery, the potential recovery of other value-added products like biopolymers, proteins, enzymes, biopesticides, and biofuels is made possible [56]. The circular economy value of sewage sludge is significant, offering phosphorus, biochar, methane, bio-oil, and biohydrogen for applications in soil fertilization, energy self-sufficiency, and clean fuel. Furthermore, the volatile fatty acids produced by anaerobic digestion can be transformed into environmentally friendly bioplastics known as polyhydroxyalkanoates (PHAs). Additionally, the organic compounds, specifically extracellular polymeric substances found in microbial aggregates containing proteins, polysaccharides, lipids, and humic acids, utilized in the treatment process can be extracted from microbial biomass and repurposed for various industrial applications [36]. Utilizing sludge for construction materials also aligns with the principles of the circular economy, offering a solution to the considerable challenges associated with sludge disposal while maintaining safety standards [45]. The water industry's growing focus on sustainability and circular economy principles has led to increased exploration resource recovery from wastewater, sewage sludge and microalgae, with biorefinery defined as a comprehensive approach for renewables production and environmental solutions.

3.4 Waste Utilization in Treatment Processes

The valorization of waste into valuable bioproducts in an eco-friendly manner is crucial for environmental protection and human health [59]. The utilization of waste to synthesize materials that are capable of treating wastewater has been explored extensively. Particularly, biochar produced from various waste streams has captured enormous attention for application as adsorbent, membrane and electrode materials in WWT technologies (Fig. 5). For example, carbonized coconut shell was used as cathode in electrochemical cells to investigate the removal of endocrine disrupting chemicals [60]. This sustainable cathode material was observed to have better removal efficiency when compared to commercial graphite plates and metal-based cathode materials. Such successful demonstrations pave the pathway for waste-based material usage in advanced WWT to support sustainability. Detailed reviews on the state of the art for biochar application in WWT are available in the literature [61, 62].

Other instances of using sustainable material include the application of dried sludge powder in sequencing batch to aid the startup process of aerobic granular sludge process [64]. Furthermore, clay-based geopolymers as adsorbent materials are proposed as alternate sustainable source in a circular economy [65]. A comprehensive and integrated approach is needed to finalize the use of waste-derived materials in WWT [63]. This includes the assessing performance of treatment process, reuse/regeneration of waste-derived materials, and their subsequent treatment/application based on accumulated contaminants in potential applications such as soil improvement, cement enhancement, and biofuels [66]. Additionally, assessing economic feasibility through energy balance of the overall integrated process, as well as

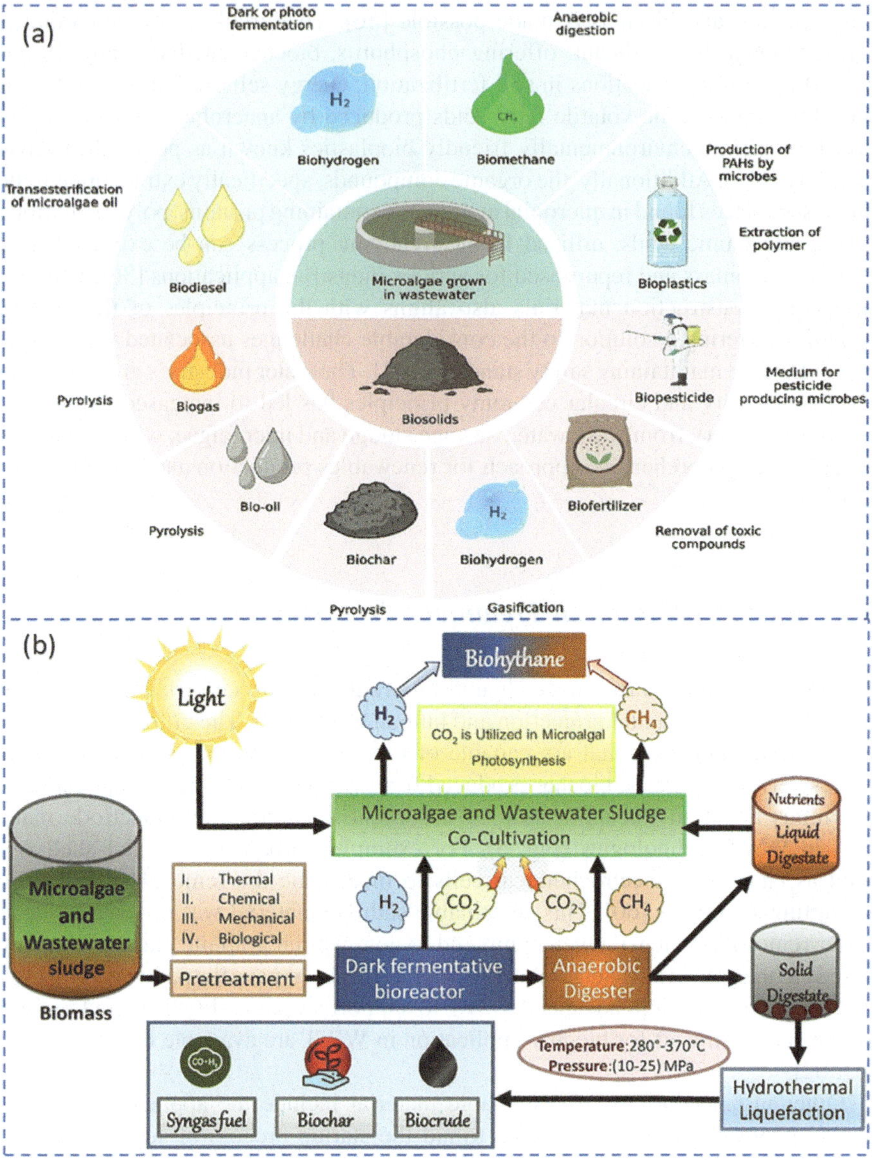

Fig. 4 Opportunities of resource recovery from microalgae and sewage sludge **a** potential energy products and value-added products **b** biorefinary demonstrations [57, 58]

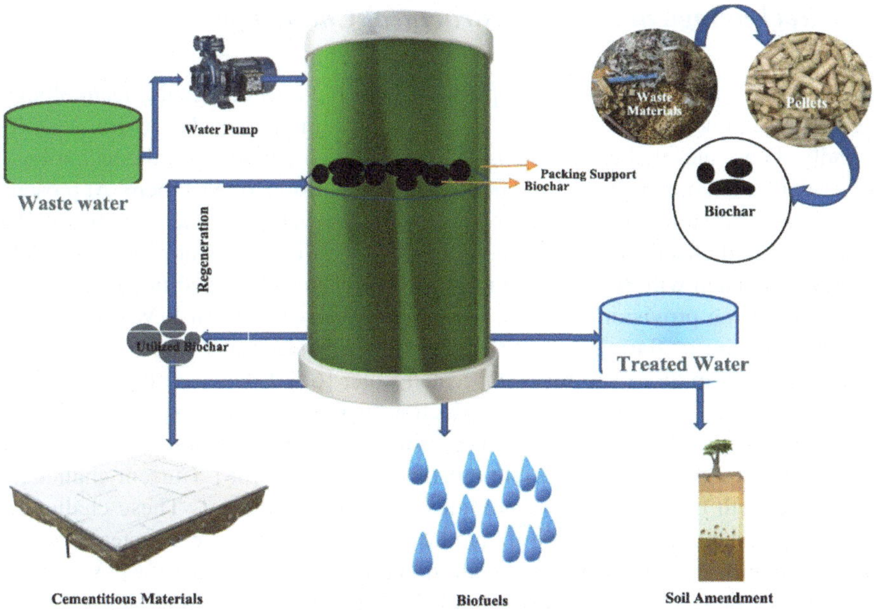

Fig. 5 Wastewater treatment by waste materials and applications of end-product [63]

economic and environmental impacts, along with sensitivity analysis, is essential to determine the economic viability of the waste-based processing scheme.

4 Challenges in Enabling Circular Economy

The endeavour to create a circular economy in the wastewater sector is confronting several challenges at multiple facets. The enormous quantity, variable compositions and emergence of newer contaminants pose uncertainty in achieving appropriate treatment. Alongside, selection from numerous alternative schemes for the treatment and management of water resources leads to a complex process of decision-making. Moreover, when suitable treatment process is selected from a technical perspective, challenges remain to overcome other barriers such as regulatory obstacles, market penetration, customer acceptance and public perception goal [53]. Hence marketability, socioeconomic and political factors play significant role in drafting a circular road to recover resources from wastewater. In this section, a concise overview of challenges encompassing technology and infrastructure, market and cost considerations, as well as public perception, policy, and regulations is provided.

4.1 Technological and Infrastructural Challenges

In the era of an ever-increasing population with high affluence, the technologies are required to be high performance, low cost and simultaneously environmentally friendly. These elevated expectations place substantial demands on the technology, necessitating ongoing research and implementation efforts to meet them. At present, there is no one stop solution for treatment of wastewater and technologies suffer from a few bottlenecks. For instance, anaerobic digestion is slower processing technology associated with higher retention times of reactor [45]. Additionally, in most cases there is no provision to utilize biogas when extracted from WWT, leading to its direct release or flaring into the atmosphere, alongside a fraction of produced biogas gets lost in dissolved forms in treated water [67]. Likewise, several microbe integrated technologies for energy production or material recovery perform superior in controlled lab environment, however maintaining the stability of a microbial consortium at a pilot and field scale remains challenging [68]. Current challenges associated with the available technologies are listed in Table 1. These challenges further extend to the production of biodiesel from sludge, where water residuals present economic hurdles, requiring additional pretreatment steps [56]. Moreover, the goals of process integration in small WWTPs are difficult due to insufficient sludge mass for onsite biorefinery processing. Furthermore, need for downstream processing remains a significant obstacle in producing chemicals or fuels from waste streams, especially for chemical feedstocks requiring high purity [2].

The direct potable usage of advanced treated water further exacerbates the challenging situation, where the retention of emerging contaminants could adversely impact plant operations [12, 69]. Hence, when omitting the environmental buffer, an equivalent engineered buffer will be required to provide supplementary measures. Furthermore, toxicological and epidemiological studies are necessary for emerging contaminants to regulate and maintain the health and safety aspects of potable reuse. The present infrastructure struggles to efficiently treat wastewater with changing chemical and biological compositions, emphasizing the need for retrofitting with process modifications [35, 70]. Regardless of the treatment technology used, enhancing the energy efficiency can be achieved by replacing outdated and inefficient equipment (such as pumps, blowers, and diffusers) with energy-efficient alternatives [53]. Additionally, the decision to create centralized or decentralized systems involves a complex tradeoff between the economy of scale and advanced onsite management technologies [5].

Achieving the primary objective of SDG6 and provide universal access to clean water and sanitation required not only enhancing and extending "hard-infrastructure" such as sewer networks and WWTPs but also minimizing pollutant emissions at their source [71]. Overall, addressing these technology and infrastructure challenges is crucial for advancing WWT towards a sustainable and circular future.

4.2 Marketability and Cost Considerations

Implementing commercial-scale biorefineries necessitates addressing technical, commercial, and sustainability challenges, including optimizing feedstock logistics and establishing reliable supply chains for organic feedstocks. The application of circular economy principles often requires a significant upfront investment. This is a significant barrier for municipalities or companies, especially in regions with limited financial resources. The market for recovered materials from wastewater is also uncertain. It is a challenge to identify products that offer a significant advantage over its conventional production from petrochemical or sugar-based refinery processes [2]. For example, when using sludge as raw material, higher production costs are associated with the production of enzymes and proteins, whereas low production yield of biohydrogen limits its economic viability [56]. The intentionally low price of water for certain purposes, like drinking or irrigation, often subsidized in many countries, hinders the competitiveness of water reuse [14]. Furthermore, when a contaminant of concern is present in wastewater, expensive advanced treatments are required such as reverse osmosis. Price regulations and a stable market are demands yet to be fulfilled to establish a robust and secured supply chain and production line for biobased material and energy from WWT.

4.3 Public Perception, Policy and Regulations

Public acceptance and regulatory considerations are crucial aspects of sustainable systems. The viability of circular wastewater schemes can be assured by integrating regulatory instruments, removing legislative barriers and conducting awareness programs to positively turn public attitudes [56]. Currently, the innovative small-scale technologies are able to produce on-site treated water with good quality, however legislative barriers are hindering the closure of loops [72]. For example, there is a lack of specific regulations for handling sewage sludge in decentralized WWT systems. Developing and implementing supportive policies is crucial to incentivize circular practices in the wastewater sector. These policies should clearly specify the responsibilities of all the stakeholders including academia, government, industries, and communities for their effective collaboration. Encouragement to public–private model and subsidies can motivate businesses in this sector [73]. Furthermore, consumer acceptance can be enhanced by publicity programs by referring to successful examples of Singapore for the publicity of NEWater (treated potable water) [33]. Clear communication and education regarding circular economy and its benefit in wastewater sector will continue to play a key role in the overall shift towards the sustainable change.

Another challenge is associated with monitoring the global progress in achieving SDG 6 in absence of consistent and comprehensive datasets [16]. Performance comparison across countries is not straightforward, with inconsistent reporting definitions and protocols concerning diverse geography and income levels [7, 74]. The delay in planning, implementation and reporting is also not recorded, alongside use of non-standard and unreliable methods to gather data adds considerable skepticism for the available data [8]. These challenges can be addressed by establishing international standardized definitions in WWT at the utility level. This would aid in normalization and aggregation, facilitating a comprehensive assessment of national performance. Furthermore, centralized and transparent data reporting and regressions must be performed to enhance reliability and certainty. On local levels, decision-making is further challenged by the complexity of evaluating different alternatives, which involves substantial scientific and procedural steps [73]. Various tools are available to aid the decision-making process, such as LCA, optimization modelling etc. which are discussed in next section. Overall, the circularity challenge is subject to local implementation and monitoring administered by global planning and directions.

5 Decision Making Tools to Promote Circular Economy

Making sustainable decisions involves consideration of various factors such as determining the degree of decentralization, assessing environmental impacts of technologies, maximizing resource recovery, evaluating installation and operational costs, selecting the plant's location, and addressing other socioeconomic aspects. Careful analysis of these factors contributes to the complexity in decision making process, however such analysis is crucial to enable circularity within the wastewater sector and integrating it with other sectors. To simplify the complex problem and take appropriate actions, decision makers can be equipped with tools and knowledge of available frameworks that promote a holistic assessment for a circular economy. This section provides an overview of commonly used tools, each varying in mathematical complexity, yet contributing to the assessment and decision-making process. The discussion covers the fundamentals of tools including LCA, multi-criteria decision modelling, SWOT analysis, cost–benefit analysis mathematical modelling and hybrid modelling. More extensive reviews of these decision support tools applied in the wastewater sector are available in the literature [10, 75].

5.1 Life Cycle Assessment

Life cycle assessment is a framework which evaluates the environmental impacts of a product or process in its lifespan, including raw material extraction, production, distribution, usage and end-of-life management. LCA comprises 4 stages as per ISO:14040 framework namely, Goal and Scope definition, Life cycle Inventory, Life cycle impact assessment, and interpretation of results. While performing an LCA study, great care has to be taken in each step including, the selection of functional

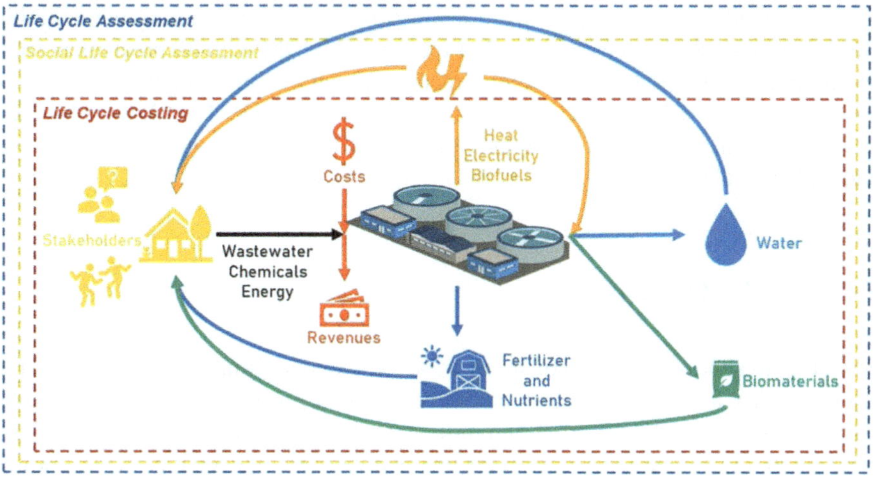

Fig. 6 Overview of life cycle sustainability assessment boundary and its relation to circular economies within the WWTPs [78]

units, setting the system boundaries, creating inventory, and finalizing categorization and characterization, as the results are sensitive to all these processes and parameters.

As a quantitative environmental assessment method, LCA serves as a useful decision-support tool for examining alternative future operational scenarios, during strategic planning within the wastewater sector [76]. LCA is playing an important role in evaluating the environmental sustainability of new technologies and processes by capturing trade-offs across various categories of environmental concern. The use of LCA was started in early 1990's in the wastewater field, following 20 + years of significant interest from researchers in this area [77]. The constraint with this method is its exclusive focus on quantifying environmental impact, and the assessment heavily relies on the quality and availability of data. To extend the analysis towards an integrated assessment, there is a growing demand for combining social life cycle assessment (SLCA) and life cycle costing (LCC) alongside conventional LCA. Figure 6 illustrates the dimensions considered in a wastewater biofactory encompassing LCA, SLCA, and LCC.

5.2 *Multi Criteria Decision Modelling*

Multi-criteria decision modelling (MCDM) is an effective decision-making tool for evaluating and selecting the appropriate alternative based on weighing several criteria. In case of the wastewater sector, the major criteria are environmental, social and economic, which are further divided into factor and assigned weights for assessing each available alternative technology or process scenario. This approach enhances decision-making by providing a systematic framework to accommodate various considerations in making a holistic approach in determining the complex choices in wastewater management. For example, MCDM with 30 parameters such as effluent BOD, energy for resource recovery, years of plant operation etc. was applied to compare four of the available treatment technologies [79]. It was observed that mixotrophic algal systems outperformed the MBR, conventional activated sludge and high-rate algal ponds.

5.3 *SWOT Analysis*

SWOT is an abbreviated form of strength, weakness, opportunities and threats. This is a qualitative assessment tool which can highlight key aspects involved for capitalizing strengths and opportunities and mitigating the weaknesses and strengths. This conceptual tool can be applied on overviewing diverse set of decisions in wastewater management. For example, [51] performed a SWOT analysis on implementation of circular economy in the wastewater sector in European countries. They highlighted decreasing pressure on natural resources and global trend to promote greener processes as few of the strengths and weakness of circular change, while financial outlay and political changes as weakness and threat to implementation. One another instance of applying SWOT is to identify key drivers of heat recovery from WWT [80]. They underscored the reduction in carbon emission as a major strength for heat recovery in WWT systems. Overall, this tool sets the ground for a more comprehensive analysis if required, and based on these factors decision-makers can formulate strategies to enhance treatment efficiency, address infrastructure challenges, and align with evolving environmental standards, ensuring a sustainable and resilient wastewater management system.

5.4 *Cost–Benefit Analysis*

Cost–benefit analysis (CBA) evaluates the economic viability of a project by contrasting monetary costs with achieved benefits through WWT. Costs involve capital and operational expenses, while benefits include improved water quality, public health, and environmental impacts etc. Monetization of the process allows for direct comparisons, and discounting methods adjusts future values to present worth. A positive net present value (NPV) indicates economic viability, guiding decision-makers in resource allocation for WWT infrastructure. CBA facilitates informed decisions, ensuring that the benefits of a project outweigh its costs and promoting sustainable water management. A model study for such analysis was performed by Kordana-Obuch et al. [81] to close the loop in water sector in Jordan. They used shadow price approach for monetization of benefits, which implies calculating the avoided cost associated with non-recovery of resources and direct discharge of wastewater. Net-zero energy models can also be seen as performing the similar activity in a non-monetized circular evaluation of energy and resources [52].

5.5 Mathematical Modelling

Mathematical modelling in WWT employs mathematical equations to simulate and optimize complex processes within treatment systems. These models represent physical, chemical, and biological interactions, aiding in system understanding and prediction. By inputting variables like contaminant concentrations and flow rates, models simulate the system's behaviour over time, facilitating the prediction of outcomes under different conditions. The mathematical models are utilized further in optimization exercise, which involves defining variables, constraints, and an objective function to maximize or minimize [10].

Moreover, mathematical models play a crucial role in the design and modification of WWTPs by guiding decisions on infrastructure, component sizing, and layout. They also enable sensitivity analysis, allowing for the assessment of how variations in input parameters impact overall system performance. In essence, mathematical modelling contributes to the advancement of efficient and sustainable WWT practices, offering a systematic approach to analyse, design, and optimize treatment processes for improved water quality and environmental impact towards a circular economy.

5.6 Hybrid Models

Hybrid modelling in WWT combines process-based models, which use mathematical equations to represent treatment mechanisms, with data-driven models like machine learning algorithms [82]. Process-based models provide a solid foundation for understanding treatment processes, while data-driven models enhance adaptability and accommodate real-world variability. The integration of these models in hybrid approaches aims to capitalize on their complementary strengths. This synergy improves the overall accuracy, robustness, and predictive capacity of the models. Hybrid modelling finds applications in optimizing WWT processes, developing control strategies, and conducting scenario analyses. By offering a more comprehensive and adaptable modelling framework, hybrid models contribute to the enhancement of treatment efficiency and reliability. Furthermore, more intelligent data driven tools alone, like virtual reality, and cyber-physical systems will prepare WWTPs for Industry 4.0 [83].

6 Conclusion

The concept of a circular economy has become a cornerstone for economic development, aligning with environmental concerns. This approach is intricately linked with SDGs, and the wastewater sector stands out as a key player in its implementation. Globally, various projects focus on circular resource extraction, and a few WWTPs have attained energy self-sufficiency. The evolving landscape of WWT emphasizes the imperative of resource recovery. From innovative technologies like anaerobic MBR and MFCs to the integration of microalgae-based solutions, the journey towards sustainable wastewater treatment is intrinsically linked with harnessing resources from the treatment processes. However, challenges hinder the widespread deployment of circular measures. Issues related to technology, infrastructure, cost, policy, and public perception pose significant obstacles. Presently, only 63% of wastewater is collected, with 52% undergoing treatment. The sheer quantity, diverse compositions, and the emergence of new contaminants add uncertainty to achieving appropriate treatment. Decision-making becomes complex with numerous alternative schemes for water resource treatment and management, coupled with barriers like regulatory obstacles, market penetration, customer acceptance, and public perception. To make sustainable decisions, factors such as decentralization, environmental impacts, resource recovery, costs, location selection, and socioeconomic considerations must be carefully analyzed.

A sustainable and circular approach is deemed essential for water resource management, aligning with governments' agendas worldwide to ensure clean water. To enforce the narratives of sustainable and circular economy, wastewater must be seen as raw material and wastewater treatment as a simultaneous production process. An integrated approach towards research in technology improvement is required which not only emphasizes the efficiency of treatment, but an overall reduced burden on the environment. This could be achieved by including the use of several sustainability analysis tools, such as life cycle assessment, sustainability index etc. Decision-makers should utilize the results from these tools and draft frameworks that promote a holistic assessment for a circular economy. Furthermore, each stakeholder should collaborate on open platforms for implementing ground-level solutions enhanced by advanced research and transition through guiding policies. Combating the circularity challenge is possible only through local implementation and monitoring administered by global planning and directions.

References

1. Tonelli M, Cristoni N (2019) An introduction to the circular economy. https://doi.org/10.4324/9781315102641-3
2. Puyol D, Batstone DJ, Hülsen T, Astals S, Peces M, Krömer JO (2017) Resource recovery from wastewater by biological technologies: opportunities, challenges, and prospects. Front Microbiol 7:1–23. https://doi.org/10.3389/fmicb.2016.02106

3. Steffen W, Richardson K, Rockström J et al (2015) Planetary boundaries: Guiding human development on a changing planet. Science (80-.), 347. https://doi.org/10.1126/science.1259855
4. Robles Á, Aguado D, Barat R et al (2020) New frontiers from removal to recycling of nitrogen and phosphorus from wastewater in the circular economy. Bioresour Technol 300:122673. https://doi.org/10.1016/j.biortech.2019.122673
5. Ali M, Hong PY, Mishra H, Vrouwenvelder J, Saikaly PE (2022) Adopting the circular model: opportunities and challenges of transforming wastewater treatment plants into resource recovery factories in Saudi Arabia. Water Reuse 12:346–365. https://doi.org/10.2166/wrd.2022.038
6. Colella M, Ripa M, Cocozza A, Panfilo C, Ulgiati S (2021) Challenges and opportunities for more efficient water use and circular wastewater management. The case of Campania Region, Italy. J Environ Manage 297:113171. https://doi.org/10.1016/j.jenvman.2021.113171
7. Malik OA, Hsu A, Johnson LA, de Sherbinin A (2015) A global indicator of wastewater treatment to inform the sustainable (SDGs). Environ Sci Policy 48:172–185. https://doi.org/10.1016/j.envsci.2015.01.005
8. Guven H, Ersahin ME, Ozgun H, Ozturk I, Koyuncu I (2023) Energy and material refineries of future: wastewater treatment plants. J Environ Manage 329:117130. https://doi.org/10.1016/j.jenvman.2022.117130
9. Makarigakis AK, Jimenez-Cisneros BE (2019) UNESCO's contribution to face global water challenges. Water (Switzerland) 11. https://doi.org/10.3390/w11020388
10. Pasciucco F, Pecorini I, Iannelli R (2022) Planning the centralization level in wastewater collection and treatment: a review of assessment methods. J Clean Prod 375:134092. https://doi.org/10.1016/j.jclepro.2022.134092
11. Adhikari S, Halden RU (2022) Opportunities and limits of wastewater-based epidemiology for tracking global health and attainment of UN sustainable development goals. Environ Int 163:107217. https://doi.org/10.1016/j.envint.2022.107217
12. Tortajada C (2020) Contributions of recycled wastewater to clean water and sanitation Sustainable Development Goals. npj Clean Water 3. https://doi.org/10.1038/s41545-020-0069-3
13. Obaideen K, Shehata N, Sayed ET, Abdelkareem MA, Mahmoud MS, Olabi AG (2022) The role of wastewater treatment in achieving sustainable development goals (SDGs) and sustainability guideline. Energy Nexus 7:100112. https://doi.org/10.1016/j.nexus.2022.100112
14. Tzanakakis VA, Capodaglio AG, Angelakis AN (2023) Insights into global water reuse opportunities. Sustainable 15. https://doi.org/10.3390/su151713007
15. Muga HE, Mihelcic JR (2008) Sustainability of wastewater treatment technologies. J Environ Manage 88:437–447. https://doi.org/10.1016/j.jenvman.2007.03.008
16. Bilalova S, Newig J, Tremblay-Lévesque LC, Roux J, Herron C, Crane S (2023) Pathways to water sustainability? A global study assessing the benefits of integrated water resources management. J Environ Manage 343. https://doi.org/10.1016/j.jenvman.2023.118179
17. Rani A, Snyder SW, Kim H, Lei Z, Pan SY (2022) Pathways to a net-zero-carbon water sector through energy-extracting wastewater technologies. npj Clean Water 5:1–17. https://doi.org/10.1038/s41545-022-00197-8
18. Pandit A, Minné EA, Li F et al (2017) Infrastructure ecology: an evolving paradigm for sustainable urban development. J Clean Prod 163:S19–S27. https://doi.org/10.1016/j.jclepro.2015.09.010
19. G20 Presidency (2023)
20. Manisha M, Verma K, Ramesh N, Anirudha TP, Santrupt RM, Das R, Mohan Kumar MS, Chanakya HN, Rao L (2023) Socio-economic impact assessment of large-scale recycling of treated municipal wastewater for indirect groundwater recharge. Sci Total Environ 859:160207. https://doi.org/10.1016/j.scitotenv.2022.160207
21. Avellán T, Nagabhatla N, Jalan I, Liao D (2021) Integrating circularity to achieve sustainability: examples of various wastewater treatment systems. Elsevier Inc. https://doi.org/10.1016/B978-0-12-821664-4.00002-9

22. Brucculeri M, Bolzonella D, Battistoni P, Cecchi F (2005) Treatment of mixed municipal and winery wastewaters in a conventional activated sludge process: a case study. Water Sci Technol 51:89–98. https://doi.org/10.2166/wst.2005.0011
23. Akunna JC (2018) Biological treatment processes. https://doi.org/10.1201/9781351170529-1
24. Iorhemen OT, Hamza RA, Tay JH (2016) Membrane bioreactor (Mbr) technology for wastewater treatment and reclamation: membrane fouling. Membranes (Basel) 6:13–16. https://doi.org/10.3390/membranes6020033
25. Uddin MM, Wright MM (2023) Anaerobic digestion fundamentals, challenges, and technological advances. Phys Sci Rev 8:2819–2837. https://doi.org/10.1515/psr-2021-0068
26. Parde D, Patwa A, Shukla A, Vijay R, Killedar DJ, Kumar R (2021) A review of constructed wetland on type, treatment and technology of wastewater. Environ Technol Innov 21:101261. https://doi.org/10.1016/j.eti.2020.101261
27. Bhatt P, Bhandari G, Turco RF, Aminikhoei Z, Bhatt K, Simsek H (2022) Algae in wastewater treatment, mechanism, and application of biomass for production of value-added product. Environ Pollut 309:119688. https://doi.org/10.1016/j.envpol.2022.119688
28. Ghangrekar MM, Kumar S, Ahmad A, Das S (2022) Concomitant bioenergy production and wastewater treatment employing microbial electrochemical technologies. Elsevier Inc. https://doi.org/10.1016/B978-0-323-85269-2.00015-0
29. Priyadarshini M, Das I, Ghangrekar MM, Blaney L (2022) Advanced oxidation processes: performance, advantages, and scale-up of emerging technologies. J Environ Manage 316:115295. https://doi.org/10.1016/j.jenvman.2022.115295
30. Saúco C, Cano R, Marín D, Lara E, Rogalla F, Arbib Z (2021) Hybrid wastewater treatment system based in a combination of high rate algae pond and vertical constructed wetland system at large scale. J Water Process Eng 43:102311. https://doi.org/10.1016/j.jwpe.2021.102311
31. Djandja OS, Liew RK, Liu C et al (2023) Catalytic hydrothermal carbonization of wet organic solid waste: a review. Sci Total Environ. 873. https://doi.org/10.1016/j.scitotenv.2023.162119
32. Mandari V, Devarai SK (2022) Biodiesel production using homogeneous, heterogeneous, and enzyme catalysts via transesterification and esterification reactions: a critical Review. Bioenergy Res 15:935–961. https://doi.org/10.1007/s12155-021-10333-w
33. Kog YC (2020) Water reclamation and reuse in Singapore. J Environ Eng 146:1–8. https://doi.org/10.1061/(asce)ee.1943-7870.0001675
34. Shehata N, Egirani D, Olabi AG, Inayat A, Abdelkareem MA, Chae KJ, Sayed ET (2023) Membrane-based water and wastewater treatment technologies: Issues, current trends, challenges, and role in achieving sustainable development goals, and circular economy. Chemosphere 320:137993. https://doi.org/10.1016/j.chemosphere.2023.137993
35. Singh S, Kumar V, Anil AG et al (2021) Adsorption and detoxification of pharmaceutical compounds from wastewater using nanomaterials: a review on mechanism, kinetics, valorization and circular economy. J Environ Manage 300:113569. https://doi.org/10.1016/j.jenvman.2021.113569
36. Semaha P, Lei Z, Yuan T, Zhang Z, Shimizu K (2023) Transition of biological wastewater treatment from flocculent activated sludge to granular sludge systems towards circular economy. Bioresour Technol Reports 21:101294. https://doi.org/10.1016/j.biteb.2022.101294
37. Simon D, Adam-Bradford A (2016) Archaeology and contemporary dynamics for more sustainable, resilient cities in the peri-urban interface. https://doi.org/10.1007/978-3-319-28112-4_5
38. Garduño-Jiménez AL, Durán-Álvarez JC, Ortori CA, Abdelrazig S, Barrett DA, Gomes RL (2023) Delivering on sustainable development goals in wastewater reuse for agriculture: Initial prioritization of emerging pollutants in the Tula Valley, Mexico. Water Res 238. https://doi.org/10.1016/j.watres.2023.119903
39. Alabiso A, Frasca S, Cantelmo V, Andrea MMD, Braglia R, Costa F, Savio S, Congestri R, Migliore L (2023) A dirty job : dishwasher wastewater reuse and upcycle through an ad hoc engineered microbial consortium. npj Clean Water 6:1–14. https://doi.org/10.1038/s41545-023-00280-8

40. Dominguez S, Laso J, Margallo M, Aldaco R, Rivero MJ, Irabien Á, Ortiz I (2018) LCA of greywater management within a water circular economy restorative thinking framework. Sci Total Environ 621:1047–1056. https://doi.org/10.1016/j.scitotenv.2017.10.122
41. Ghimire U, Sarpong G, Gude VG (2021) Transitioning wastewater treatment plants toward circular economy and energy sustainability. ACS Omega 6:11794–11803. https://doi.org/10.1021/acsomega.0c05827
42. Zhao Q, Peng Y, Li J, Gao R, Jia T, Deng L, Du R (2022) Sustainable upgrading of biological municipal wastewater treatment based on anammox: from microbial understanding to engineering application. Sci Total Environ 813:152468. https://doi.org/10.1016/j.scitotenv.2021.152468
43. Jadhav DA, Ghosh Ray S, Ghangrekar MM (2017) Third generation in bio-electrochemical system research—A systematic review on mechanisms for recovery of valuable by-products from wastewater. Renew Sustain Energy Rev 76:1022–1031. https://doi.org/10.1016/j.rser.2017.03.096
44. Zhang X, Liu Y (2022) Circular economy is game-changing municipal wastewater treatment technology towards energy and carbon neutrality. Chem Eng J 429:132114. https://doi.org/10.1016/j.cej.2021.132114
45. Gherghel A, Teodosiu C, De Gisi S (2019) A review on wastewater sludge valorisation and its challenges in the context of circular economy. J Clean Prod 228:244–263. https://doi.org/10.1016/j.jclepro.2019.04.240
46. Kurniawan TA, Othman MHD, Liang X, Ayub M, Goh HH, Kusworo TD, Mohyuddin A, Chew KW (2022) Microbial Fuel Cells (MFC): a potential game-changer in renewable energy development. Sustain 14:1–20. https://doi.org/10.3390/su142416847
47. Venkatramanan V, Shah S, Prasad R (2021) A critical review on microbial fuel cells technology: perspectives on wastewater treatment. Open Biotechnol J 15:131–141. https://doi.org/10.2174/1874070702115010131
48. Femina CC, Kamalesh T, Kumar PS, Rangasamy G (2023) A critical review on the sustainable approaches for the removal of toxic heavy metals from water systems. Ind Eng Chem Res 62:8575–8601. https://doi.org/10.1021/acs.iecr.3c00709
49. Goswami RK, Mehariya S, Verma P, Lavecchia R, Zuorro A (2021) Microalgae-based biorefineries for sustainable resource recovery from wastewater. J Water Process Eng 40:101747. https://doi.org/10.1016/j.jwpe.2020.101747
50. Das S, Raj R, Ghangrekar MM (2023) Efficient algal lipid extraction via a green bio-electro-Fenton process and its conversion into biofuel and bioelectricity with concurrent wastewater treatment in a photosynthetic microbial fuel cell. Green Chem 7166–7182. https://doi.org/10.1039/d3gc01548c
51. Guerra-Rodríguez S, Oulego P, Rodríguez E, Singh DN, Rodríguez-Chueca J (2020) Towards the implementation of circular economy in the wastewater sector: Challenges and opportunities. Water (Switzerland) 12. https://doi.org/10.3390/w12051431
52. Yan P, Qin RC, Guo JS, Yu Q, Li Z, Chen YP, Shen Y, Fang F (2017) Net-zero-energy model for sustainable wastewater treatment. Environ Sci Technol 51:1017–1023. https://doi.org/10.1021/acs.est.6b04735
53. Hernández-Chover V, Castellet-Viciano L, Fuentes R, Hernández-Sancho F (2023) Circular economy and efficiency to ensure the sustainability in the wastewater treatment plants. J Clean Prod 384. https://doi.org/10.1016/j.jclepro.2022.135563
54. Walsh M, Schenk GSS (2023) Realising the circular phosphorus economy delivers for sustainable development goals. AuthoreaCom 1–15. https://doi.org/10.1038/s44264-023-00002-0
55. Djandja OS, Salami AA, Wang ZC, Duo J, Yin LX, Duan PG (2022) Random forest-based modeling for insights on phosphorus content in hydrochar produced from hydrothermal carbonization of sewage sludge. Energy 245. https://doi.org/10.1016/j.energy.2022.123295
56. Capodaglio AG (2023) Biorefinery of sewage sludge: overview of possible value-added products and applicable process technologies. Water (Switzerland) 15. https://doi.org/10.3390/w15061195

57. Awasthi MK, Ganeshan P, Gohil N et al (2023) Advanced approaches for resource recovery from wastewater and activated sludge: a review. Bioresour Technol 384:129250. https://doi.org/10.1016/j.biortech.2023.129250
58. Bin KS, Khalekuzzaman M, Hossain N, Jamal M, Alam MA, Abomohra AEF (2022) Progress in biohythane production from microalgae-wastewater sludge co-digestion: an integrated biorefinery approach. Biotechnol Adv 57:107933. https://doi.org/10.1016/j.biotechadv.2022.107933
59. Leong HY, Chang CK, Khoo KS, Chew KW, Chia SR, Lim JW, Chang JS, Show PL (2021) Waste biorefinery towards a sustainable circular bioeconomy: a solution to global issues. Biotechnol Biofuels 14:1–15. https://doi.org/10.1186/s13068-021-01939-5
60. Raj R, Tripathi A, Das S, Ghangrekar MM (2023) Waste coconut shell-derived carbon monolith as an efficient binder-free cathode for electrochemical advanced oxidation treatment of endocrine-disrupting compounds. J Environ Manage 348:119328. https://doi.org/10.1016/j.jenvman.2023.119328
61. Xiang W, Zhang X, Chen J, Zou W, He F, Hu X, Tsang DCW, Ok YS, Gao B (2020) Biochar technology in wastewater treatment: a critical review. Chemosphere 252:126539. https://doi.org/10.1016/j.chemosphere.2020.126539
62. Weidner E, Karbassiyazdi E, Altaee A, Jesionowski T, Ciesielczyk F (2022) Hybrid metal oxide/biochar materials for wastewater treatment technology: a review. ACS Omega 7:27062–27078. https://doi.org/10.1021/acsomega.2c02909
63. Hossain N, Bhuiyan MA, Pramanik BK, Nizamuddin S, Griffin G (2020) Waste materials for wastewater treatment and waste adsorbents for biofuel and cement supplement applications: a critical review. J. Clean. Prod. 255. https://doi.org/10.1016/j.jclepro.2020.120261
64. Sellamuthu B, Gandhi V, Yadav B, Tyagi RD (2021) Sustainable production of bioadsorbents from municipal and industrial wastes in a circular bioeconomy context. Elsevier Inc. https://doi.org/10.1016/B978-0-12-821878-5.00019-2
65. Maged A, El-Fattah HA, Kamel RM, Kharbish S, Elgarahy AM (2023) A comprehensive review on sustainable clay-based geopolymers for wastewater treatment: circular economy and future outlook. Springer International Publishing. https://doi.org/10.1007/s10661-023-11303-9
66. Baskar AV, Bolan N, Hoang SA et al (2022) Recovery, regeneration and sustainable management of spent adsorbents from wastewater treatment streams: a review. Sci Total Environ 822:153555. https://doi.org/10.1016/j.scitotenv.2022.153555
67. Stazi V, Tomei MC (2021) Dissolved methane in anaerobic effluents: a review on sustainable strategies for optimization of energy recovery or internal process reuse. J Clean Prod 317:128359. https://doi.org/10.1016/j.jclepro.2021.128359
68. Sharma M, Agarwal S, Agarwal Malik R et al (2023) Recent advances in microbial engineering approaches for wastewater treatment: a review. Bioengineered 14:2184518. https://doi.org/10.1080/21655979.2023.2184518
69. Letsoalo MR, Sithole T, Mufamadi S, Mazhandu Z, Sillanpaa M, Kaushik A, Mashifana T (2023) Efficient detection and treatment of pharmaceutical contaminants to produce clean water for better health and environmental. J Clean Prod 387:135798. https://doi.org/10.1016/j.jclepro.2022.135798
70. Tian X, Richardson RE, Tester JW, Lozano JL, You F (2020) Retrofitting municipal wastewater treatment facilities toward a greener and circular economy by virtue of resource recovery: techno-economic analysis and life cycle assessment. ACS Sustain Chem Eng 8:13823–13837. https://doi.org/10.1021/acssuschemeng.0c05189
71. Jones ER, Bierkens MFP, Wanders N, Sutanudjaja EH, van Beek LPH, van Vliet MTH (2022) Current wastewater treatment targets are insufficient to protect surface water quality. Commun Earth Environ 3:1–8. https://doi.org/10.1038/s43247-022-00554-y
72. Cipolletta G, Ozbayram EG, Eusebi AL, Akyol Ç, Malamis S, Mino E, Fatone F (2021) Policy and legislative barriers to close water-related loops in innovative small water and wastewater systems in Europe: a critical analysis. J Clean Prod 288. https://doi.org/10.1016/j.jclepro.2020.125604

73. Kumar R, Maurya A, Raj A (2023) Emerging technological solutions for the management of paper mill wastewater: Treatment, nutrient recovery and fourth industrial revolution (IR 4.0). J Water Process Eng 53:103715. https://doi.org/10.1016/j.jwpe.2023.103715
74. Huang D, Jin L, Liu J, Han G, Liu W, Huber-Lee A (2023) Tsthe Beijing Case. Water (Switzerland) 15. https://doi.org/10.3390/w15040630
75. Mannina G, Rebouças TF, Cosenza A, Sànchez-Marrè M, Gibert K (2019) Decision support systems (DSS) for wastewater treatment plants—A review of the state of the art. Bioresour Technol 290:121814. https://doi.org/10.1016/j.biortech.2019.121814
76. Corominas L, Byrne DM, Guest JS, Hospido A, Roux P, Shaw A, Short MD (2020) The application of life cycle assessment (LCA) to wastewater treatment: a best practice guide and critical review. Water Res 184. https://doi.org/10.1016/j.watres.2020.116058
77. Singh A, Sawant M, Kamble SJ, Herlekar M, Starkl M, Aymerich E, Kazmi A (2019) Performance evaluation of a decentralized wastewater treatment system in India. Environ Sci Pollut Res 26:21172–21188. https://doi.org/10.1007/s11356-019-05444-z
78. Furness M, Bello-Mendoza R, Dassonvalle J, Chamy-Maggi R (2021) Building the 'Bio-factory": a bibliometric analysis of circular economies and life cycle sustainability assessment in wastewater treatment'. J Clean Prod 323:129127. https://doi.org/10.1016/j.jclepro.2021.129127
79. Kanchanamala Delanka-Pedige HM, Munasinghe-Arachchige SP, Abeysiriwardana-Arachchige ISA, Nirmalakhandan N (2021) Evaluating wastewater treatment infrastructure systems based on UN sustainable development goals and targets. J Clean Prod 298:126795. https://doi.org/10.1016/j.jclepro.2021.126795
80. Kordana-Obuch S, Wojtoń M, Starzec M, Piotrowska B (2023) opportunities and challenges for research on heat recovery from wastewater: bibliometric and strategic analyses. Energies 16. https://doi.org/10.3390/en16176370
81. Abu-Ghunmi D, Abu-Ghunmi L, Kayal B, Bino A (2016) Circular economy and the opportunity cost of not "closing the loop" of water industry: the case of Jordan. J Clean Prod 131:228–236. https://doi.org/10.1016/j.jclepro.2016.05.043
82. Schneider MY, Quaghebeur W, Borzooei S, Froemelt A, Li F, Saagi R, Wade MJ, Zhu JJ, Torfs E (2022) Hybrid modelling of water resource recovery facilities: status and opportunities. Water Sci Technol 85:2503–2524. https://doi.org/10.2166/wst.2022.115
83. Matheri AN, Mohamed B, Ntuli F, Nabadda E, Ngila JC (2022) Sustainable circularity and intelligent data-driven operations and control of the wastewater treatment plant. Phys Chem Earth 126:103152. https://doi.org/10.1016/j.pce.2022.103152

Life Cycle Assessment of Hybrid-Biological Wastewater Treatment Plants: Challenges and Opportunities

Rajarshi Bhar, Joydeepa Taran, Brajesh Kumar Dubey, and Makarand Madhao Ghangrekar

Abstract Primarily propelled by the growing population, wastewater generation has been continuously increasing. Rapid urbanization and industrial development are also increasing the load on the existing wastewater treatment plants (WWTPs). Henceforth, the pursuit of sustainable development goals, aided by technological advancements, has fuelled the construction of new WWTPs, along with the transformation of the existing conventional WWTPs into hybrid-biological treatment plants, which has emphasized the domains of resource recovery and circular economy. However, it is crucial to ensure that the overall environmental impact of the modified WWTP is lower than that of the conventional one. Life Cycle Assessment (LCA) serves as a tool that quantitatively assesses the anticipated environmental impact of WWTPs and facilitates comparisons with their conventional counterparts. This book chapter provides an overview of the LCA methodology within the context of WWTPs, highlighting the iterative nature of the different phases of an LCA, along with a detailed discussion of the individual phase. The environmental impact of employing hybrid-biological wastewater treatment units was discussed with the aid of case studies on nutrient removal, constructed wetlands, moving bed biofilm reactors and membrane bioreactors. Apart from highlighting the trade-offs inherent to different hybrid systems, the chapter also emphasizes the challenges associated with data availability, quality and acceptability of the LCA results while underscoring the associated methodological challenges. Therefore, this chapter can aid researchers, policymakers and academicians alike in understanding the intricacies associated with the LCA of WWTPs and support better decision-making.

R. Bhar · J. Taran · B. K. Dubey (✉) · M. M. Ghangrekar
Department of Civil Engineering, Indian Institute of Technology Kharagpur, Kharagpur, West Bengal 721302, India
e-mail: bkdubey@civil.iitkgp.ac.in

B. K. Dubey
School of Water Resources, Indian Institute of Technology Kharagpur, Kharagpur, West Bengal 721302, India

M. M. Ghangrekar et al. (eds.), *Biological and Hybrid Wastewater Treatment Technology*, Earth and Environmental Sciences Library,
https://doi.org/10.1007/978-3-031-63046-0_21

Keywords Hybrid-biological wastewater treatment · Life cycle assessment · Environmental footprint · Environmental performance · Sustainability

1 Introduction

Wastewater management has been an integral part of human civilization, which has undergone several changes based on social and technological developments. The increased urbanization and industrialization have also led to the continuous rise in wastewater generation. Furthermore, aligning with the sustainable development goals, the market for wastewater treatment was estimated to be 281 billion USD in the year 2021 and was predicted to grow by about 74% by 2029 [1]. However, conventional wastewater treatment plants aren't designed to remove emerging contaminants (and/or micropollutants) from the incoming effluent [2]. This inefficiency can also be owed to their recalcitrant nature. Moreover, the presence of such contaminants in the treated effluent increases the risk of their potential accumulation and subsequent consequences.

With increasing freshwater demand, the use of treated wastewater for potable and non-potable applications has garnered increasing interest due to reduced environmental load [3, 4]. This is particularly evident from the increasing employment of various nutrient recovery units, which can also reduce the eutrophication risks from the discharged effluents [5]. In this regard, hybrid-biological processes address some major limitations of biological wastewater treatment and present a more effective approach [6]. The selection of a hybrid process is primarily dictated by the nature of the targeted contaminant (or contaminants) and typically employs a combination of physicochemical processes along with biological processes. Some of the commonly employed complementary techniques can include two or more stage biological treatments, electrochemical oxidation processes, adsorption processes, and electro-coagulation. Apart from increasing the wastewater treatment efficiency, a hybrid approach can also be aimed at increasing the yield of biogas, bioelectricity, biohydrogen or other valuable by-products. However, the choice of technologies and strategies should account for the overall environmental sustainability of the wastewater treatment plant.

2 LCA Methodology from a Wastewater Treatment Perspective

The primary objective of wastewater treatment plants (WWTPs) is to reduce the environmental impact associated with the direct discharge of untreated wastewater into water bodies or land. Therefore, it is vital to ensure that the adverse environmental impacts associated with the material and energy inputs for operating the WWTPs

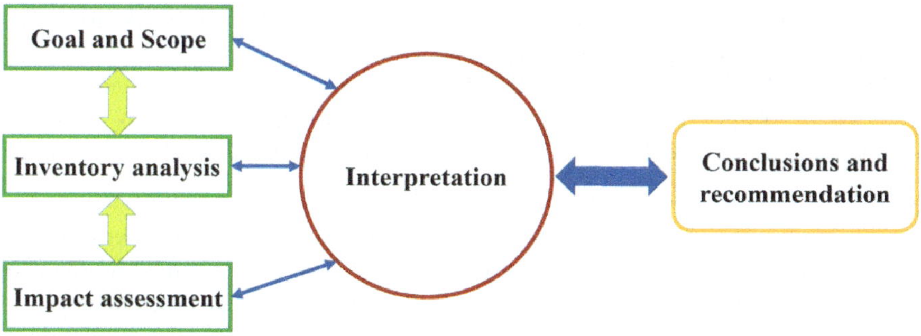

Fig. 1 Four major phases during an LCA

don't outweigh that of the direct discharge of untreated wastewater. Furthermore, it should also be ascertained that the WWTPs don't exert a larger environmental burden than the benefits associated with remediation. This demands a comprehensive environmental assessment of different scenarios to get better clarity. Life cycle assessment (LCA) has been one of the most commonly used tools to analyze the environmental footprint of WWTPs, to identify process hotspots and recommend relevant measures to improve the overall environmental performance [7]. It can be defined as a systematic and standardized approach to quantify the environmental footprint of any product, process or service throughout [8]. Figure 1 depicts the major stages of an LCA, which tend to be iterative in nature.

LCA played a significant role in assessing the environmental sustainability in the transition of conventional WWTPs from just treating wastewater towards resource recovery. A comprehensive environmental assessment can reveal the environmental trade-offs associated with the implementation of newer technologies and aid in evaluating the overall environmental sustainability. Moreover, based on the desired objectives, an LCA can consider the complete life cycle or a certain portion of it. A more detailed discussion of the methodology of LCA has been provided in the following sections. The International Organization of Standards (ISO) can be credited for the formal standardization of the LCA process. The principles and framework of LCA were first addressed by ISO 14040, followed by ISO 14041, defining the goal and scope; ISO 14042, defining life cycle impact assessment (LCIA); and ISO 14043, highlighting the interpretation of LCIA results. Later on, ISO 14041, 14042 and 14043 were compiled into ISO 14044. Thereafter, the International Life Cycle Data (ILCD) system was developed by the European Commission, which contained the methodological guidelines and a database for the life cycle inventory (LCI) data. However, whether to consider the entire life cycle or a certain part of it depends on the objective of the assessment. From the perspective of municipal wastewater treatment, the following phases can be considered:

Phase 1: Generation of sewage
Phase 2: Collection of sewage and conveyance to the WWTP
Phase 3: Wastewater treatment in the WWTP

Phase 4: Reuse of the treated wastewater for different applications (like irrigation, groundwater recharge, cooling water or toilet flushing)
Phase 5: Sludge management in the WWTP
Phase 6: Reuse of the treated sludge for different applications (like soil amendment, land reclamation, solid fuel, construction material or fish/animal feed).

Considering the above phases and with an objective to assess the environmental footprint of treating the wastewater alone, the LCA will be performed by focusing particularly on Phase 3 (i.e., wastewater treatment in the WWTP). Such an approach analyses a certain span in the life cycle of wastewater. Conversely, when examining an industrial effluent treatment plant, the situation can exhibit variability. Consider a specific industry producing 'Product X' and concurrently generating a wastewater stream. The subsequent stages can be contemplated to elucidate the LCA of the industrial process.

Phase 1: Procurement of raw materials and chemicals
Phase 2: Production of **Product X** (also considering the energy consumption)
Phase 3: Effluent treatment
Phase 4: Reuse of the treated effluent for different applications (like irrigation, groundwater recharge, cooling water or toilet flushing)
Phase 5: Use phase of **Product X**
Phase 6: End-of-life management of **Product X**.

From a commercial perspective and aiming to assess the environmental footprint for the production of one unit of Product X, Phases 1–4 may be taken into account. However, adopting a life cycle perspective necessitates considering all six phases mentioned above to estimate the potential environmental impact throughout the entire life cycle, from cradle-to-grave. Furthermore, when the objective is to compare different effluent treatment approaches and determine the one with the highest environmental efficiency, an LCA should be performed by particularly focusing on Phase 3 and, if required, Phase 4.

2.1 Goal and Scope

The first step of an LCA involves the articulation of the goal and objective of the study. The phase highlights the purpose of the assessment, also underlining the intended application and audience. Furthermore, the scope of an LCA study is also dictated by the intended goal. On the other hand, defining the scope of the study is equally vital as it delineates the functional unit (FU), impact assessment method, allocation approach, system boundaries, and data description (or quality). Herein, the function defines the selection of FU and, therefore, also quantifies the function, laying the basis for comparison between different systems. For instance, for a WWTP with a function to treat the influent and reduce the chemical oxygen demand (COD)

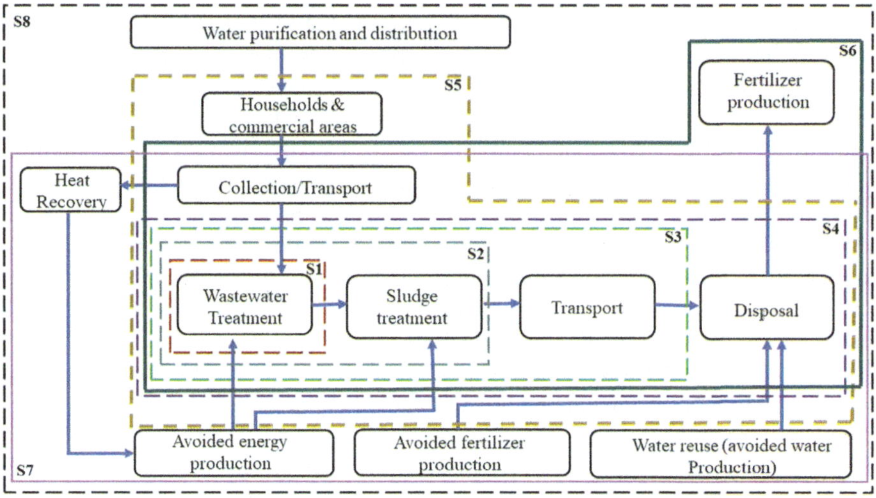

Fig. 2 Typical system boundaries adopted for LCA of wastewater treatment plants. Adopted and modified from Corominas et al. [10]. S1–S8 denote different system boundaries

of the influent, the corresponding FU can be the volume of wastewater treated or mass removal of COD or population equivalent [9]. Resultingly, the environmental performance of different WWTPs serving the same function can be compared by using the above-mentioned FUs. Additionally, defining the scope also frequently includes the description of data quality, geographical reference, time reference, and technological reference.

The system boundary is another indispensable term used during any LCA. As the name suggests, it highlights the processes that are to be analyzed for the quantification of the overall environmental footprint. Thus, the processes lying outside the boundary aren't considered for the LCA. Furthermore, the nature of the system boundary is typically characterized as cradle-to-grave, cradle-to-gate, gate-to-gate, gate-to-grave, or cradle-to-cradle, based on the phases of the product life cycle or selected activities. Figure 2 depicts the system boundaries assessed by different studies. Most of the LCA studies exclude the wastewater generation and conveyance part, owing to the greater emphasis on assessing and comparing different wastewater treatment technologies. However, the inclusion of the wastewater generation and distribution phase by expanding the system boundary (merging S5, S6 and S7) can provide a more comprehensive outlook on the environmental footprint of the entire system.

2.2 *Inventory*

LCI analysis comprises of a quantified compilation of material and energy inputs and outputs for a pre-defined system boundary. In other words, it is typically a table that contains the quantities of materials and energy consumed, along with the products (mass or energy) and emissions generated from the system. It is also the most important phase of an LCA, as the quality of the data constituting the inventory will define the precision and certainty of the estimated environmental footprint. As a result, this phase is iterative in nature to increase the data accuracy and accounting. The data used for preparing the LCI can be either primary or secondary in nature. However, it is vital to clearly report the assumptions and calculations made while adopting any secondary data. Finally, the collected (or adopted) data is represented in an LCI table, clearly indicating and quantifying the inputs and outputs from a system. The quality and reliability of the collected data include addressing the associated uncertainties. This is typically performed through sensitivity analysis.

With respect to the operational phase of wastewater treatment, the inputs often comprise of influent volume, influent COD and other macro- or micro-contaminants, chemical consumptions (like NaCl for electro-chlorination or O_3 as an oxidizing agent) in case of advanced oxidation processes, and energy consumption. The outputs from the system can include effluent volume, effluent COD and other macro- or micro-contaminants, aerial emissions (like CO_2, CH_4 and N_2O), and sludge production. Among the system outputs and based on the adopted technology, treated effluent (as irrigation water) can be considered as an avoided burden for conventional freshwater and fertilizer, while biogas (as an energy source) for fossil energy, and sludge (as fertilizer or fish-feed) for fertilizer.

2.3 *Impact Assessment*

The potential environmental footprint is evaluated during life cycle impact assessment (LCIA) by using the data compiled during LCI. The impacts are quantified based on the characterization factors defined in different LCIA methods. Some of the commonly used methods are ReCiPe 2016, Impact 2002+, CML method and TRACI. Based on the selection of the LCIA method, the environmental footprint can be classified into different impact categories. Furthermore, the characterization results can be normalized relative to a reference amount to judge the magnitude of threat on different impact categories. From the perspective of WWTPs, some of the most impacted and studied impact categories include fossil resource depletion (FRD), eutrophication potential (EP), global warming potential (GWP), acidification potential (AP), ozone layer depletion potential (OLDP) and human toxicity potential (HTP) [11]. Additionally, some of the other impact categories that are gaining increasing attention include water footprint, land use impact, and energy consumption [11].

2.4 Interpretation

The potential environmental impacts estimated during LCIA are evaluated and scrutinized to ensure their alignment with the initially defined goal and scope of LCA. The four major components of this phase include (1) identification of system hotspots and significant issues, (2) evaluating the reliability of the obtained results, (3) conclusions and recommendations, and (4) reporting. This phase can involve deriving meaningful insights from the derived results. This phase goes beyond the quantification of environmental impacts and aims to provide a comprehensive understanding on the implications of the assessment. The process hotspots can be different unit processes in an LCA and these are often identified by the LCI parameters, which make a major contribution to the overall environmental footprint. This impact from different inventory parameters can also be clubbed as different unit processes, which can highlight the process that has the highest environmental impacts. Another aspect can be to underline the most affected impact categories and also emphasize the major contributors in different categories.

The reliability of the results is usually evaluated through completeness, sensitivity and consistency checks. As the name suggests, a completeness check is performed to identify any incomplete or missing information required to achieve the goal of the LCA study and also fall within its scope. If found, the missing data has to be accounted for, and the associated impacts need to be added or subtracted from the previously calculated values. Sensitivity checks deal with the uncertainty associated with the various adopted factors like assumptions, data, allocation methods, and LCIA methods. Resultingly, sensitivity analysis becomes particularly important to understand the significance of different factors and also to compare the impact on the estimated footprint due to any variation in the factors. Finally, consistency checks include confirmation that the data quality, system boundary, allocation method and LCIA method are aligned with the goal and scope of the LCA study.

While drawing conclusions and providing recommendations, the LCA practitioner also highlights the associated limitations or gaps of the study. The goal of the study is summarized, and a more detailed discussion is presented, addressing the identified issues and providing recommendations for exploring options to reduce the environmental footprint of the system. Finally, a comprehensive report is to be prepared, presenting the adopted methodology and obtained results in a transparent and well-structured manner. Sufficient details about the compilation of LCI should incorporate assumptions and calculations, as well as highlight possible limitations.

3 LCA of Hybrid-Biological WWTPs

3.1 Case Studies on Ammonia Removal

Nutrient removal from wastewater often requires additional chemical and energy requirements, which can significantly contribute to the total environmental footprint of a WWTP. However, recovering valuable fertilizers from the wastewater can be an environmentally benign approach. Hauck et al. assessed the environmental performance when a two-step anaerobic ammonia oxidation (ANAMMOX), followed by stable high-rate ammonia removal over nitrite (SHARON) was integrated into a conventional WWTP, treating wastewater via activated sludge process (ASP) [12]. Owing to the reduced nitrogen emissions by introducing the hybrid system, a 16% reduction in the marine EP was observed. However, this benefit came with a trade-off with other impact categories, among which the carbon footprint of the process rose by 9% due to increased N_2O emissions and electricity use.

Later on, van Zelm et al. compared the Nijhuis Ammonia Recovery (AECO-NAR) system's nitrogen removal (as ammonia) from digestate and recovering nitrogen fertilizer (as ammonium sulfate) with the hybrid nitrogen removal system assessed by Hauck et al. [12, 13]. The AECO-NAR was primarily developed for nitrogen removal (as ammonia) from digestate supernatant and recovering nitrogen fertilizer (as ammonium sulfate) [13]. van Zelm et al. [13] focused only on nutrient removal (excluding primary and tertiary treatment of wastewater) and endpoint impacts on human health and the ecosystem. Excluding the environmental benefits of ammonium sulfate recovery, sulfuric acid contributed to about 84% and 27% of the total impacts on human health and ecosystem, respectively. In the same way, 65% of the impact of direct emissions on the ecosystem was mainly attributed to the EP due to the release of phosphates into water bodies.

3.2 Case Studies on Constructed Wetlands (CWs)

CWs mimic natural wetlands with plantations indigenous to moist regions, which can be combined with microbes and substrates to remove nutrients and contaminants from wastewater. Hybrid CWs can further improve wastewater treatment efficiency by combining conventional CWs with other approaches. For instance, Resende et al. assessed the environmental footprint for treating the effluent of septic tank by employing CWs and considering two scenarios [14]. In scenario 1, the septic tank effluent was treated through a vertical flow CW, followed by a horizontal flow CW. In scenario 2, the septic tank effluent sequentially passes through a free-flow CW (with artificial aeration), secondary clarifier and vertical sub-surface flow CW. Furthermore, the study also considered the construction phase of the components of WWTP.

The LCIA results of scenario 1 revealed that the construction phase had more than 80% of the total impacts on ecotoxicity, human toxicity and acidification, while the operation phase dominated other impact categories (Fig. 3). During the construction phase, the GWP of scenario 1 was about four times higher than that of scenario 2, with about 1.5–4 times higher impact in other categories as well. In the case of scenario 2, the operation phase posed the major environmental impact, with the exception of freshwater eutrophication and GWP, wherein the former scenario 1 was 1.28 times higher, while GWP was almost equal. For the rest of the discussed impact categories, the impact of scenario 2 was about ten times higher than scenario 1. The study also assessed the impact of the decommissioning phase and again found the impact of scenario 1 to be about 2.7 times higher across the discussed categories. The relative contributions of different parameters have been depicted in Figs. 4 and 5.

Rufí-Salís et al. studied four different scenarios for treating sewage and used 1 m^3 of treated wastewater as the FU (Fig. 6) [15]. The study focused on GWP, freshwater EP (FEP), marine EP (MEP) and ecotoxicity to determine the environmental performance of different scenarios, among which struvite recovery (SR) and total resource recovery (TRR) emerged as more environmentally benign scenarios than the baseline (BS) and savings-E (SE) scenario (See Fig. 7). Particularly focusing on the GWP, the hierarchy of the net GWP was SE > TRR > BS > SR. Moreover, as characteristic of WWTPs, direct aerial emissions had a 48–75% contribution to the GWP, with almost the entirety of the eutrophication impacts. Additionally, the autotrophic biological nitrogen removal (BNR) was the major contributor to the SE and TRR scenario,

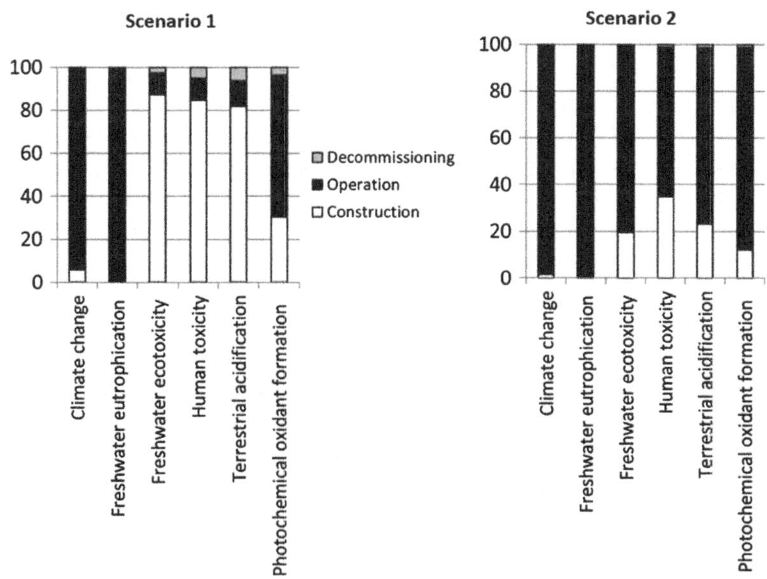

Fig. 3 Contribution analysis using the LCIA results of construction, operation and decommissioning phase for scenarios 1 and 2. Adopted and modified from Resende et al. [14]

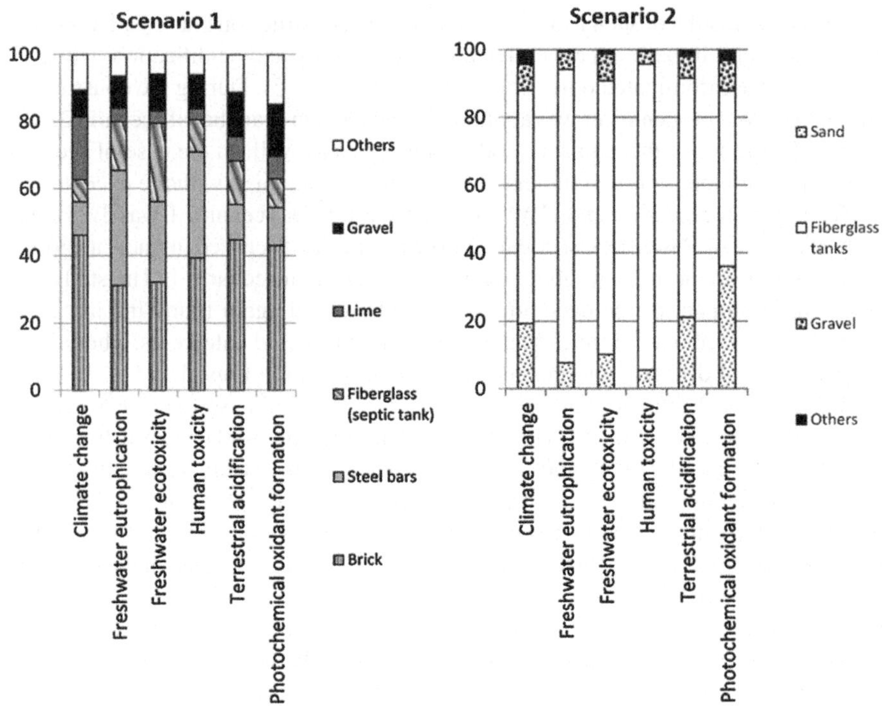

Fig. 4 Contribution analysis using the LCIA results of different LCI parameters involved in the construction phase for scenarios 1 and 2. Adopted and modified from Resende et al. [14]

which even though reduces the discharge of nutrients into the water bodies, unfortunately also increases associated N_2O, which is a very potent greenhouse gas (almost 273 times CO_2) [16]. Furthermore, the MEP of the SR scenario was high due to the absence of the BNR phase. In the case of ecotoxicity, wastewater reagents were the major hotspots for the BS and SE scenarios, while the infrastructural requirements had a significant footprint in the SE and TRR scenarios. The environmental benefits for all the scenarios were attributed to either or both the energy production and fertilizer substitution.

The environmental implications of the integration of CWs with microbial fuel cell (MFC) were assessed by considering the impact of gravel-based anode MFC (enclosed in stainless steel mesh) and graphite-based anode MFC individually [17]. When compared to the employment of only CWs, the introduction of MFCs had similar impacts for both anode types, except for abiotic depletion potential, wherein the gravel-based MFC system was estimated to have about two times higher impact. Furthermore, introducing graphite-based anode MFCs can reduce the environmental impacts of CWs by about 20%.

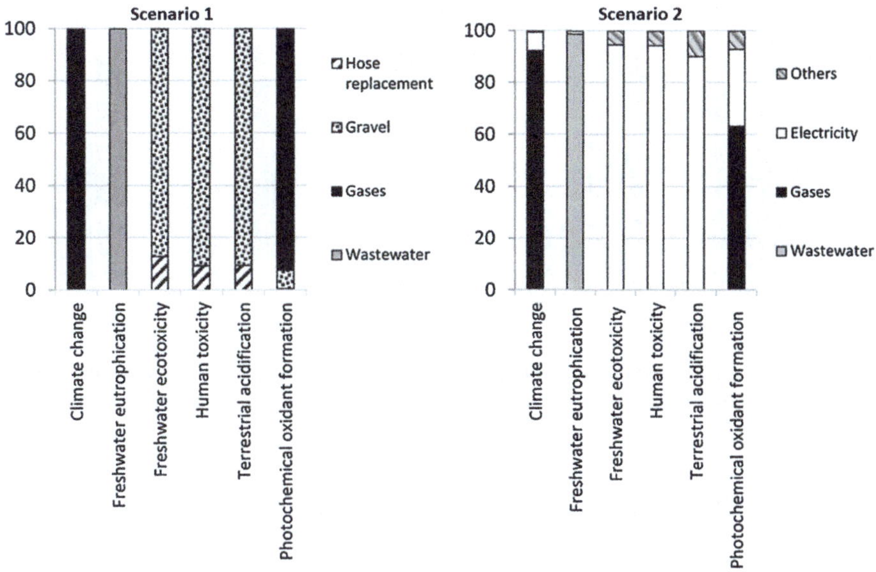

Fig. 5 Contribution analysis using the LCIA results of different materials and energy inputs used in the operation phase for scenarios 1 and 2. Adopted and modified from Resende et al. [14]

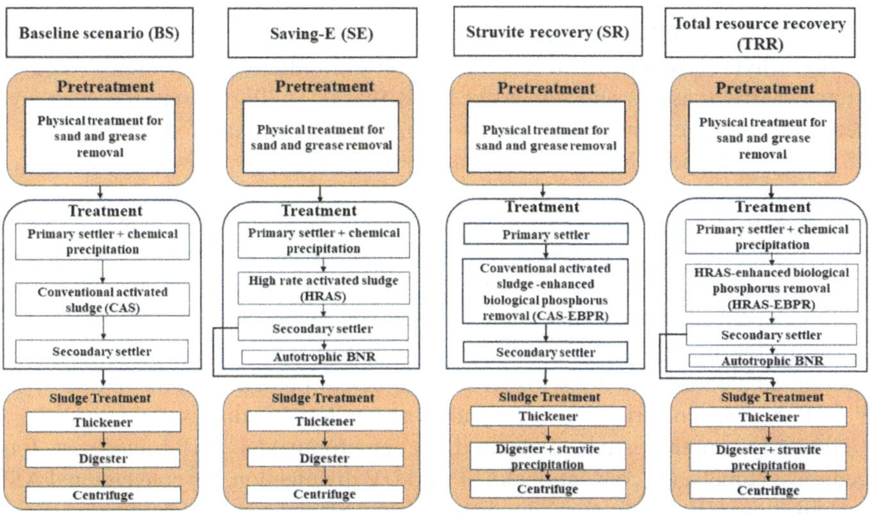

Fig. 6 Different scenarios considered for a comparative LCA of WWTPs. Modified from Rufí-Salís et al. [15]

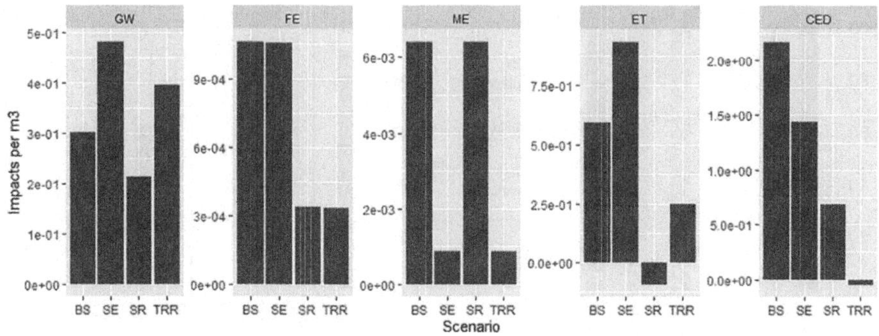

Fig. 7 The net environmental footprint of different scenarios investigated by comparing baseline scenarios with resource recovery-based scenarios [15]. Copyright with the permission from Rufí-Salís et al. [15]

3.3 Case Studies on Membrane Bioreactors (MBRs)

MBRs have been known for their high wastewater treatment efficiencies, especially reducing the EP of the treated effluent when it is discharged to water bodies [18]. However, the higher energy consumption due to aeration and membrane cleaning reduces their environmental efficiency, especially in areas that rely more on fossil sources of energy [19]. Ortiz et al. compared stand-alone conventional ASP (case 1) with a hybrid system when integrated with ultrafiltration membranes (case 2), external MBR (case 3) and immersed MBR (case 4) [20]. Among the compared scenarios, case 2 exhibited the highest environmental burden, followed by case 3 (78.6% lower as per CML 2 baseline method), case 4 (62.3% lower as per CML 2 baseline method), and case 1 (48.5% lower as per CML 2 baseline method). The external MBR system had a higher environmental impact than the immersed MBR due to the former's higher energy consumption. If the functional unit had been with respect to the quality of the treated effluent, the environmental hierarchy of the cases might have changed due to much superior effluent quality from the MBR systems. Owing to the higher energy consumption in the operation phase, the study also highlighted a correspondingly higher environmental burden from this phase of the WWTP than the construction, maintenance and end-of-life disposal phase.

Similarly, another study compared the environmental performance of four scenarios, namely, (1) conventional ASP and UV disinfection, (2) conventional ASP and high-dose UV disinfection, (3) conventional ASP and peracetic acid-aided chemical disinfection, (4) up-flow anaerobic sludge blanket reactor (UASBR) and anaerobic MBR, and the LCA was performed considering the use of treated effluent for irrigation [21]. The study found scenario 4, which employs an ultrafiltration anaerobic MBR, to be superior to other scenarios. For instance, when compared to scenario 1, marked improvements were observed in GWP (28% reduction), fossil resource (31% reduction), mineral resource depletion (52% reduction), and freshwater ecotoxicity (35% reduction). However, the trade-off was accompanied by an increased risk of freshwater eutrophication (23% increase) due to the discharge of nutrient-rich effluent into the water bodies. Human toxicity and terrestrial ecotoxicity also faced higher risks in scenario 4. The environmental performances of scenarios 2 and 3 were similar, with scenario 3 showing 9%, 18% and 6% reduction in the GWP, fossil depletion and mineral resource depletion, respectively.

On similar lines, another study performed a comparative LCA of a conventional centralized water supply and wastewater treatment system and decentralized-hybrid greywater treatment systems [22]. The decentralized greywater treatment employed submerged MBRs, coupled with a preliminary filtration unit and then a microfiltration unit after MBR to remove microbes. Furthermore, a fraction of the treated greywater was also reclaimed for non-potable applications. On the other hand, the centralized wastewater treatment included a primary clarifier, biological nutrient removal, secondary clarifier, filtration and UV disinfection units for wastewater treatment, while sludge was treated through anaerobic digestion (AD) followed by sludge dewatering and incineration. As compared to treating the greywater in a centralized WWTP, the decentralized system led up to 49% reduction in both non-potable water demand and electricity. Resultingly, the overall environmental efficiency of treating the greywater in a centralized system was 20–41% lower than the decentralised hybrid treatment approach.

3.4 Case Studies on Moving Bed Biofilm Reactors (MBBRs)

Jabri et al. performed an LCA for different scenarios of a hybrid decentralized grey treatment unit comprising two units of MBBRs, followed by a UV disinfection system [23]. Additionally, the impacts of both the construction and operational phases were considered. The study revealed a promising benefit of 3.9, 16.2 and 17% reduction in OLDP, marine aquatic ecotoxicity and EP, respectively. Furthermore, shifting towards renewable energy sources was also reported to reduce the environmental footprint with marked benefits in GWP and AP.

In another case-study, the wastewater generated from an industrial park undergoes primary, secondary, and tertiary influent treatment and sludge management [24]. The secondary treatment employs ASP and sequencing batch reactor (SBR) for wastewater treatment, while the tertiary treatment comprises an integrated anoxic/oxic (AO) process and an MBBR. Furthermore, a fraction of the treated effluent from the WWTP was diverted to a water recycling plant (WRP), which cleaned the effluent via reverse osmosis (RO) and used the treated effluent as deionized (DI) water. The avoided scenario for the application of treated sewage as DI water included DI water synthesis through RO and mixed bed (RO-MB) technologies. Among the different scenarios explored by the study, the most environmentally benign one corresponded to 99% reuse of the treated water for industrial applications.

The normalized results revealed marine ecotoxicity potential to be the most impacted category, with OLDP exhibiting negligible effect compared to other impact categories. In fact, the 99% reuse scenario predicted an overall positive impact on the environment for most of the studied indicators. For instance, the highest avoided abiotic depletion was attributed to the chemical-intensive processes (RO-MB) conventionally employed by the industries. The avoided DI water supply also had a considerable positive impact in most of the impact categories, with the negative impacts mainly attributed to the tertiary treatment of the wastewater. However, in the case of EP, sludge treatment and disposal also pose significant risks.

Another study performed the environmental assessment of an integrated system, wherein residential wastewater and digestate were treated using a hybrid MBBR-MBR process, while the generated sludge and food waste were further treated through AD [25]. The biogas was assumed to substitute LPG demand in canteens, while the treated wastewater was assumed to be used for irrigation purposes. Resultingly, compared to conventional systems where the food waste and sludge were assumed to be landfilled and wastewater to be treated via ASP (with no reuse of treated effluent), the integrated zero-waste system exhibited a net environmental benefit on GWP, AP, EP, photochemical oxidant formation and fossil depletion. The marked reduction in freshwater EP can be owed to the effective ammonia and nitrate removal by the MBBR-MBR system, with the treated effluent being used for irrigation purposes. In fact, the system turned out to have a negative carbon footprint with a value ranging between − 10.3 and − 7.9 kg CO_2eq $kg_{AD\ substrate}^{-1}$, with major benefits coming from the use of biogas as an LPG substitute. The MBBR-MBR system alone had a GWP of 0.6 kg CO_2eq m^{-3}.

4 Challenges and Recommendations Associated with the LCA of Hybrid-Biological WWTPs

Even though LCA can be a promising tool for the environmental assessment of WWTPs, the absence of a standardized procedure tailored to this specific domain poses numerous challenges. For instance, variations in the functional units, LCIA methods, and significant disparities in the LCI data are some of the typical hurdles for comparison between different studies. Assessing the environmental performance of WWTPs becomes even more intricate due to the complexities inherent in the interdependent processes of wastewater treatment. Furthermore, the paradigm shift towards resource recovery also adds to these challenges. These issues are more severe in developing countries primarily due to the lack of appropriate data (1) across different geographical and temporal distributions, (2) lack of technical expertise for wastewater and sludge treatment, and (3) release of contaminants from WWTPs [26].

Moreover, a methodological challenge arises in impact allocation when various resources (such as recycled water, biogas, biosolids, and biochemicals) are recovered during wastewater treatment. Additionally, the impact of pathogens on human health is not firmly established, especially concerning the potable use of treated wastewater [10]. Other major challenges include the disparity in using the upscaling factors and the absence of well-developed characterization and normalization factors for metals [10].

Corominas et al. discussed the challenges associated with the communication of the outcomes of LCA studies, given the complexity of environmental processes and mechanisms [27]. The complexity is often compounded by the reliance on models, which are based on assumptions that are often concealed within databases. The involvement of utility personnel early in the process has been recommended to foster greater confidence in the validity of LCA and its assumptions. While end-point impact categories may be deemed more relevant by stakeholders, their use introduces additional assumptions and greater uncertainty compared to mid-point indicators.

5 Recommendations Associated with the LCA of Hybrid-Biological WWTPs

Based on the challenges highlighted in the above section, it is essential to develop a standard protocol suited for the LCA of both conventional and hybrid WWTPs. This will ensure better comparability between different studies and also increase process transparency. Necessary efforts should also be made in developing region-specific background databases and emission factors.

Employing skilled manpower is also encouraged for better plant operation and systematic collection of operational data. In this regard, continuous monitoring and maintaining the performance of the WWTPs is also recommended to minimize plant's downtime and reduce environmental pollution, that can further aid in the collection of valuable data on lifecycle performance of WWTPs. Installation of smart sensors, internet of things (IoT) devices, and artificial intelligence-driven anomaly detection and optimization algorithms can also be adopted to further improving plant efficiency and ease the data collection process.

From the perspective of LCA, involvement of utility personnel early in the process is suggested to foster greater confidence in the validity of LCA and its assumptions [27]. Moreover, the integration of LCA with life cycle costing and social LCA can provide a more comprehensive view of overall sustainability. The above discussions, therefore, underline the importance of LCA and also emphasize its role as an aid or a complementary tool for decision-making, supported by other sources of evaluations and information.

6 Conclusions

The introduction of hybrid-biological systems for wastewater treatment can be an effective approach for not only reducing the contaminant loading in the treated effluent but can also be tuned to recover various resources. However, the overall environmental performance of such units should be determined to judge their overall sustainability. In that regard, LCA has proven to be a useful tool for comparing different wastewater treatment approaches and evaluating their environmental performance. Furthermore, with the increasing interest in integrating resource recovery and circular economy with conventional wastewater treatment approaches, the importance of LCA has increased significantly.

In accordance with the characteristics and intended use of the hybrid-biological unit, previous research has indicated the potential for trade-offs among various impact categories. While certain impact categories may show favourable enhancements when compared with conventional wastewater treatment approaches, others could experience adverse effects. Furthermore, the construction phase of the WWTP also tends to have a significant impact on the overall environmental performance, at times overshadowing the operational phase. However, data availability and methodological refinements are some of the major aspects that have been identified and need to be addressed to increase the reliability of the LCIA results and reduce the associated uncertainties. The reliability of the environmental footprint assessment can be further increased by addressing the current challenges and issues associated with the methodology and data quality.

This chapter can assist policymakers, academics, and researchers in gaining a more profound comprehension of the requisite methodological consistency and precision in LCA studies. Moreover, it can also contribute to decision-making processes by considering geographical-specific conditions and environmental risks before implementing modifications to existing WWTPs or introducing new units.

References

1. Tsangas M, Papamichael I, Banti D, Samaras P, Zorpas AA (2023) LCA of municipal wastewater treatment. Chemosphere 341:139952. https://doi.org/10.1016/J.CHEMOSPHERE.2023.139952
2. Bhar R, Mondal A, Dubey BK, Ghangrekar MM (2023) A review on the scope of remediating chlorinated paraffin contaminated water bodies and soils/sediments. Sci Total Environ 885:163941. https://doi.org/10.1016/J.SCITOTENV.2023.163941
3. Singh A, Kamble SJ, Sawant M, Chakravarthy Y, Kazmi A, Aymerich E, Starkl M, Ghangrekar M, Philip L (2018) Technical, hygiene, economic, and life cycle assessment of full-scale moving bed biofilm reactors for wastewater treatment in India. Environ Sci Pollut Res Int 25:2552–2569. https://doi.org/10.1007/S11356-017-0605-Y
4. Mehmeti A, Canaj K (2022) Environmental assessment of wastewater treatment and reuse for irrigation: a mini-review of LCA studies. Resources 11:94. https://doi.org/10.3390/RESOURCES11100094
5. Ghangrekar MM, Kumar S, Ahmad A, Das S (2022) Concomitant bioenergy production and wastewater treatment employing microbial electrochemical technologies. In: Biofuels and bioenergy: opportunities and challenges, pp 359–385. https://doi.org/10.1016/B978-0-323-85269-2.00015-0
6. Grandclément C, Seyssiecq I, Piram A, Wong-Wah-Chung P, Vanot G, Tiliacos N, Roche N, Doumenq P (2017) From the conventional biological wastewater treatment to hybrid processes,

the evaluation of organic micropollutant removal: a review. Water Res 111:297–317. https://doi.org/10.1016/J.WATRES.2017.01.005
7. Bhar R, Tiwari BR, Sarmah AK, Brar SK, Dubey BK (2022) A comparative life cycle assessment of different pyrolysis-pretreatment pathways of wood biomass for levoglucosan production. Bioresour Technol 356:127305. https://doi.org/10.1016/J.BIORTECH.2022.127305
8. Vanapalli KR, Bhar R, Maity SK, Dubey BK, Kumar S, Kumar V (2023) Life cycle assessment of fermentative production of lactic acid from bread waste based on process modelling using pinch technology. Sci Total Environ 905:167051. https://doi.org/10.1016/J.SCITOTENV.2023.167051
9. Gallego-Schmid A, Tarpani RRZ (2019) Life cycle assessment of wastewater treatment in developing countries: a review. Water Res 153:63–79. https://doi.org/10.1016/J.WATRES.2019.01.010
10. Corominas L, Byrne DM, Guest JS, Hospido A, Roux P, Shaw A, Short MD (2020) The application of life cycle assessment (LCA) to wastewater treatment: a best practice guide and critical review. Water Res 184:116058. https://doi.org/10.1016/j.watres.2020.116058
11. Rashid SS, Harun SN, Hanafiah MM, Razman KK, Liu YQ, Tholibon DA (2023) Life cycle assessment and its application in wastewater treatment: a brief overview. Processes 11:208. https://doi.org/10.3390/PR11010208
12. Hauck M, Maalcke-Luesken FA, Jetten MSM, Huijbregts MAJ (2016) Removing nitrogen from wastewater with side stream anammox: what are the trade-offs between environmental impacts? Resour Conserv Recycl 107:212–219. https://doi.org/10.1016/J.RESCONREC.2015.11.019
13. van Zelm R, de Paiva Seroa da Motta R, Lam WY, Menkveld W, Broeders E (2020) Life cycle assessment of side stream removal and recovery of nitrogen from wastewater treatment plants. J Ind Ecol 24:913–922. https://doi.org/10.1111/JIEC.12993
14. Resende JD, Nolasco MA, Pacca SA (2019) Life cycle assessment and costing of wastewater treatment systems coupled to constructed wetlands. Resour Conserv Recycl 148:170–177. https://doi.org/10.1016/J.RESCONREC.2019.04.034
15. Rufí-Salís M, Petit-Boix A, Leipold S, Villalba G, Rieradevall J, Moliné E, Gabarrell X, Carrera J, Suárez-Ojeda ME (2022) Increasing resource circularity in wastewater treatment: environmental implications of technological upgrades. Sci Total Environ 838:156422. https://doi.org/10.1016/J.SCITOTENV.2022.156422
16. US EPA (2023) Understanding global warming potentials. US EPA. https://www.epa.gov/ghgemissions/understanding-global-warming-potentials. Accessed 29 Dec 2023
17. Corbella C, Puigagut J, Garfí M (2017) Life cycle assessment of constructed wetland systems for wastewater treatment coupled with microbial fuel cells. Sci Total Environ 584–585:355–362. https://doi.org/10.1016/J.SCITOTENV.2016.12.186
18. Saidulu D, Majumder A, Gupta AK (2021) A systematic review of moving bed biofilm reactor, membrane bioreactor, and moving bed membrane bioreactor for wastewater treatment: comparison of research trends, removal mechanisms, and performance. J Environ Chem Eng 9:106112. https://doi.org/10.1016/J.JECE.2021.106112
19. Shahid MK, Kashif A, Rout PR, Aslam M, Fuwad A, Choi Y, Banu JR, Park JH, Kumar G (2020) A brief review of anaerobic membrane bioreactors emphasizing recent advancements, fouling issues and future perspectives. J Environ Manage 270:110909. https://doi.org/10.1016/J.JENVMAN.2020.110909
20. Ortiz M, Raluy RG, Serra L (2007) Life cycle assessment of water treatment technologies: wastewater and water-reuse in a small town. Desalination 204:121–131. https://doi.org/10.1016/J.DESAL.2006.04.026
21. Foglia A, Andreola C, Cipolletta G, Radini S, Akyol Ç, Eusebi AL, Stanchev P, Katsou E, Fatone F (2021) Comparative life cycle environmental and economic assessment of anaerobic membrane bioreactor and disinfection for reclaimed water reuse in agricultural irrigation: a case study in Italy. J Clean Prod 293:126201. https://doi.org/10.1016/J.JCLEPRO.2021.126201

22. Jeong H, Broesicke OA, Drew B, Crittenden JC (2018) Life cycle assessment of small-scale greywater reclamation systems combined with conventional centralized water systems for the City of Atlanta, Georgia. J Clean Prod 174:333–342. https://doi.org/10.1016/J.JCLEPRO.2017.10.193
23. Masmoudi Jabri K, Nolde E, Ciroth A, Bousselmi L (2020) Life cycle assessment of a decentralized greywater treatment alternative for non-potable reuse application. Int J Environ Sci Technol 17:433–444. https://doi.org/10.1007/S13762-019-02511-3
24. Tong L, Liu X, Liu X, Yuan Z, Zhang Q (2013) Life cycle assessment of water reuse systems in an industrial park. J Environ Manage 129:471–478. https://doi.org/10.1016/J.JENVMAN.2013.08.018
25. Sakcharoen T, Ratanatamskul C, Chandrachai A (2021) Factors affecting technology selection, techno-economic and environmental sustainability assessment of a novel zero-waste system for food waste and wastewater management. J Clean Prod 314:128103. https://doi.org/10.1016/J.JCLEPRO.2021.128103
26. Parra-Saldivar R, Bilal M, Iqbal HMN (2020) Life cycle assessment in wastewater treatment technology. Curr Opin Environ Sci Health 13:80–84. https://doi.org/10.1016/J.COESH.2019.12.003
27. Corominas L, Foley J, Guest JS, Hospido A, Larsen HF, Morera S, Shaw A (2013) Life cycle assessment applied to wastewater treatment: state of the art. Water Res 47:5480–5492. https://doi.org/10.1016/J.WATRES.2013.06.049

The manufacturer's authorised representative in the EU is Springer Nature Customer Service Centre GmbH, Europaplatz 3, 69115 Heidelberg, Germany. If you have any concerns regarding our products, please contact ProductSafety@springernature.com

Printed and bound by CPI Group (UK) Ltd, Croydon, CR0 4YY
28/09/2026
02224065-0012